Trends in Mathematics is a series devoted to the publication of volumes arising from conferences and lecture series focusing on a particular topic from any area of mathematics. Its aim is to make current developments available to the community as rapidly as possible without compromise to quality and to archive these for reference.

Proposals for volumes can be sent to the Mathematics Editor at either

Birkhäuser Verlag
P.O. Box 133
CH-4010 Basel
Switzerland

or

Birkhäuser Boston Inc.
675 Massachusetts Avenue
Cambridge, MA 02139
USA

Material submitted for publication must be screened and prepared as follows:

All contributions should undergo a reviewing process similar to that carried out by journals and be checked for correct use of language which, as a rule, is English. Articles without proofs, or which do not contain any significantly new results, should be rejected. High quality survey papers, however, are welcome.

We expect the organizers to deliver manuscripts in a form that is essentially ready for direct reproduction. Any version of TeX is acceptable, but the entire collection of files must be in one particular dialect of TeX and unified according to simple instructions available from Birkhäuser.

Furthermore, in order to guarantee the timely appearance of the proceedings it is essential that the final version of the entire material be submitted no later than one year after the conference. The total number of pages should not exceed 350. The first-mentioned author of each article will receive 25 free offprints. To the participants of the congress the book will be offered at a special rate.

C*-algebras
and Elliptic Theory II

Dan Burghelea
Richard Melrose
Alexander S. Mishchenko
Evgenij V. Troitsky
Editors

Birkhäuser
Basel · Boston · Berlin

Editors:

Dan Burghelea
Department of Mathematics
Ohio State University
231 West 18th Avenue
Columbus, OH 43210
USA
e-mail: burghele@math.ohio-state.edu

Richard B. Melrose
Department of Mathematics
Massachusetts Institute of Technology
Room 2-174
77 Massachusetts Avenue
Cambridge, MA 02139
USA
e-mail: rbm@math.mit.edu

Alexander S. Mishchenko
Evgenij V. Troitsky
Department of Mechanics and Mathematics
Moscow State University
Leninskie Gory
119992 Moscow
Russia
e-mail: asmish@mech.math.msu.su
        troitsky@mech.math.msu.su

2000 Mathematical Subject Classification: 19Kxx, 58Jxx, 57Rxx, 58J32, 19D55

Library of Congress Control Number: 2007942640

Bibliographic information published by Die Deutsche Bibliothek. Die Deutsche Bibliothek lists
this publication in the Deutsche Nationalbibliografie; detailed bibliographic data is available in
the Internet at http://dnb.ddb.de

ISBN 978-3-7643-8603-0 Birkhäuser Verlag AG, Basel - Boston - Berlin

© 2008 Birkhäuser Verlag AG
Basel · Boston · Berlin
P.O. Box 133, CH-4010 Basel, Switzerland
Part of Springer Science+Business Media
Printed on acid-free paper produced from chlorine-free pulp. TCF ∞
Cover Design: Alexander Faust, CH-4051 Basel, Switzerland
Printed in Germany
ISBN 978-3-7643-8603-0   e-ISBN 978-3-7643-8604-7

9 8 7 6 5 4 3 2 1   www.birkhauser.ch

Contents

Editors' Introduction

The conference "C^*-algebras and elliptic theory, II" was held at the Stefan Banach International Mathematical Center in Będlewo, Poland, in January 2006, one of a series of meetings in Poland and Russia. This volume is a collection of original and refereed research and expository papers related to the meeting. Although centered on the K-theory of operator algebras, a broad range of topics is covered including geometric, L^2- and spectral invariants, such as the analytic torsion, signature and index, of differential and pseudo-differential operators on spaces which are possibly singular, foliated or non-commutative. This material should be of interest to researchers in Mathematical Physics, Differential Topology and Analysis.

The series of conferences including this one originated with an idea of Professor Bogdan Bojarski, namely, to strengthen collaboration between mathematicians from Poland and Russia on the basis of common scientific interests, particularly in the field of Non-commutative Geometry. This led to the first meeting, in 2004, which brought together about 60 mathematicians not only from Russia and Poland, but from other leading centers. It was supported by the European program "Geometric Analysis Research Training Network". Since then there have been annual meetings alternating between Będlewo and Moscow. The second conference was organized in Moscow in 2005 and was dedicated to the memory of Yu.P. Solovyov. The proceedings will appear in the *Journal of K-Theory*. The conference on which this volume is based was the third conference in the overall series with the fourth being held in Moscow in 2007. A further meeting in Będlewo is planned for 2009.

D. Burghelea, R.B. Melrose, A. Mishchenko, E. Troitsky

Contents

Pseudo-differential operators

In two papers *"Dual manifolds and pseudo-differential operators"* and *"Homotopy classification and K-homology"* **V. Nazaykinskiy, A. Savin and B. Sternin** examine index questions and the homotopy classification of pseudo-differential operators on manifolds with corners.

The paper *"Dixmier traceability for general pseudo-differential operators"* by **F. Nicola and L. Rodino** generalizes previous results about the finiteness of the Dixmier trace of pseudo-differential operators.

In *"Boundaries, Eta invariant and the determinant bundle"*, **R. Melrose and F. Rochon** show that the exponentiated η invariant gives a section of the determinant bundle over the boundary for cusp pseudo-differential operators, generalizing a theorem of Dai and Freed in the Dirac setting.

K-theory

The paper *"K-theory of twisted group algebras"* by **S. Echterhoff** presents applications of the Baum-Connes conjecture to the study of the K-theory of twisted group algebras.

A geometric formulation of the description of the dual of a finite group is extended to discrete infinite groups in the paper *"Twisted Burnside theorem for two-step torsion-free nilpotent groups"* by **A. Felshtyn, F. Indukaev and E. Troitsky**.

The paper *"Group bundle duality, invariants for certain C^*-algebras, and twisted equivariant K-theory"* by **E. Vasselli** describes a general duality for Lie group bundles and its the relation with twisted K-theory.

In the paper *"Topological invariants of bifurcation"*, **J. Pejsachowicz** uses the J-functor in K-theory to describe bifurcation for some nonlinear Fredholm operator families.

Torsion and determinants

"Torsion, as a function on the space of representations" is a survey by **D. Burghelea and S. Haller** of their results on three complex-valued invariants of a smooth closed manifold arising from combinatorial topology, from regularized determinants and from the counting instantons and closed trajectories.

The Ihara zeta function for infinite periodic simple graphs, involving a "determinant" in the setting of von Neumann linear algebra, is defined and studied in the paper *"Ihara zeta function for periodic simple graphs"* by **D. Guido, T. Isola and M. Lapidus**.

Operator algebras

Ch. Wahl, in *"A new topology on the space of unbounded selfadjoint operators and the spectral flow"*, revisits the relationship between the space of Fredholm operators and the classical K^1 and K^0 functors.

In the paper *"L^2-invariants and rank metric"*, **A. Thom** gives results about L^2-Betti numbers for tracial algebras.

A positive answer to a conjecture on non-commutative spheres, is provided by **U. Krähmer** in *"On the non-standard Podleś spheres"*.

The paper *"Modified Hochschild and periodic cyclic homology"* by **N. Teleman** proposes a modification in the definition of these two homologies to better relate them to the Alexander-Spanier homology.

Foliated manifolds

Lefschetz theory associated to a "transverse" action of a Lie group on a foliated manifold is examined in the paper *"Lefschetz distribution of Lie foliation"* by **J. Alvarez Lopez and Yu. Kordyukov**.

The paper *"Adiabatic limits and the spectrum of the Laplacian on foliated manifolds"* by **Yu. Kordyukov and A. Yakovlev** presents results on the spectrum of the Laplacian on differential forms as the Riemannian metric is expanded normal to the leaves.

C^*-algebras and Elliptic Theory II

Trends in Mathematics, 1–40

© 2008 Birkhäuser Verlag Basel/Switzerland

Lefschetz Distribution of Lie Foliations

Jesús A. Álvarez López and Yuri A. Kordyukov

Abstract. Let $\mathcal{F}$ be a Lie foliation on a closed manifold M with structural Lie group G. Its transverse Lie structure can be considered as a transverse action Φ of G on $(M, \mathcal{F})$; *i.e.*, an "action" which is defined up to leafwise homotopies. This Φ induces an action Φ^* of G on the reduced leafwise cohomology $\overline{H}(\mathcal{F})$. By using leafwise Hodge theory, the supertrace of Φ^* can be defined as a distribution $L_{\mathrm{dis}}(\mathcal{F})$ on G called the Lefschetz distribution of $\mathcal{F}$. A distributional version of the Gauss-Bonett theorem is proved, which describes $L_{\mathrm{dis}}(\mathcal{F})$ around the identity element. On any small enough open subset of G, $L_{\mathrm{dis}}(\mathcal{F})$ is described by a distributional version of the Lefschetz trace formula.

Mathematics Subject Classification (2000). 58J22, 57R30, 58J42.

Keywords. Lie foliation, Riemannian foliation, leafwise reduced cohomology, distributional trace, Lefschetz distribution, Λ-Euler characteristic, Λ-Lefschetz number, Lefschetz trace formula.

Contents

J.A.L. was partially supported by MEC (Spain), grant MTM2004-08214. Y.K. was partially supported by the RFBR grant 06-01-00208 and by the joint RFBR-DFG grant 07-01-91555-NNIO_a.

1. Introduction

Let $\mathcal{F}$ be a C^∞ foliation on a manifold M. Let $\mathrm{Diff}(M, \mathcal{F})$ be the group of foliated diffeomorphisms $(M, \mathcal{F}) \to (M, \mathcal{F})$. The elements of $\mathrm{Diff}(M, \mathcal{F})$ that are C^∞ leafwisely homotopic to id_M form a normal subgroup $\mathrm{Diff}_0(\mathcal{F})$, and let $\overline{\mathrm{Diff}}(M, \mathcal{F})$ denote the corresponding quotient group. A *right transverse action* of a group G on $(M, \mathcal{F})$ is an anti-homomorphism $\Phi : G \to \overline{\mathrm{Diff}}(M, \mathcal{F})$. A *local representation* of Φ on some open subset $O \subset G$ is a map $\phi : M \times O \to M$ such that $\phi_g = \phi(\cdot, g)$ is a foliated diffeomorphism representing Φ_g for all $g \in G$. Then Φ is said to be of *class* C^∞ if it has a C^∞ local representation on each small enough open subset of G.

Recall that the leafwise de Rham complex $(\Omega(\mathcal{F}), d_\mathcal{F})$ consists of the differential forms on the leaves which are C^∞ on M, endowed with the de Rham derivative of the leaves. Its cohomology $H(\mathcal{F})$ is called the leafwise cohomology. This becomes a topological vector space with the topology induced by the C^∞ topology, and its maximal Hausdorff quotient is the reduced leafwise cohomology $\overline{H}(\mathcal{F})$.

Consider the canonical right action of $\mathrm{Diff}(M, \mathcal{F})$ on $\overline{H}(\mathcal{F})$ defined by pulling-back leafwise differential forms. Since $\mathrm{Diff}_0(\mathcal{F})$ acts trivially, we get a canonical right action of $\overline{\mathrm{Diff}}(M, \mathcal{F})$ on $\overline{H}(\mathcal{F})$. Then any right transverse action Φ of a group G on $(M, \mathcal{F})$ induces a left action Φ^* of G on $\overline{H}(\mathcal{F})$.

Suppose from now on that $\mathcal{F}$ is a Lie foliation and the manifold M is closed. It is shown that its transverse Lie structure can be described as a right transverse action Φ of its structural Lie group G on $(M, \mathcal{F})$. Consider the induced left action Φ^* of G on $\overline{H}(\mathcal{F})$. For each $g \in G$, we would like to define the supertrace $\mathrm{Tr}^s \Phi_g^*$, which could be called the *leafwise Lefschetz number* $L(\Phi_g)$ of Φ_g. This can be achieved when $\overline{H}(\mathcal{F})$ is of finite dimension, obtaining a C^∞ function $L(\mathcal{F})$ on G defined by $L(\mathcal{F})(g) = L(\Phi_g)$; the value of $L(\mathcal{F})$ at the identity element e of G is the Euler characteristic $\chi(\mathcal{F})$ of $\overline{H}(\mathcal{F})$, which can be called the *leafwise Euler* *characteristic* of $\mathcal{F}$. But $\overline{H}(\mathcal{F})$ may be of infinite dimension, even when the leaves are dense [1], and thus $L(\mathcal{F})$ is not defined in general.

The first goal of this paper is to show that, in general, the role of the function $L(\mathcal{F})$ can be played by a distribution $L_{\mathrm{dis}}(\mathcal{F})$ on G, called the *Lefschetz distribution* of $\mathcal{F}$, whose singularities are motivated by the infinite dimension of $\overline{H}(\mathcal{F})$.

The first ingredient to define $L_{\mathrm{dis}}(\mathcal{F})$ is the leafwise Hodge theory studied in [2] for Riemannian foliations; recall that Lie foliations form a specially important class of Riemannian foliations [19]. Fix a bundle-like metric on M whose transverse part is induced by a left invariant Riemannian metric on G. For the induced Riemannian structure on the leaves, let $\Delta_{\mathcal{F}}$ be the Laplacian of the leaves operating in $\Omega(\mathcal{F})$. The kernel $\mathcal{H}(\mathcal{F})$ of Δ_F is the space of harmonic forms on the leaves that are C^{∞} on M. The metric induces an L^2 inner product on $\Omega(\mathcal{F})$, obtaining a Hilbert space $\boldsymbol{\Omega}(\mathcal{F})$. Then $\Delta_{\mathcal{F}}$ is an essentially self-adjoint operator in $\boldsymbol{\Omega}(\mathcal{F})$ whose closure is denoted by $\boldsymbol{\Delta}_{\mathcal{F}}$. The kernel of $\boldsymbol{\Delta}_{\mathcal{F}}$ is denoted by $\boldsymbol{\mathcal{H}}(\mathcal{F})$, and let $\boldsymbol{\Pi} : \boldsymbol{\Omega}(\mathcal{F}) \to \boldsymbol{\mathcal{H}}(\mathcal{F})$ denote the orthogonal projection. In [2], it is proved that $\boldsymbol{\Pi}$ has a restriction $\Pi : \Omega(\mathcal{F}) \to \mathcal{H}(\mathcal{F})$ that induces an isomorphism $\overline{H}(\mathcal{F}) \cong \mathcal{H}(\mathcal{F})$, which can be called the *leafwise Hodge isomorphism*.

Let Λ be the volume form of G, and let $\phi : M \times O \to M$ be a C^{∞} local representation of Φ. For each $f \in C_c^{\infty}(O)$, consider the operator

$$P_f = \int_G \phi_g^* \cdot f(g) \, \Lambda(g) \circ \Pi$$

in $\Omega(\mathcal{F})$. Our first main result is the following.

Proposition 1.1. *P_f is of trace class, and the functional $f \mapsto \mathrm{Tr}^s P_f$ defines a distribution on O.*

It can be easily seen that $\mathrm{Tr}^s P_f$ is independent of the choice of ϕ, and thus the distributions given by Proposition 1.1 can be combined to define a distribution $L_{\mathrm{dis}}(\mathcal{F})$ on G; this is the *Lefschetz distribution* of $\mathcal{F}$.

Observe that $L_{\mathrm{dis}}(\mathcal{F}) \equiv L(\mathcal{F}) \cdot \Lambda$ when $\overline{H}(\mathcal{F})$ is of finite dimension. This justifies the consideration of $L_{\mathrm{dis}}(\mathcal{F})$ as a generalization of $L(\mathcal{F})$; in particular, the germ of $L_{\mathrm{dis}}(\mathcal{F})$ at e generalizes $\chi(\mathcal{F})$.

If the operators P_f are restricted to $\Omega^i(\mathcal{F})$ for each degree i, its trace defines a distribution $\mathrm{Tr}_{\mathrm{dis}}^i(\mathcal{F})$, called *distributional trace*, whose germ at e generalizes the *leafwise Betti number* $\beta^i(\mathcal{F}) = \dim \overline{H}^i(\mathcal{F})$.

The distributions $L_{\mathrm{dis}}(\mathcal{F})$ and $\mathrm{Tr}_{\mathrm{dis}}^i(\mathcal{F})$ depend on Λ and $\mathcal{F}$, endowed with the transverse Lie structure. If the leaves are dense, then the transverse Lie structure is determined by the foliation, and thus these distributions depend only on Λ and the foliation. On the other hand, the dependence on Λ can be avoided by using top-dimensional currents instead of distributions, in the obvious way.

Our second goal is to prove a distributional version of the Gauss-Bonnett theorem, which describes $L_{\mathrm{dis}}(\mathcal{F})$ around e. Let $R_{\mathcal{F}}$ be the curvature of the leafwise metric. Suppose for simplicity that $\mathcal{F}$ is oriented. Then $\mathrm{Pf}(R_{\mathcal{F}}/2\pi) \in \Omega^p(\mathcal{F})$ ($p = \dim \mathcal{F}$) can be called the *leafwise Euler form*. This form can be paired with Λ, considered as a transverse invariant measure, to give a differential form $\omega_\Lambda \wedge$ $\mathrm{Pf}(R_\mathcal{F}/2\pi)$ of top degree on M. In particular, if $\dim \mathcal{F} = 2$, then

$$\omega_\Lambda \wedge \mathrm{Pf}(R_\mathcal{F}/2\pi) = \frac{1}{2\pi} K_\mathcal{F} \,\omega_M \,,$$

where $K_\mathcal{F}$ is the Gauss curvature of the leaves and ω_M is the volume form of M. Let δ_e denote the Dirac measure at e.

Theorem 1.2 (Distributional Gauss-Bonett theorem). *We have*

$$L_{\mathrm{dis}}(\mathcal{F}) = \int_M \omega_\Lambda \wedge \mathrm{Pf}(R_\mathcal{F}/2\pi) \cdot \delta_e$$

on some neighborhood of e.

To prove Theorem 1.2, we really prove that

$$L_{\mathrm{dis}}(\mathcal{F}) = \chi_\Lambda(\mathcal{F}) \cdot \delta_e \tag{1.1}$$

around e, where Λ is considered as a transverse invariant measure of $\mathcal{F}$, and $\chi_\Lambda(\mathcal{F})$ is the Λ-Euler characteristic of $\mathcal{F}$ introduced by Connes [9]. Then Theorem 1.2 follows from the index theorem of [9].

The third goal is to prove a distributional version of the Lefschetz trace formula, which describes $L_{\mathrm{dis}}(\mathcal{F})$ on any small enough open subset of G. For a C^∞ local representation $\phi : M \times O \to M$ of Φ, let $\phi' : M \times O \to M \times O$ be the map defined by $\phi'(x, g) = (\phi_g(x), g)$. The fixed point set of ϕ', $\mathrm{Fix}(\phi')$, consists of the points (x, g) such that $\phi_g(x) = x$. A point $(x, g) \in \mathrm{Fix}(\phi')$ is said to be *leafwise simple* when $\phi_{g*} - \mathrm{id} : T_x\mathcal{F} \to T_x\mathcal{F}$ is an isomorphism; in this case, the sign of the determinant of this isomorphism is denoted by $\epsilon(x, g)$. The set of leafwise simple fixed points of ϕ' is denoted by $\mathrm{Fix}_0(\phi')$. Let $\mathrm{pr}_1 : M \times O \to M$ and $\mathrm{pr}_2 : M \times O \to O$ be the factor projections. It is proved that $\mathrm{Fix}_0(\phi')$ is a C^∞ manifold of dimension equal to $\mathrm{codim}\,\mathcal{F}$. Moreover the restriction $\mathrm{pr}_1 : \mathrm{Fix}_0(\phi') \to M$ is a local embedding transverse to $\mathcal{F}$. So Λ defines a measure $\Lambda'_{\mathrm{Fix}_0(\phi')}$ on $\mathrm{Fix}_0(\phi')$. Observe that $\mathrm{pr}_2 : \mathrm{Fix}(\phi') \to O$ is a proper map.

Theorem 1.3 (Distributional Lefschetz trace formula). *Suppose that every fixed point of ϕ' is leafwise simple. Then*

$$L_{\mathrm{dis}}(\mathcal{F}) = \mathrm{pr}_{2*}(\epsilon \cdot \Lambda'_{\mathrm{Fix}(\phi')})$$

on O.

To prove Theorem 1.3, we consider certain submanifold $M_1' \subset M \times O$ endowed with a foliation $\mathcal{F}_1'$, whose leaves are of the form $L \times \{g\}$, where L is a leaf of $\mathcal{F}$ and $g \in G$. It is proved that $\mathrm{pr}_2(M_1')$ is open in some orbit of the adjoint action of G on itself, $\mathrm{pr}_1 : M_1' \to M$ is a local diffeomorphism, and $\mathcal{F}_1' = \mathrm{pr}_1^* \mathcal{F}$. So Λ lifts to a transverse invariant measure Λ_1' of $\mathcal{F}_1'$. Moreover the restriction ϕ_1' of ϕ' to M_1' is defined and maps each leaf of $\mathcal{F}_1'$ to itself. For each $f \in C_c^\infty(O)$ supported in an appropriate open subset $O_1 \subset O$, the transverse invariant measure $\Lambda_{1,f}' = \mathrm{pr}_2^* f \cdot \Lambda_1'$ is compactly supported. Then the $\Lambda_{1,f}'$-Lefschetz number $L_{\Lambda_{1,f}'}(\phi_1')$ is defined according to [14]. Without assuming any condition on the fixed point set, we show that

$$\langle L_{\text{dis}}(\mathcal{F}), f \rangle = L_{\Lambda'_{1,f}}(\phi'_1) . \tag{1.2}$$

We have that $\text{Fix}(\phi'_1)$ is a C^∞ local transversal of $\mathcal{F}'_1$. Hence Theorem 1.3 follows from (1.2) and the foliation Lefschetz theorem of [14, 24].

The numbers $\chi_\Lambda(\mathcal{F})$ and $L_{\Lambda'_{1,f}}(\phi'_1)$ are defined by using L^2 differential forms on the leaves, whilst $L_{\text{dis}}(\mathcal{F})$ is defined by using leafwise differential forms that are C^∞ on M. These are sharply different conditions when the leaves are not compact. So (1.1) and (1.2) are surprising relations.

By (1.2), $L_{\text{dis}}(\mathcal{F})$ is supported in the union of a discrete set of orbits of the adjoint action. Therefore, when $\text{codim}\,\mathcal{F} > 0$, $L_{\text{dis}}(\mathcal{F})$ is C^∞ just when it is trivial, obtaining the following.

Corollary 1.4. *If* $\overline{H}(\mathcal{F})$ *is of finite dimension and* $\text{codim}\,\mathcal{F} > 0$, *then* $L_{\text{dis}}(\mathcal{F}) \equiv L(\mathcal{F}) = 0$.

By Corollary 1.4, $\chi(\mathcal{F})$ is useless: it vanishes just when it can be defined. Moreover $\chi_\Lambda(\mathcal{F}) = 0$ in this case by (1.1). So, when $\text{codim}\,\mathcal{F} > 0$, the condition $\chi_\Lambda(\mathcal{F}) \neq 0$ yields $\dim \overline{H}(\mathcal{F}) = \infty$. More precise results of this type would be desirable.

Let $\dim \mathcal{F} = p$. When the leaves are dense, $\beta^0(\mathcal{F})$ and $\beta^p(\mathcal{F})$ are finite, and thus $\text{Tr}^0_{\text{dis}}(\mathcal{F})$ and $\text{Tr}^p_{\text{dis}}(\mathcal{F})$ are C^∞. On the other hand, when the leaves are not compact, the Λ-Betti numbers of [9] satisfy $\beta^0_\Lambda(\mathcal{F}) = \beta^p_\Lambda(\mathcal{F}) = 0$. Then the following result follows from (1.1) and Corollary 1.4.

Corollary 1.5. *If* $\text{codim}\,\mathcal{F} > 0$, $\dim \mathcal{F} = 2$ *and the leaves are dense, then* $\text{Tr}^1_{\text{dis}}(\mathcal{F}) - \beta^1_\Lambda(\mathcal{F}) \cdot \delta_e$ *is* C^∞ *around* e.

In Corollary 1.5, we could say that $\beta^1_\Lambda(\mathcal{F}) \cdot \delta_e$ is the "singular part" of $\text{Tr}^1_{\text{dis}}(\mathcal{F})$ around e.

Corollary 1.6. *Suppose that* $\text{codim}\,\mathcal{F} > 0$ *and* $\dim \mathcal{F} = 2$. *If there is a nontrivial harmonic* L^2 *differential form of degree one on some leaf, then* $\dim \overline{H}^1(\mathcal{F}) = \infty$.

It would be nice to generalize Corollary 1.6 for arbitrary dimension. Thus we conjecture the following.

Conjecture 1.7. *If* $\text{codim}\,\mathcal{F} > 0$ *and the leaves are dense, then* $\text{Tr}^i_{\text{dis}}(\mathcal{F}) - \beta^i_\Lambda(\mathcal{F}) \cdot \delta_e$ *is* C^∞ *around* e *for each degree* i.

The main results were proved in [3] for the case of codimension one. Our results also overlap the corresponding results of [20].

We hope to prove elsewhere another version of Theorem 1.3 with a more general condition on the fixed points, always satisfied by some local representation ϕ of Φ defined around any point of G. By (1.2), what is needed is another version of the Lefschetz theorem of [14], which holds for more general fixed point sets when the transverse measure is C^∞.

The idea of using such type of trace class operators to define distributional spectral invariants is due to Atiyah and Singer [5, 30]. They consider transversally elliptic operators with respect to compact Lie group actions. Further generalizations to foliations and non-compact Lie group actions were given in [21, 10, 15, 17]. In our case, $\Delta_{\mathcal{F}}$ is not transversally elliptic with respect to any Lie group action or any foliation, but it can be considered as being "transversely elliptic" with $\mathcal{G}$ respect to the structural transverse action; this simply means that it is elliptic along the leaves of $\mathcal{F}$.

2. Transverse actions

Recall that a foliation $\mathcal{F}$ on a manifold M can be described by a *foliated cocycle*, which is a collection $\{U_i, f_i\}$, where $\{U_i\}$ is an open cover of X and each f_i is a topological submersion of U_i onto some manifold T_i whose fibers are connected open subsets of $\mathbb{R}^n$, such that the following *compatibility condition* is satisfied: for every $x \in U_i \cap U_j$, there is an open neighborhood $U_{i,j}^x$ of x in $U_i \cap U_j$ and a homeomorphism $h_{i,j}^x : f_i(U_{i,j}^x) \to f_j(U_{i,j}^x)$ such that $f_j = h_{i,j}^x \circ f_i$ on $U_{i,j}^x$. Two foliated cocycles describe the same foliation $\mathcal{F}$ when their union is a foliated cocycle. The *leaf topology* on M is the topology with a base given by the open sets of the fibers of all the submersions f_i. The *leaves* of $\mathcal{F}$ are the connected components of M with the leaf topology. The leaf through each point $x \in M$ is denoted by L_x. The pseudogroup on $\bigsqcup_i T_i$ generated by the maps $h_{i,j}^x$, given by the compatibility condition, is called (a representative of) the *holonomy pseudogroup* of $\mathcal{F}$, and describes the "transverse dynamics" of $\mathcal{F}$. Different foliated cocycles of $\mathcal{F}$ induce equivalent pseudogroups in the sense of [12, 13].

Another representative of the holonomy pseudogroup is defined on any transversal of $\mathcal{F}$ that meets every leaf. It is generated by "sliding" small open subsets (local transversals) along the leaves; its precise definition is given in [12].

When M is a C^∞ manifold, it is said that $\mathcal{F}$ is C^∞ if it is described by a foliated cocycle $\{U_i, f_i\}$ which is C^∞ in the sense that each f_i is a C^∞ submersion to some C^∞ manifold.

Let Γ be a group of homeomorphisms of a manifold T. A foliated cocycle (U_i, f_i) of $\mathcal{F}$, with $f_i : U_i \to T_i$, is said to be (T, Γ)-*valued* when each T_i is an open subset of T, and the maps $h_{i,j}^x$, given by the compatibility condition, are restrictions of maps in Γ. A *transverse* (T, Γ)-*structure* of $\mathcal{F}$ is given by a (T, Γ)-valued foliated cocycle, and two (T, Γ)-valued foliated cocycles define the same transverse (T, Γ)-structure when their union is a (T, Γ)-valued foliated cocycle. When $\mathcal{F}$ is endowed with a transverse (T, Γ)-structure, it is called a (T, Γ)-*foliation*.

Let $\mathcal{F}$ and $\mathcal{G}$ be foliations on manifolds M and N, respectively. Recall the following concepts. A *foliated map* $f : (M, \mathcal{F}) \to (N, \mathcal{G})$ is a map $f : M \to N$ that maps each leaf of $\mathcal{F}$ to a leaf of $\mathcal{G}$; the simpler notation $f : \mathcal{F} \to \mathcal{G}$ will be also used. A *leafwise homotopy* (or *integrable homotopy*) between two continuous foliated maps $f, f' : (M, \mathcal{F}) \to (N, \mathcal{G})$ is a continuous map $H : M \times I \to N$

($I = [0, 1]$) such that the path $H(x, \cdot) : I \to N$ lies in a leaf of $\mathcal{G}$ for each $x \in M$; in this case, it is said that f and f' are *leafwisely homotopic* (or *integrably homotopic*).

Suppose from now on that $\mathcal{F}$ and $\mathcal{G}$ are C^∞. Two C^∞ foliated maps are said to be C^∞ *leafwisely homotopic* when there is a C^∞ leafwise homotopy between them. As usual, $T\mathcal{F} \subset TM$ denotes the subbundle of vectors tangent to the leaves of $\mathcal{F}$, $\mathfrak{X}(M, \mathcal{F})$ denotes the Lie algebra of infinitesimal transformations of $(M, \mathcal{F})$, and $\mathfrak{X}(\mathcal{F}) \subset \mathfrak{X}(M, \mathcal{F})$ is the normal Lie subalgebra of vector fields tangent to the leaves of $\mathcal{F}$ (C^∞ sections of $T\mathcal{F} \to M$). Then we can consider the quotient Lie algebra $\overline{\mathfrak{X}}(M, \mathcal{F}) = \mathfrak{X}(M, \mathcal{F})/\mathfrak{X}(\mathcal{F})$, whose elements are called *transverse vector fields*. Observe that, for each $x \in M$, the evaluation map $\mathrm{ev}_x : \mathfrak{X}(M, \mathcal{F}) \to T_x M$ induces a map $\overline{\mathrm{ev}}_x : \overline{\mathfrak{X}}(M, \mathcal{F}) \to T_x M/T_x \mathcal{F}$, which can be also called *evaluation map*. For any Lie algebra $\mathfrak{g}$, a homomorphism $\mathfrak{g} \to \overline{\mathfrak{X}}(M, \mathcal{F})$ is called an *infinitesimal transverse action* of $\mathfrak{g}$ on $(M, \mathcal{F})$. In particular, we have a canonical infinitesimal transverse action of $\overline{\mathfrak{X}}(M, \mathcal{F})$ on $(M, \mathcal{F})$.

Let $\mathrm{Diff}(M, \mathcal{F})$ be the group of C^∞ foliated diffeomorphisms $(M, \mathcal{F}) \to (M, \mathcal{F})$ with the operation of composition, let $\mathrm{Diff}(\mathcal{F}) \subset \mathrm{Diff}(M, \mathcal{F})$ be the normal subgroup C^∞ foliated diffeomorphisms that preserve each leaf of $\mathcal{F}$, and let $\mathrm{Diff}_0(\mathcal{F}) \subset \mathrm{Diff}(\mathcal{F})$ be the normal subgroup of C^∞ foliated diffeomorphisms that are C^∞ leafwisely homotopic to the identity map. Then we can consider the quotient group $\overline{\mathrm{Diff}}(M, \mathcal{F}) = \mathrm{Diff}(M, \mathcal{F})/\mathrm{Diff}_0(\mathcal{F})$, whose operation is also denoted by "$\circ$". The elements of $\overline{\mathrm{Diff}}(M, \mathcal{F})$ can be called *transverse transformations* of $(M, \mathcal{F})$. For any group G, an anti-homomorphism $\Phi : G \to \overline{\mathrm{Diff}}(M, \mathcal{F})$, $g \mapsto \Phi_g$, is called a *right transverse action* of G on $(M, \mathcal{F})$. For an open subset $O \subset G$, a map $\phi : M \times O \to M$ is called a *local representation* of Φ on O if $\phi_g = \phi(\cdot, g) \in \Phi_g$ for all $g \in O$. For any leaf L of $\mathcal{F}$ and any $g \in O$, the leaf $\phi_g(L)$ is independent of the local representative ϕ, and thus it will be denoted by $\Phi_g(L)$. When G is a Lie group, Φ is said to be of *class* C^∞ if it has a C^∞ local representation around each element of G.

Somehow, we can think of $\overline{\mathrm{Diff}}(M, \mathcal{F})$ as a Lie group whose Lie algebra is $\overline{\mathfrak{X}}(M, \mathcal{F})$; indeed, it will be proved elsewhere that, if G is a simply connected Lie group and $\mathfrak{g}$ is its Lie algebra of left invariant vector fields, then there is a canonical bijection between infinitesimal transverse actions of $\mathfrak{g}$ on $(M, \mathcal{F})$ and C^∞ right transverse actions of G on $(M, \mathcal{F})$.

The *leafwise de Rham complex* $(\Omega(\mathcal{F}), d_\mathcal{F})$ is the space of differential forms on the leaves smooth on M (C^∞ sections of $\bigwedge T\mathcal{F}^* \to M$) endowed with the leafwise de Rham differential. It is also a topological vector space with the C^∞ topology, and $d_\mathcal{F}$ is continuous. The cohomology $H(\mathcal{F})$ of $(\Omega(\mathcal{F}), d_\mathcal{F})$ is called the *leafwise cohomology* of $\mathcal{F}$, which is a topological vector space with the induced topology. Its maximal Hausdorff quotient $\overline{H}(\mathcal{F}) = H(\mathcal{F})/\overline{0}$ is called the *reduced leafwise cohomology*.

By pulling back leafwise differential forms, any C^∞ foliated map $f : (M, \mathcal{F}) \to (N, \mathcal{G})$ induces a continuous homomorphism of complexes, $f^* : \Omega(\mathcal{G}) \to \Omega(\mathcal{F})$, obtaining a continuous homomorphism $f^* : \overline{H}(\mathcal{G}) \to \overline{H}(\mathcal{F})$. Moreover, if f is

C^∞ leafwisely homotopic to another C^∞ foliated map $f' : (M, \mathcal{F}) \to (M, \mathcal{F})$, then $f^* = f'^* : \overline{H}(\mathcal{G}) \to \overline{H}(\mathcal{F})$ by standard arguments [7]. Therefore, for any $F \in \overline{\mathrm{Diff}}(M, \mathcal{F})$ and any $f \in F$, the endomorphism f^* of $\overline{H}(\mathcal{F})$ can be denoted by F^*. So any right transverse action Φ of a group G on $(M, \mathcal{F})$ induces a left action Φ^* of G on $\overline{H}(\mathcal{F})$ given by $(g, \xi) \mapsto \Phi_g^* \xi$.

3. Lie foliations

Let $\mathcal{F}$ be a C^∞ foliation of codimension q on a C^∞ closed manifold M. Let G be a simply connected Lie group of dimension q, and $\mathfrak{g}$ its Lie algebra of left invariant vector fields. A *transverse Lie structure* of $\mathcal{F}$, with *structural Lie group* G and *structural Lie algebra* $\mathfrak{g}$, can be described with any of the following objects that determine each other [11, 19]:

(L.1) A transverse (G, G)-structure of $\mathcal{F}$, where G is identified with the group of its left translations.

(L.2) A $\mathfrak{g}$-valued 1-form ω on M such that $\omega_x : T_x M \to \mathfrak{g}$ is surjective with kernel $T_x \mathcal{F}$ for every $x \in M$, and

$$d\omega + \frac{1}{2}[\omega, \omega] = 0 \ .$$

(L.3) A homomorphism $\theta : \mathfrak{g} \to \overline{\mathfrak{X}}(M, \mathcal{F})$ such that the composite

$$\mathfrak{g} \xrightarrow{\ \theta\ } \overline{\mathfrak{X}}(M, \mathcal{F}) \xrightarrow{\ \overline{\mathrm{ev}}_x\ } T_x M / T_x \mathcal{F}$$

is an isomorphism for every $x \in M$.

In (L.1), the elements of G whose corresponding left translations are involved in the definition of the transverse (G, G)-structure form a subgroup Γ, which is called the *holonomy group* of $\mathcal{F}$. So the transverse (G, G)-structure is a transverse (G, Γ)-structure. In (L.2) and (L.3), ω and θ can be respectively called the *structural form* and the *structural infinitesimal transverse action*.

A C^∞ foliation endowed with a transverse Lie structure is called a *Lie foliation*; the terms *Lie G-foliation* or *Lie $\mathfrak{g}$-foliation* are used too. If the leaves are dense, then the transverse Lie structure is unique, and thus it is determined by the foliation.

A Lie G-foliation $\mathcal{F}$ on a C^∞ closed manifold M has the following description due to Fedida [11, 19]. There exists a regular covering $\pi : \widetilde{M} \to M$, a fibre bundle $D : \widetilde{M} \to G$ and an injective homomorphism $h : \mathrm{Aut}(\pi) \to G$ such that the leaves of $\widetilde{\mathcal{F}} = \pi^* \mathcal{F}$ are the fibres of D, and D is h-equivariant; *i.e.*,

$$D \circ \sigma(\tilde{x}) = h(\sigma) \cdot D(\tilde{x})$$

for all $\tilde{x} \in \widetilde{M}$ and $\sigma \in \mathrm{Aut}(\pi)$. This h is called the *holonomy homomorphism*. By using the covering space $\ker(h) \backslash \widetilde{M}$ of M if necessary, we can assume that h is injective, and thus π restricts to diffeomorphisms of the leaves of $\widetilde{\mathcal{F}}$ to the leaves of $\mathcal{F}$. The leaf of $\widetilde{\mathcal{F}}$ through each point $\tilde{x} \in \widetilde{M}$ will be denoted by $\widetilde{L}_{\tilde{x}}$.

Given a (G,G)-valued foliated cocycle $\{U_i, f_i\}$ defining the transverse Lie structure according to (L.1), the $\mathfrak{g}$-valued 1-form ω of (L.2) and the infinitesimal transverse action θ of (L.3) can be defined as follows. For $x \in U_i$ and $v \in T_x M$, $\omega_x(v)$ is the left invariant vector field on G whose value at $f_i(x)$ is $f_{i*}(v)$. To define θ, fix an auxiliary vector subbundle $\nu \subset TM$ complementary of $T\mathcal{F}$ ($TM = \nu \oplus T\mathcal{F}$). Each $X \in \mathfrak{g}$ defines a C^∞ vector field $X^\nu \in \mathfrak{X}(M, \mathcal{F})$ by the conditions $X^\nu(x) \in \nu_x$ and $f_{i*}(X^\nu(x)) = X(f_i(x))$ if $x \in U_i$. Then $\theta(X)$ is the class of X^ν in $\overline{\mathfrak{X}}(M, \mathcal{F})$, which is independent of the choice of ν.

By using Fedida's geometric description of $\mathcal{F}$, the definitions of ω and X^ν can be better understood:

- Let ω_G be the canonical $\mathfrak{g}$-valued 1-form on G defined by $\omega_G(X(g)) = X$ for any $X \in \mathfrak{g}$ and any $g \in G$. Then ω is determined by the condition $\pi^* \omega = D^* \omega_G$.
- Let $\tilde{\nu} = \pi_*^{-1}(\nu) \subset T\widetilde{M}$, which is a vector subbundle complementary of $T\widetilde{\mathcal{F}}$. Then, for any $X \in \mathfrak{g}$, there is a unique $\widetilde{X}^\nu \in \mathfrak{X}(\widetilde{M}, \widetilde{\mathcal{F}})$ which is a section of $\tilde{\nu}$ and satisfies $D_* \circ \widetilde{X}^\nu = X \circ D$. Since D is h-equivariant, $\widetilde{X}^\nu$ is $\mathrm{Aut}(\pi)$-invariant. Then X^ν is the projection of $\widetilde{X}^\nu$ to M.

4. Structural transverse action

Let G be a simply connected Lie group, and let $\mathcal{F}$ be a Lie G-foliation on a closed manifold M. According to Section 2, the structural infinitesimal transverse action corresponds to a unique right transverse action of G on $(M, \mathcal{F})$, obtaining another description of the transverse Lie structure:

(L.4) A C^∞ right transverse action Φ of G on $(M, \mathcal{F})$ which has a C^∞ local representation ϕ around the identity element e of G such that the composite

$$T_e G \xrightarrow{\phi^x_*} T_x M \longrightarrow T_x M / T_x \mathcal{F}$$

is an isomorphism for all $x \in M$, where $\phi^x = \phi(x, \cdot)$ and the second map is the canonical projection. This condition is independent of the choice of ϕ. This Φ is called the *structural transverse action*.

To describe Φ, consider Fedida's geometric description of $\mathcal{F}$ (Section 3). For any $g \in G$, take a continuous, piecewise C^∞ path $c : I \to G$ with $c(0) = e$ and $c(1) = g$. For any $\tilde{x} \in \widetilde{M}$, there exists a unique continuous piecewise C^∞ path $\tilde{c}^\nu_{\tilde{x}} : I \to \widetilde{M}$ such that

- $\tilde{c}^\nu_{\tilde{x}}(0) = \tilde{x}$,
- $\tilde{c}^\nu_{\tilde{x}}$ is tangent to $\tilde{\nu}$ at every $t \in I$ where it is C^∞, and
- $D \circ \tilde{c}^\nu_{\tilde{x}}(t) = D(\tilde{x}) \cdot c(t)$ for any $t \in I$.

It is easy to see that such a $\tilde{c}^\nu_{\tilde{x}}$ depends smoothly on $\tilde{x}$.

Lemma 4.1. *We have* $\sigma \circ \tilde{c}^\nu_{\tilde{x}} = \tilde{c}^\nu_{\sigma(\tilde{x})}$ *for* $\tilde{x} \in \widetilde{M}$ *and* $\sigma \in \mathrm{Aut}(\pi)$.

Proof. This is a direct consequence of the h-equivariance of D and the unicity of the paths $\tilde{c}_{\tilde{x}}^{\nu}$. □

For each $g \in G$, let $\tilde{\phi}_g : (\widetilde{M}, \widetilde{\mathcal{F}}) \to (\widetilde{M}, \widetilde{\mathcal{F}})$ be the C^∞ foliated diffeomorphism given by $\tilde{\phi}_g(\tilde{x}) = \tilde{c}_{\tilde{x}}^{\nu}(1)$. For any $\tilde{x} \in \widetilde{M}$ and $\sigma \in \mathrm{Aut}(\pi)$, we have

$$\sigma \circ \tilde{\phi}_g(\tilde{x}) = \sigma \circ \tilde{c}_{\tilde{x}}^{\nu}(1) = \tilde{c}_{\sigma(\tilde{x})}^{\nu}(1) = \tilde{\phi}_g \circ \sigma(\tilde{x})$$

by Lemma 4.1, yielding $\sigma \circ \tilde{\phi}_g = \tilde{\phi}_g \circ \sigma$. Therefore, there exists a unique C^∞ foliated diffeomorphism $\phi_g : (M, \mathcal{F}) \to (M, \mathcal{F})$ such that $\pi \circ \tilde{\phi}_g = \phi_g \circ \pi$.

Lemma 4.2. *The C^∞ leafwise homotopy class of ϕ_g is independent of the choice of c.*

Proof. Let $d : I \to G$ be another continuous and piecewise smooth path with $d(0) = e$ and $d(1) = g$, which defines a C^∞ foliated map $\varphi_g : (M, \mathcal{F}) \to (M, \mathcal{F})$ as above. Since G is simply connected, there exists a family of continuous and piecewise smooth paths $c_s : I \to G$, depending smoothly on $s \in I$, with $c_s(0) = e$, $c_s(1) = g$, $c_0 = c$ and $c_1 = d$. The paths c_s induce a family of C^∞ foliated maps $\phi_{g,s} : (M, \mathcal{F}) \to (M, \mathcal{F})$ as above, defining a C^∞ leafwise homotopy between ϕ_g and φ_g. □

Lemma 4.3. *The C^∞ leafwise homotopy class of ϕ_g is independent of the choice of ν.*

Proof. Let $\nu' \subset TM$ be another vector subbundle complementary of $T\mathcal{F}$, which can be used to define a C^∞ foliated map ϕ'_g as above. It is easy to find a C^∞ deformation of vector subbundles of $\nu_s \subset TM$ complementary of $T\mathcal{F}$, $s \in I$, with $\nu_0 = \nu$ and $\nu_1 = \nu'$. Then the foliated maps $\phi_{g,s}$, induced by the vector bundles ν_s as above, define a C^∞ leafwise homotopy between ϕ_g and ϕ'_g. □

Therefore, for each g, the C^∞ leafwise homotopy class Φ_g of ϕ_g depends only on g, $\mathcal{F}$ and its transverse Lie structure. So a map $\Phi : G \to \overline{\mathrm{Diff}}(M, \mathcal{F})$ is given by $g \mapsto \Phi_g$.

Lemma 4.4. *Φ is a right transverse action of G in $(M, \mathcal{F})$.*

Proof. Given $g_1, g_2 \in G$, let $c_1, c_2 : I \to G$ be continuous, piecewise smooth paths such that $c_1(0) = c_2(0) = e$, $c_1(1) = g_1$ and $c_2(1) = g_2$, which are used to define ϕ_{g_1} and ϕ_{g_2} as above. Let $c : I \to G$ be the path product of c_1 and $L_{g_1} \circ c_2$, where L_{g_1} denotes the left translation by g_1. We have $c(0) = e$ and $c(1) = g_1 g_2$. We can use this c to define $\phi_{g_1 g_2}$, obtaining $\phi_{g_1 g_2} = \phi_{g_2} \circ \phi_{g_1}$, and thus $\Phi_{g_1 g_2} = \Phi_{g_2} \circ \Phi_{g_1}$. □

Lemma 4.5. *Φ is C^∞.*

Proof. It is easy to prove that each element of G has a neighbourhood O such that there is a C^∞ map $c : I \times O \to G$ so that each $c_g = c(\cdot, g)$ is a path from e to g. The corresponding foliated diffeomorphisms ϕ_g form a C^∞ representation of Φ on O. □

This construction defines the structural transverse action Φ. According to Section 2, Φ induces a left action Φ^* of G on $\overline{H}(\mathcal{F})$.

Lemma 4.6. *There is a local representation $\varphi : M \times O \to M$ of Φ around the identity element e such that $\varphi_e = \mathrm{id}_M$.*

Proof. Construct ϕ like in the proof of Lemma 4.5 such that $e \in O$ and c_e is the constant path at e. $\qquad\square$

Let $\varphi : M \times O \to M$ be a local representation of Φ. A map $\tilde{\varphi} : \widetilde{M} \times O \to \widetilde{M}$ is called a *lift* of φ if $\pi \circ \tilde{\varphi}_g = \varphi_g \circ \pi$ for all $g \in O$, where $\tilde{\varphi}_g = \tilde{\varphi}(\cdot, g)$. In particular, the above construction of ϕ also gives a lift $\tilde{\phi}$. Let $R_g : G \to G$ denote the right translation by any $g \in G$.

Lemma 4.7. *Any C^∞ lift $\tilde{\varphi} : \widetilde{M} \times O \to \widetilde{M}$ of each C^∞ local representation $\varphi : M \times O \to M$ of Φ, such that O is connected, satisfies $D \circ \tilde{\varphi}_g = R_g \circ D$ for all $g \in O$.*

Proof. It is enough to prove the result when O is as small as desired. It is clear that the property of the statement is satisfied by the maps $\tilde{\phi}$ constructed above for connected O.

For an arbitrary φ, if O is small enough and connected, there is some $\phi : M \times O \to M$ defined by the above construction and some homotopy $H : M \times O \times I \to M$ between φ and ϕ such that each path $t \mapsto H(x, g, t)$ is contained in a leaf of $\mathcal{F}$. This H lifts to a homotopy $\widetilde{H} : \widetilde{M} \times O \times I \to \widetilde{M}$ between $\tilde{\varphi}$ and $\tilde{\phi}$ so that each path $t \mapsto \widetilde{H}(\tilde{x}, g, t)$ is contained in a leaf of $\widetilde{\mathcal{F}}$. Then $D \circ \tilde{\varphi} = D \circ \tilde{\phi}$, completing the proof. $\qquad\square$

Corollary 4.8. *$\tilde{\varphi} : \widetilde{L} \times O \to \widetilde{M}$ is a C^∞ embedding for each leaf $\widetilde{L}$ of $\widetilde{\mathcal{F}}$.*

The transverse Lie structure of $\mathcal{F}$ lifts to a transverse Lie structure of $\widetilde{\mathcal{F}}$, whose structural right transverse action is locally represented by the C^∞ lifts of C^∞ local representations of Φ.

5. The Hodge isomorphism

Recall that any Lie foliation is Riemannian [23]. Then fix a bundle-like metric on M [23], and equip the leaves of $\mathcal{F}$ with the induced Riemannian metric. Let $\delta_\mathcal{F}$ denote the leafwise coderivative on the leaves operating in $\Omega(\mathcal{F})$, and set $D_\mathcal{F} = d_\mathcal{F} + \delta_\mathcal{F}$. Then $\Delta_\mathcal{F} = D_\mathcal{F}^2 = d_\mathcal{F} \circ d_\mathcal{F} + d_\mathcal{F} \circ \delta_\mathcal{F}$ is the leafwise Laplacian operating in $\Omega(\mathcal{F})$. Let $\mathcal{H}(\mathcal{F}) = \ker \Delta_\mathcal{F}$ (the space of leafwise harmonic forms which are smooth on M). Since the metric is bundle-like, the transverse volume element is holonomy invariant, which implies that $D_\mathcal{F}$ and $\Delta_\mathcal{F}$ are symmetric, and thus they have the same kernel.

Let $\mathbf{\Omega}(\mathcal{F})$ be the Hilbert space of square integrable leafwise differential forms on M. The metric of M induces a Hilbert structure in $\mathbf{\Omega}(\mathcal{F})$. For any C^∞ foliated map $f : (M, \mathcal{F}) \to (M, \mathcal{F})$, the endomorphism f^* of $\Omega(\mathcal{F})$ is obviously L^2-bounded, and thus extends to a bounded operator f^* in $\mathbf{\Omega}(\mathcal{F})$. Consider $D_\mathcal{F}$ and $\Delta_\mathcal{F}$ as unbounded operators in $\mathbf{\Omega}(\mathcal{F})$, which are essentially self-adjoint [8], and whose closures are denoted by $\mathbf{D}_\mathcal{F}$ and $\mathbf{\Delta}_\mathcal{F}$ (see, *e.g.*, [4, 16]). By [2], $\mathcal{H}(\mathcal{F}) = \ker \mathbf{\Delta}_\mathcal{F}$ is the closure of $\mathcal{H}(\mathcal{F})$ in $\mathbf{\Omega}(\mathcal{F})$, and the orthogonal projection $\mathbf{\Pi} : \mathbf{\Omega}(\mathcal{F}) \to \mathcal{H}(\mathcal{F})$ has a restriction $\Pi : \Omega(\mathcal{F}) \to \mathcal{H}(\mathcal{F})$, which induces a leafwise Hodge isomorphism

$$\overline{H}(\mathcal{F}) \cong \mathcal{H}(\mathcal{F}) .$$

For any C^∞ foliated map $f : (M, \mathcal{F}) \to (M, \mathcal{F})$, the homomorphism $f^* : \overline{H}(\mathcal{F}) \to \overline{H}(\mathcal{F})$ corresponds to the operator $\Pi \circ f^*$ in $\mathcal{H}(\mathcal{F})$ via the Hodge isomorphism. So the left G-action on $\overline{H}(\mathcal{F})$, defined in Section 4, corresponds to the left G-action on $\mathcal{H}(\mathcal{F})$ given by $(g, \alpha) \mapsto \Pi \circ \phi_g^* \alpha$ for any $\phi_g \in \Phi_g$.

Since the left action of G on $\mathcal{H}(\mathcal{F})$ is L^2-continuous, we get an extended left action of G on $\mathcal{H}(\mathcal{F})$ given by $(g, \alpha) \mapsto \mathbf{\Pi} \circ \phi_g^* \alpha$ for any $\phi_g \in \Phi_g$.

These actions on $\mathcal{H}(\mathcal{F})$ and $\mathcal{H}(\mathcal{F})$ are continuous on G since Φ is C^∞.

6. A class of smoothing operators

6.1. Preliminaries on smoothing and trace class operators

Let ω_M denote the volume forms of M. A *smoothing operator* in $\Omega(\mathcal{F})$ is a linear map $P : \Omega(\mathcal{F}) \to \Omega(\mathcal{F})$, continuous with respect to the C^∞ topology, given by

$$(P\alpha)(x) = \int_M k(x, y) \, \alpha(y) \, \omega_M(y)$$

for some C^∞ section k of $\bigwedge T\mathcal{F}^* \boxtimes \bigwedge T\mathcal{F}$ over $M \times M$; thus

$$k(x, y) \in \bigwedge T\mathcal{F}_x^* \otimes \bigwedge T\mathcal{F}_y \equiv \mathrm{Hom}(\bigwedge T\mathcal{F}_y^*, \bigwedge T\mathcal{F}_x^*)$$

for any $x, y \in M$. This k is called the *smoothing kernel* or *Schwartz kernel* of P. Such a P defines a trace class operator in $\mathbf{\Omega}(\mathcal{F})$, and we have

$$\mathrm{Tr}\, P = \int_M \mathrm{Tr}\, k(x, x) \, \omega_M(x) .$$

The supertrace formalism will be also used. For any homogeneous operator T in $\Omega(\mathcal{F})$ or in $\bigwedge T_x \mathcal{F}^*$, let $T^\pm$ denote its restriction to the even and odd degree part, and let $T^{(i)}$ denote its restriction to the part of degree i. If T is of trace class, then its supertrace is

$$\mathrm{Tr}^s\, T = \mathrm{Tr}\, T^+ - \mathrm{Tr}\, T^- = \sum_i (-1)^i \, \mathrm{Tr}\, T^{(i)} .$$

Thus

$$\mathrm{Tr}^s\, P = \int_M \mathrm{Tr}^s\, k(x, x) \, \omega_M(x) .$$

Let $W^k \Omega(\mathcal{F})$ denote the Sobolev space of order k of leafwise differential forms on M, and let $\|\cdot\|_k$ denote a norm of $W^k \Omega(\mathcal{F})$. A continuous operator P in $\Omega(\mathcal{F})$ is smoothing if and only if P extends to a bounded operator $P : W^k \Omega(\mathcal{F}) \to W^l \Omega(\mathcal{F})$ for any k and l.

If an operator P in $\Omega(\mathcal{F})$ has an extension $P : W^k\Omega(\mathcal{F}) \to W^\ell\Omega(\mathcal{F})$, then $\|P\|_{k,\ell}$ denotes the norm of this extension; the notation $\|P\|_k$ is used when $k = \ell$. By the Sobolev embedding theorem, the trace of a smoothing operator P in $\Omega(\mathcal{F})$ can be estimated in the following way: for any $k > \dim M$, there is some $C > 0$ independent of P such that

$$| \operatorname{Tr} P| \leq C \, \|P\|_{0,k} \ . \tag{6.1}$$

6.2. The class $\mathcal{D}$

Let $\mathcal{A}$ be the set of all functions $\psi : \mathbb{R} \to \mathbb{C}$, extending to an entire function ψ on $\mathbb{C}$ such that, for each compact set $K \subset \mathbb{R}$, the set of functions $\{(x \mapsto \psi(x+iy)) \mid y \in K\}$ is bounded in the Schwartz space $\mathcal{S}(\mathbb{R})$. This $\mathcal{A}$ has a structure of Fréchet algebra, and, in fact, it is a module over $\mathbb{C}[z]$. This algebra contains all functions with compactly supported Fourier transform, and the functions $x \mapsto e^{-tx^2}$ with $t > 0$.

By [25, Proposition 4.1], there exists a "functional calculus map" $\mathcal{A} \to \operatorname{End}(\Omega(\mathcal{F}))$, $\psi \mapsto \psi(D_\mathcal{F})$, which is a continuous homomorphism of $\mathbb{C}[z]$-modules and of algebras. Any operator $\psi(D_\mathcal{F})$, $\psi \in \mathcal{A}$, extends to a bounded operator in $W^k\Omega(\mathcal{F})$ for any k with the following estimate for its norm: there is some $C > 0$, independent of ψ, such that

$$\|\psi(D_\mathcal{F})\|_k \leq \int |\hat{\psi}(\xi)| \, e^{C\,|\xi|} \, d\xi \ , \tag{6.2}$$

where $\hat{\psi}$ denotes the Fourier transform of ψ. Therefore, for any natural N, the operator $(\operatorname{id} + \Delta_\mathcal{F})^N \psi(D_\mathcal{F})$ extends to a bounded operator in $W^k\Omega(\mathcal{F})$ for any k whose norm can be estimated as follows: there is some $C > 0$, independent of ψ, such that

$$\|(\operatorname{id} + \Delta_\mathcal{F})^N \psi(D_\mathcal{F})\|_k \leq \int |(\operatorname{id} - \partial_\xi^2)^N \hat{\psi}(\xi)| \, e^{C\,|\xi|} \, d\xi \ . \tag{6.3}$$

Fix a left-invariant Riemannian metric on G, and let Λ denote its volume form. We can assume that the metrics on M and G agree in the sense that the maps f_i of (L.1) are Riemannian submersions (Section 3). Thus $D : \widetilde{M} \to G$ is a Riemannian submersion with respect to the lift of the bundle-like metric to $\widetilde{M}$.

A *leafwise differential* operator in $\Omega(\mathcal{F})$ is a differential operator which involves only leafwise derivatives; for instance, $d_\mathcal{F}$, $\delta_\mathcal{F}$, $D_\mathcal{F}$ and $\Delta_\mathcal{F}$ are leafwise differential operators. A family of leafwise differential operators in $\Omega(\mathcal{F})$, $A = \{A_v \mid v \in V\}$, is said to be *smooth* when V is a C^∞ manifold and, with respect to C^∞ local coordinates, the local coefficients of each A_v depend smoothly on v in the C^∞-topology. We also say that A is *compactly supported* when there is some compact subset $K \subset V$ such that $A_v = 0$ if $v \notin K$. Given another smooth family of leafwise differential operators in $\Omega(\mathcal{F})$ with the same parameter manifold, $B = \{B_v \mid v \in V\}$, the *composite* $A \circ B$ is the family defined by $(A \circ B)_v = A_v \circ B_v$. Similarly, we can define the *sum* $A + B$ and the *product* $\lambda \cdot A$ for some $\lambda \in \mathbb{R}$.

We introduce the class $\mathcal{D}$ of operators $P : \Omega(\mathcal{F}) \to \Omega(\mathcal{F})$ of the form

$$P = \int_O \phi_g^* \circ A_g \, \Lambda(g) \circ \psi(D_{\mathcal{F}}) \,,$$

where O is some open subset of G, $\phi : M \times O \to M$ is a C^∞ local representation of Φ, $A = \{A_g \mid g \in O\}$ is a smooth compactly supported family of leafwise differential operators in $\Omega(\mathcal{F})$, and $\psi \in \mathcal{A}$.

Proposition 6.1. *Any operator $P \in \mathcal{D}$ is a smoothing operator in $\Omega(\mathcal{F})$.*

Proof. Let $P \in \mathcal{D}$ as above. By (6.3) and since the operator ϕ_g^* preserves any Sobolev space, P defines a bounded operator in $W^k\Omega(\mathcal{F})$ for any k.

Let $\varphi : M \times O_0 \to M$ be a C^∞ local representation of Φ on some open neighborhood O_0 of the identity element e; we can assume that $\varphi_e = \mathrm{id}_M$ by Corollary 4.8. For any $Y \in \mathfrak{g}$, let $\widehat{Y}$ be the first-order differential operator in $\Omega(\mathcal{F})$ defined by

$$\widehat{Y}u = \frac{d}{dt} \varphi_{\exp tY}^* u \Big|_{t=0} \,,$$

which makes sense because $\exp tY \in O_0$ for any $t > 0$ small enough.

Fix a base $Y_1, \ldots, Y_q$ of $\mathfrak{g}$. Then the second-order differential operator $L = -\sum_{j=1}^q \widehat{Y}_j^2$ in $\Omega(\mathcal{F})$ is transversely elliptic. Moreover $\Delta_{\mathcal{F}}$ is leafwise elliptic. By the elliptic regularity theorem, it suffices to prove that $L^N \circ P$ and $\Delta_{\mathcal{F}}^N \circ P$ belong to $\mathcal{D}$ for any natural N. In turn, this follows by showing that $Q \circ P$ and $\widehat{Y} \circ P$ are in $\mathcal{D}$ for any leafwise differential operator Q and any $Y \in \mathfrak{g}$.

We have

$$Q \circ P = \int_O \phi_g^* \circ B_g \, \Lambda(g) \circ \psi(D_{\mathcal{F}}) \,,$$

where $B_g = (\phi_g^*)^{-1} \circ Q \circ \phi_g^* \circ A_g$. Since ϕ_g is a foliated map, it follows that $\{B_g \mid g \in O\}$ is a smooth family of leafwise differential operators, yielding $Q \circ P \in \mathcal{D}$.

For $g \in O$ and $a \in O_0$ close enough to e, let

$$F_{a,g} = \phi_{ag} \circ \varphi_a \circ \phi_g^{-1} \,.$$

Observe that $F_{e,g} = \mathrm{id}_M$ because $\varphi_e = \mathrm{id}_M$. For each $Y \in \mathfrak{g}$, we get a smooth family $V_Y = \{V_{Y,g} \mid g \in O\}$ of first-order leafwise differential operators in $\Omega(\mathcal{F})$ given by

$$V_{Y,g}u = \frac{d}{dt} F_{\exp tY,g}^* u \Big|_{t=0} \,.$$

Let also $L_Y A = \{(L_Y A)_g \mid g \in O\}$ be the smooth family of leafwise differential operators given by

$$(L_Y A)_g u = \frac{d}{dt} A_{\exp(-tY) \cdot g} u \Big|_{t=0} \,.$$

In particular, if A_g is given by multiplication by $f(g)$ for some $f \in C_c^\infty(G)$, then $(L_Y A)_g$ is given by multiplication by $(Yf)(g)$.

We proceed as follows:

$$\int_O \varphi^*_{\exp tY} \circ \phi^*_g \circ A_g \Lambda(g) = \int_O \phi^*_{\exp tY \cdot g} \circ F^*_{\exp tY, \exp(-tY) \cdot g} \circ A_g \Lambda(g)$$

$$= \int_O \phi^*_g \circ F^*_{\exp tY, g} \circ A_{\exp tY \cdot g} \Lambda(g) \, ,$$

yielding

$$\widehat{Y} \circ P = \lim_{t \to 0} \frac{1}{t} \left(\int_O \varphi^*_{\exp tY} \circ \phi^*_g \circ A_g \, dg - \int_O \phi^*_g \circ A_g \, dg \right) \circ \psi(D_{\mathcal{F}})$$

$$= \lim_{t \to 0} \frac{1}{t} \left(\int_O \phi^*_g \circ F^*_{\exp tY, g} \circ A_{\exp tY \cdot g} \, dg - \int_O \phi^*_g \circ A_g \, dg \right) \circ \psi(D_{\mathcal{F}})$$

$$= \int_O \phi^*_g \circ (V_Y \circ A + L_Y A)_g \, dg \circ \psi(D_{\mathcal{F}}) \, .$$

So $\widehat{Y} \circ P \in \mathcal{D}$. □

With the above notation, by the proof of Proposition 6.1 and (6.3), it can be easily seen that, for integers $k \leq \ell$, there are some $C, C' > 0$ and some natural N such that

$$\|P\|_{k,\ell} \leq C' \int |(\mathrm{id} - \partial^2_\xi)^N \hat{\psi}(\xi)| \, e^{C|\xi|} \, d\xi \, . \tag{6.4}$$

Here, C depends on k and ℓ, and C' depends on k, ℓ and A.

6.3. A norm estimate

Let

$$P = \int_O \phi^*_g \cdot f(g) \Lambda(g) \circ \psi(D_{\mathcal{F}}) \in \mathcal{D} \, ,$$

where ϕ and ψ are like in Section 6.2, and $f \in C^\infty_c(O)$. In this case, (6.4) is improved by the following result, where Δ_G denotes the Laplacian of G.

Proposition 6.2. *Let $K \subset O$ be a compact subset containing $\mathrm{supp} \, f$. For naturals $k \leq \ell$, there are some $C, C'' > 0$ and some natural N, depending only on K, k and ℓ, such that*

$$\|P\|_{k,\ell} \leq C'' \max_{g \in K} |(\mathrm{id} + \Delta_G)^N f(g)| \int |(\mathrm{id} - \partial^2_\xi)^N \hat{\psi}(\xi)| \, e^{C|\xi|} \, d\xi \, .$$

Proof. Fix an orthonormal frame $Y_1, \ldots, Y_q$ of $\mathfrak{g}$. Consider any multi-index $J = (j_1, \ldots, j_k)$ with $j_1, \ldots, j_k \in \{1, \ldots, q\}$. We use the standard notation $|J| = k$, and, with the notation of the proof of Proposition 6.1, let:

- $Y_J = Y_{j_1} \circ \cdots \circ Y_{j_k}$ (operating in $C^\infty(G)$);
- $\widehat{Y}_J = \widehat{Y}_{j_1} \circ \cdots \circ \widehat{Y}_{j_k}$;
- $V_J = V_{Y_{j_1}} \circ \cdots \circ V_{Y_{j_k}}$; and
- $L_J A = L_{Y_{j_1}} \cdots L_{Y_{j_k}} A$ for any smooth family A of leafwise differential operators in $\Omega(\mathcal{F})$.

Consider the empty multi-index $\emptyset$ too, with $|\emptyset| = 0$, and define:

- $Y_\emptyset = \mathrm{id}_{C^\infty(G)}$;
- $\widehat{Y}_\emptyset = \mathrm{id}_{\Omega(\mathcal{F})}$;
- $V_{\emptyset,g} = \mathrm{id}_{\Omega(\mathcal{F})}$ for all $g \in O$, defining a smooth family $V_\emptyset$; and
- $L_\emptyset A = A$ for any smooth family A of leafwise differential operators in $\Omega(\mathcal{F})$.

Given any natural N, there is some $C_1 > 0$ such that

$$\|\phi_g^*\|_k \le C_1 \,, \qquad \|(L_J V_{J'})_g\| \le C_1 \,,$$

$$\|(Y_J f)(g)\| \le C_1 \max_{g \in K} |(\mathrm{id} + \Delta_G)^N f(g)| \,,$$

$$\|(\mathrm{id} + \phi_g^{*-1} \circ \Delta_\mathcal{F} \circ \phi_g^*)^N \circ \psi(\Delta_\mathcal{F})\|_k \le C_1 \|(\mathrm{id} + \Delta_\mathcal{F})^N \circ \psi(D_\mathcal{F})\|_k$$

for all $g \in K$ and all multi-indices J and J' with $|J|, |J'| \le N$.

For any multi-index J, we have

$$\widehat{Y}_J \circ P = \int_O \phi_g^* \circ A_{J,g} \, \Lambda(g) \circ \psi(D_\mathcal{F}) \,,$$

where $A_J = \{A_{J,g} \mid g \in G\}$ is the smooth family of leafwise differential operators inductively defined by setting

$$A_{\emptyset,g} = \mathrm{id}_{\Omega(\mathcal{F})} \cdot f(g) \,,$$
$$A_{(j,J)} = V_j \circ A_J + L_j A_J \,.$$

By induction on $|J|$, we easily get that A_J is a sum of smooth families of leafwise differential operators of the form

$$L_{J_1} V_{J_1'} \circ \cdots \circ L_{J_\ell} V_{J_\ell'} \cdot Y_{J''} f \,,$$

where $J_1, J_1', \ldots, J_\ell, J_\ell', J''$ are possibly empty multi-indices satisfying

$$|J_1| + |J_1'| + \cdots + |J_\ell| + |J_\ell'| + |J''| = |J| \,.$$

So there is some $C_2 > 0$ such that

$$\|A_{J,g}\|_k \le C_2 \max_{g \in K} |(\mathrm{id} + \Delta_G)^N f(g)|$$

for all $g \in K$ and every multi-index J with $|J| \le N$. Hence

$$\|\widehat{Y}_J \circ P\|_k \le \int_O \|\phi_g^*\|_k \|A_{J,g}\|_k \, dg \, \|\psi(D_\mathcal{F})\|_k$$

$$\le C_1 C_2 \max_{g \in K} |(\mathrm{id} + \Delta_G)^N f(g)| \int |\hat\psi(\xi)| \, e^{C|\xi|} \, d\xi$$

for some $C > 0$ by (6.2). On the other hand,

$$\|(\mathrm{id} + \Delta_\mathcal{F})^N \circ P\|_k \le \int_O \|(\mathrm{id} + \phi_g^{*-1} \circ \Delta_\mathcal{F} \circ \phi_g^*)^N \circ \psi(\Delta_\mathcal{F})\|_k \, |f(g)| \, \Lambda(g)$$

$$\le C_1 \int_O \|(\mathrm{id} + \Delta_\mathcal{F})^N \circ \psi(\Delta_\mathcal{F})\|_k \, |f(g)| \, \Lambda(g)$$

$$\le C_1 \max_{g \in K} |f(g)| \int |(\mathrm{id} - \partial_\xi^2)^N \hat\psi(\xi)| \, e^{C|\xi|} \, d\xi$$

for some $C > 0$ by (6.3). Now, the result follows because $-\sum_{j=1}^{q} \widehat{Y_j}^2$ is transversely elliptic, and $\Delta_{\mathcal{F}}$ is leafwise elliptic. □

6.4. Parameter independence of the supertrace

Choose an even function in $\mathcal{A}$, which can be written as $x \mapsto \psi(x^2)$. Take also a C^∞ local representation $\phi : M \times O \to M$ of Φ and some $f \in C_c^\infty(O)$. Then consider the one parameter family of operators $P_t \in \mathcal{D}$, $t > 0$, defined by

$$P_t = \int_O \phi_g^* \cdot f(g) \, \Lambda(g) \circ \psi(t\Delta_{\mathcal{F}})^2 \, .$$

Lemma 6.3. $\mathrm{Tr}^s P_t$ *is independent of* t.

Proof. The proof is similar to the proof of the corresponding result in the heat equation proof of the Lefschetz trace formula (see, *e.g.*, [28]). We have

$$\frac{d}{dt} \, \mathrm{Tr}^s P_t = 2 \, \mathrm{Tr}^s \int_O \phi_g^* \cdot f(g) \, \Lambda(g) \circ \Delta_{\mathcal{F}} \circ \psi'(t\Delta_{\mathcal{F}}) \circ \psi(t\Delta_{\mathcal{F}})$$

$$= 2 \, \mathrm{Tr} \int_O \phi_g^* \cdot f(g) \, \Lambda(g) \circ d_{\mathcal{F}}^- \circ \delta_{\mathcal{F}}^+ \circ \psi'(t\Delta_{\mathcal{F}}^+) \circ \psi(t\Delta_{\mathcal{F}}^+)$$

$$- 2 \, \mathrm{Tr} \int_O \phi_g^* \cdot f(g) \, \Lambda(g) \circ d_{\mathcal{F}}^+ \circ \delta_{\mathcal{F}}^- \circ \psi'(t\Delta_{\mathcal{F}}^-) \circ \psi(t\Delta_{\mathcal{F}}^-)$$

$$+ 2 \, \mathrm{Tr} \int_O \phi_g^* \cdot f(g) \, \Lambda(g) \circ \delta_{\mathcal{F}}^- \circ d_{\mathcal{F}}^+ \circ \psi'(t\Delta_{\mathcal{F}}^+) \circ \psi(t\Delta_{\mathcal{F}}^+)$$

$$- 2 \, \mathrm{Tr} \int_O \phi_g^* \cdot f(g) \, \Lambda(g) \circ \delta_{\mathcal{F}}^+ \circ d_{\mathcal{F}}^- \circ \psi'(t\Delta_{\mathcal{F}}^-) \circ \psi(t\Delta_{\mathcal{F}}^-) \, .$$

On the other hand, since the function $x \mapsto \psi'(x^2)$ is in $\mathcal{A}$, we have

$$\mathrm{Tr} \int_O \phi_g^* \cdot f(g) \, \Lambda(g) \circ d_{\mathcal{F}}^\mp \circ \delta_{\mathcal{F}}^\pm \circ \psi'(t\Delta_{\mathcal{F}}^\pm) \circ \psi(t\Delta_{\mathcal{F}}^\pm)$$

$$= \mathrm{Tr} \, d_{\mathcal{F}}^\mp \circ \int_O \phi_t^* \cdot f(g) \, \Lambda(g) \circ \psi'(t\Delta_{\mathcal{F}}^\pm) \circ \psi(t\Delta_{\mathcal{F}}^\pm) \circ \delta_{\mathcal{F}}^\pm$$

$$= \mathrm{Tr} \, \psi(t\Delta_{\mathcal{F}}^\pm) \circ \delta_{\mathcal{F}}^\pm \circ d_{\mathcal{F}}^\mp \circ \int_O \phi_g^* \cdot f(g) \, \Lambda(g) \circ \psi'(t\Delta_{\mathcal{F}}^\pm)$$

$$= \mathrm{Tr} \int_O \phi_g^* \cdot f(g) \, \Lambda(g) \circ \psi'(t\Delta_{\mathcal{F}}^\pm) \circ \psi(t\Delta_{\mathcal{F}}^\pm) \circ \delta_{\mathcal{F}}^\pm \circ d_{\mathcal{F}}^\mp$$

$$= \mathrm{Tr} \int_O \phi_g^* \cdot f(g) \, \Lambda(g) \circ \delta_{\mathcal{F}}^\pm \circ d_{\mathcal{F}}^\mp \circ \psi'(t\Delta_{\mathcal{F}}^\pm) \circ \psi(t\Delta_{\mathcal{F}}^\pm) \, ,$$

where we have used the well-known fact that, if A is a trace class operator and B is bounded, then AB and BA are trace class operators with the same trace. Therefore $\frac{d}{dt} \, \mathrm{Tr}^s P_t = 0$ as desired. □

6.5. The global action on the leafwise complex

Let $\mathfrak{G}$ be the holonomy groupoid of $\mathcal{F}$. Since the leaves of Lie foliations have trivial holonomy groups, we have

$$\mathfrak{G} \equiv \{(x, y) \in M \times M \mid x \text{ and } y \text{ lie in the same leaf of } \mathcal{F}\} \ .$$

This is a C^∞ submanifold of $M \times M$ which contains the diagonal Δ_M. Let $d_\mathcal{F}$ be the distance function of the leaves of $\mathcal{F}$. For each $r > 0$, the *r-penumbra* of Δ_M in $\mathfrak{G}$ is defined by

$$\mathrm{Pen}_\mathfrak{G}(\Delta_M, r) = \{(x, y) \in \mathfrak{G} \mid d_\mathcal{F}(x, y) < r\} \ .$$

Observe that a subset of $\mathfrak{G}$ has compact closure if and only if it is contained in some penumbra of Δ_M. The product of two elements $(x_1, y_1), (x_2, y_2) \in \mathfrak{G}$ is defined when $y_1 = x_2$, and it is equal to (x_1, y_2). The space of units of $\mathfrak{G}$ is $\Delta_M \equiv M$. The source and target projections $s, r : \mathfrak{G} \to M$ are the restrictions of the first and second factor projections $M \times M \to M$; thus

$$r^{-1}(x) = L_x \times \{x\} \ , \quad s^{-1}(x) = \{x\} \times L_x$$

for each $x \in M$.

Let S denote the C^∞ vector bundle

$$s^* \bigwedge T\mathcal{F}^* \otimes r^* \bigwedge T\mathcal{F}$$

over $\mathfrak{G}$; thus

$$S_{(x,y)} \equiv \bigwedge T_x\mathcal{F}^* \otimes \bigwedge T_y\mathcal{F} \equiv \mathrm{Hom}(\bigwedge T_y\mathcal{F}^*, \bigwedge T_x\mathcal{F}^*)$$

for each $(x, y) \in \mathfrak{G}$. Let $\omega_\mathcal{F}$ be the volume form of the leaves of $\mathcal{F}$ (we assume that $\mathcal{F}$ is oriented). Recall that $C_c^\infty(S)$ is an algebra with the convolution product given by

$$(k_1 \cdot k_2)(x, y) = \int_{L_x} k_1(x, z) \circ k_2(z, y) \, \omega_\mathcal{F}(z)$$

for $k_1, k_2 \in C_c^\infty(S)$ and $(x, y) \in \mathfrak{G}$. Recall also that the *global action* of $C_c^\infty(S)$ in $\Omega(\mathcal{F})$ is defined by

$$(k \cdot \alpha)(x) = \int_{L_x} k(x, y) \, \alpha(y) \, \omega_\mathcal{F}(y)$$

for $k \in C_c^\infty(S)$, $\alpha \in \Omega(\mathcal{F})$ and $x \in M$.

Consider the lift to $\widetilde{M}$ of the bundle-like metric of M, and its restriction to the leaves of $\widetilde{\mathcal{F}}$. Let $U\Omega(\widetilde{\mathcal{F}}) \subset \Omega(\widetilde{\mathcal{F}})$ be the subcomplex of differential forms α whose covariant derivatives $\nabla^r \alpha$ of arbitrary order r are uniformly bounded; this is a Fréchet space with the metric induced by the seminorms

$$\|\|\alpha\|\|_r = \sup\{\nabla^r \alpha(\tilde{x}) \mid \tilde{x} \in \widetilde{M}\} \ .$$

Observe that $\pi^*(\Omega(\mathcal{F})) \subset U\Omega(\widetilde{\mathcal{F}})$.

The holonomy groupoid $\widetilde{\mathfrak{G}}$ of $\widetilde{\mathcal{F}}$ satisfies the same properties as $\mathfrak{G}$, except that, in $\widetilde{\mathfrak{G}}$, the penumbras of the diagonal $\Delta_{\widetilde{M}}$ have compact closure if and only $\widetilde{M}$ is compact.

The map $\pi \times \pi : \widetilde{M} \times \widetilde{M} \to M \times M$ restricts to a covering map $\widetilde{\mathfrak{G}} \to \mathfrak{G}$, whose group of deck transformations is isomorphic to $\mathrm{Aut}(\pi)$: for each $\sigma \in \mathrm{Aut}(\pi)$, the corresponding element in $\mathrm{Aut}(\widetilde{\mathfrak{G}} \to \mathfrak{G})$ is the restriction $\sigma \times \sigma : \widetilde{\mathfrak{G}} \to \widetilde{\mathfrak{G}}$.

Let $\widetilde{S}$ denote the C^∞ vector bundle

$$\widetilde{s}^* \bigwedge T\widetilde{\mathcal{F}}^* \otimes \widetilde{r}^* \bigwedge T\widetilde{\mathcal{F}}$$

over $\widetilde{\mathfrak{G}}$, and let $C_\Delta^\infty(\widetilde{S}) \subset C^\infty(\widetilde{S})$ denote the subspace of sections supported in some penumbra of $\Delta_{\widetilde{M}}$. As above, this set becomes an algebra with the convolution product, and there is a *global action* of $C_\Delta^\infty(\widetilde{S})$ in $U\Omega(\widetilde{\mathcal{F}})$.

Any $k \in C^\infty(S)$ lifts via $\pi \times \pi$ to a section $\widetilde{k} \in C^\infty(\widetilde{S})$. Since π restricts to diffeomorphisms of the leaves of $\widetilde{\mathcal{F}}$ to the leaves of $\mathcal{F}$, it follows that $\widetilde{k} \in C_\Delta^\infty(\widetilde{S})$ if $k \in C_c^\infty(S)$.

Take any $\psi \in \mathcal{A}$. For each leaf L of $\mathcal{F}$, denoting by Δ_L the Laplacian of L, the spectral theorem defines a smoothing operator $\psi(\Delta_L)$ in $\mathbf{\Omega}(L)$, and the family

$$\{\psi(\Delta_L) \mid L \text{ is a leaf of } \mathcal{F}\}$$

is also denoted by $\psi(\Delta_{\mathcal{F}})$. By [26, Proposition 2.10], the Schwartz kernels k_L of the operators $\psi(\Delta_L)$ can be combined to define a section $k \in C^\infty(S)$, called the *leafwise smoothing kernel* or *leafwise Schwartz kernel* of $\psi(\Delta_{\mathcal{F}})$.

Suppose that the Fourier transform $\hat{\psi}$ of ψ is supported in $[-R, R]$ for some $R > 0$. Then, according to the proof of Assertion 1 in [25, page 461], k is supported in the R-penumbra of Δ_M, and thus $k \in C_c^\infty(S)$. Moreover the operator $\psi(D_{\mathcal{F}})$ in $\Omega(\mathcal{F})$, defined by the spectral theorem, equals the operator given by the global action of k.

Consider also the lift $\widetilde{k} \in C_\Delta^\infty(\widetilde{S})$, whose global action in $U\Omega(\widetilde{\mathcal{F}})$ defines an operator denoted by $\psi(D_{\widetilde{\mathcal{F}}})$. It is clear that the diagram

$$
\begin{array}{ccc}
U\Omega(\widetilde{\mathcal{F}}) & \xrightarrow{\psi(D_{\widetilde{\mathcal{F}}})} & U\Omega(\widetilde{\mathcal{F}}) \\
{\scriptstyle \pi^*}\uparrow & & \uparrow{\scriptstyle \pi^*} \\
\Omega(\mathcal{F}) & \xrightarrow{\psi(D_{\mathcal{F}})} & \Omega(\mathcal{F})
\end{array}
\tag{6.5}
$$

commutes.

Any function $\psi \in \mathcal{A}$ with compactly supported Fourier transform can be modified as follows to achieve the condition of being supported in $[-R, R]$. For each $t > 0$, let $\psi_t \in \mathcal{A}$ be the function defined by $\psi_t(x) = \psi(tx)$.

Lemma 6.4. *If $\hat{\psi}$ is compactly supported for some $\psi \in \mathcal{A}$, then $\widehat{\psi_t}$ is supported in $[-R, R]$ for t small enough.*

Proof. This holds because $\widehat{\psi_t}(\xi) = \frac{1}{t}\hat{\psi}(\frac{\xi}{t})$. $\square$

6.6. Schwartz kernels

Let ϕ, f, ψ and P be like in Section 6.3 such that $\hat{\psi}$ is compactly supported. Take some $R > 0$ so that $\operatorname{supp} \hat{\psi} \subset [-R, R]$. Let $k \in C_c^\infty(S)$ be the leafwise kernel of $\psi(D_{\mathcal{F}})$, and let $\tilde{k} \in C_\Delta^\infty(\tilde{S})$ be the lift of k, whose action in $\Omega(\widetilde{\mathcal{F}})$ defines the operator $\psi(D_{\widetilde{\mathcal{F}}})$ (Section 6.5).

Let $\tilde{\phi} : \widetilde{M} \times O \to \widetilde{M}$ be a C^∞ lift of ϕ. Define $\widetilde{P} : U\Omega(\widetilde{\mathcal{F}}) \to U\Omega(\widetilde{\mathcal{F}})$ by

$$\widetilde{P} = \int_O \tilde{\phi}_g^* \cdot f(g)\, \Lambda(g) \circ \psi(D_{\widetilde{\mathcal{F}}}).$$

The commutativity of the diagram

$$
\begin{array}{ccc}
U\Omega(\widetilde{\mathcal{F}}) & \xrightarrow{\ \widetilde{P}\ } & U\Omega(\widetilde{\mathcal{F}}) \\
{\scriptstyle \pi^*}\big\uparrow & & \big\uparrow{\scriptstyle \pi^*} \\
\Omega(\mathcal{F}) & \xrightarrow{\ P\ } & \Omega(\mathcal{F})
\end{array}
$$

follows from the commutativity of (6.5).

Let $\omega_{\widetilde{\mathcal{F}}}$ be the volume form of the leaves of $\widetilde{\mathcal{F}}$, which can be also considered as a differential form on M that vanishes when some vector is orthogonal to the leaves. Thus the volume form of $\widetilde{M}$ is $\omega_{\widetilde{M}} = D^*\Lambda \wedge \omega_{\widetilde{\mathcal{F}}}$ with the right choice of orientations. For $\tilde{x} \in \widetilde{M}$ and $\alpha \in U\Omega(\widetilde{\mathcal{F}})$, we have

$$
\begin{aligned}
(\widetilde{P}\alpha)(\tilde{x}) &= \Big(\int_O \tilde{\phi}_g^* \cdot f(g)\, \Lambda(g) \circ \psi(D_{\widetilde{\mathcal{F}}})\alpha\Big)(\tilde{x}) \\
&= \int_O \tilde{\phi}_g^*((\psi(D_{\widetilde{\mathcal{F}}})\alpha)(\tilde{\phi}_g(\tilde{x}))) \cdot f(g)\, \Lambda(g) \\
&= \int_O \int_{\widetilde{L}_{\tilde{x}}} \tilde{\phi}_g^* \circ \tilde{k}(\tilde{\phi}_g(\tilde{x}), \tilde{y})(\alpha(\tilde{y}))\, \omega_{\widetilde{\mathcal{F}}}(\tilde{y}) \cdot f(g)\, \Lambda(g) \\
&= \int_{\phi(\widetilde{L}_{\tilde{x}} \times O)} \tilde{\phi}_g^* \circ \tilde{k}(\tilde{\phi}_g(\tilde{x}), \tilde{y})(\alpha(\tilde{y})) \cdot f(g)\, \omega_{\widetilde{M}}(\tilde{y})
\end{aligned}
$$

by Corollary 4.8, where $g \in O$ is determined by the condition $\tilde{y} \in \tilde{\phi}_g(\widetilde{L}_{\tilde{x}})$, which means $g = D(\tilde{x})^{-1}D(\tilde{y})$ by Lemma 4.7. So we can say that $\widetilde{P}$ is given by the Schwartz kernel $\tilde{p}$ defined by

$$
\tilde{p}(\tilde{x}, \tilde{y}) = \begin{cases} \tilde{\phi}_g^* \circ \tilde{k}(\tilde{\phi}_g(\tilde{x}), \tilde{y}) \cdot f(g) & \text{if } \tilde{y} \in \tilde{\phi}(\widetilde{L}_{\tilde{x}} \times O) \\ 0 & \text{otherwise} \end{cases} \tag{6.6}
$$

for $g \in O$ as above. It follows that

$$
p(x, y) = \sum_{\sigma \in \operatorname{Aut}(\pi)} \tilde{p}(\tilde{x}, \sigma(\tilde{y})), \tag{6.7}
$$

where $\tilde{x} \in \pi^{-1}(x)$, $\tilde{y} \in \pi^{-1}(y)$, and we use identifications $T_{\tilde{x}}\widetilde{\mathcal{F}} \equiv T_x\mathcal{F}$ and $T_{\sigma(\tilde{y})}\widetilde{\mathcal{F}} \equiv T_y\mathcal{F}$ given by π_*.

For each $x \in M$, $\tilde{x} \in \widetilde{M}$ and $r > 0$, let $B_{\mathcal{F}}(x, r)$ and $B_{\widetilde{\mathcal{F}}}(\tilde{x}, r)$ be the r-balls of centers x and $\tilde{x}$ in L_x and $\tilde{L}_{\tilde{x}}$, respectively. Let O_1 be an open subset of G whose closure is compact and contained in O. By the compactness of $M \times \overline{O}_1$, there is some $R_1 > 0$ such that

$$B_{\mathcal{F}}(\phi_g(x), R) \subset \phi_g(B_{\mathcal{F}}(x, R_1)) \tag{6.8}$$

for all $x \in M$ and all $g \in O_1$. So

$$B_{\widetilde{\mathcal{F}}}(\tilde{\phi}_g(\tilde{x}), R) \subset \tilde{\phi}_g(B_{\widetilde{\mathcal{F}}}(\tilde{x}, R_1)) \tag{6.9}$$

for all $\tilde{x} \in \widetilde{M}$ and all $g \in O_1$ because π restricts to isometries of the leaves of $\widetilde{\mathcal{F}}$ to the leaves of $\mathcal{F}$.

Lemma 6.5. *Each $g \in O$ has a neighborhood O_1 as above such that*

$$\pi : \tilde{\phi}(\overline{B_{\widetilde{\mathcal{F}}}(\tilde{x}, R_1)} \times O_1) \to M$$

is injective for any $\tilde{x} \in \widetilde{M}$.

Proof. Since M is compact, there exists a compact subset $K \subset \widetilde{M}$ with $\pi(K) = M$. Notice that, if the statement holds for some $\tilde{x} \in \widetilde{M}$, then it also holds for all points in the $\mathrm{Aut}(\pi)$-orbit of $\tilde{x}$. So, if the statement fails, there exist sequences $\tilde{x}_i, \tilde{y}_i \in \widetilde{M}$ and $\sigma_i \in \mathrm{Aut}(\pi)$ such that $\tilde{x}_i \in K$, $\sigma_i \neq \mathrm{id}_{\widetilde{M}}$, and

$$d_{\widetilde{M}}(\{\tilde{y}_i, \sigma_i(\tilde{y}_i)\}, \tilde{\phi}_g(B_{\widetilde{\mathcal{F}}}(\tilde{x}_i, R_1))) \to 0$$

as $i \to \infty$; observe that $D(\tilde{x}_i)^{-1} D(\tilde{y}_i) \to g$ by Lemma 4.7. Since K is compact, we can assume that there exists $\lim_i \tilde{x}_i = \tilde{x} \in \widetilde{M}$, where $d_{\widetilde{M}}$ denotes the distance function of $\widetilde{M}$. Hence $\tilde{y}_i$ and $\sigma_i(\tilde{y}_i)$ approach $\tilde{\phi}_g(B_{\widetilde{\mathcal{F}}}(\tilde{x}, R_1))$. Since $\tilde{\phi}_g(B_{\widetilde{\mathcal{F}}}(\tilde{x}, R_1))$ has compact closure, it follows that $\tilde{y}_i$ and $\sigma_i(\tilde{y}_i)$ lie in some compact neighborhood Q of $\tilde{\phi}_g(B_{\widetilde{\mathcal{F}}}(\tilde{x}, R_1))$ for infinitely many indices i, yielding $\sigma_i(Q) \bigcap Q \neq \emptyset$. So there is some $\sigma \in \mathrm{Aut}(\pi)$ such that $\sigma_i = \sigma$ for infinitely many indices i. In particular, $\sigma \neq \mathrm{id}_{\widetilde{M}}$.

On the other hand, since $\tilde{y}_i$ and $\sigma_i(\tilde{y}_i)$ approach $\tilde{\phi}_g(B_{\widetilde{\mathcal{F}}}(\tilde{x}, R_1))$, which has compact closure, we can assume that there exist $\lim_i \tilde{y}_i = \tilde{y}$ and $\lim_i \sigma_i(\tilde{y}_i) = \sigma(\tilde{y})$ in $\tilde{\phi}_g(B_{\widetilde{\mathcal{F}}}(\tilde{x}, R_1))$, which is contained in the leaf $\tilde{\phi}_g(\tilde{L}_{\tilde{x}})$ (a fiber of D). So

$$D(\tilde{y}) = D(\sigma(\tilde{y})) = h(\sigma) \cdot D(\tilde{y}) ,$$

yielding $h(\sigma) = e$, and thus $\sigma = \mathrm{id}_{\widetilde{M}}$ because h is injective. This contradiction concludes the proof. $\qquad\square$

From now on, assume that ϕ satisfies (6.8) and the property of the statement of Lemma 6.5 with some fixed open subset $O_1 \subset O$ which contains the support of f.

Corollary 6.6. *The map π is injective on the support of $\tilde{p}(\tilde{x}, \cdot)$ for any $\tilde{x} \in \widetilde{M}$.*

Proof. By (6.6), (6.9) and since $\tilde{k}$ is supported in the R-penumbra of $\Delta_{\widetilde{M}}$, we get

$$\text{supp}(\tilde{p}(\tilde{x}, \cdot)) \subset \tilde{\phi}(B_{\widetilde{\mathcal{F}}}(\tilde{x}, R_1) \times O_1)$$

for any $\tilde{x} \in \widetilde{M}$, and the result follows from Lemma 6.5. □

Corollary 6.7. *We have*

$$p(x, y) = \begin{cases} \phi_g^* \circ k(\phi_g(x), y) \cdot f(g) & \text{if } y \in \phi(B_{\mathcal{F}}(x, R_1) \times O_1) \\ 0 & \text{otherwise,} \end{cases}$$

where $g \in O_1$ is determined by the condition $y \in \phi_g(B_{\mathcal{F}}(x, R_1))$.

Proof. This is a consequence of (6.6), (6.7), Corollary 6.6 and Lemma 6.5. □

Corollary 6.8. *If $e \in O_1$ and $\phi_e = \text{id}_M$, then*

$$p(x, x) = k(x, x) \cdot f(e) .$$

Proof. Since $\phi_e = \text{id}_M$, the result follows from Corollary 6.7 and the following assertion.

Claim 1. For all $g \in O_1$ and $x \in M$, if $x \in \phi_g(B_{\mathcal{F}}(x, R_1))$, then $g = e$.

By Lemma 6.5,

$$\pi : \tilde{\phi}(B_{\widetilde{\mathcal{F}}}(\tilde{x}, R_1) \times O_1) \to \phi(B_{\mathcal{F}}(x, R_1) \times O_1)$$

is a diffeomorphism. On the other hand,

$$\tilde{\phi} : \widetilde{L}_{\tilde{x}} \times O_1 \to \tilde{\phi}(\widetilde{L}_{\tilde{x}} \times O_1)$$

is a diffeomorphism as well by Corollary 4.8. It follows that

$$\phi : B_{\mathcal{F}}(x, R_1) \times O_1 \to \phi(B_{\mathcal{F}}(x, R_1) \times O_1)$$

is also a diffeomorphism, which implies Claim 1 because $\phi_e(x) = x$. □

Lemma 6.9. *For $i \in \{1, 2\}$, suppose that $x_i \in \phi_{g_i}(B_{\mathcal{F}}(x_i, R_1))$ for some $(x_i, g_i) \in M \times O_1$. If x_2 is close enough to x_1, then there is some $a \in G$ such that $x_2 \in \Phi_a(L_{x_1})$ and $g_2 = a^{-1}g_1 a$.*

Proof. We have

$$\Phi_a(L_{x_1}) = \Phi_a \circ \Phi_{g_1}(L_{x_1}) = \Phi_{a^{-1}g_1 a} \circ \Phi_a(L_{x_1})$$

for all $a \in G$. Therefore, if x_2 is close enough to x_1, there is some $a \in G$ such that $a^{-1}g_1 a \in O_1$ and

$$x_2 \in \Phi_a(L_{x_1}) \cap \phi_{a^{-1}g_1 a}(B_{\mathcal{F}}(x_2, R_1)) .$$

Then the result follows because the condition $x_2 \in \phi_{g_2}(B_{\mathcal{F}}(x_2, R_1))$ determines g_2 in O_1 by Lemma 6.5. □

7. Lefschetz distribution

Let $\phi : M \times O \to M$ be a C^∞ local representation of the structural transverse action Φ on some open subset $O \subset G$. For any $f \in C_c^\infty(O)$ and $t > 0$, let P_f and $Q_{t,f}$ be the operators in $\Omega(\mathcal{F})$ defined by

$$P_f = \int_O \phi_g^* \cdot f(g) \, \Lambda(g) \circ \Pi \, ,$$

$$Q_{t,f} = \int_O \phi_g^* \cdot f(g) \, \Lambda(g) \circ e^{-t\Delta_\mathcal{F}} \, .$$

The operator $Q_{t,f}$ is in the class $\mathcal{D}$, and thus it is smoothing by Proposition 6.1.

Proposition 7.1. P_f is a smoothing operator.

Proof. By [2], Π defines a bounded operator in each Sobolev space $W^k \Omega^i(\mathcal{F})$. Hence, $P_f = Q_{t,f} \circ \Pi$ is smoothing because so is $Q_{t,f}$. $\square$

By Proposition 7.1, P_f is a trace class operator in the space $\Omega(\mathcal{F})$, and thus so is $P_f^{(i)}$.

Proposition 7.2. *The functional $f \mapsto \mathrm{Tr}\, P_f^{(i)}$ is a distribution on O.*

Proof. Since Π is a projection in $\Omega(\mathcal{F})$ and $P_f = Q_{t,f} \circ \Pi$, we have

$$\|P_f^{(i)}\|_{0,k} \leq \|Q_{t,f}^{(i)}\|_{0,k} \, ,$$

and the result follows by (6.1) and Proposition 6.2. $\square$

Proposition 1.1 is given by Propositions 7.1 and 7.2.

Because the endomorphism Φ_g^* of $\overline{H}(\mathcal{F})$ corresponds to the operator $\Pi \circ \phi_g^*$ in $\mathcal{H}(\mathcal{F})$ by the leafwise Hodge isomorphism, the composite $\Pi \circ P_f$ is independent of the choice of ϕ. Moreover $\mathrm{Tr}\, P_f^{(i)} = \mathrm{Tr}(\Pi \circ P_f^{(i)})$. Hence the distributions given by Proposition 7.2 can be combined to form a global distribution $\mathrm{Tr}_{\mathrm{dis}}^i(\mathcal{F})$ on G; in this notation, $\mathcal{F}$ refers to the foliation endowed with the given transverse Lie structure, which indeed is determined by the foliation when the leaves are dense. Each $\mathrm{Tr}_{\mathrm{dis}}^i(\mathcal{F})$ is called a *distributional trace* of $\mathcal{F}$, and define the *Lefschetz distribution* of $\mathcal{F}$ by the formula

$$L_{\mathrm{dis}}(\mathcal{F}) = \sum_i (-1)^i \, \mathrm{Tr}_{\mathrm{dis}}^i(\mathcal{F}) \, .$$

Lemma 7.3. *For any $f \in C_c^\infty(O)$, $\mathrm{Tr}\, Q_{t,f}^{(i)} \to \mathrm{Tr}\, P_f^{(i)}$ as $t \to \infty$.*

Proof. Since $Q_{1,f}$ is smoothing, it defines a bounded operator $W^{-1}\Omega^i(\mathcal{F}) \to W^k \Omega^i(\mathcal{F})$ for any k. By [2], $e^{-(t-1)\Delta_\mathcal{F}} - \Pi$ is bounded in $W^{-1}\Omega^i(\mathcal{F})$ for $t > 1$ and converges strongly to 0 as $t \to \infty$. From the compactness of the canonical embedding $\Omega^i(\mathcal{F}) \to W^{-1}\Omega^i(\mathcal{F})$, it follows that $e^{-(t-1)\Delta_\mathcal{F}} - \Pi$ converges uniformly to 0 as $t \to \infty$ as an operator $\mathbf{\Omega}^i(\mathcal{F}) \to W^{-1}\Omega^i(\mathcal{F})$. Therefore $\|Q_{t,f} - P_f\|_{0,k} \to 0$ as $t \to \infty$ for any k because

$$Q_{t,f} - P_f = Q_{1,f} \circ \left(e^{-(t-1)\Delta_{\mathcal{F}}} - \Pi \right).$$

Then the result follows from (6.1). $\qquad\qquad\qquad\qquad\qquad\qquad\qquad\qquad\qquad\qquad\square$

Corollary 7.4. $\operatorname{Tr}^s Q_{t,f} = \operatorname{Tr}^s P_f$ *for all* t.

Proof. This follows from Lemmas 6.3 and 7.3. $\qquad\qquad\qquad\qquad\qquad\qquad\square$

8. The distributional Gauss-Bonett theorem

The holonomy pseudogroup of $\mathcal{F}$ is represented by the pseudogroup on G generated by the left translations given by elements of Γ. Thus Λ can be considered as a holonomy invariant transverse measure of $\mathcal{F}$. To be more precise, take a (G, Γ)-valued foliated cocycle $\{U_i, f_i\}$ defining the given transverse Lie structure (Section 3). The differential forms $f_i^* \Lambda$ can be combined to get the transverse volume form ω_Λ of $\mathcal{F}$. We can also describe ω_Λ by the condition $D^*\Lambda = \pi^*\omega_\Lambda$. The restriction of ω_Λ to smooth local transversals is the precise interpretation of Λ as a holonomy invariant measure on local transversals.

By non-commutative integration theory [9], the holonomy invariant transverse measure Λ defines a trace $\operatorname{Tr}_\Lambda$ on the twisted foliation von Neumann algebra $W^*(M, \mathcal{F}, \bigwedge T\mathcal{F}^*)$. Consider also the corresponding supertrace $\operatorname{Tr}_\Lambda^s$, equal to $\pm \operatorname{Tr}_\Lambda$, depending on whether the even-odd bigrading is preserved or interchanged.

With the notation of Section 6.5, we have $C_c^\infty(S) \subset W^*(M, \mathcal{F}, \bigwedge T\mathcal{F}^*)$; here, each $k \in C_c^\infty(S)$ is identified to the family of operators on the leaves whose Schwartz kernels are the restrictions of k, and moreover

$$\operatorname{Tr}_\Lambda(k) = \int_M \operatorname{Tr} k(x, x)\, \omega_M(x)\,, \quad \operatorname{Tr}_\Lambda^s(k) = \int_M \operatorname{Tr}^s k(x, x)\, \omega_M(x)\,.$$

For each leaf L, let $\mathbf{\Omega}(L)$ denote the Hilbert space of L^2 differential forms on L, let $\mathcal{H}(L) \subset \mathbf{\Omega}(L)$ be the subspace of harmonic L^2 forms, and let Π_L be the orthogonal projection $\mathbf{\Omega}(L) \to \mathcal{H}(L)$. The family

$$\Pi_{\mathcal{F}} = \{\Pi_L \mid L \text{ is a leaf of } \mathcal{F}\}$$

defines a projection in $W^*(M, \mathcal{F}, \bigwedge T\mathcal{F}^*)$. The notation $\Pi_L^{(i)}$ and $\Pi_{\mathcal{F}}^{(i)}$ is used when we are only considering differential forms of degree i. For each leaf L, let $S_L = S|_{L \times L}$, and let $k_L, k_L^{(i)} \in C^\infty(S_L)$ denote the Schwartz kernels of Π_L and $\Pi_L^{(i)}$. These sections can be combined to define measurable sections k and $k^{(i)}$ of S, called the *leafwise Schwartz kernels* of $\Pi_{\mathcal{F}}$ and $\Pi_{\mathcal{F}}^{(i)}$. Since k and $k^{(i)}$ are C^∞ along the fibers of the source and target projections, their restrictions to the diagonal Δ_M are measurable, and we have

$$\operatorname{Tr}_\Lambda(\Pi_{\mathcal{F}}^{(i)}) = \int_M \operatorname{Tr} k^{(i)}(x, x)\, \omega_M(x)\,, \quad \operatorname{Tr}_\Lambda^s(\Pi_{\mathcal{F}}) = \int_M \operatorname{Tr}^s k(x, x)\, \omega_M(x)\,.$$

According to [9], the ith Λ-*Betti number* is defined by

$$\beta_\Lambda^i(\mathcal{F}) = \mathrm{Tr}_\Lambda(\Pi_\mathcal{F}^{(i)}) \, ,$$

and the Λ-*Euler characteristic* is given by the formula

$$\chi_\Lambda(\mathcal{F}) = \mathrm{Tr}_\Lambda^s(\Pi_\mathcal{F}) = \sum_i (-1)^i \beta_\Lambda^i(\mathcal{F}) \, .$$

Theorem 8.1. $L_{\mathrm{dis}}(\mathcal{F}) = \chi_\Lambda(\mathcal{F}) \cdot \delta_e$ *in some neighborhood of* e.

Like in [25, p. 463], choose a sequence of smooth even functions on $\mathbb{R}$, written as $x \mapsto \psi_m(x^2)$ with $\psi_m(0) = 1$, whose Fourier transforms are compactly supported and which tend to the function $x \mapsto e^{-x^2/2}$ in the Schwartz space $\mathcal{S}(\mathbb{R})$. Let $k_{m,t}$ be the leafwise Schwartz kernel of $\psi_m(t\Delta_\mathcal{F})^2$, which is in $C_c^\infty(S)$ according to [25]. In [25, p. 463], it is proved that

$$\mathrm{Tr}_\Lambda^s \, \psi_m(t\Delta_\mathcal{F})^2 = \chi_\Lambda(\mathcal{F}) \, . \tag{8.1}$$

Let $\phi : M \times O \to M$ be any C^∞ local representation of Φ on some neighborhood O of e such that $\phi_e = \mathrm{id}_M$, whose existence is given by Lemma 4.6. Given $R > 0$, take $R_1 > 0$ and some open subset O_1 of O containing e such that (6.8) and Lemma 6.5 are satisfied.

For every $f \in C_c^\infty(O)$ supported in O_1, let

$$Q_{m,t,f} = \int_O \phi_g^* \cdot f(g) \, \Lambda(g) \circ \psi_m(t\Delta_\mathcal{F})^2 \in \mathcal{D} \, .$$

Lemma 8.2. $\mathrm{Tr}^s \, Q_{m,t,f} = \chi_\Lambda(\mathcal{F}) \cdot f(e)$.

Proof. By Lemma 6.4, we can apply Corollary 6.8 to $Q_{m,t,f}$ when t is small enough, obtaining

$$\mathrm{Tr}^s \, Q_{m,t,f} = \int_M \mathrm{Tr}^s \, k_{m,t}(x,x) \cdot f(e) \, \omega_M(x)$$
$$= \mathrm{Tr}_\Lambda^s \, \psi_m(t\Delta_\mathcal{F})^2 \cdot f(e) \, .$$

Then the result follows by (8.1). $\qquad\square$

Consider the operators $Q_{t,f}$ and P_f of Section 7.

Lemma 8.3. *We have*

$$\lim_{m\to\infty} \mathrm{Tr}^s \, Q_{m,t,f} = \mathrm{Tr}^s \, Q_{t,f}$$

for each t.

Proof. Since the function $x \mapsto \psi_m(tx^2) - e^{-\frac{t}{2}x^2}$ tends to zero in $\mathcal{A}$ as $m \to \infty$, we get

$$\lim_{m\to\infty} \|Q_{m,t,f} - Q_{t,f}\|_{0,k} = 0$$

for all k by (6.4) (or Lemma 6.2), and the result follows from (6.1). $\qquad\square$

Theorem 8.1 follows from Lemmas 8.2 and 8.3, and Corollary 7.4.

9. The distributional Lefschetz trace formula

Let $\mathcal{F}'$ be the foliation of $M \times G$ whose leaves are the sets $L \times \{g\}$ for leaves L of $\mathcal{F}$ and points $g \in G$. Lemma 6.9 suggests the following definition: for each $x \in M$ and $g \in G$, let

$$M'_{(x,g)} = \bigcup_{a \in G} (\Phi_a(L_x) \times \{a^{-1}ga\}) .$$

Observe that $M'_{(x,e)} = M \times \{e\}$. Moreover $M'_{(x_1,g_1)} = M'_{(x_2,g_2)}$ if and only if $(x_2, g_2) \in M'_{(x_1,g_1)}$; thus these sets form a partition of $M \times G$.

Proposition 9.1. *The sets $M'_{(x,g)}$ are the leaves of a C^∞ foliation $\mathcal{G}$ on $M \times G$.*

Proof. Consider the canonical identity $T_{(x,g)}(M \times G) \equiv T_x M \oplus T_g G$ for each $(x, g) \in M \times G$, and let $\mathrm{Ad} : G \to \mathrm{Aut}(\mathfrak{g})$ denote the adjoint representation of G. With the notation of Section 4, consider the C^∞ vector subbundles $\mathcal{V}, \mathcal{W} \subset T(M \times G)$ given by

$$\mathcal{V}_{(x,g)} = \{(X^\nu(x), (X - \mathrm{Ad}_{g^{-1}}(X))(g)) \mid X \in \mathfrak{g}\} ,$$
$$\mathcal{W}_{(x,g)} = \mathcal{V}_{(x,g)} + T_{(x,g)}\mathcal{F}' .$$

The distribution defined by $\mathcal{V}$ is not completely integrable. Nevertheless, since $[X^\nu, Y^\nu] - [X, Y]^\nu \in \mathfrak{X}(\mathcal{F}')$ for all $X, Y \in \mathfrak{g}$, it follows that the distribution defined by $\mathcal{W}$ is completely integrable. Thus there is a C^∞ foliation $\mathcal{G}$ on $M \times G$ so that $T\mathcal{G} = \mathcal{W}$. It is easy to check that the leaves of $\mathcal{G}$ are the sets $M'_{(x,g)}$. $\square$

Let pr_1 and pr_2 denote the first and second factor projections of $M \times G$ onto M and G, respectively.

Proposition 9.2. *For each leaf M' of $\mathcal{G}$, we have the following:*

(i) *the restriction $\mathrm{pr}_1 : M' \to M$ is a covering map; and*
(ii) *pr_2 restricts to a fiber bundle map of M' to some orbit of the adjoint action of G on itself.*

Proof. For any $x \in M$, there is some open neighborhood P of x in L_x, and some local representation $\varphi : M \times O \to M$ of Φ on some open neighborhood O of e such that φ restricts to a diffeomorphism of $P \times O$ onto some neighborhood U of x. For any $g \in G$ such that $(x, g) \in M'$, the set

$$\widetilde{U}_g = \{(\varphi_a(y), a^{-1}ga) \mid y \in P, \ a \in O\}$$

is an open neighborhood of (x, g) in M', and the restriction $\mathrm{pr}_1 : \widetilde{U}_g \to U$ is a diffeomorphism. Therefore property (i) follows.

It is clear that $\mathrm{pr}_2(M')$ is an orbit of the adjoint action of G on itself, and that $\mathrm{pr}_2 : M' \to \mathrm{pr}_2(M')$ is a C^∞ submersion; thus its fibers are C^∞ submanifolds. If $(x, g) \in M'$, it can be easily seen that

$$\mathrm{pr}_2^{-1}(g) \cap M' = \{(\phi_a(y), g) \mid y \in L_x, \ a \in G_g, \ \phi_g \in \Phi_g\} ,$$

where G_g is the centralizer of g in G. For $\varphi : M \times O \to M$ as above, the set $O' = \{b^{-1}gb \mid b \in O\}$ is an open neighborhood of g in $\mathrm{pr}_2(M')$. Let

$$F : O' \times (\mathrm{pr}_2^{-1}(g) \cap M') \to \mathrm{pr}_2^{-1}(O') \cap M'$$

be the map defined by

$$F(b^{-1}gb; \varphi_a(y), g) = (\varphi_{b^{-1}ab} \circ \varphi_b(y), b^{-1}gb)$$

for $y \in L_x$, $a \in G_g$ and $b \in O'$. It is easy to see that F is a C^∞ diffeomorphism, which shows property (ii). $\qquad \square$

Observe that $\mathcal{F}'$ is a subfoliation of $\mathcal{G}$, and, for each leaf M' of $\mathcal{G}$, the restriction $\mathcal{F}'|_{M'}$ is equal to the lift of $\mathcal{F}$ by $\mathrm{pr}_1 : M' \to M$.

Let $\phi : M \times O \to M$ be any C^∞ local representation of Φ. Given $R > 0$, take $R_1 > 0$ and some open subset O_1 of O containing e such that (6.8) and Lemma 6.5 are satisfied. Let

$$\mathcal{S} = \{(x, g) \in M \times O_1 \mid x \in \phi_g(B_{\mathcal{F}}(x, R_1))\} .$$

Proposition 9.3. *We have:*

(i) *$\mathcal{S}$ is contained in a finite union of leaves of $\mathcal{G}$; and*
(ii) *the restriction $\mathrm{pr}_1 : \mathcal{S} \to M$ is injective.*

Proof. Property (i) is a consequence of Lemma 6.9 and the compactness of M. Property (ii) follows from Lemma 6.5. $\qquad \square$

Let $\phi' : M \times O \to M \times O$ be the C^∞ diffeomorphism defined by $\phi'(x, g) = (\phi(x, g), g)$. Observe that ϕ' is a foliated map $\mathcal{F}'|_{M \times O} \to \mathcal{F}'|_{M \times O}$.

Proposition 9.4. *Let M' be a leaf of $\mathcal{G}$. If ϕ' preserves some leaf of $\mathcal{F}'|_{M' \cap (M \times O)}$, then it preserves every leaf of $\mathcal{F}'|_{M' \cap (M \times O)}$.*

Proof. Take some point (x, g) in a leaf L' of $\mathcal{F}'|_{M' \cap (M \times O)}$; thus $L' = L_x \times \{g\}$. Suppose $\phi'(L') \subset L'$, which means $\Phi_g(L_x) = L_x$. Any leaf of $\mathcal{F}'|_{M' \cap (M \times O)}$ is of the form $\Phi_a(L_x) \times \{a^{-1}ga\}$ for some $a \in G$. We have

$$\Phi_{a^{-1}ga} \circ \Phi_a(L_x) = \Phi_{ga}(L_x) = \Phi_a \circ \Phi_g(L_x) = \Phi_a(L_x) .$$

So ϕ' preserves $\Phi_a(L_x) \times \{a^{-1}ga\}$. $\qquad \square$

According to Proposition 9.3, if O_1 is small enough, then $\mathcal{S}$ is contained in a leaf M' of $\mathcal{G}$; this property is assumed from now on. Let $M_1' = M' \cap (M \times O_1)$ and $\mathcal{F}_1' = \mathcal{F}|_{M_1'}$. By Proposition 9.4, ϕ' maps each leaf of M' to itself, and thus can be restricted to a map $\phi_1' : M_1' \to M_1'$, which is a foliated map $(M_1', \mathcal{F}_1') \to (M_1', \mathcal{F}_1')$.

Consider the volume form Λ of G as a transverse invariant measure of $\mathcal{F}$. By Proposition 9.2-(i), Λ lifts to a transverse invariant measure Λ_1' of $\mathcal{F}'$. Similarly, the Riemannian metric of M lifts to a Riemannian metric of M', which can be restricted to M_1'; the volume form of this restriction is denoted by $\omega_{M_1'}$.

Even though the foliated manifolds of [14] are compact, it is clear that its Lefschetz theorem for foliations with transverse invariant measures generalizes to the non-compact case when the transverse invariant measure is compactly supported.

In our case, M_1' may not be compact, but, for every $f \in C^\infty(O)$ supported in O_1, $\Lambda_{1,f}' = \mathrm{pr}_2^* f \cdot \Lambda_1'$ of $\mathcal{F}_1'$ is a compactly supported transverse invariant measure of $\mathcal{F}_1'$. Therefore, according to [14], the $\Lambda_{1,f}'$-Lefschetz number $L_{\Lambda_{1,f}'}(\phi_1')$ of ϕ_1' can be defined.

Theorem 9.5. *With the above notation and conditions, we have*

$$\langle L_{\mathrm{dis}}(\mathcal{F}), f \rangle = L_{\Lambda_{1,f}'}(\phi_1')$$

for every $f \in C_c^\infty(O)$ supported in O_1.

The proof of Theorem 9.5 is analogous to the proof of Theorem 8.1. The holonomy groupoid $\mathfrak{G}_1'$ of $\mathcal{F}_1'$ can be described like $\mathfrak{G}$ in Section 6.5 as a C^∞ submanifold of $M_1' \times M_1'$ containing the diagonal. Its penumbras of the diagonal can be also defined like those of $\mathfrak{G}$. Its source and target projections are denoted by $s_1', r_1' : \mathfrak{G}_1' \to M_1'$. The restriction $\mathrm{pr}_1 \times \mathrm{pr}_1 : \mathfrak{G}_1' \to \mathfrak{G}$ is a covering map by Proposition 9.2-(i).

Let S_1' be the C^∞ vector bundle

$$s_1'^* \bigwedge T\mathcal{F}_1'^* \otimes r_1'^* \bigwedge T\mathcal{F}_1'$$

over $\mathfrak{G}_1'$, which can be identified with $(\mathrm{pr}_1 \times \mathrm{pr}_1)^* S$. The space of C^∞ sections of S_1' supported in penumbras of the diagonal will be denoted by $C_\Delta^\infty(S_1')$. Like in Section 6.5, there is a global action of $C_\Delta^\infty(S_1')$ in $\Omega(\mathcal{F}_1')$.

For each leaf L' of $\mathcal{F}_1'$, the composite $\phi_1'^* \circ \Pi_{L'}$ is a smoothing operator on L', and let $k_{\phi,L'}'$ denote its smoothing kernel. All of these smoothing kernels can be combined to define a measurable section k_ϕ of S_1' with C^∞ restrictions to the fibers of s_1'; k_ϕ can be called the *leafwise smoothing kernel* or *leafwise Schwartz kernel* of $\phi_1'^* \circ \Pi_{\mathcal{F}_1'}$. So the restriction of k_ϕ to the diagonal $\Delta_{M_1'}$ is measurable too. Then $\phi_1'^* \circ \Pi_{\mathcal{F}_1'}$ defines an element of the von Neumann algebra $W^*(M_1', \mathcal{F}_1', \bigwedge T\mathcal{F}_1'^*)$, and we have

$$L_{\Lambda_{1,f}'}(\phi_1') = \mathrm{Tr}_{\Lambda_1'}^s(\phi_1'^* \circ \Pi_{\mathcal{F}_1'}) = \int_{M_1'} \mathrm{Tr}^s k_\phi(x,x)\, \omega_{M_1'} \ . \tag{9.1}$$

For any $\psi \in \mathcal{A}$ with $\mathrm{supp}\,\psi \subset [-R, R]$, we have defined the leafwise Schwartz kernels $k \in C^\infty(S)$ and $\tilde{k} \in C_\Delta^\infty(S)$ of $\psi(D_\mathcal{F})$ and $\psi(D_{\tilde{\mathcal{F}}})$ in Section 8. Similarly, we can define the leafwise Schwartz kernels $k_1', k_\phi' \in C_\Delta^\infty(S_1')$ of $\psi(D_{\mathcal{F}_1'})$ and $\phi_1'^* \circ \psi(\Delta_{\mathcal{F}_1'})$, respectively. It is easy to see that k_1' can be identified with the lift of k via $\mathrm{pr}_1 \times \mathrm{pr}_1$. Therefore k_ϕ' is given by

$$k_\phi'((x,g),(y,g)) = \phi_1'^* \circ k_1'(\phi_1'(x,g),(y,g)) \equiv \phi_g^* \circ k(\phi_g(x), y) \ . \tag{9.2}$$

Choose a sequence of functions ψ_m like in Section 8. Let k and $k_{m,t}$ be the leafwise Schwartz kernels of $\Pi_\mathcal{F}$ and $\psi_m(t\Delta_\mathcal{F})^2$, respectively. By [27, Lemma 1.2], $k_{m,t}$ tends to k as $t \to \infty$, and moreover $k_{m,t}$ is uniformly bounded for large m and t. Hence, by (9.2), the leafwise Schwartz kernel $k_{\phi,m,t}'$ of $\phi_1'^* \circ \psi_m(t\Delta_{\mathcal{F}_1'})^2$ tends to

k'_ϕ as $t \to \infty$, and $k_{m,t}$ is uniformly bounded for large m and t. Therefore

$$\lim_{t \to \infty} \mathrm{Tr}^s_{\Lambda'_{1,f}}(\phi_1'^* \circ \psi_m(t\Delta_{\mathcal{F}'_1})^2) = L_{\Lambda'_{1,f}}(\phi'_1)$$

for each m by (9.1) and the dominated convergence theorem. Furthermore

$$\mathrm{Tr}^s_{\Lambda'_{1,f}}(\phi_1'^* \circ \psi_m(t\Delta_{\mathcal{F}'_1})^2)$$

is independent of t (see [14, Theorem 5.1]). Therefore

$$\mathrm{Tr}^s_{\Lambda'_{1,f}}(\phi_1'^* \circ \psi_m(t\Delta_{\mathcal{F}'_1})^2) = L_{\Lambda'_{1,f}}(\phi'_1) \tag{9.3}$$

for all m and t.

Let $Q_{m,t,f}$ be defined like in Section 8.

Lemma 9.6. *We have*

$$\mathrm{Tr}^s Q_{m,t,f} = L_{\Lambda'_{1,f}}(\phi'_1) \, .$$

Proof. By Lemma 6.4, the Schwartz kernel $q_{m,t,f}$ of $Q_{m,t,f}$ is given by Corollary 6.7 when t is small enough. So, if $(x, x) \in \mathrm{supp}\, q_{m,t,f}$ for some $x \in M$, we have

$$q_{m,t,f}(x,x) = \phi_g^* \circ k_{m,t}(\phi_g(x), x) \cdot f(g) \, ,$$

where $g \in O$ is determined by the condition $x \in \phi_g(B_{\mathcal{F}}(x, R_1))$; thus $(x, g) \in \mathcal{S} \subset M'_1$. Therefore, since $\mathrm{pr}_1 : \mathcal{S} \to M$ is injective (Proposition 9.3-(ii)),

$$\mathrm{Tr}^s Q_{m,t,f} = \int_{\mathcal{S}} \mathrm{Tr}^s(\phi_g^* \circ k_{m,t}(\phi_g(x), x)) \cdot f(g) \, \omega_{M'_1}(x, g)$$

$$= \int_{M'_1} \mathrm{Tr}^s k'_{\phi,m,t}((x, g), (x, g)) \cdot f(g) \, \omega_{M'_1}(x, g)$$

by (9.2)

$$= \mathrm{Tr}^s_{\Lambda'_{1,f}}(\phi_1'^* \circ \psi_m(t\Delta_{\mathcal{F}'_1}))$$

for t small enough. Then the result follows by (9.3). $\qquad\square$

Theorem 9.5 follows from Lemmas 9.6 and 8.3, and Corollary 7.4.

Now, let us prove Theorem 1.3. Let $\mathrm{Fix}(\phi')$ and $\mathrm{Fix}(\phi'_1)$ denote the fixed point sets of ϕ' and ϕ'_1. Observe that $\mathrm{Fix}(\phi') \subset M'$, and thus

$$\mathrm{Fix}(\phi'_1) = \mathrm{Fix}(\phi') \cap (M \times O_1) \, . \tag{9.4}$$

It is clear that $\mathrm{pr}_2 : \mathrm{Fix}(\phi') \to O$ is a proper map because M is compact and $\mathrm{Fix}(\phi')$ is closed in $M \times O$. Then $\mathrm{pr}_2 : \mathrm{Fix}(\phi'_1) \to O_1$ is proper too by (9.4).

A fixed point (x, g) of ϕ' is said to be *leafwise simple* if $\phi_{g*} - \mathrm{id} : T_x\mathcal{F} \to T_x\mathcal{F}$ is an isomorphism. The set of simple fixed points of ϕ' is denoted by $\mathrm{Fix}_0(\phi')$. Define $\epsilon : \mathrm{Fix}_0(\phi') \to \{\pm 1\}$ by

$$\epsilon(x, g) = \mathrm{sign}\det(\phi_{g*} - \mathrm{id} : T_x\mathcal{F} \to T_x\mathcal{F}) \, .$$

Lemma 9.7. $\mathrm{Fix}_0(\phi')$ *is a C^∞ regular submanifold of M' whose dimension is equal to* $\mathrm{codim}\, \mathcal{F}$.

Proof. Let $\hat{\phi} : M \times O \to M \times M$ be the C^∞ map defined by $\hat{\phi}(x, g) = (x, \phi_g(x))$, and let Δ_M denote the diagonal in $M \times M$. Then $\text{Fix}(\phi') = \hat{\phi}^{-1}(\Delta_M)$.

There is some open subset $U \subset M \times O$ such that $\text{Fix}_0(\phi') = \text{Fix}(\phi') \cap U$. Then the result follows by showing that the restriction $\hat{\phi} : U \to M \times M$ is transverse to Δ_M.

Pick any $(x, g) \in \text{Fix}_0(\phi')$. Let Δ_{L_x} denote the diagonal in $L_x \times L_x$. Consider the canonical identity $T_{(x,x)}(M \times M) \equiv T_x M \oplus T_x M$. The fact that x is a simple fixed point of ϕ_g means that

$$T_x L_x \oplus T_x L_x = \hat{\phi}_*(T_{(x,g)}(L_x \times \{g\})) + T_{(x,x)}\Delta_{L_x} . \tag{9.5}$$

Observe that

$$\mu_x = \phi_*(T_{(x,g)}(\{x\} \times G))$$

is complementary of $T_x \mathcal{F}$, and

$$\hat{\phi}_*(T_{(x,g)}(\{x\} \times G)) = 0_x \oplus \mu_x ,$$

where 0_x denotes the zero subspace of $T_x M$. So

$$\begin{aligned}
T_x M \oplus T_x M &= (T_x L_x \oplus T_x L_x) + T_{(x,x)}\Delta_M \\
&= (T_x L_x \oplus T_x L_x) + (0_x \oplus \mu_x) + T_{(x,x)}\Delta_M \\
&= \hat{\phi}_*(T_{(x,g)}(M \times G)) + T_{(x,x)}\Delta_M
\end{aligned}$$

by (9.5). $\qquad\qquad\square$

Proposition 9.8. $\text{Fix}_0(\phi')$ *is a* C^∞ *transversal of* $\mathcal{F}'|_{M'}$.

Proof. By Lemma 9.7, it is enough to prove that $\text{Fix}_0(\phi')$ is transverse to $\mathcal{F}'|_{M'}$, which follows from the following claim for any point $(x, g) \in \text{Fix}_0(\phi')$.

Claim 2. We have

$$T_{(x,g)}(\text{Fix}_0(\phi')) \cap T_{(x,g)}\mathcal{F}' = 0 .$$

The proof of Claim 2 involves another assertion:

Claim 3. We have

$$T_{(x,g)}(\text{Fix}_0(\phi')) = \ker(\phi_* - \text{pr}_{1*} : T_{(x,g)}M' \to T_x M) .$$

For any $v \in T_{(x,g)}(\text{Fix}_0(\phi'))$, there is a C^∞ curve (x_t, g_t) in $\text{Fix}_0(\phi')$, with $-\epsilon < t < \epsilon$ for some $\epsilon > 0$, such that $(x_0, g_0) = (x, g)$ and $\frac{d}{dt}(x_t, g_t)\big|_{t=0} = v$. We have $\phi(x_t, g_t) = x_t = \text{pr}_1(x_t, g_t)$, yielding $\phi_*(v) = \text{pr}_{1*}(v)$. So

$$v \in \ker(\phi_* - \text{pr}_{1*} : T_{(x,g)}M' \to T_x M) ,$$

obtaining the inclusion "$\subset$" of Claim 3.

Since $\phi_{g*} - \text{id} : T_x \mathcal{F} \to T_x \mathcal{F}$ is an isomorphism, so is $\phi_* - \text{pr}_{1*} : T_{(x,g)}\mathcal{F}' \to T_x \mathcal{F}$. Hence

$$\ker(\phi_* - \text{pr}_{1*} : T_{(x,g)}M' \to T_x M) \cap T_{(x,g)}\mathcal{F}' = 0 ,$$

yielding Claims 2 and 3 because the inclusion "$\subset$" of Claim 3 is already proved. $\qquad\square$

Proposition 9.9. $\mathrm{pr}_2 : \mathrm{Fix}_0(\phi') \to \mathrm{pr}_2(M')$ *is a* C^∞ *submersion.*

Proof. Since the leaves of $\mathcal{F}'$ are contained in the fibers of pr_2, the tangent map pr_{2*} induces a homomorphism $\overline{\mathrm{pr}_{2*}} : T(M \times G)/T\mathcal{F}' \to TG$. Take any $(x, g) \in \mathrm{Fix}_0(\phi')$. By Proposition 9.8 and according to the proof of Proposition 9.1, the restrictions

$$T_{(x,g)}\mathrm{Fix}_0(\phi') \longrightarrow T_{(x,g)}M'/T_{(x,g)}\mathcal{F}' \longleftarrow \mathcal{V}_{(x,g)} \ .$$

of the quotient map $T(M \times G) \to T(M \times G)/T\mathcal{F}'$ are isomorphisms. Moreover pr_{2*} corresponds to $\overline{\mathrm{pr}_{2*}}$ by these isomorphisms. So

$$\mathrm{pr}_{2*}(T_{(x,g)}(\mathrm{Fix}_0(\phi'))) = \{(X - \mathrm{Ad}_{g^{-1}}(X))(g)) \mid X \in \mathfrak{g}\}$$
$$= T_g(\mathrm{pr}_2(M'))$$

by the proof of Proposition 9.1. $\qquad\square$

According to Proposition 9.8, the measure given by Λ' on $\mathrm{Fix}_0(\phi')$ is denoted by $\Lambda'_{\mathrm{Fix}_0(\phi')}$. The direct image $\mathrm{pr}_{2*}(\epsilon \cdot \Lambda'_{\mathrm{Fix}_0(\phi')})$ is supported in $\mathrm{pr}_2(M') \cap O$.

Let ω_Λ be the transverse volume form of $\mathcal{F}$ defined by Λ. Then the transverse volume form of $\mathcal{F}'|_{M'}$ defined by Λ' is $\omega_{\Lambda'} = \mathrm{pr}_1^* \omega_\Lambda$. The restriction of $\omega_{\Lambda'}$ to the C^∞ local transversal $\mathrm{Fix}_0(\phi')$ is a volume form, which can be identified to the measure $\Lambda'_{\mathrm{Fix}_0(\phi')}$. According to Proposition 9.9, $\mathrm{pr}_{2*}(\epsilon \cdot \Lambda'_{\mathrm{Fix}_0(\phi')})$ is given by the top degree differential form on $\mathrm{pr}_2(M') \cap O$ defined by the integration along the fibers

$$\int_{\mathrm{pr}_2} \epsilon \cdot \omega_{\Lambda'}|_{\mathrm{Fix}_0(\phi')} \ .$$

By (9.4), Theorem 9.5 and the Lefschetz theorem of [14], we have

$$\langle L_{\mathrm{dis}}(\mathcal{F}), f \rangle = L_{\Lambda'_{1,f}}(\phi'_1)$$
$$= \int_{\mathrm{Fix}(\phi'_1)} \epsilon(x, g)\, f(g)\, \Lambda'_{\mathrm{Fix}(\phi')}(x)$$
$$= \int_{O_1 \cap \mathrm{pr}_2(M')} f(g)\, \mathrm{pr}_{2*}(\epsilon \cdot \Lambda'_{\mathrm{Fix}(\phi')})(g)$$
$$= \langle \mathrm{pr}_{2*}(\epsilon \cdot \Lambda'_{\mathrm{Fix}(\phi')}), f \rangle \ ,$$

completing the proof of Theorem 1.3.

10. Examples

10.1. Codimension one foliations

Consider the case when $\mathcal{F}$ is a codimension one Lie foliation. So we have $\mathfrak{g} = \mathbb{R}$, $G = \mathbb{R}$, and $\mathcal{F}$ is defined by a closed nonsingular 1-form ω. The leaves of $\mathcal{G}$ in $M \times \mathbb{R}$ are $M'_s = M \times \{s\}$, $s \in \mathbb{R}$. A global C^∞ representation of Φ is given by the flow $\phi : M \times \mathbb{R} \to M$ of an arbitrary vector field X on M such that $\omega(X) = 1$. Then

$$\mathrm{Fix}(\phi') = \{(x, s) \in M \times \mathbb{R} \mid \phi_s(x) = x\} \ .$$

So we have $\text{Fix}(\phi') \cap M'_s \neq \emptyset$ if and only if either $s = 0$ or s is the period of a closed orbit of the flow ϕ. In the latter case, we have

$$\text{Fix}(\phi') \cap M'_s = \bigcup_c \mathcal{O}_c \times \{s\} \,,$$

where c runs over the set of all closed orbits of period s, and $\mathcal{O}_c$ is the corresponding primitive closed orbit:

$$\mathcal{O}_c = \{\phi_t(x) \in M \mid t \in [0, \ell(c)]\}$$

where $x \in c$ is an arbitrary point, and $\ell(c)$ is the length of $\mathcal{O}_c$. Assume that all closed orbits of ϕ are simple. Then $\epsilon : \text{Fix}(\phi') \to \{\pm 1\}$ is constant on each $\mathcal{O}_c \times \{s\} \subset \text{Fix}(\phi') \cap M'_s$, and its value on $\mathcal{O}_c \times \{s\}$ will be denoted by $\epsilon_s(c)$.

The Lebesgue measure $\Lambda = dt$ on $\mathbb{R}$ can be considered as an invariant transverse measure of $\mathcal{F}$. So we have

$$L_{\text{dis}}(\mathcal{F}) = \chi_\Lambda(\mathcal{F}) \cdot \delta_0$$

in some neighborhood of 0. The restriction of the transverse volume form ω'_Λ to $\text{Fix}(\phi') \cap M'_s$ coincides with ω_Λ on each $\mathcal{O}_c$. For any component $\mathcal{O}_c \times \{s\} \subset \text{Fix}(\phi') \cap M'_s$, one can write $s = k\,\ell(c)$ for some $k \neq 0$, and we see that, on $\mathbb{R} \setminus \{0\}$,

$$L_{\text{dis}}(\mathcal{F}) = \text{pr}_2^*(\epsilon \cdot \Lambda'|_{\text{Fix}(\phi')}) = \sum_c \ell(c) \sum_{k \neq 0} \epsilon_{k\,\ell(c)}(c) \cdot \delta_{k\,\ell(c)} \,,$$

where c runs over all primitive closed orbits of the flow ϕ [3].

10.2. Suspensions

Let X be a connected compact manifold, $\widetilde{X}$ its universal cover, G a compact Lie group, and $h : \Gamma = \pi_1(X) \to G$ a homomorphism. Consider the canonical right action of Γ on $\widetilde{X}$, and the diagonal right action of Γ on $\widetilde{M} = \widetilde{X} \times G$:

$$(x, a) \cdot \gamma = (x \cdot \gamma, h(\gamma^{-1}) \cdot a) \,.$$

Let $M = \widetilde{M}/\Gamma$ (usually denoted by $\widetilde{X} \times_\Gamma G$). The canonical projection $\pi : \widetilde{M} \to M$ is a covering map. Let $[x, a]$ be the element of M represented by each $(x, a) \in \widetilde{M}$. The foliation $\widetilde{\mathcal{F}}$ on $\widetilde{M}$ given by the fibers of the second factor projection $\text{pr}_2 : \widetilde{M} \to G$ gives rise to a foliation $\mathcal{F}$ on M. Let Λ be a left invariant volume form on G, which can be considered as an invariant transverse measure of $\mathcal{F}$ because its holonomy pseudogroup can be represented by the pseudogroup generated by the left translations by elements of $h(\Gamma)$. The corresponding transverse volume form ω_Λ is defined by the condition $\pi^*\omega_\Lambda = \text{pr}_2^* \Lambda$ of $\widetilde{\mathcal{F}}$, whose restriction to local transversals is another interpretation of Λ as transverse invariant measure of $\mathcal{F}$. It is easy to see that

$$\chi_\Lambda(\mathcal{F}) = \text{vol}(G) \cdot \chi_\Gamma(\widetilde{X}) \,,$$

where $\chi_\Gamma(\widetilde{X})$ is the Γ-Euler characteristic of the covering manifold $\widetilde{X}$ of X defined by Atiyah [6]. By Atiyah's Γ-index theorem [6], we have $\chi_\Gamma(\widetilde{X}) = \chi(X)$, where $\chi(X)$ is the Euler characteristic of X.

There is a C^∞ global representation $\phi : M \times G \to M$ of the structural transverse action Φ, defined by

$$\phi([x, a], g) = [x, ag] .$$

This ϕ is a free action. Therefore

$$L_{\mathrm{dis}}(\mathcal{F}) = \mathrm{vol}(G) \cdot \chi(X) \cdot \delta_e \qquad (10.1)$$

on the whole of G. In particular, if $\chi(X) \neq 0$, then $\dim \overline{H}(\mathcal{F}) = \infty$ for any homomorphism $h : \Gamma \to G$.

We can consider the following concrete example. Let X be a compact oriented surface of genus $g \geq 2$ endowed with a hyperbolic metric. One can show that there exists an injective homomorphism $h : \pi_1(X) \to \mathrm{SO}(3, \mathbb{R})$. One obtains a Lie $\mathrm{SO}(3, \mathbb{R})$-foliation $\mathcal{F}$ whose leaves are dense, simply connected (diffeomorphic to $\mathbb{R}^2$) and isometric to the hyperbolic plane. Assuming that $\mathrm{vol}(G) = 1$, we get

$$\beta_\Lambda^0(\mathcal{F}) = \beta_\Lambda^2(\mathcal{F}) = 0 , \qquad \beta_\Lambda^1(\mathcal{F}) = 2g - 2 .$$

Since the leaves of $\mathcal{F}$ are dense, we have $\overline{H}^0(\mathbb{R}) \cong \overline{H}^2(\mathbb{R}) \cong \mathbb{R}$, and therefore

$$\mathrm{Tr}_{\mathrm{dis}}^0(\mathcal{F}) = \mathrm{Tr}_{\mathrm{dis}}^2(\mathcal{F}) = 1 .$$

By (10.1), we get

$$L_{\mathrm{dis}}(\mathcal{F}) = (2 - 2g) \cdot \delta_e ,$$

and

$$\mathrm{Tr}_{\mathrm{dis}}^1(\mathcal{F}) = (2g - 2) \cdot \delta_e + 2 .$$

One can also take any homomorphism of Γ to the n-torus $\mathbb{R}^n / \mathbb{Z}^n$ to produce a foliation, which has infinite-dimensional reduced cohomology of degree one (see [1, Example 2.11]). In this case, we have

$$\mathrm{Tr}_{\mathrm{dis}}^i(\mathcal{F}) \neq \beta_\Lambda^i(\mathcal{F}) \cdot \delta_e ,$$

but $\mathrm{Tr}_{\mathrm{dis}}^i(\mathcal{F}) - \beta_\Lambda^i(\mathcal{F}) \cdot \delta_e$ is C^∞.

10.3. Bundles over homogeneous spaces and the Selberg trace formula

Let G be a simply connected Lie group, Γ a discrete cocompact subgroup in G, and α an injective homomorphism of Γ to the diffeomorphism group $\mathrm{Diff}(X)$ of some compact connected C^∞ manifold X. Consider a left action of Γ on $\widetilde{M} = G \times X$ given by

$$\gamma \cdot (a, x) = (\gamma a, \alpha(\gamma)(x)) .$$

Let $M = \Gamma \backslash (G \times X)$, and let $[a, x]$ be the element of M represented by any $(a, x) \in \widetilde{M}$. The canonical projection $\pi : \widetilde{M} \to M$ is a covering map. The first factor projection $G \times X \to G$ defines a fiber bundle map $M \to \Gamma \backslash G$, whose fibers are the leaves of a foliation $\mathcal{F}$. For each $a \in G$, the leaf of $\mathcal{F}$ through Γa is

$$L_{\Gamma a} = \{[a, x] \mid x \in X\} ,$$

which is diffeomorphic to X because α is injective. Consider a left-invariant volume form Λ on G. It induces a volume form on $\Gamma \backslash G$, denoted by $\Lambda_{\Gamma \backslash G}$, whose pull-back to M via the map $M \to \Gamma\backslash G$ defines a transverse volume form ω_Λ of $\mathcal{F}$. Since $M \to \Gamma\backslash G$ is a fiber bundle map with typical fiber X, we get

$$\chi_\Lambda(\mathcal{F}) = \mathrm{vol}(\Gamma\backslash G) \cdot \chi(X) \,,$$

where $\chi(X)$ is the Euler characteristic of X.

The structural transverse action Φ_g of an element $g \in G$ is given by the leafwise homotopy class of diffeomorphisms $\phi_g : M \to M$ of the form

$$\phi_g([a,x]) = [ag, \beta(x)] \,,$$

where β is any diffeomorphism of X homotopic to id_X.

The leaf of the foliation $\mathcal{G}$ through a point $([a,x],b) \in M \times G$ is

$$\{([ag,y], g^{-1}bg) \mid y \in X, \ g \in G\} \,.$$

So the leaves of $\mathcal{G}$ are

$$M_b' = \{([g,y], g^{-1}bg) \mid y \in X, \ g \in G\} \,, \quad b \in G \,,$$

with $M_{b_1}' = M_{b_2}'$ when $b_2 \in \mathrm{Ad}(\Gamma)b_1$; thus the leaves of $\mathcal{G}$ are parameterized by the Γ-conjugacy classes in G.

Let pr_1 and pr_2 denote the factor projections of $M \times G$ to M and G, respectively. The restriction $\mathrm{pr}_2 : M_b' \to G$ is a bundle map over the orbit

$$\mathcal{O}_b = \{g^{-1}bg \mid g \in G\} \equiv G_b\backslash G$$

of the adjoint representation of G on G, where

$$G_b = \{g \in G \mid gb = bg\}$$

is the centralizer of b in G.

For each $b \in G$, the restriction $\mathrm{pr}_1 : M_b' \to M$ is a covering map. Indeed, we have $M_b' \equiv \Gamma_b\backslash(G \times X)$, where

$$\Gamma_b = \{\gamma \in \Gamma \mid \gamma b = b\gamma\} = \Gamma \cap G_b \,.$$

The leaves of the foliation $\mathcal{F}_b' = \mathrm{pr}_1^* \mathcal{F}$ on M_b' are described as

$$L_a = \{([a,y], a^{-1}ba) \mid y \in X\} \,, \quad a \in G \,,$$

with $L_{a_1} = L_{a_2}$ if and only if $\Gamma_b a_1 = \Gamma_b a_2$. Therefore the leaves of $\mathcal{F}'$ are the fibers of the natural map

$$M_b' \equiv \Gamma_b\backslash(G \times X) \to \Gamma_b\backslash G \,, \quad ([a,y], a^{-1}ba) \mapsto \Gamma_b a \,.$$

Take a C^∞ global representation $\phi : M \times G \to M$ of Φ defined by

$$\phi([a,x], g) = [ag, x] \,.$$

We have

$$\mathrm{Fix}(\phi') = \{([a,x], g) \in M \times G \mid [ag, x] = [a,x]\} \,.$$

The identity $[ag, x] = [a, x]$ holds if and only if there exists $\gamma \in \Gamma$ such that $ag = \gamma a$ and $\alpha(\gamma)(x) = x$. Hence

$$\mathrm{Fix}(\phi') = \bigcup_{\gamma \in \Gamma} \{([a,x], a^{-1}\gamma a) \mid x \in X, \ \alpha(\gamma)x = x, \ a \in G\} \,.$$

We see that if $\mathrm{Fix}(\phi') \cap M'_b \neq \emptyset$, then one can assume that $b = \gamma \in \Gamma$ and $\alpha(\gamma)$ has a fixed point in X. In this case,

$$\mathrm{Fix}(\phi') \bigcap M'_\gamma = \{([a, x], a^{-1}\gamma a) \mid x \in X, \ \alpha(\gamma)x = x, \ a \in G\} \ .$$

A point $([a, x], a^{-1}\gamma a) \in \mathrm{Fix}(\phi') \cap M'_\gamma$ is simple if and only if x is a simple fixed point of $\alpha(\gamma)$; in this case, we have

$$\epsilon([a, x], a^{-1}\gamma a) = \mathrm{sign}\det(\alpha(\gamma)_* - \mathrm{id} : T_x X \to T_x X) \ ,$$

which is denoted by $\epsilon_{\alpha(\gamma)}(x)$. Assume that, for any $\gamma \in \Gamma \backslash \{e\}$, all the fixed points of the diffeomorphism $\alpha(\gamma)$, denoted by $x_1(\gamma), x_2(\gamma), \ldots, x_{d(\gamma)}(\gamma)$, are simple. Then

$$\mathrm{Fix}(\phi') \bigcap M'_\gamma = \bigcup_{k=1}^{d(\gamma)} \{([a, x_k(\gamma)], a^{-1}\gamma a) \mid a \in G\} \ .$$

The transverse volume form $\omega'_\Lambda = \mathrm{pr}_1^* \omega_\Lambda$ of $\mathcal{F}'_\gamma$ is, by definition, the pull-back of $\Lambda_{\Gamma_\gamma \backslash G}$ via the map $M'_\gamma \to \Gamma_\gamma \backslash G$. Let Σ be a complete set of representatives of the Γ-conjugacy classes in Γ. For $f \in C_c^\infty(G \backslash \{e\})$, we get

$$\langle \mathrm{pr}_{2*}(\epsilon \cdot \Lambda'|_{\mathrm{Fix}(\phi')}), f \rangle$$

$$= \int_{\mathrm{Fix}(\phi')} f \circ \mathrm{pr}_2 \cdot \epsilon \, \omega'_\Lambda$$

$$= \sum_{\gamma \in \Sigma \backslash \{e\}} \sum_{k=1}^{d(\gamma)} \int_{\Gamma_\gamma \backslash G} f(a^{-1}\gamma a) \cdot \epsilon_{\alpha(\gamma)}(x_k(\gamma)) \, \Lambda_{\Gamma_\gamma \backslash G}(\Gamma_\gamma a) \ .$$

By the classical Lefschetz theorem, we have

$$\sum_{k=1}^{d(\gamma)} \epsilon_{\alpha(\gamma)}(x_k(\gamma)) = L(\alpha(\gamma)) \ ,$$

where

$$L(\alpha(\gamma)) = \sum_{i=1}^{\dim X} (-1)^i \, \mathrm{Tr}(\alpha(\gamma)^* : H^i(X) \to H^i(X))$$

is the Lefschetz number of the diffeomorphism $\alpha(\gamma)$. It can be easily seen that $L(\alpha(\gamma))$ depends only on the conjugacy class of γ. Take a left invariant Riemannian metric on G whose volume form is Λ. Consider the Riemannian metric on $G_\gamma \backslash G$ so that the canonical projection $G \to G_\gamma \backslash G$ is a Riemannian submersion, and let $\Lambda_{G_\gamma \backslash G}$ be the corresponding volume form. Then

$$\langle \mathrm{pr}_{2*}(\epsilon\Lambda'|_{\mathrm{Fix}(\phi')}), f \rangle = \sum_{\gamma \in \Sigma \backslash \{e\}} L(\alpha(\gamma)) \int_{\Gamma_\gamma \backslash G} f(a^{-1}\gamma a) \, \Lambda_{\Gamma_\gamma \backslash G}(\Gamma_\gamma a)$$

$$= \sum_{\gamma \in \Sigma \backslash \{e\}} L(\alpha(\gamma)) \, \mathrm{vol}(\Gamma_\gamma \backslash G_\gamma) \int_{G_\gamma \backslash G} f(a^{-1}\gamma a) \, \Lambda_{G_\gamma \backslash G}(G_\gamma a) \ .$$

Finally, we get the following Selberg type trace formula (cf. [29]):

$$\langle L_{\mathrm{dis}}(\mathcal{F}), f \rangle = \mathrm{vol}(\Gamma \backslash G) \, \chi(X) \, f(e)$$
$$+ \sum_{\gamma \in \Sigma \backslash \{e\}} L(\alpha(\gamma)) \, \mathrm{vol}(\Gamma_\gamma \backslash G_\gamma) \int_{G_\gamma \backslash G} f(a^{-1} \gamma a) \, \Lambda_{G_\gamma \backslash G}(G_\gamma a) \, .$$

In the particular case when $G = \mathbb{R}$, $\Gamma = \mathbb{Z}$ and the homomorphism α is given by a diffeomorphism F of a compact manifold X, the manifold M is the mapping torus of F and the foliation $\mathcal{F}$ is given by the fibers of the natural map $M \to S^1$. Then the formula gives

$$L_{\mathrm{dis}}(\mathcal{F}) = \chi(X) \cdot \delta_0 + \sum_{k \in \mathbb{Z} \backslash \{0\}} L(F^k) \cdot \delta_k \, .$$

10.4. Homogeneous foliations

Let H and G be simply connected Lie groups, Γ a uniform discrete subgroup in H, and $D : H \to G$ a surjective homomorphism so that $\Gamma_1 = D(\Gamma)$ is dense in G. Then $M = \Gamma \backslash H$ is a compact manifold, and let $\mathcal{F}$ be the foliation on M whose leaves are the projections of the fibers of D. If $K = \ker D$, then the leaves of $\mathcal{F}$ are the orbits of the right action of K on M induced by the right action on H defined by right translations.

This $\mathcal{F}$ is a Lie G-foliation whose structural transverse action Φ is given as follows: for each $g \in G$, Φ_g is represented by the foliated map $\mathcal{F} \to \mathcal{F}$ induced by the right multiplication by any element of $D^{-1}(g)$.

The leaf of the foliation $\mathcal{G}$ on $M \times G$ through a point $(\Gamma h, a) \in M \times G$ is, by definition,

$$M'_{(\Gamma h, a)} = \{ (\Gamma h_1, g^{-1} a g) \mid g \in G, \ h_1 \in D^{-1}(D(\Gamma h) \, g) \} \, .$$

It is easy to see that there is a bijection between the set of leaves of $\mathcal{G}$ and the orbit space $G / \mathrm{Ad}(\Gamma_1)$ of the adjoint action of Γ_1 on G so that, for $\mathrm{Ad}(\Gamma_1) g_0 \in G / \mathrm{Ad}(\Gamma_1)$, the corresponding leaf is described as

$$M'_{\mathrm{Ad}(\Gamma_1) g_0} = \{ (\Gamma h, g) \in M \times G \mid D(h) \, g \, D(h)^{-1} \in \mathrm{Ad}(\Gamma_1) \, g_0 \} \, .$$

The first factor projection $\mathrm{pr}_1 : M'_{\mathrm{Ad}(\Gamma_1) g_0} \to M$ is a covering map; indeed, $M'_{g_0} \equiv \Gamma_{g_0} \backslash H$, where $\Gamma_{g_0} = \Gamma \cap D^{-1}(\Gamma_{1, g_0})$, denoting by Γ_{1, g_0} the centralizer of g_0 in Γ_1.

The leaves of $\mathcal{F}$ can be described as

$$L_{\Gamma_1 g_1} = \{ \Gamma h \in M \mid D(h) \in \Gamma_1 g_1 \} \, , \quad g_1 \in \Gamma_1 \backslash G \, .$$

By definition, the leaf $L'_{\Gamma_1 g_1} = \mathrm{pr}_1^*(L_{\Gamma_1 g_1})$ of the foliation $\mathcal{F}' = \mathrm{pr}_1^* \mathcal{F}$ on M'_{g_0} consists of all $(\Gamma h, g) \in M \times G$ such that $D(h) \, g \, D(h)^{-1} \in \mathrm{Ad}(\Gamma_1) \, g_0$ and $D(h) \in \Gamma_1 g_1$. So it can be parameterized by the elements of $(\Gamma_1 \backslash G) \times (G / \mathrm{Ad}(\Gamma_1))$, and it can be described as

$$L'_{\Gamma_1 g_1} = \{ (\Gamma h, g) \in M \times G : D(h) \in \Gamma_1 g_1, \ g \in \mathrm{Ad}(g_1) \mathrm{Ad}(\Gamma_1) \, g_0 \} \, .$$

We also see that $\mathrm{pr}_2(M'_{g_0})$ is the orbit $\mathcal{O}_{g_0}$ of the adjoint action of G on G through g_0. Moreover, $\mathrm{pr}_2 : M'_{g_0} \to \mathcal{O}_{g_0}$ is a bundle map, and the fiber of this bundle over $y \in \mathcal{O}_{g_0}$ can be identified with $\Gamma_x \backslash H_x$, where $x \in H$ is any element such that $D(x) = y$.

Denote by $\mathfrak{h}$, $\mathfrak{g}$ and $\mathfrak{k}$ the Lie algebras of H, G and K, respectively. We have a short exact sequence

$$0 \longrightarrow \mathfrak{k} \longrightarrow \mathfrak{h} \overset{D_*}{\longrightarrow} \mathfrak{g} \longrightarrow 0 \ .$$

To construct C^∞ local representations of Φ, we choose a splitting of this short exact sequence; that is, a linear map $s : \mathfrak{g} \to \mathfrak{h}$ such that $D_* \circ s = \mathrm{id}_{\mathfrak{g}}$. So s is injective and $s(\mathfrak{g}) \oplus \mathfrak{k} = \mathfrak{h}$. Let $U \subset \mathfrak{g}$ be an open neighborhood of 0 in $\mathfrak{g}$ such that the restriction $\exp : U \to \exp(U) \subset G$ of the exponential map to U is a diffeomorphism. Then, for any $g \in G$, a C^∞ local representation $\phi : M \times O \to M$ of Φ is defined on the open neighborhood $O = g \exp(U)$ of g as

$$\phi(\Gamma h, g \exp Y) = \Gamma h h_1 \exp s(Y) \ , \quad h \in H \ , \quad Y \in U \ ,$$

where $h_1 \in H$ is any element such that $D(h_1) = g$.

Now fix $g \in G$ and $h_1 \in H$ such that $D(h_1) = g$. By definition,

$$(\Gamma h, g \exp Y) \in \mathrm{Fix}(\phi') \Leftrightarrow \Gamma h h_1 \exp s(Y) = \Gamma h \Leftrightarrow h h_1 \exp s(Y) \, h^{-1} \in \Gamma \ .$$

We have

$$D(h) \, g \exp Y \, D(h)^{-1} = D(h h_1 \exp s(Y) \, h^{-1}) \in \Gamma_1 \ ,$$

therefore, we get $\mathrm{Fix}(\phi') \cap M'_{g_0} \neq \emptyset$ iff $g_0 \in \Gamma_1$. In particular, it follows that

$$\mathrm{pr}_2(\mathrm{Fix}(\phi')) = \bigcup_{\gamma \in \Sigma} \mathcal{O}_\gamma \ ,$$

where Σ is a complete set of representatives of the Γ_1-conjugacy classes in Γ_1. For a fixed class $\gamma \in \Sigma$, let $[D^{-1}(\gamma)]$ be the Γ-conjugacy class of the unique element $\gamma_1 \in \Gamma$ such that $D(\gamma_1) = \gamma$. Then we have

$$\mathrm{Fix}(\phi') \cap M'_\gamma = \{(\Gamma h, g \exp Y) \in (\Gamma \backslash H) \times G \mid h h_1 \exp s(Y) \, h^{-1} \in [D^{-1}(\gamma)]\} \ .$$

For any $(\Gamma h, g \exp Y) \in \mathrm{Fix}(\phi')$, the left translation by h determines an isomorphism of the tangent space $T_{\Gamma h} \mathcal{F}$ with $\mathfrak{k}$, and, under this isomorphism, the induced map $(\phi_{h_1 \exp s(Y)})_* : T_{\Gamma h} \mathcal{F} \to T_{\Gamma h} \mathcal{F}$ corresponds to the restriction $\mathrm{Ad}(h_1 \exp s(Y))_*|_{\mathfrak{k}} : \mathfrak{k} \to \mathfrak{k}$ of the differential of the adjoint action of $g \exp Y \in G$ on G to $\mathfrak{k}$. In particular, $(\Gamma h, g \exp Y) \in \mathrm{Fix}(\phi)$ is simple if and only if $\mathrm{Ad}(h_1 \exp s(Y))_*|_{\mathfrak{k}} : \mathfrak{k} \to \mathfrak{k}$ is an isomorphism. It should be noted that this condition depends only on $g \exp Y$ and is independent of the choice h_1 and s.

Assume that $\mathrm{Ad}(h \gamma h^{-1})_*|_{\mathfrak{k}} : \mathfrak{k} \to \mathfrak{k}$ is an isomorphism for any $\gamma \in \Gamma$ and $h \in H$. Then the value

$$\epsilon(\Gamma h, g \exp Y) = \mathrm{sign} \det \left((\phi_{h_1 \exp s(Y)})_* - \mathrm{id} : T_{\Gamma h} \mathcal{F} \to T_{\Gamma h} \mathcal{F} \right)$$
$$= \mathrm{sign} \det \left(\mathrm{Ad}(h_1 \exp s(Y))_*|_{\mathfrak{k}} - \mathrm{id} : \mathfrak{k} \to \mathfrak{k} \right)$$

is the same for any $(\Gamma h, g \exp Y) \in \mathrm{Fix}(\phi') \cap M'_\gamma$, and equals

$$\epsilon(\gamma) = \mathrm{sign} \det \left(\mathrm{Ad}(\gamma)_*|_{\mathfrak{k}} - \mathrm{id} : \mathfrak{k} \to \mathfrak{k} \right) \ .$$

Let Λ be a left invariant volume form on G, which can be identified with a transverse volume form of $\mathcal{F}$. Fix $\gamma \in \Sigma$. Then the transverse volume form $\Lambda' = \mathrm{pr}_1^* \Lambda$ of $\mathcal{F}'$ is given by the lift of Λ to M'_γ by the restriction of the map

$$(\Gamma h, g) \in (\Gamma \backslash H) \times G \mapsto D(h) \in G \ .$$

to

$$M'_\gamma = \{(\Gamma h, g) \in (\Gamma \backslash H) \times G \mid D(h)\, g\, D(h)^{-1} \in \mathrm{Ad}(\Gamma_1)\, \gamma\} \ .$$

As above, take a left invariant Riemannian metric on G whose volume form is Λ. Consider the Riemannian metric on $G_\gamma \backslash G$ so that the canonical projection $G \to G_\gamma \backslash G$ is a Riemannian submersion, and let $\Lambda_{G_\gamma \backslash G}$ be the corresponding volume form. Restricting the form Λ' to $\mathrm{Fix}(\phi') \cap M'_\gamma$ and integrating it along the fibers of pr_2, for any $f \in C_c^\infty(G)$, we get

$$\langle \chi_{\mathrm{dis}}(\mathcal{F}), f \rangle = \langle \mathrm{pr}_{2*}(\epsilon \Lambda'), f \rangle$$

$$= \sum_{\gamma \in \Sigma} \epsilon(\gamma)\, \mathrm{vol}(\Gamma_{\gamma_0} \backslash H_{\gamma_0}) \int_{G_\gamma \backslash G} f(g^{-1} \gamma g)\, \Lambda_{G_\gamma \backslash G}(G_\gamma g) \ ,$$

where $\gamma_0 \in \Gamma$ is the unique element such that $D(\gamma_0) = \gamma$.

10.5. Nilpotent homogeneous foliations

Let G be a nontrivial simply connected nilpotent Lie group and let $\Gamma_1 \subset G$ be a finitely generated dense subgroup. By Malcev's theory [18], there exists a simply connected nilpotent Lie group H, an embedding $i : \Gamma_1 \to H$ and a surjective homomorphism $D : H \to G$ such that $\Gamma = i(\Gamma_1)$ is discrete and uniform in H, and $D \circ i = \mathrm{id}_{\Gamma_1}$. Consider the corresponding homogeneous foliation on the closed nilmanifold $M = \Gamma \backslash H$. As above, K denotes the kernel of D, which is a normal connected Lie subgroup in H, and $\mathfrak{k}$ denotes the Lie algebra of K. As shown in [1, Theorem 2.10], there is a canonical isomorphism $\overline{H}(\mathcal{F}) \cong H(\mathfrak{k})$ (c.f. [22]), and thus $L_{\mathrm{dis}}(\mathcal{F}) = 0$ by Corollary 1.4. Let us check this triviality in another way. It can be easily seen that, under this isomorphism, the action of an element $g \in G$ on $\overline{H}(\mathcal{F})$ induced by the structural action Φ corresponds to the action $\mathrm{Ad}_*(h)$ on $H(\mathfrak{k})$ induced by the adjoint action of any element $h \in D^{-1}(g)$. So $\mathrm{Tr}_{\mathrm{dis}}^i(\mathcal{F})$ is a smooth function on G, whose value at $g \in G$ is the trace of $\mathrm{Ad}_*(h)$ on $H^i(\mathfrak{k})$ with $h \in D^{-1}(g)$. Since H is nilpotent, $\mathrm{Ad}_*(h)$ has a triangular matrix representation whose diagonal entries are equal to 1. So

$$\mathrm{Tr}_{\mathrm{dis}}^i(\mathcal{F}) \equiv \dim H^i(\mathfrak{k}) \ ,$$

yielding

$$L_{\mathrm{dis}}(\mathcal{F}) \equiv \sum_i (-1)^i \dim H^i(\mathfrak{k}) = \sum_i (-1)^i \dim \bigwedge^i \mathfrak{k} = 0 \ .$$

Any local section $g \mapsto h_g$ of D on some open subset $O \subset G$ induces a C^∞ local representation $\phi : M \times O \to M$ of the structural action Φ, where each ϕ_g is induced by the right multiplication by h_g. All the fixed points of ϕ are not simple.

References

[1] J.A. Álvarez López and G. Hector. The dimension of the leafwise reduced cohomology. *Amer. J. Math.* **123** (2001), 607–646.

[2] J.A. Álvarez López and Y.A. Kordyukov. Long time behavior of leafwise heat flow for Riemannian foliations. *Compositio Math.* **125** (2001), 129–153.

[3] J.A. Álvarez López and Y.A. Kordyukov. Distributional Betti numbers of transitive foliations of codimension one. In *Foliations: Geometry and Dynamics (Warsaw, 2000)*, ed. P. Walczak et al. World Scientific, Singapore, 2002, pp. 159–183.

[4] J.A. Álvarez López and P. Tondeur. Hodge decomposition along the leaves of a Riemannian foliation. *J. Funct. Anal.* **99** (1991), 443–458.

[5] M.F. Atiyah. Elliptic operators and compact groups. In *Lecture Notes in Mathematics.* Vol. 401, pages 1–93. Springer, Berlin, Heidelberg, New York, 1974.

[6] M.F. Atiyah. Elliptic operators, discrete groups and von Neumann algebras. *Astérisque* **32** (1976), 43–72.

[7] R. Bott and L.W. Tu. *Differential Forms in Algebraic Topology.* Graduate Texts in Mathematics, Vol. 82. Springer-Verlag, Berlin, Heidelberg, New York, 1982.

[8] P.R. Chernoff. Finite propagation speed, kernel estimates for functions of the Laplace operator, and the geometry of complete Riemannian manifolds. *J. Funct. Anal.* **12** (1973), 401–414.

[9] A. Connes. Sur la théorie non commutative de l'intégration. In *Algèbres d'opérateurs (Sém., Les Plans-sur-Bex, 1978)*, Lecture Notes in Math. Vol. 725, pp. 19–143. Springer, Berlin, Heidelberg, New York, 1979.

[10] A. Connes. Noncommutative differential geometry. *Publ. Math.* **62** (1986), 41–144.

[11] E. Fedida. Sur les feuilletages de Lie. *C. R. Acad. Sci. Paris. Ser. A-B* **272** (1971), A999–A1001.

[12] A. Haefliger. Some remarks on foliations with minimal leaves. *J. Diff. Geom.* **15** (1980), 269–284.

[13] A. Haefliger. Foliations and compactly generated pseudogroups. In *Foliations: Geometry and Dynamics (Warsaw, 2000)*, pages 275–295. World Sci. Publishing, River Edge, NJ, 2002.

[14] J.L. Heitsch and C. Lazarov. A Lefschetz theorem for foliated manifolds. *Topology* **29** (1990), 127–162.

[15] Y.A. Kordyukov. Transversally elliptic operators on G-manifolds of bounded geometry. *Russ. J. Math. Ph.* **2** (1994), 175–198.

[16] Y.A. Kordyukov. Functional calculus for tangentially elliptic operators on foliated manifolds. In *Analysis and Geometry in Foliated Manifolds, Proceedings of the VII International Colloquium on Differential Geometry, Santiago de Compostela, 1994*, pp. 113–136. World Scientific, Singapore, 1995.

[17] Y.A. Kordyukov. Noncommutative spectral geometry of Riemannian foliations. *Manuscripta Math.* **94** (1997), 45–73.

[18] A.I. Mal'cev. On a class of homogeneous spaces. *Transl. Amer. Math. Soc.* **39** (1951), 276–307.

[19] P. Molino. Géométrie globale des feuilletages Riemanniens. *Proc. Nederl. Acad. A1* **85** (1982), 45–76.

[20] B. Mümken. On tangential cohomology of Riemannian foliations. *Amer. J. Math.* **128** (2006), 1391–1408.

[21] A. Nestke and P. Zuckermann. The index of transversally elliptic complexes. *Rend. Circ. Mat. Palermo* **34**, Suppl. 9 (1985), 165–175.

[22] K. Nomizu. On the cohomology of compact homogeneous spaces of nilpotent lie groups. *Ann. of Math.* **59** (1954), 531–538.

[23] B.L. Reinhart. Foliated manifolds with bundle-like metrics. *Ann. of Math.* **69** (1959), 119–132.

[24] A.F. Rich. A Lefschetz theorem for foliated manifolds. PhD Thesis, The University of Chicago, 1989.

[25] J. Roe. Finite propagation speed and Connes' foliation algebra. *Math. Proc. Cambridge Philos. Soc.* **102** (1987), 459–466.

[26] J. Roe. An index theorem on open manifolds. I. *J. Diff. Geom.* **27** (1988), 87–113.

[27] J. Roe. An index theorem on open manifolds. II. *J. Diff. Geom.* **27** (1988), 115–136.

[28] J. Roe. *Elliptic Operators, Topology and Asymptotic Methods. Second Edition*. Pitman Research Notes in Mathematics Series, 395. Longman, Harlow, 1998.

[29] A. Selberg. Harmonic analysis and discontinuous groups in weakly symmetric Riemannian spaces with applications to Dirichlet series. *J. Indian Math. Soc. (N.S.)* **20** (1956), 47–87.

[30] I.M. Singer. Recent applications of index theory for elliptic operators. In *Proc. Symp. Pure Appl. Math.* 23, pp. 11–31. Amer. Math. Soc., Providence, R. I., 1973.

Jesús A. Álvarez López
Departamento de Xeometría e Topoloxía
Facultade de Matemáticas
Universidade de Santiago de Compostela
15782 Santiago de Compostela
Spain
e-mail: jalvarez@usc.es

Yuri A. Kordyukov
Institute of Mathematics
Russian Academy of Sciences
112 Chernyshevsky str.
450077 Ufa
Russia
e-mail: yurikor@matem.anrb.ru

C^*-algebras and Elliptic Theory II
Trends in Mathematics, 41–66
ⓒ 2008 Birkhäuser Verlag Basel/Switzerland

Torsion, as a Function on the Space of Representations

Dan Burghelea and Stefan Haller

Abstract. Riemannian Geometry, Topology and Dynamics permit to introduce partially defined holomorphic functions on the variety of representations of the fundamental group of a manifold. The functions we consider are the complex-valued Ray–Singer torsion, the Milnor–Turaev torsion, and the dynamical torsion. They are associated essentially to a closed smooth manifold equipped with a (co)Euler structure and a Riemannian metric in the first case, a smooth triangulation in the second case, and a smooth flow of type described in Section 2 in the third case. In this paper we define these functions, describe some of their properties and calculate them in some case. We conjecture that they are essentially equal and have analytic continuation to rational functions on the variety of representations. We discuss what we know to be true. As particular cases of our torsion, we recognize familiar rational functions in topology such as the Lefschetz zeta function of a diffeomorphism, the dynamical zeta function of closed trajectories, and the Alexander polynomial of a knot. A numerical invariant derived from Ray–Singer torsion and associated to two homotopic acyclic representations is discussed in the last section.

Mathematics Subject Classification (2000). 57R20, 58J52.

Keywords. Euler structure; coEuler structure; combinatorial torsion; analytic torsion; theorem of Bismut–Zhang; Chern–Simons theory; geometric regularization; mapping torus; rational function.

Part of this work was done while both authors enjoyed the hospitality of the Max Planck Institute for Mathematics in Bonn. A previous version was written while the second author enjoyed the hospitality of the Ohio State University. The second author was partially supported by the *Fonds zur Förderung der wissenschaftlichen Forschung* (Austrian Science Fund), project number P17108-N04.

Contents

1. Introduction

For a finitely presented group Γ denote by $\mathrm{Rep}(\Gamma; V)$ the algebraic set of all complex representations of Γ on the complex vector space V. For a closed base pointed manifold (M, x_0) with $\Gamma = \pi_1(M, x_0)$ denote by $\mathrm{Rep}^M(\Gamma; V)$ the algebraic closure of $\mathrm{Rep}_0^M(\Gamma; V)$, the Zariski open set of representations $\rho \in \mathrm{Rep}(\Gamma; V)$ so that $H^*(M; \rho) = 0$. The manifold M is called V-acyclic iff $\mathrm{Rep}^M(\Gamma; V)$, or equivalently $\mathrm{Rep}_0^M(\Gamma; V)$, is non-empty. If M is V-acyclic then the Euler–Poincaré characteristic $\chi(M)$ vanishes. There are plenty of V-acyclic manifolds.

If $\dim V = 1$ then $\mathrm{Rep}(\Gamma; V) = (\mathbb{C} \setminus 0)^k \times F$, where k denotes the first Betty number of M, and F is a finite Abelian group. If in addition M is V-acyclic and $H_1(M; \mathbb{Z})$ is torsion free, then $\mathrm{Rep}^M(\Gamma; V) = (\mathbb{C} \setminus 0)^k$. There are plenty of V-acyclic ($\dim V = 1$) manifolds M with $H_1(M; \mathbb{Z})$ torsion free.

In this paper, to a V-acyclic manifold and an Euler or coEuler structure we associate three partially defined holomorphic functions on $\mathrm{Rep}^M(\Gamma; V)$, the complex-valued Ray–Singer torsion, the Milnor–Turaev torsion, and the dynamical torsion, and describe some of their properties. They are defined with the help of a Riemannian metric, resp. smooth triangulation resp. a vector field with the properties listed in Section 2.4, but are independent of these data.

We expect that they are essentially equal, and have analytic continuation to rational functions on $\mathrm{Rep}^M(\Gamma; V)$ (if $\dim V = 1$ genuine rational functions of k variables). This is not entirely true and we discuss what we know so far.

We calculate them in some cases and recognize familiar rational functions in topology (Lefschetz zeta function of a diffeomorphism, dynamical zeta function of some flows, Alexander polynomial of a knot) as particular cases of our torsion, cf. Section 7.

The results answer the question

(Q) *Is the Ray–Singer torsion the absolute value of a holomorphic function on the space of representations?*[1]

and establish the analytic continuation of the dynamical torsion (for a related result consult [4]). Both issues are subtle when formulated inside the field of spectral geometry or of dynamical systems and can hardly be decided using internal techniques of these fields. There are interesting dynamical implications on the growth of the number of instantons and of closed trajectories, some of them improving on a conjecture formulated by S.P. Novikov about the gradients of closed Morse one form, cf. Section 8.

This paper surveys results from [7], [9], [10] [11] and reports on additional work in progress on these lines. Its contents is the following.

In Section 2, for the reader's convenience, we recall some less familiar characteristic forms used in this paper and describe the class of vector fields we use to define the dynamical torsion. These vector fields have finitely many rest points but infinitely many instantons and closed trajectories. However, despite their infiniteness, they can be counted by appropriate counting functions which can be related to the topology and the geometry of the underlying manifold cf. [8]. The dynamical torsion is derived from them.

All torsion functions referred to above involve some additional topological data; the Milnor–Turaev and dynamical torsion involve an Euler structure while the complex Ray–Singer torsion a coEuler structure, a sort of Poincaré dual of the first. In Section 3 we define Euler and coEuler structures and discuss some of their properties. Although they can be defined for arbitrary base pointed manifolds (M, x_0) we present the theory only in the case $\chi(M) = 0$ when the base point is irrelevant.

While the complex Ray–Singer torsion and dynamical torsion are new concepts the Milnor–Turaev torsion is not, however our presentation is somehow different from the traditional one. In Section 4 we discuss the algebraic variety of cochain complexes of finite-dimensional vector spaces and introduce the Milnor torsion as a rational function on this variety. The Milnor–Turaev torsion is obtained as a pull back by a characteristic map of this rational function.

Section 5 is about analytic torsion. In Section 5.1, we recall the familiar Ray–Singer torsion slightly modified with the help of a coEuler structure. This is a positive real-valued function defined on $\mathrm{Rep}_0^M(\Gamma; V)$, the variety of the acyclic representations. We show that this function is independent of the Riemannian metric, and that it is the absolute value of a complex valued rational function, provided the coEuler structure is integral. In Section 5.2 we introduce the complex-valued Ray–Singer torsion, and show the relation to the first. The complex Ray–Singer torsion, denoted $\mathcal{ST}$, is a meromorphic function on a finite cover of the space

---

[1]A similar question was considered in [25] and a positive answer provided.

of representations and is defined analytically using regularized determinants of elliptic operators but not self adjoint.

The Milnor–Turaev torsion, defined in Section 6.1, is associated with a smooth manifold, a given Euler structure and a homology orientation and is constructed using a smooth triangulation. Its square was conjectured to be equal to the complex Ray–Singer torsion as defined in Section 5.2, when the coEuler structure for Ray–Singer corresponds, by Poincaré duality map, to the Euler structure for Milnor–Turaev. The conjecture was verified in [11] for odd-dimensional manifolds[2] and in full generality in [28].

Up to sign the dynamical torsion, introduced in Section 6.2, is associated to a smooth manifold and a given Euler structure and is constructed using a smooth vector field in the class described in Section 2.4. The sign can be fixed with the help of an equivalence class of orderings of the rest points of X, cf. Section 6.2. A priori the dynamical torsion is only a partially defined holomorphic function on $\mathrm{Rep}^M(\Gamma; V)$ and is defined using the instantons and the closed trajectories of X. For a representation ρ the dynamical torsion is expressed as a series which might not be convergent for each ρ but should be convergent for ρ in a subset U of $\mathrm{Rep}^M(\Gamma; V)$ with non-empty interior. At present this convergence was established only in the case of rank one representations. The existence of U is guaranteed by the exponential growth property (EG) required from the vector field (cf. Section 2.4 for the definition).

The main results, Theorems 1, 2 and 3, establish the relationship between these torsion functions. Theorem 3 is expected to hold for V of arbitrary dimension.

One can calculate the Milnor–Turaev torsion when M has a structure of mapping torus of a diffeomorphism ϕ as the "twisted Lefschetz zeta function" of the diffeomorphism ϕ, cf. Section 7.1. The Alexander polynomial as well as the twisted Alexander polynomials of a knot can also be recovered from this torsion cf. Section 7.3. If the vector field has no rest points but admits a closed Lyapunov cohomology class, cf. Section 7.2, the dynamical torsion can be expressed in terms of closed trajectories only, and the dynamical zeta function of the vector field (including all its twisted versions) can be recovered from the dynamical torsion described here.

In Section 8.1 we express the phase difference of the Milnor–Turaev torsion of two representations in the same connected component of $\mathrm{Rep}_0^M(\Gamma; V)$ in terms of the Ray–Singer torsion. This invariant is analogous to the Atiyah–Patodi–Singer spectral flow but has not been investigated so far. Section 8 discusses some progress towards a conjecture of Novikov which came out from the work on dynamical torsion.

---

[2]with comments on how derive it for even-dimensional manifolds

2. Characteristic forms and vector fields

2.1. Euler, Chern–Simons, and Mathai–Quillen form

Let M be smooth closed manifold of dimension n. Let $\pi : TM \to M$ denote the tangent bundle, and $\mathcal{O}_M$ the orientation bundle, which is a flat real line bundle over M. For a Riemannian metric g denote by

$$e(g) \in \Omega^n(M; \mathcal{O}_M)$$

its Euler form, and for two Riemannian metrics g_1 and g_2 by

$$cs(g_1, g_2) \in \Omega^{n-1}(M; \mathcal{O}_M)/d(\Omega^{n-2}(M; \mathcal{O}_M))$$

their Chern–Simons class. The following properties follow from (4) and (5) below.

$$d \, cs(g_1, g_2) = e(g_2) - e(g_1) \tag{1}$$
$$cs(g_2, g_1) = -cs(g_1, g_2) \tag{2}$$
$$cs(g_1, g_3) = cs(g_1, g_2) + cs(g_2, g_3) \tag{3}$$

If the dimension of M is odd both $e(g)$ and $cs(g_1, g_2)$ vanish.

Denote by ξ the Euler vector field on TM which assigns to a point $x \in TM$ the vertical vector $-x \in T_x TM$. A Riemannian metric g determines the Levi–Civita connection in the bundle $\pi : TM \to M$. There is a canonic n-form $\mathrm{vol}(g) \in \Omega^n(TM; \pi^* \mathcal{O}_M)$, which assigns to an n-tuple of vertical vectors *their volume times their orientation* and vanishes when contracted with horizontal vectors. The global angular form, see for instance [3], is the differential form

$$A(g) := \frac{\Gamma(n/2)}{(2\pi)^{n/2}|\xi|^n} i_\xi \, \mathrm{vol}(g) \in \Omega^{n-1}(TM \setminus M; \pi^* \mathcal{O}_M).$$

In [21] Mathai and Quillen have introduced a differential form

$$\Psi(g) \in \Omega^{n-1}(TM \setminus M; \pi^* \mathcal{O}_M).$$

When pulled back to the fibers of $TM \setminus M \to M$ the form $\Psi(g)$ coincides with $A(g)$. If $U \subseteq M$ is an open subset on which the curvature of g vanishes, then $\Psi(g)$ coincides with $A(g)$ on $TU \setminus U$. In general we have the equalities:

$$d\Psi(g) = \pi^* e(g) \tag{4}$$
$$\Psi(g_2) - \Psi(g_1) = \pi^* cs(g_1, g_2) \mod d\Omega^{n-2}(TM \setminus M; \pi^* \mathcal{O}_M) \tag{5}$$

2.2. Euler and Chern–Simons chains

For a vector field X with non-degenerate rest points we have the singular 0-chain $e(X) \in C_0(M; \mathbb{Z})$ defined by $e(X) := \sum_{x \in \mathcal{X}} \mathrm{IND}(x)x$, with $\mathrm{IND}(x)$ the Hopf index.

For two vector fields X_1 and X_2 with non-degenerate rest points we have the singular 1-chain rel. boundaries $cs(X_1, X_2) \in C_1(M; \mathbb{Z})/\partial C_2(M; \mathbb{Z})$ defined via the zero set of a homotopy from X_1 to X_2, cf. [7]. They are related by the formulas, see [7],

$$\partial \operatorname{cs}(X_1, X_2) = \operatorname{e}(X_2) - \operatorname{e}(X_1) \tag{6}$$
$$\operatorname{cs}(X_2, X_1) = -\operatorname{cs}(X_1, X_2) \tag{7}$$
$$\operatorname{cs}(X_1, X_3) = \operatorname{cs}(X_1, X_2) + \operatorname{cs}(X_2, X_3). \tag{8}$$

2.3. Kamber–Tondeur one form

Let E be a real or complex vector bundle over M. For a connection ∇ and a Hermitian structure μ on E define a real-valued one form $\omega(\nabla, \mu) \in \Omega^1(M; \mathbb{R})$ by

$$\omega(\nabla, \mu)(Y) := -\frac{1}{2} \operatorname{tr}(\mu^{-1} \cdot (\nabla_Y \mu)), \quad Y \in TM. \tag{9}$$

Here we consider μ as an element in $\Omega^0(M; \hom(E, \bar{E}^*))$, where $\bar{E}^*$ denotes the dual of the complex conjugate bundle. With respect to the induced connection on $\hom(E, \bar{E}^*)$ we have $\nabla_Y \mu \in \Omega^1(M; \hom(E, \bar{E}^*))$ and therefore $\mu^{-1} \cdot \nabla_Y \mu \in \Omega^1(M; \operatorname{end}(E, E))$. Actually the latter one form has values in the endomorphisms of E which are symmetric with respect to μ, and thus the (complex) trace, see (9), will indeed be real. Since any two Hermitian structures μ_1 and μ_2 are homotopic, the difference $\omega(\nabla, \mu_2) - \omega(\nabla, \mu_1)$ will be exact. If ∇ is flat then $\omega(\nabla, \mu)$ is closed and its cohomology class independent of μ.

Replacing the Hermitian structure by a non-degenerate symmetric bilinear form b, we define a complex-valued one form $\omega(\nabla, b) \in \Omega^1(M; \mathbb{C})$ by a similar formula

$$\omega(\nabla, b)(Y) := -\frac{1}{2} \operatorname{tr}(b^{-1} \cdot (\nabla_Y b)), \quad Y \in TM. \tag{10}$$

Here we regard b as an element in $\Omega^0(M; \hom(E, E^*))$. If two non-degenerate symmetric bilinear forms b_1 and b_2 are homotopic, then $\omega(\nabla, b_2) - \omega(\nabla, b_1)$ is exact. If ∇ is flat, then $\omega(\nabla, b)$ is closed. Note that $\omega(\nabla, b) \in \Omega^1(M; \mathbb{C})$ depends holomorphically on ∇.

2.4. Vector fields, instantons and closed trajectories

Consider a vector field X which satisfies the following properties:

  (H) All rest points are of hyperbolic type.

(EG) The vector field has exponential growth at all rest points.

  (L) The vector field is of Lyapunov type.

(MS) The vector field satisfies Morse–Smale condition.

(NCT) The vector field has all closed trajectories non-degenerate.

Precisely this means that:

  (H) The differential of X at each rest point x has all eigenvalues with non-trivial real part; the number of eigenvalues with negative real part is called the index and denoted by $\operatorname{ind}(x)$; as a consequence the stable and unstable stable sets are images of one-to-one immersions $i_x^\pm : W_x^\pm \to M$ with $W_x^\pm$ diffeomorphic to $\mathbb{R}^{n-\operatorname{ind}(x)}$ resp. $\mathbb{R}^{\operatorname{ind}(x)}$.

(EG) With respect to one and then any Riemannian metric g on M, the volume of the disk of radius r in W_x^- (w.r. to the induced Riemannian metric) is $\leq e^{Cr}$, for some constant $C > 0$.

(L) There exists a real-valued closed one form ω so that $\omega(X)_x < 0$ for x not a rest point.[3]

(MS) For any two rest points x and y the mappings i_x^- and i_y^+ are transversal and therefore the space of non-parameterized trajectories form x to y, $\mathcal{T}(x, y)$, is a smooth manifold of dimension $\mathrm{ind}(x) - \mathrm{ind}(y) - 1$. Instantons are exactly the elements of $\mathcal{T}(x, y)$ when this is a smooth manifold of dimension zero, i.e., $\mathrm{ind}(x) - \mathrm{ind}(y) - 1 = 0$.

(NCT) Any closed trajectory is non-degenerate, i.e., the differential of the return map in normal direction at one and then any point of a closed trajectory does not have non-zero fixed points.

Recall that a trajectory θ is an equivalence class of parameterized trajectories and two parameterized trajectories θ_1 and θ_2 are equivalent iff $\theta_1(t + c) = \theta_2(t)$ for some real number c. Recall that a closed trajectory $\hat{\theta}$ is a pair consisting of a trajectory θ and a positive real number T so that $\theta(t + T) = \theta(t)$.

Property (L), (H), (MS) imply that for any real number R the set of instantons θ from x to y with $-\omega([\theta]) \leq R$ is finite and properties (L), (H), (MS), (NCT) imply that for any real number R the set of the closed trajectories $\hat{\theta}$ with $-\omega([\hat{\theta}]) \leq R$ is finite. Here $[\theta]$ resp $[\hat{\theta}]$ denote the homotopy class of instantons resp. closed trajectories.[4]

Denote by $\mathcal{P}_{x,y}$ the set of homotopy classes of paths from x to y and by $\mathcal{X}_q$ the set of rest points of index q. Suppose a collection $\mathcal{O} = \{\mathcal{O}_x \mid x \in \mathcal{X}\}$ of orientations of the unstable manifolds is given and (MS) is satisfied. Then any instanton θ has a sign $\epsilon(\theta) = \pm 1$ and therefore, if (L) is also satisfied, for any two rest points $x \in \mathcal{X}_{q+1}$ and $y \in \mathcal{X}_q$ we have the counting function of instantons $\mathbb{I}_{x,y}^{X,\mathcal{O}} : \mathcal{P}_{x,y} \to \mathbb{Z}$ defined by

$$\mathbb{I}_{x,y}^{X,\mathcal{O}}(\alpha) := \sum_{\theta \in \alpha} \epsilon(\theta). \tag{11}$$

Under the hypothesis (NCT) any closed trajectory $\hat{\theta}$ has a sign $\epsilon(\hat{\theta}) = \pm 1$ and a period $p(\hat{\theta}) \in \{1, 2, \dots\}$, cf. [17]. If (H), (L), (MS), (NCT) are satisfied, as the set of closed trajectories in a fixed homotopy class $\gamma \in [S^1, M]$ is compact, we have the counting function of closed trajectories $\mathbb{Z}_X : [S^1, M] \to \mathbb{Q}$ defined by

$$\mathbb{Z}_X(\gamma) := \sum_{\hat{\theta} \in \gamma} \epsilon(\hat{\theta})/p(\hat{\theta}). \tag{12}$$

Here are a few properties about vector fields which satisfy (H) and (L).

---

[3]This ω has no relation with the Kamber–Tondeur form $\omega(\nabla, b)$ considered in the previous section.

[4]For a closed trajectory the map whose homotopy class is considered is $\hat{\theta} : \mathbb{R}/T\mathbb{Z} \to M$.

Proposition 1.

1. *Given a vector field X which satisfies (H) and (L) arbitrary close in the C^r-topology for any $r \geq 0$ there exists a vector field Y which agrees with X on a neighborhood of the rest points and satisfies (H), (L), (MS) and (NCT).*

2. *Given a vector field X which satisfies (H) and (L) arbitrary close in the C^0-topology there exists a vector field Y which agrees with X on a neighborhood of the rest points and satisfies (H), (EG), (L), (MS) and (NCT).*

3. *If X satisfies (H), (L) and (MS) and a collection $\mathcal{O}$ of orientations is given then for any $x \in \mathcal{X}_q$, $z \in \mathcal{X}_{q-2}$ and $\gamma \in \mathcal{P}_{x,z}$ one has[5]*

$$\sum_{\substack{y \in \mathcal{X}_{q-1}, \alpha \in \mathcal{P}_{x,y}, \beta \in \mathcal{P}_{y,z} \\ \alpha * \beta = \gamma}} \mathbb{I}^{X,\mathcal{O}}_{x,y}(\alpha) \cdot \mathbb{I}^{X,\mathcal{O}}_{y,z}(\beta) = 0. \tag{13}$$

This proposition is a recollection of some of the main results in [8], see Proposition 3, Theorem 1 and Theorem 5 in there.

3. Euler and coEuler structures

Although not always necessary in this section as in fact always in this paper M is supposed to be closed connected smooth manifold.

3.1. Euler structures

Euler structures have been introduced by Turaev [30] for manifolds M with $\chi(M) = 0$. If one removes the hypothesis $\chi(M) = 0$ the concept of Euler structure can still be considered for any connected base pointed manifold (M, x_0) cf. [5] and [7]. Here we will consider only the case $\chi(M) = 0$. The set of Euler structures, denoted by $\mathfrak{Eul}(M; \mathbb{Z})$, is equipped with a free and transitive action

$$m : H_1(M; \mathbb{Z}) \times \mathfrak{Eul}(M; \mathbb{Z}) \to \mathfrak{Eul}(M; \mathbb{Z})$$

which makes $\mathfrak{Eul}(M; \mathbb{Z})$ an affine version of $H_1(M; \mathbb{Z})$. If $\mathfrak{e}_1, \mathfrak{e}_2$ are two Euler structure we write $\mathfrak{e}_2 - \mathfrak{e}_1$ for the unique element in $H_1(M; \mathbb{Z})$ with $m(\mathfrak{e}_2 - \mathfrak{e}_1, \mathfrak{e}_1) = \mathfrak{e}_2$.

To define the set $\mathfrak{Eul}(M; \mathbb{Z})$ we consider pairs (X, c) with X a vector field with non-degenerate zeros and $c \in C_1(M; \mathbb{Z})$ so that $\partial c = e(X)$. We make (X_1, c_1) and (X_2, c_2) equivalent iff $c_2 - c_1 = cs(X_1, X_2)$ and write $[X, c]$ for the equivalence class represented by (X, c). The action m is defined by $m([c'], [X, c]) := [X, c' + c]$.

Observation 1. *Suppose X is a vector field with non-degenerate zeros, and assume its zero set $\mathcal{X}$ is non-empty. Moreover, let $\mathfrak{e} \in \mathfrak{Eul}(M; \mathbb{Z})$ be an Euler structure and $x_0 \in M$. Then there exists a collection of paths $\{\sigma_x \mid x \in \mathcal{X}\}$ with $\sigma_x(0) = x_0$, $\sigma_x(1) = x$ and such that $\mathfrak{e} = [X, c]$ where $c = \sum_{x \in \mathcal{X}} \mathrm{IND}(x)\sigma_x$.*

---

[5]It is understood that only finitely many terms from the left side of the equality are not zero. Here $*$ denotes juxtaposition.

A remarkable source of Euler structures is the set of homotopy classes of nowhere vanishing vector fields. Any nowhere vanishing vector field X provides an Euler structure $[X, 0]$ which only depends on the homotopy class of X. Still assuming $\chi(M) = 0$, every Euler structure can be obtained in this way provided $\dim(M) > 2$. Be aware, however, that different homotopy classes may give rise to the same Euler structure.

To construct such a homotopy class one can proceed as follows. Represent the Euler structure $\mathfrak{e}$ by a vector field X and a collection of paths $\{\sigma_x \mid x \in \mathcal{X}\}$ as in Observation 1. Since $\dim(M) > 2$ we may assume that the interiors of the paths are mutually disjoint. Then the set $\bigcup_{x \in \mathcal{X}} \sigma_x$ is contractible. A smooth regular neighborhood of it is the image by a smooth embedding $\varphi : (D^n, 0) \to (M, x_0)$. Since $\chi(M) = 0$, the restriction of the vector field X to $M \backslash \mathrm{int}(D^n)$ can be extended to a non-vanishing vector field $\tilde{X}$ on M. It is readily checked that $[\tilde{X}, 0] = \mathfrak{e}$. For details see [7].

If M has dimension larger than 2 an alternative description of $\mathfrak{Eul}(M; \mathbb{Z})$ with respect to a base point x_0 is $\mathfrak{Eul}(M; \mathbb{Z}) = \pi_0(\mathfrak{X}(M, x_0))$, where $\mathfrak{X}(M, x_0)$ denotes the space of vector fields of class C^r, $r \geq 0$, which vanish at x_0 and are non-zero elsewhere. We equip this space with the C^r-topology and note that the result $\pi_0(\mathfrak{X}(M, x_0))$ is the same for all r, and since $\chi(M) = 0$, canonically identified for different base points.

Let τ be a smooth triangulation of M and consider the function $f_\tau : M \to \mathbb{R}$ linear on any simplex of the first barycentric subdivision and taking the value $\dim(s)$ on the barycenter x_s of the simplex $s \in \tau$. A smooth vector field X on M with the barycenters as the only rest points all of them hyperbolic and f_τ strictly decreasing on non-constant trajectories is called an Euler vector field of τ. By an argument of convexity two Euler vector fields are homotopic by a homotopy of Euler vector fields.[6] Therefore, a triangulation τ, a base point x_0 and a collection of paths $\{\sigma_s \mid s \in \tau\}$ with $\sigma_s(0) = x_0$ and $\sigma_s(1) = x_s$ define an Euler structure $[X_\tau, c]$, where $c := \sum_{s \in \tau} (-1)^{n+\dim(s)} \sigma_s$, X_τ is any Euler vector field for τ, and this Euler structure does not depend on the choice of X_τ. Clearly, for fixed τ and x_0, every Euler structure can be realized in this way by an appropriate choice of $\{\sigma_s \mid s \in \tau\}$, cf. Observation 1.

3.2. CoEuler structures

Again, suppose $\chi(M) = 0$.[7] Consider pairs (g, α) where g is a Riemannian metric on M and $\alpha \in \Omega^{n-1}(M; \mathcal{O}_M)$ with $d\alpha = \mathrm{e}(g)$ where $\mathrm{e}(g) \in \Omega^n(M; \mathcal{O}_M)$ denotes the Euler form of g, see Section 2.1. We call two pairs (g_1, α_1) and (g_2, α_2) equivalent if

$$\mathrm{cs}(g_1, g_2) = \alpha_2 - \alpha_1 \in \Omega^{n-1}(M; \mathcal{O}_M)/d\Omega^{n-2}(M; \mathcal{O}_M).$$

---

[6] Any Euler vector field X satisfies (H), (EG), (L) and has no closed trajectory, hence also satisfies (NCT). The counting functions of instantons are exactly the same as the incidence numbers of the triangulation hence take the values 1, −1 or 0.

[7] The hypothesis is not necessary and the theory of coEuler structure can be pursued for an arbitrary base pointed smooth manifold (M, x_0), cf. [7].

We will write $\mathfrak{Eul}^*(M;\mathbb{R})$ for the set of equivalence classes and $[g,\alpha]$ for the equivalence class represented by the pair (g,α). Elements of $\mathfrak{Eul}^*(M;\mathbb{R})$ are called *coEuler structures*.

There is a natural action

$$m^* : H^{n-1}(M;\mathcal{O}_M) \times \mathfrak{Eul}^*(M;\mathbb{R}) \to \mathfrak{Eul}^*(M;\mathbb{R})$$

given by

$$m^*([\beta],[g,\alpha]) := [g,\alpha - \beta]$$

for $[\beta] \in H^{n-1}(M;\mathcal{O}_M)$. This action is obviously free and transitive. In this sense $\mathfrak{Eul}^*(M;\mathbb{R})$ is an affine version of $H^{n-1}(M;\mathcal{O}_M)$. If $\mathfrak{e}_1^*$ and $\mathfrak{e}_2^*$ are two coEuler structures we write $\mathfrak{e}_2^* - \mathfrak{e}_1^*$ for the unique element in $H^{n-1}(M;\mathcal{O}_M)$ with $m^*(\mathfrak{e}_2^* - \mathfrak{e}_1^*, \mathfrak{e}_1^*) = \mathfrak{e}_2^*$.

Observation 2. *Given a Riemannian metric g on M any coEuler structure can be represented as a pair (g,α) for some $\alpha \in \Omega^{n-1}(M;\mathcal{O}_M)$ with $d\alpha = \mathrm{e}(g)$.*

There is a natural map $\mathrm{PD} : \mathfrak{Eul}(M;\mathbb{Z}) \to \mathfrak{Eul}^*(M;\mathbb{R})$ which combined with the Poincaré duality map $D : H_1(M;\mathbb{Z}) \to H_1(M;\mathbb{R}) \to H^{n-1}(M;\mathcal{O}_M)$, the composition of the coefficient homomorphism for $\mathbb{Z} \to \mathbb{R}$ with the Poincaré duality isomorphism,[8] makes the diagram below commutative:

$$
\begin{array}{ccc}
H_1(M;\mathbb{Z}) \times \mathfrak{Eul}(M;\mathbb{Z}) & \xrightarrow{\ m\ } & \mathfrak{Eul}(M;\mathbb{Z}) \\
{\scriptstyle D\times\mathrm{PD}}\downarrow & & \downarrow{\scriptstyle \mathrm{PD}} \\
H^{n-1}(M;\mathcal{O}_M) \times \mathfrak{Eul}^*(M;\mathbb{R}) & \xrightarrow{\ m^*\ } & \mathfrak{Eul}^*(M;\mathbb{R})
\end{array}
$$

There are many ways to define the map PD, cf. [7]. For example, assuming $\chi(M) = 0$ and $\dim M > 2$ one can proceed as follows. Represent the Euler structure by a nowhere vanishing vector field $\mathfrak{e} = [X,0]$. Choose a Riemannian metric g, regard X as mapping $X : M \to TM \setminus M$, set $\alpha := X^*\Psi(g)$, put $\mathrm{PD}(\mathfrak{e}) := [g,\alpha]$ and check that this does indeed only depend on $\mathfrak{e}$.

A coEuler structure $\mathfrak{e}^* \in \mathfrak{Eul}^*(M;\mathbb{R})$ is called *integral* if it belongs to the image of PD. Integral coEuler structures constitute a lattice in the affine space $\mathfrak{Eul}^*(M;\mathbb{R})$.

Observation 3. *If $\dim M$ is odd, then there is a canonical coEuler structure $\mathfrak{e}_0^* \in \mathfrak{Eul}^*(M;\mathbb{R})$; it is represented by the pair $[g,0]$, with any g Riemannian metric. In general this coEuler structure is not integral.*

---

[8]We will use the same notation D for the Poincaré duality isomorphism $D : H_1(M;\mathbb{R}) \to H^{n-1}(M;\mathcal{O}_M)$.

4. Complex representations and cochain complexes

4.1. Complex representations

Let Γ be a finitely presented group with generators $g_1, \ldots, g_r$ and relations

$$R_i(g_1, g_2, \ldots, g_r) = e, \quad i = 1, \ldots, p,$$

and V be a complex vector space of dimension N. Let $\mathrm{Rep}(\Gamma; V)$ be the set of linear representations of Γ on V, i.e., group homomorphisms $\rho : \Gamma \to \mathrm{GL}_\mathbb{C}(V)$. By identifying V to $\mathbb{C}^N$ this set is, in a natural way, an algebraic set inside the space $\mathbb{C}^{rN^2+1}$ given by $pN^2 + 1$ equations. Precisely if $A_1, \ldots, A_r, z$ represent the coordinates in $\mathbb{C}^{rN^2+1}$ with $A := (a^{ij})$, $a^{ij} \in \mathbb{C}$, so $A \in \mathbb{C}^{N^2}$ and $z \in \mathbb{C}$, then the equations defining $\mathrm{Rep}(\Gamma; V)$ are

$$\begin{aligned}
z \cdot \det(A_1) \cdot \det(A_2) \cdots \det(A_r) &= 1 \\
R_i(A_1, \ldots, A_r) &= \mathrm{id}, \quad i = 1, \ldots, p
\end{aligned}$$

with each of the equalities R_i providing N^2 polynomial equations.

Suppose $\Gamma = \pi_1(M, x_0)$, M a closed manifold. Denote by $\mathrm{Rep}_0^M(\Gamma; V)$ the set of representations ρ with $H^*(M; \rho) = 0$ and notice that they form a Zariski open set in $\mathrm{Rep}(\Gamma; V)$. Denote the closure of this set by $\mathrm{Rep}^M(\Gamma; V)$. This is an algebraic set which depends only on the homotopy type of M, and is a union of irreducible components of $\mathrm{Rep}(\Gamma; V)$.

Recall that every representation $\rho \in \mathrm{Rep}(\Gamma; V)$ induces a canonical vector bundle F_ρ equipped with a canonical flat connection ∇_ρ. They are obtained from the trivial bundle $\tilde{M} \times V \to \tilde{M}$ and the trivial connection by passing to the Γ quotient spaces. Here $\tilde{M}$ is the canonical universal covering provided by the base point x_0. The Γ-action is the diagonal action of deck transformations on $\tilde{M}$ and of the action ρ on V. The fiber of F_ρ over x_0 identifies canonically with V. The holonomy representation determines a right Γ-action on the fiber of F_ρ over x_0, i.e., an anti homomorphism $\Gamma \to \mathrm{GL}(V)$. When composed with the inversion in $\mathrm{GL}(V)$ we get back the representation ρ. The pair (F_ρ, ∇_ρ) will be denoted by $\mathbb{F}_\rho$.

If ρ_0 is a representation in the connected component $\mathrm{Rep}_\alpha(\Gamma; V)$ one can identify $\mathrm{Rep}_\alpha(\Gamma; V)$ to the connected component of ∇_{ρ_0} in the complex analytic space of flat connections of the bundle F_{ρ_0} modulo the group of bundle isomorphisms of F_{ρ_0} which fix the fiber above x_0.

Remark 1. An element $a \in H_1(M; \mathbb{Z})$ defines a holomorphic function

$$\det_a : \mathrm{Rep}^M(\Gamma; V) \to \mathbb{C}_*.$$

The complex number $\det_a(\rho)$ is the evaluation on $a \in H_1(M; \mathbb{Z})$ of $\det(\rho) : \Gamma \to \mathbb{C}_*$ which factors through $H_1(M; \mathbb{Z})$. Note that for $a, b \in H_1(M; \mathbb{Z})$ we have $\det_{a+b} = \det_a \det_b$. If a is a torsion element, then $\det_a$ is constant equal to a root of unity of order, the order of a.

4.2. The space of cochain complexes

Let $k = (k_0, k_1, \ldots, k_n)$ be a string of non-negative integers. The string is called admissible, and we will write $k \geq 0$ in this case, if the following requirements are satisfied

$$k_0 - k_1 + k_2 \mp \cdots + (-1)^n k_n = 0 \tag{14}$$

$$k_i - k_{i-1} + k_{i-2} \mp \cdots + (-1)^i k_0 \geq 0 \quad \text{for any } i \leq n-1. \tag{15}$$

Denote by $\mathbb{D}(k) = \mathbb{D}(k_0, \ldots, k_n)$ the collection of cochain complexes of the form

$$C = (C^*, d^*) : 0 \to C^0 \xrightarrow{d^0} C^1 \xrightarrow{d^1} \cdots \xrightarrow{d^{n-2}} C^{n-1} \xrightarrow{d^{n-1}} C^n \to 0$$

with $C^i := \mathbb{C}^{k_i}$, and by $\mathbb{D}_{ac}(k) \subseteq \mathbb{D}(k)$ the subset of acyclic complexes. Note that $\mathbb{D}_{ac}(k)$ is non-empty iff $k \geq 0$. The cochain complex C is determined by the collection $\{d^i\}$ of linear maps $d^i : \mathbb{C}^{k_i} \to \mathbb{C}^{k_{i+1}}$. If regarded as the subset of those $\{d^i\} \in \bigoplus_{i=0}^{n-1} L(\mathbb{C}^{k_i}, \mathbb{C}^{k_{i+1}})$, with $L(V, W)$ the space of linear maps from V to W, which satisfy the quadratic equations $d^{i+1} \cdot d^i = 0$, the set $\mathbb{D}(k)$ is an affine algebraic set given by degree two homogeneous polynomials and $\mathbb{D}_{ac}(k)$ is a Zariski open set. The map $\pi_0 : \mathbb{D}_{ac}(k) \to \text{Emb}(C^0, C^1)$ which associates to $C \in \mathbb{D}_{ac}(k)$ the linear map d^0, is a bundle whose fiber is isomorphic to $\mathbb{D}_{ac}(k_1 - k_0, k_2, \ldots, k_n)$.

This can be easily generalized as follows. Consider a string $b = (b_0, \ldots, b_n)$. We will write $k \geq b$ if $k - b = (k_0 - b_0, \ldots, k_n - b_n)$ is admissible, i.e., $k - b \geq 0$. Denote by $\mathbb{D}_b(k) = \mathbb{D}_{(b_0, \ldots, b_n)}(k_0, \ldots, k_n)$ the subset of cochain complexes $C \in \mathbb{D}(k)$ with $\dim(H^i(C)) = b_i$. Note that $\mathbb{D}_b(k)$ is non-empty iff $k \geq b$. The obvious map $\pi_0 : \mathbb{D}_b(k) \to L(C^0, C^1; b_0)$, $L(C^0, C^1; b_0)$ the space of linear maps in $L(C^0, C^1)$ whose kernel has dimension b_0, is a bundle whose fiber is isomorphic to $\mathbb{D}_{b_1, \ldots, b_n}(k_1 - k_0 + b_0, k_2, \ldots, k_n)$. Note that $L(C^0, C^1; b_0)$ is the total space of a bundle $\text{Emb}(\mathbb{C}^{k_0}/L, \mathbb{C}^{k_1}) \to \text{Gr}_{b_0}(k_0)$ with $L \to \text{Gr}_{b_0}(k_0)$ the tautological bundle over $\text{Gr}_{b_0}(k_0)$ and $\mathbb{C}^{k_0}$ resp. $\mathbb{C}^{k_1}$ the trivial bundles over $\text{Gr}_{b_0}(k_0)$ with fibers of dimension k_0 resp. k_1. As a consequence we have

Proposition 2. 1. $\mathbb{D}_{ac}(k)$ and $\mathbb{D}_b(k)$ are connected smooth quasi affine algebraic sets whose dimension is

$$\dim \mathbb{D}_b(k) = \sum_j (k^j - b^j) \cdot \left(k^j - \sum_{i \leq j} (-1)^{i+j} (k^i - b^i) \right).$$

2. *The closures* $\hat{\mathbb{D}}_{ac}(k)$ *and* $\hat{\mathbb{D}}_b(k)$ *are irreducible algebraic sets, hence affine algebraic varieties, and* $\hat{\mathbb{D}}_b(k) = \bigsqcup_{k \geq b' \geq b} \mathbb{D}_{b'}(k)$.

For any cochain complex in $C \in \mathbb{D}_{ac}(k)$ denote by $B^i := \text{img}(d^{i-1}) \subseteq C^i = \mathbb{C}^{k_i}$ and consider the short exact sequence $0 \to B^i \xrightarrow{\text{inc}} C^i \xrightarrow{d^i} B^{i+1} \to 0$. Choose a base $\mathfrak{b}_i$ for each B_i, and choose lifts $\overline{\mathfrak{b}}_{i+1}$ of $\mathfrak{b}_{i+1}$ in C^i using d^i, i.e., $d^i(\overline{\mathfrak{b}}_{i+1}) = \mathfrak{b}_{i+1}$. Clearly $\{\mathfrak{b}_i, \overline{\mathfrak{b}}_{i+1}\}$ is a base of C^i. Consider the base $\{\mathfrak{b}_i, \overline{\mathfrak{b}}_{i+1}\}$ as a collection of vectors in $C^i = \mathbb{C}^{k_i}$ and write them as columns of a matrix $[\mathfrak{b}_i, \overline{\mathfrak{b}}_{i+1}]$. Define the torsion of the acyclic complex C, by

$$\tau(C) := (-1)^{N+1} \prod_{i=0}^{n} \det[\mathfrak{b}_i, \overline{\mathfrak{b}}_{i+1}]^{(-1)^i}$$

where $(-1)^N$ is Turaev's sign, see [15]. The result is independent of the choice of the bases $\mathfrak{b}_i$ and of the lifts $\overline{\mathfrak{b}}_i$ cf. [22] [15], and leads to the function

$$\tau : \mathbb{D}_{\mathrm{ac}}(k) \to \mathbb{C}_*.$$

Turaev provided a simple formula for this function, cf. [31], which permits to recognize τ as the restriction of a rational function on $\hat{\mathbb{D}}_{\mathrm{ac}}(k)$.

For $C \in \hat{\mathbb{D}}_{\mathrm{ac}}(k)$ denote by $(d^i)^t : \mathbb{C}^{k_{i+1}} \to \mathbb{C}^{k_i}$ the transpose of $d^i : \mathbb{C}^{k_i} \to \mathbb{C}^{k_{i+1}}$, and define $P_i = d^{i-1} \cdot (d^{i-1})^t + (d^i)^t \cdot d^i$. Define $\Sigma(k)$ as the subset of cochain complexes in $\hat{\mathbb{D}}_{\mathrm{ac}}(k)$ where $\ker P \neq 0$, and consider $S\tau : \hat{\mathbb{D}}_{\mathrm{ac}}(k) \backslash \Sigma(k) \to \mathbb{C}_*$ defined by

$$S\tau(C) := \left(\prod_{i \text{ even}} (\det P_i)^i / \prod_{i \text{ odd}} (\det P_i)^i \right)^{-1}.$$

One can verify

Proposition 3. *Suppose $k = (k_0, \dots, k_n)$ is admissible.*

1. *$\Sigma(k)$ is a proper subvariety containing the singular set of $\hat{\mathbb{D}}_{\mathrm{ac}}(k)$.*
2. *$S\tau = \tau^2$ and implicitly $S\tau$ has an analytic continuation to $\mathbb{D}_{\mathrm{ac}}(k)$.*

In particular τ defines a square root of $S\tau$. We will not use explicitly $S\tau$ in this writing however it justifies the definition of complex Ray–Singer torsion.

5. Analytic torsion

Let M be a closed manifold, g Riemannian metric and (g, α) a representative of a coEuler structure $\mathfrak{e}^* \in \mathfrak{Eul}^*(M; \mathbb{R})$. Suppose $E \to M$ is a complex vector bundle and denote by $\mathcal{C}(E)$ the space of connections and by $\mathcal{F}(E)$ the subset of flat connections. $\mathcal{C}(E)$ is a complex affine (Fréchet) space while $\mathcal{F}(E)$ a closed complex analytic subset (Stein space) of $\mathcal{C}(E)$. Let b be a non-degenerate symmetric bilinear form and μ a Hermitian (fiber metric) structure on E. While Hermitian structures always exist, non-degenerate symmetric bilinear forms exist iff the bundle is the complexification of some real vector bundle, and in this case $E \simeq E^*$.

The connection $\nabla \in \mathcal{C}(E)$ can be interpreted as a first-order differential operator $d^\nabla : \Omega^*(M; E) \to \Omega^{*+1}(M; E)$ and g and b resp. g and μ can be used to define the formal b-adjoint resp. μ-adjoint $\delta^\nabla_{q;g,b}$ resp. $\delta^\nabla_{q;g,\mu} : \Omega^{q+1}(M; E) \to \Omega^q(M; E)$ and therefore the Laplacians

$$\Delta^\nabla_{q;g,b} \text{ resp. } \Delta^\nabla_{q;g,\mu} : \Omega^q(M; E) \to \Omega^q(M; E).$$

They are elliptic second-order differential operators with principal symbol $\sigma_\xi = |\xi|^2$. Therefore they have a unique well-defined zeta regularized determinant (modified determinant) $\det(\Delta^\nabla_{q;g,b}) \in \mathbb{C}$ ($\det'(\Delta^\nabla_{q;g,b}) \in \mathbb{C}_*$) resp. $\det(\Delta^\nabla_{q;g,\mu}) \in \mathbb{R}_{\geq 0}$ ($\det'(\Delta^\nabla_{q;g,\mu}) \in \mathbb{R}_{>0}$) calculated with respect to a non-zero Agmon angle avoiding the spectrum cf. [10]. Recall that the zeta regularized determinant (modified determinant) is the zeta regularized product of all (non-zero) eigenvalues.

Denote by

$$\Sigma(E, g, b) := \left\{ \nabla \in \mathcal{C}(E) \mid \ker(\Delta^\nabla_{*;g,b}) \neq 0 \right\}$$
$$\Sigma(E, g, \mu) := \left\{ \nabla \in \mathcal{C}(E) \mid \ker(\Delta^\nabla_{*;g,\mu}) \neq 0 \right\}$$

and by

$$\Sigma(E) := \left\{ \nabla \in \mathcal{F}(E) \mid H^*(\Omega^*(M; E), d^\nabla) \neq 0 \right\}.$$

Note that $\Sigma(E, g, \mu) \cap \mathcal{F}(E) = \Sigma(E)$ for any μ, and $\Sigma(E, g, b) \cap \mathcal{F}(E) \supseteq \Sigma(E)$. Both, $\Sigma(E)$ and $\Sigma(E, g, b) \cap \mathcal{F}(E)$, are closed complex analytic subsets of $\mathcal{F}(E)$, and $\det(\Delta^\nabla_{q;g,\dots}) = \det'(\Delta^\nabla_{q;g,\dots})$ on $\mathcal{F}(E) \setminus \Sigma(E, g, \dots)$.

We consider the real analytic functions: $T^{\text{even}}_{g,\mu} : \mathcal{C}(E) \to \mathbb{R}_{\geq 0}$, $T^{\text{odd}}_{g,\mu} : \mathcal{C}(E) \to \mathbb{R}_{\geq 0}$, $R_{\alpha,\mu} : \mathcal{C}(E) \to \mathbb{R}_{>0}$ and the holomorphic functions $T^{\text{even}}_{g,b} : \mathcal{C}(E) \to \mathbb{C}$, $T^{\text{odd}}_{g,b} : \mathcal{C}(E) \to \mathbb{C}$, $R_{\alpha,b} : \mathcal{C}(E) \to \mathbb{C}_*$ defined by:

$$T^{\text{even}}_{g,\dots}(\nabla) := \prod_{q \text{ even}} (\det \Delta^\nabla_{q;g,\dots})^q,$$

$$T^{\text{odd}}_{g,\dots}(\nabla) := \prod_{q \text{ odd}} (\det \Delta^\nabla_{q;g,\dots})^q, \qquad (16)$$

$$R_{\alpha,\dots}(\nabla) := e^{\int_M \omega(\dots,\nabla) \wedge \alpha}.$$

We also write $T'^{\text{even}}_{g,\dots}$ resp. $T'^{\text{odd}}_{g,\dots}$ for the same formulas with $\det'$ instead of $\det$. These functions are discontinuous on $\Sigma(E, g, \dots)$ and coincide with $T^{\text{even}}_{g,\dots}$ resp. $T^{\text{odd}}_{g,\dots}$ on $\mathcal{F}(E) \setminus \Sigma(E, g, \dots)$. Here $\dots$ stands for either b or μ. For the definition of real or complex analytic space/set, holomorphic/meromorphic function/map in infinite dimension the reader can consult [13] and [19], although the definitions used here are rather straightforward.

Let $E_r \to M$ be a smooth real vector bundle equipped with a non-degenerate symmetric positive definite bilinear form b_r. Let $\mathcal{C}(E_r)$ resp. $\mathcal{F}(E_r)$ the space of connections resp. flat connections in E_r. Denote by $E \to M$ the complexification of E_r, $E = E_r \otimes \mathbb{C}$, and by b resp. μ the complexification of b_r resp. the Hermitian structure extension of b_r. We continue to denote by $\mathcal{C}(E_r)$ resp. $\mathcal{F}(E_r)$ the subspace of $\mathcal{C}(E)$ resp. $\mathcal{F}(E)$ consisting of connections which are complexification of connections resp. flat connections in E_r, and by ∇ the complexification of the connection $\nabla \in \mathcal{C}(E_r)$. If $\nabla \in \mathcal{C}(E_r)$, then

$$\text{Spect } \Delta^\nabla_{q;g,b} = \text{Spect } \Delta^\nabla_{q;g,\mu} \subseteq \mathbb{R}_{\geq 0}$$

and therefore

$$T_{g,b}^{\text{even/odd}}(\nabla) = \left|T_{g,b}^{\text{even/odd}}(\nabla)\right| = T_{g,\mu}^{\text{even/odd}}(\nabla),$$
$$T_{g,b}'^{\text{even/odd}}(\nabla) = \left|T_{g,b}'^{\text{even/odd}}(\nabla)\right| = T_{g,\mu}'^{\text{even/odd}}(\nabla), \tag{17}$$
$$R_{\alpha,b}(\nabla) = \left|R_{\alpha,b}(\nabla)\right| = R_{\alpha,\mu}(\nabla).$$

Observe that $\Omega^*(M; E)(0)$ the (generalized) eigenspace of $\Delta_{*;g,b}^\nabla$ corresponding to the eigenvalue zero is a finite-dimensional vector space of dimension the multiplicity of 0. The restriction of the symmetric bilinear form induced by b remains non-degenerate and defines for each component $\Omega^q(M; E)(0)$ an equivalence class of bases. Since d^∇ commutes with $\Delta_{*;g,b}^\nabla$, $\left(\Omega^*(M; E)(0), d^\nabla\right)$ is a finite-dimensional complex. When acyclic, i.e., $\nabla \in \mathcal{F}(E) \setminus \Sigma(E)$, denote by

$$T_{\text{an}}(\nabla, g, b)(0) \in \mathbb{C}_*$$

the Milnor torsion associated to the equivalence class of bases induced by b.

5.1. The modified Ray–Singer torsion

Let $E \to M$ be a complex vector bundle, and let $\mathfrak{e}^* \in \mathfrak{Eul}^*(M; \mathbb{R})$ be a coEuler structure. Choose a Hermitian structure (fiber metric) μ on E, a Riemannian metric g on M and $\alpha \in \Omega^{n-1}(M; \mathcal{O}_M)$ so that $[g, \alpha] = \mathfrak{e}^*$, see Section 3.2. For $\nabla \in \mathcal{F}(E) \setminus \Sigma(E)$ consider the quantity

$$T_{\text{an}}(\nabla, \mu, g, \alpha) := \left(T_{g,\mu}^{\text{even}}(\nabla)/T_{g,\mu}^{\text{odd}}(\nabla)\right)^{-1/2} \cdot R_{\alpha,\mu}(\nabla) \in \mathbb{R}_{>0}$$

referred to as the *modified Ray–Singer torsion*. The following proposition is a reformulation of one of the main theorems in [2], cf. also [6] and [7].

Proposition 4. *If $\nabla \in \mathcal{F}(E) \setminus \Sigma(E)$, then $T_{\text{an}}(\nabla, \mu, g, \alpha)$ is gauge invariant and independent of μ, g, α.*

When applied to $\mathbb{F}_\rho$ the number $T_{\text{an}}^{\mathfrak{e}^*}(\rho) := T_{\text{an}}(\nabla_\rho, \mu, g, \alpha)$ defines a real analytic function $T_{\text{an}}^{\mathfrak{e}^*} : \text{Rep}_0^M(\Gamma; V) \to \mathbb{R}_{>0}$. It is natural to ask if $T_{\text{an}}^{\mathfrak{e}^*}$ is the absolute value of a holomorphic function.

The answer is no as one can see on the simplest possible example $M = S^1$ equipped with the the canonical coEuler structure $\mathfrak{e}_0^*$. In this case $\text{Rep}^M(\Gamma; \mathbb{C}) = \mathbb{C} \setminus 0$, and $T_{\text{an}}^{\mathfrak{e}_0^*}(z) = |\frac{(1-z)}{z^{1/2}}|$, cf. [10]. However, Theorem 2 in Section 6.1 below provides the following answer to the question (Q) from the introduction.

Observation 4. *If $\mathfrak{e}^*$ is an integral coEuler structure, then $T_{\text{an}}^{\mathfrak{e}^*}$ is the absolute value of a holomorphic function on $\text{Rep}_0^M(\Gamma; V)$ which is the restriction of a rational function on $\text{Rep}^M(\Gamma; V)$. For a general coEuler structure $T_{\text{an}}^{\mathfrak{e}^*}$ still locally is the absolute value of a holomorphic function.*

5.2. Complex Ray–Singer torsion

Let E be a complex vector bundle equipped with a non-degenerate symmetric bilinear form b. Suppose (g, α) is a pair consisting of a Riemannian metric g and a differential form $\alpha \in \Omega^{n-1}(M; \mathcal{O}_M)$ with $d\alpha = \mathrm{e}(g)$. For any $\nabla \in \mathcal{F}(E) \setminus \Sigma(E)$ consider the complex number

$$\mathcal{ST}_{\mathrm{an}}(\nabla, b, g, \alpha) := \left(T_{g,b}^{\prime\,\mathrm{even}}(\nabla)/T_{g,b}^{\prime\,\mathrm{odd}}(\nabla)\right)^{-1} \cdot R_{\alpha,b}(\nabla)^2 \cdot T_{\mathrm{an}}(\nabla, g, b)(0)^2 \in \mathbb{C}_*$$

(18)

referred to as the *complex-valued Ray–Singer torsion*.[9]

It is possible to provide an alternative definition of $\mathcal{ST}_{\mathrm{an}}(\nabla, b, g, \alpha)$. Suppose $R > 0$ is a positive real number so that the Laplacians $\Delta_{q;g,b}^{\nabla}$ have no eigenvalues of absolute value R. In this case denote by $\det^R \Delta_{q;g,b}^{\nabla}$ the regularized product of all eigenvalues larger than R w.r. to a non-zero Agmon angle disjoint from the spectrum and by $T_{g,b}^{R,\mathrm{even}}$ resp. $T_{g,b}^{R,\mathrm{even}}$ the quantities defined by the formulae (16) with $T^{R,\mathrm{even/odd}}(\Delta)$ instead of $T^{\prime\,\mathrm{even/odd}}(\Delta)$. Consider $\Omega^*(M; E)(R)$ to be the sum of generalized eigenspaces of $\Delta_{*;g,b}^{\nabla}$ corresponding to eigenvalues smaller in absolute value than R. $(\Omega^*(M; E)(R), d^{\nabla})$ is a finite-dimensional complex. As before b remains non-degenerate and when acyclic (and this is the case iff $(\Omega^*(M; E), d^{\nabla})$ is acyclic) denote by $T_{\mathrm{an}}(\nabla, g, b)(R)$ the Milnor torsion associated to the equivalence class of bases induced by b. It is easy to check that

$$\mathcal{ST}_{\mathrm{an}}(\nabla, b, g, \alpha) = \left(T_{g,b}^{R,\mathrm{even}}(\nabla)/T_{g,b}^{R,\mathrm{odd}}(\nabla)\right)^{-1} \cdot R_{\alpha,b}(\nabla)^2 \cdot T_{\mathrm{an}}(\nabla, g, b)(R)^2 \quad (19)$$

Proposition 5.

1. $\mathcal{ST}_{\mathrm{an}}(\nabla, b, g, \alpha)$ *is a holomorphic function on* $\mathcal{F}(E) \setminus \Sigma(E)$ *and the restriction of a meromorphic function on* $\mathcal{F}(E)$ *with poles and zeros in* $\Sigma(E)$.
2. *If* b_1 *and* b_2 *are two non-degenerate symmetric bilinear forms which are homotopic, then* $\mathcal{ST}_{\mathrm{an}}(\nabla, b_1, g, \alpha) = \mathcal{ST}_{\mathrm{an}}(\nabla, b_2, g, \alpha)$.
3. *If* (g_1, α_1) *and* (g_2, α_2) *are two pairs representing the same coEuler structure, then* $\mathcal{ST}_{\mathrm{an}}(\nabla, b, g_1, \alpha_1) = \mathcal{ST}_{\mathrm{an}}(\nabla, b, g_2, \alpha_2)$.
4. *We have* $\mathcal{ST}_{\mathrm{an}}(\gamma\nabla, \gamma b, g, \alpha) = \mathcal{ST}_{\mathrm{an}}(\nabla, b, g, \alpha)$ *for every gauge transformation* γ *of* E.
5. $\mathcal{ST}_{\mathrm{an}}(\nabla_1 \oplus \nabla_2, b_1 \oplus b_2, g, \alpha) = \mathcal{ST}_{\mathrm{an}}(\nabla_1, b_1, g, \alpha) \cdot \mathcal{ST}_{\mathrm{an}}(\nabla_2, b_2, g, \alpha)$.

To check the first part of this proposition, one shows that for $\nabla_0 \in \mathcal{F}(E)$ one can find $R > 0$ and an open neighborhood U of $\nabla_0 \in \mathcal{F}(E)$ such that no eigenvalue of $\Delta_{q;g,b}^{\nabla}$, $\nabla \in U$, has absolute value R. The function $\left(T_{g,b}^{R,\mathrm{even}}(\nabla)/T_{g,b}^{R,\mathrm{odd}}(\nabla)\right)^{-1}$ is holomorphic in $\nabla \in U$. Moreover, on U the function $T_{\mathrm{an}}(\nabla, g, b)(R)^2$ is meromorphic in ∇, and holomorphic when restricted to $U \setminus \Sigma(E)$. The statement thus follows from (19).

The second and third part of Proposition 5 are derived from formulas for $d/dt(\mathcal{ST}_{\mathrm{an}}(\nabla, b(t), g, \alpha))$ resp. $d/dt(\mathcal{ST}_{\mathrm{an}}(\nabla, b, g(t), \alpha)$ which are similar to such

---

[9]The idea of considering b-Laplacians for torsion was brought to the attention of the first author by W. Müller [23]. The second author came to it independently.

formulas for Ray–Singer torsion in the case of a Hermitian structure instead of a non-degenerate symmetric bilinear form, cf. [10]. The proof of 4) and 5) require a careful inspection of the definitions. The full arguments are contained in [10].

As a consequence to each homotopy class of non-degenerate symmetric bilinear forms $[b]$ and coEuler structure $\mathfrak{e}^*$ we can associate a meromorphic function on $\mathcal{F}(E)$. The reader unfamiliar with the basic concepts of complex analytic geometry on Banach/Fréchet manifolds can consult [13] and [19]. Changing the coEuler structure our function changes by multiplication with a non-vanishing holomorphic function as one can see from (18). Changing the homotopy class $[b]$ is actually more subtle. However $\mathcal{ST}$ remains unchanged when the coEuler structure is integral. This fact was conjectured in [10] and is implied by the result of [28] or [11].

Denote by $\mathrm{Rep}^{M,E}(\Gamma; V)$ the union of components of $\mathrm{Rep}^M(\Gamma; V)$ which consists of representations equivalent to holonomy representations of flat connections in the bundle E. Suppose E admits non-degenerate symmetric bilinear forms and let $[b]$ be a homotopy class of such forms. Let $x_0 \in M$ be a base point and denote by $\mathcal{G}(E)_{x_0,[b]}$ the group of gauge transformations which leave fixed E_{x_0} and the class $[b]$. In view of Proposition 5, $\mathcal{ST}_{\mathrm{an}}(\nabla, b, g, \alpha)$ defines a meromorphic function $\mathcal{ST}_{\mathrm{an}}^{\mathfrak{e}^*,[b]}$ on $\pi^{-1}(\mathrm{Rep}^{M,E}(\Gamma; V) \subseteq \mathcal{F}(E)/\mathcal{G}_{x_0,[b]}$. Note that $\pi : \mathcal{F}(E)/\mathcal{G}_{x_0,[b]} \to \mathrm{Rep}(\Gamma; V)$ is a principal holomorphic covering of its image which contains $\mathrm{Rep}^{M,E}(\Gamma; V)$. The results in [28] and [11] imply that the absolute value of this function is the square of modified Ray–Singer torsion.

We summarize this in the following theorem.

Theorem 1. *With the hypotheses above we have.*

1. *If $\mathfrak{e}_1^*$ and $\mathfrak{e}_2^*$ are two coEuler structures then*

$$\mathcal{ST}_{\mathrm{an}}^{\mathfrak{e}_1^*,[b]} = \mathcal{ST}_{\mathrm{an}}^{\mathfrak{e}_2^*,[b]} \cdot e^{2([\omega(\nabla,b)], D^{-1}(\mathfrak{e}_1^* - \mathfrak{e}_2^*))}$$

   with $D : H_1(M; \mathbb{R}) \to H^{n-1}(M; \mathcal{O}_M)$ the Poincaré duality isomorphism.

2. *If $\mathfrak{e}^*$ is integral then $\mathcal{ST}_{\mathrm{an}}^{\mathfrak{e}^*,[b]}$ is independent of $[b]$ and descends to a rational function on $\mathrm{Rep}^{M,E}(\Gamma; V)$ denoted $\mathcal{ST}_{\mathrm{an}}^{\mathfrak{e}^*}$.*

3. *We have*

$$\left| \mathcal{ST}_{\mathrm{an}}^{\mathfrak{e}^*,[b]} \right| = (T_{\mathrm{an}}^{\mathfrak{e}^*} \cdot \pi)^2. \tag{20}$$

Observation 5. *Property 5) in Proposition 5 shows that up to multiplication with a root of unity the complex Ray–Singer torsion can be defined on all components of $\mathrm{Rep}^M(\Gamma; V)$, since $F = \oplus_k E$ is trivial for sufficiently large k.*

6. Milnor–Turaev and dynamical torsion

6.1. Milnor–Turaev torsion

Consider a smooth triangulation τ of M, and choose a collection of orientations $\mathcal{O}$ of the simplices of τ. Let $x_0 \in M$ be a base point, and set $\Gamma := \pi_1(M, x_0)$. Let V be a finite-dimensional complex vector space. For a representation $\rho \in \mathrm{Rep}(\Gamma; V)$, consider the chain complex $(C^*_\tau(M; \rho), d^{\mathcal{O}}_\tau(\rho))$ associated with the triangulation τ which computes the cohomology $H^*(M; \rho)$.

Denote the set of simplices of dimension q by $\mathcal{X}_q$, and set $k_i := \sharp(\mathcal{X}_i) \cdot \dim(V)$. Choose a collection of paths $\sigma := \{\sigma_s \mid s \in \tau\}$ from x_0 to the barycenters of τ as in Section 3.1. Choose an ordering o of the barycenters and a framing ϵ of V. Using σ, o and ϵ one can identify $C^q_\tau(M; \rho)$ with $\mathbb{C}^{k_q}$. We obtain in this way a map

$$t_{\mathcal{O}, \sigma, o, \epsilon} : \mathrm{Rep}(\Gamma; V) \to \mathbb{D}(k_0, \dots, k_n)$$

which sends $\mathrm{Rep}^M_0(\Gamma; V)$ to $\mathbb{D}_{\mathrm{ac}}(k_0, \dots, k_n)$. A look at the explicit definition of $d^{\mathcal{O}}_\tau(\rho)$ implies that $t_{\mathcal{O}, \sigma, o, \epsilon}$ is actually a regular map between two algebraic sets. Change of $\mathcal{O}, \sigma, o, \epsilon$ changes the map $t_{\mathcal{O}, \sigma, o, \epsilon}$.

Recall that the triangulation τ determines Euler vector fields X_τ which together with σ determine an Euler structure $\mathfrak{e} \in \mathfrak{Eul}(M; \mathbb{Z})$, see Section 3.1. Note that the ordering o induces a cohomology orientation $\mathfrak{o}$ in $H^*(M; \mathbb{R})$. In view of the arguments of [22] or [29] one can conclude (cf. [7]):

Proposition 6. *If $\rho \in \mathrm{Rep}^M_0(\Gamma; V)$ different choices of $\tau, \mathcal{O}, \sigma, o, \epsilon$ provide the same composition $\tau \cdot t_{\mathcal{O}, \sigma, o, \epsilon}(\rho)$ provided they define the same Euler structure $\mathfrak{e}$ and homology orientation $\mathfrak{o}$.*

In view of Proposition 6 we obtain a well-defined complex-valued rational function on $\mathrm{Rep}^M(\Gamma; V)$ called the Milnor–Turaev torsion and denoted from now on by $\mathcal{T}^{\mathfrak{e}, \mathfrak{o}}_{\mathrm{comb}}$.

Theorem 2.

1. *The poles and zeros of $\mathcal{T}^{\mathfrak{e}, \mathfrak{o}}_{\mathrm{comb}}$ are contained in $\Sigma(M)$, the subvariety of representations ρ with $H^*(M; \rho) \neq 0$.*

2. *The absolute value of $\mathcal{T}^{\mathfrak{e}, \mathfrak{o}}_{\mathrm{comb}}(\rho)$ calculated on $\rho \in \mathrm{Rep}^M_0(\Gamma; V)$ is the modified Ray–Singer torsion $T^{\mathfrak{e}^*}_{\mathrm{an}}(\rho)$, where $\mathfrak{e}^* = \mathrm{PD}(\mathfrak{e})$.*

3. *If $\mathfrak{e}_1$ and $\mathfrak{e}_2$ are two Euler structures then $\mathcal{T}^{\mathfrak{e}_2, \mathfrak{o}}_{\mathrm{comb}} = \mathcal{T}^{\mathfrak{e}_1, \mathfrak{o}}_{\mathrm{comb}} \cdot \det_{\mathfrak{e}_2 - \mathfrak{e}_1}$ and $\mathcal{T}^{\mathfrak{e}, -\mathfrak{o}}_{\mathrm{comb}} = (-1)^{\dim V} \cdot \mathcal{T}^{\mathfrak{e}, \mathfrak{o}}_{\mathrm{comb}}$ where $\det_{\mathfrak{e}_2 - \mathfrak{e}_1}$ is the regular function on $\mathrm{Rep}^M(\Gamma; V)$ defined in Section 4.1.*

4. *When restricted to $\mathrm{Rep}^{M, E}(\Gamma; V)$, E a complex vector bundle equipped with a non-degenerate symmetric bilinear form b $(\mathcal{T}^{\mathfrak{e}, \mathfrak{o}}_{\mathrm{comb}})^2 = \mathcal{S}\mathcal{T}^{\mathfrak{e}^*}_{\mathrm{an}}$, where $\mathfrak{e}^* = \mathrm{PD}(\mathfrak{e})$.*

Parts 1) and 3) follow from the definition and the general properties of τ, part 2) can be derived from the work of Bismut–Zhang [2] cf. also [6], and part 4) is discussed in [10], Remark 5.11. and established in the generality stated as a consequence of the results in [28] and [11].

6.2. Dynamical torsion

Let X be a vector field on M satisfying (H), (EG), (L), (MS) and (NCT) from Section 2.4. Choose orientations $\mathcal{O}$ of the unstable manifolds. Let $x_0 \in M$ be a base point and set $\Gamma := \pi_1(M, x_0)$. Let V be a finite-dimensional complex vector space. For a representation $\rho \in \mathrm{Rep}(\Gamma; V)$ consider the associated flat bundle

(F_ρ, ∇_ρ), and set $C_X^q(M; \rho) := \Gamma(F_\rho|_{\mathcal{X}_q})$, where $\mathcal{X}_q$ denotes the set of zeros of index q. Recall that for $x \in \mathcal{X}$, $y \in \mathcal{X}$ and every homotopy class $\hat\alpha \in \mathcal{P}_{x,y}$ parallel transport provides an isomorphism $(\mathrm{pt}_{\hat\alpha}^\rho)^{-1} : (F_\rho)_y \to (F_\rho)_x$. For $x \in \mathcal{X}_q$ and $y \in \mathcal{X}_{q-1}$ consider the expression:

$$\delta_X^{\mathcal{O}}(\rho)_{x,y} := \sum_{\hat\alpha \in \mathcal{P}_{x,y}} \mathbb{I}_{x,y}^{X,\mathcal{O}}(\hat\alpha)(\mathrm{pt}_{\hat\alpha}^\rho)^{-1}. \tag{21}$$

If the right-hand side of (21) is absolutely convergent for all x and y they provide a linear mapping $\delta_X^{\mathcal{O}}(\rho) : C_X^{q-1}(M; \rho) \to C_X^q(M; \rho)$ which, in view of Proposition 1(3), makes $\big(C_X^*(M; \rho), \delta_X^{\mathcal{O}}(\rho)\big)$ a cochain complex. There is an integration homomorphism $\mathrm{Int}_X^{\mathcal{O}}(\rho) : \big(\Omega^*(M; F_\rho), d^{\nabla_\rho}\big) \to \big(C_X^*(M; \rho), \delta_X^{\mathcal{O}}(\rho)\big)$ which does not always induce an isomorphism in cohomology.

Recall that for every $\rho \in \mathrm{Rep}(\Gamma; V)$ the composition $\mathrm{tr} \cdot \rho^{-1} : \Gamma \to \mathbb{C}$ factors through conjugacy classes to a function $\mathrm{tr} \cdot \rho^{-1} : [S^1, M] \to \mathbb{C}$. Let us also consider the expression

$$P_X(\rho) := \sum_{\gamma \in [S^1, M]} \mathbb{Z}_X(\gamma)(\mathrm{tr} \cdot \rho^{-1})(\gamma). \tag{22}$$

Again, the right-hand side of (22) will in general not converge.

Proposition 7. *There exists an open set U in $\mathrm{Rep}^M(\Gamma; V)$, intersecting every irreducible component, s.t. for any representation $\rho \in U$ we have:*

a) *The differentials $\delta_X^{\mathcal{O}}(\rho)$ converge absolutely.*
b) *The integration $\mathrm{Int}_X^{\mathcal{O}}(\rho)$ converges absolutely.*
c) *The integration $\mathrm{Int}_X^{\mathcal{O}}(\rho)$ induces an isomorphism in cohomology.*
d) *If in addition $\dim V = 1$, then*

$$\sum_{\sigma \in H_1(M;\mathbb{Z})/\mathrm{Tor}(H_1(M;\mathbb{Z}))} \left| \sum_{[\gamma] \in \sigma} \mathbb{Z}_X(\gamma)(\mathrm{tr} \cdot \rho^{-1})(\gamma) \right| \tag{23}$$

converges, cf. (22). Here the inner (finite) sum is over all $\gamma \in [S^1, M]$ which give rise to $\sigma \in H_1(M;\mathbb{Z})/\mathrm{Tor}(H_1(M;\mathbb{Z}))$.

This proposition is a consequence of exponential growth property (EG) and requires (for d)) Hutchings–Lee or Pajitnov results. A proof in the case $\dim V = 1$ is presented in [9]. The convergence of (23) is derived from the interpretation of this sum as the Laplace transform of a Dirichlet series with a positive abscissa of convergence.

We expect d) to remain true for V of arbitrary dimension.[10] In this case we make (22) precise by setting

$$P_X(\rho) := \sum_{\sigma \in H_1(M;\mathbb{Z})/\mathrm{Tor}(H_1(M;\mathbb{Z}))} \sum_{[\gamma] \in \sigma} \mathbb{Z}_X(\gamma)(\mathrm{tr} \cdot \rho^{-1})(\gamma). \tag{24}$$

---

[10] Even more, we conjecture that (22) converges absolutely on an open set U as in Proposition 7.

Observation 6. *A Lyapunov closed one form* ω *for* X *permits to consider the family of regular functions* $P_{X;R}$, $R \in \mathbb{R}$, *on the variety* $\mathrm{Rep}(\Gamma; V)$ *defined by:*

$$P_{X;R}(\rho) := \sum_{\hat{\theta}, -\omega(\hat{\theta}) \leq R} (\epsilon(\hat{\theta})/p(\hat{\theta})) \operatorname{tr}(\rho(\hat{\theta})^{-1}).$$

If (23) converges then $\lim_{R \to \infty} P_{X;R}$ *exists for* ρ *in an open set of representations. We expect that by analytic continuation this can be defined for all representations except those in a proper algebraic subvariety. This is the case when* $\dim V = 1$ *or, for* V *of arbitrary dimension, when the vector field* X *has only finitely many simple closed trajectories. In this case* $\lim_{R \to \infty} P_{X;R}$ *has an analytic continuation to a rational function on* $\mathrm{Rep}(\Gamma; V)$, *see Section 8 below.*

As in Section 6.1, we choose a collection of paths $\sigma := \{\sigma_x \mid x \in \mathcal{X}\}$ from x_0 to the zeros of X, an ordering o' of $\mathcal{X}$, and a framing ϵ of V. Using σ, o, ϵ we can identify $C_X^q(M; \rho)$ with $\mathbb{C}^{k_q}$, where $k_q := \sharp(\mathcal{X}_q) \cdot \dim(V)$. As in the previous section we obtain in this way a holomorphic map

$$t_{\mathcal{O}, \sigma, o', \epsilon} : U \to \hat{\mathbb{D}}_{\mathrm{ac}}(k_0, \dots, k_n).$$

An ordering o' of $\mathcal{X}$ is given by orderings o'_q of $\mathcal{X}_q$, $q = 0, 1, \dots, n$. Two orderings o'_1 and o'_2 are equivalent if $o'_{1,q}$ is obtained from $o'_{2,q}$ by a permutation π_q so that $\prod_q \operatorname{sgn}(\pi_q) = 1$. We call an equivalence class of such orderings a *rest point orientation*. Let us write $\mathfrak{o}'$ for the rest point orientation determined by o'. Moreover, let $\mathfrak{e}$ denote the Euler structure represented by X and σ, see Observation 1. As in the previous section, the composition $\tau \cdot t_{\mathcal{O}, \sigma, o', \epsilon} : U \setminus \Sigma \to \mathbb{C}_*$ is a holomorphic map which only depends on $\mathfrak{e}$ and $\mathfrak{o}'$, and will be denoted by $\tau_X^{\mathfrak{e}, \mathfrak{o}'}$. Consider the holomorphic map $P_X : U \to \mathbb{C}$ defined by formula (22). The *dynamical torsion* is the partially defined holomorphic function

$$\mathcal{T}_X^{\mathfrak{e}, \mathfrak{o}'} := \tau_X^{\mathfrak{e}, \mathfrak{o}'} \cdot e^{P_X} : U \setminus \Sigma \to \mathbb{C}_*.$$

The following result is based on a theorem of Hutchings–Lee and Pajitnov [17] cf. [9].

Theorem 3. *If* $\dim V = 1$ *the partially defined holomorphic function* $\mathcal{T}_X^{\mathfrak{e}, \mathfrak{o}'}$ *has an analytic continuation to a rational function equal to* $\pm \mathcal{T}_{\mathrm{comb}}^{\mathfrak{e}, \mathfrak{o}}$.

It is hoped that a generalization of Hutchings–Lee and Pajitnov results which will be elaborated in subsequent work [12] might led to the proof of the above result for V of arbitrary dimension.

7. Examples

7.1. Milnor–Turaev torsion for mapping tori and twisted Lefschetz zeta function

Let Γ_0 be a group, $\alpha : \Gamma_0 \to \Gamma_0$ an isomorphism and V a complex vector space. Denote by $\Gamma := \Gamma_0 \times_\alpha \mathbb{Z}$ the group whose underlying set is $\Gamma_0 \times \mathbb{Z}$ and group operation $(g', n) * (g'', m) := (\alpha^m(g') \cdot g'', n + m)$. A representation $\rho : \Gamma \to \mathrm{GL}(V)$

determines a representation $\rho_0(\rho) : \Gamma_0 \to \mathrm{GL}(V)$ the restriction of ρ to $\Gamma_0 \times 0$ and an isomorphism of V, $\theta(\rho) \in \mathrm{GL}(V)$.

Let (X, x_0) be a based point compact space with $\pi_1(X, x_0) = \Gamma_0$ and $f : (X, x_0) \to (X, x_0)$ a homotopy equivalence. For any integer k the map f induces the linear isomorphism $f^k : H^k(X; V) \to H^k(X; V)$ and then the standard Lefschetz zeta function

$$\zeta_f(z) := \frac{\prod_{k \text{ even}} \det(I - z f^k)}{\prod_{k \text{ odd}} \det(I - z f^k)}.$$

More general if ρ is a representation of Γ then f and $\rho = (\rho_0(\rho), \theta(\rho))$ induce the linear isomorphisms $f_\rho^k : H^k(X; \rho_0(\rho)) \to H^k(X; \rho_0(\rho))$ and then the ρ-twisted Lefschetz zeta function

$$\zeta_f(\rho, z) := \frac{\prod_{k \text{ even}} \det(I - z f_\rho^k)}{\prod_{k \text{ odd}} \det(I - z f_\rho^k)}.$$

Let N be a closed connected manifold and $\varphi : N \to N$ a diffeomorphism. Without loss of generality one can suppose that $y_0 \in N$ is a fixed point of φ. Define the mapping torus $M = N_\varphi$, the manifold obtained from $N \times I$ identifying $(x, 1)$ with $(\varphi(x), 0)$. Let $x_0 = (y_0, 0) \in M$ be a base point of M. Set $\Gamma_0 := \pi_1(N, n_0)$ and denote by $\alpha : \pi_1(N, y_0) \to \pi_1(N, y_0)$ the isomorphism induced by φ. We are in the situation considered above with $\Gamma = \pi_1(M, x_0)$. The mapping torus structure on M equips M with a canonical Euler structure $\mathfrak{e}$ and canonical homology orientation $\mathfrak{o}$. The Euler structure $\mathfrak{e}$ is defined by any vector field X with $\omega(X) < 0$ where $\omega := p^* dt \in \Omega^1(M; \mathbb{R})$; all are homotopic. The Wang sequence

$$\cdots \to H^*(M; \mathbb{F}_\rho) \to H^*(N; i^*(\mathbb{F}_\rho)) \xrightarrow{\varphi_\rho^* - \mathrm{id}} H^*(N; i^*(\mathbb{F}_\rho)) \to H^{*+1}(M; \mathbb{F}_\rho) \to \cdots \tag{25}$$

implies $H^*(M; \mathbb{F}_\rho) = 0$ iff $\det(I - \varphi_\rho^k) \neq 0$ for all k. The cohomology orientation is derived from the Wang long exact sequence for the trivial one-dimensional real representation. For details see [7]. We have

Proposition 8. *With these notations* $\mathcal{T}_{\mathrm{comb}}^{\mathfrak{e},\mathfrak{o}}(\rho) = \zeta_\varphi(\rho, 1)$.

This result is known cf. [18]. A proof can be also derived easily from [7].

7.2. Vector fields without rest points and Lyapunov cohomology class

Let X be a vector field without rest points, and suppose X satisfies (L) and (NCT). As in the previous section X defines an Euler structure $\mathfrak{e}$. Consider the expression (22). By Theorem 3 we have:

Observation 7. *With the hypothesis above there exists an open set* $U \subseteq \mathrm{Rep}^M(\Gamma; V)$ *so that (24) converges, and* e^{P_X} *is a well-defined holomorphic function on* U. *The function* e^{P_X} *has an analytic continuation to a rational function on* $\mathrm{Rep}^M(\Gamma; V)$ *equal to* $\pm \mathcal{T}_{\mathrm{comb}}^{\mathfrak{e},\mathfrak{o}}$. *The set* U *intersects non-trivially each connected component of* $\mathrm{Rep}^M(\Gamma; V)$.

7.3. The Alexander polynomial

If M is obtained by surgery on a framed knot, and $\dim V = 1$, then $\mathrm{Rep}(\Gamma; V) = \mathbb{C} \setminus 0$, and the function $(z - 1)^2 \mathcal{T}_{\mathrm{comb}}^{\mathfrak{e},\mathfrak{o}}$ equals the Alexander polynomial of the knot. see [32]. One can extend the definition of $\mathcal{T}_{\mathrm{comb}}^{\mathfrak{e},\mathfrak{o}}$ to compact manifolds with boundary and recover Alexander polynomial from this function when applied to the complement of an open neighborhood of the knot, see [32]. Any twisted Alexander polynomial of the knot can be also recovered from $\mathcal{T}_{\mathrm{comb}}^{\mathfrak{e},\mathfrak{o}}$ for V of higher dimension. One expects that passing to higher-dimensional representations $\mathcal{T}_{\mathrm{comb}}^{\mathfrak{e},\mathfrak{o}}$ captures even more subtle knot invariants.

8. Applications

8.1. The invariant $A^{\mathfrak{e}^*}(\rho_1, \rho_2)$

Let M be a V-acyclic manifold and $\mathfrak{e}^*$ a coEuler structure. Using the modified Ray–Singer torsion we define a $\mathbb{R}/\pi\mathbb{Z}$-valued invariant (which resembles the Atiyah–Patodi–Singer spectral flow) for two representations ρ_1, ρ_2 in the same component of $\mathrm{Rep}_0^M(\Gamma; V)$.

By a holomorphic path in $\mathrm{Rep}_0^M(\Gamma; V)$ we understand a holomorphic map $\tilde{\rho} : U \to \mathrm{Rep}_0^M(\Gamma; V)$ where U is an open neighborhood of the segment of real numbers $[1, 2] \times \{0\} \subset \mathbb{C}$ in the complex plane. For a coEuler structure $\mathfrak{e}^*$ and a holomorphic path $\tilde{\rho}$ in $\mathrm{Rep}_0^M(\Gamma; V)$ define

$$\mathrm{arg}^{\mathfrak{e}^*}(\tilde{\rho}) := \Re\left(2/i \int_1^2 \frac{\partial(T_{\mathrm{an}}^{\mathfrak{e}^*} \circ \tilde{\rho})}{T_{\mathrm{an}}^{\mathfrak{e}^*} \circ \tilde{\rho}}\right) \quad \mathrm{mod}\ \pi. \tag{26}$$

Here, for a smooth function φ of complex variable z, $\partial\varphi$ denotes the complex-valued 1-form $(\partial\varphi/\partial z)dz$ and the integration is along the path $[1, 2] \times 0 \subset U$. Note that

Observation 8.

1. *Suppose E is a complex vector bundle with a non-degenerate bilinear form b, and suppose $\tilde{\rho}$ is a holomorphic path in $\mathrm{Rep}_0^{M,E}(\Gamma; V)$. Then*

$$\mathrm{arg}^{\mathfrak{e}^*}(\tilde{\rho}) = \mathrm{arg}\left(\mathcal{ST}_{\mathrm{an}}^{\mathfrak{e}^*,[b]}(\tilde{\rho}(2))/\mathcal{ST}_{\mathrm{an}}^{\mathfrak{e}^*,[b]}(\tilde{\rho}(1))\right) \quad \mathrm{mod}\ \pi.$$

   As consequence

2. *If $\tilde{\rho}'$ and $\tilde{\rho}''$ are two holomorphic paths in $\mathrm{Rep}_0^M(\Gamma; V)$ with $\tilde{\rho}'(1) = \tilde{\rho}''(1)$ and $\tilde{\rho}'(2) = \tilde{\rho}''(2)$ then*

$$\mathrm{arg}^{\mathfrak{e}^*}(\tilde{\rho}') = \mathrm{arg}^{\mathfrak{e}^*}(\tilde{\rho}'') \quad \mathrm{mod}\ \pi.$$

3. *If $\tilde{\rho}'$, $\tilde{\rho}''$ and $\tilde{\rho}'''$ are three holomorphic paths in $\mathrm{Rep}_0^M(\Gamma; V)$ with $\tilde{\rho}'(1) = \tilde{\rho}'''(1)$, $\tilde{\rho}'(2) = \tilde{\rho}''(1)$ and $\tilde{\rho}''(2) = \tilde{\rho}'''(2)$ then*

$$\mathrm{arg}^{\mathfrak{e}^*}(\tilde{\rho}''') = \mathrm{arg}^{\mathfrak{e}^*}(\tilde{\rho}') + \mathrm{arg}^{\mathfrak{e}^*}(\tilde{\rho}'') \quad \mathrm{mod}\ \pi.$$

Observation 8 permits to define a $\mathbb{R}/\pi\mathbb{Z}$-valued numerical invariant $A^{\mathfrak{e}^*}(\rho_1, \rho_2)$ associated to a coEuler structure $\mathfrak{e}^*$ and two representations ρ_1, ρ_2 in the same connected component of $\mathrm{Rep}_0^M(\Gamma; V)$. If there exists a holomorphic path with $\tilde{\rho}(1) = \rho_1$ and $\tilde{\rho}(2) = \rho_2$ we set

$$A^{\mathfrak{e}^*}(\rho_1, \rho_2) := \arg^{\mathfrak{e}^*}(\tilde{\rho}) \mod \pi.$$

Given any two representations ρ_1 and ρ_2 in the same component of $\mathrm{Rep}_0^M(\Gamma; V)$ one can always find a finite collection of holomorphic paths $\tilde{\rho}_i$, $1 \leq i \leq k$, in $\mathrm{Rep}_0^M(\Gamma; V)$ so that $\tilde{\rho}_i(2) = \tilde{\rho}_{i+1}(1)$ for all $1 \leq i < k$, and such that $\tilde{\rho}_1(1) = \rho_1$ and $\tilde{\rho}_k(2) = \rho_2$. Then take

$$A^{\mathfrak{e}^*}(\rho_1, \rho_2) := \sum_{i=1}^{k} \arg^{\mathfrak{e}^*}(\tilde{\rho}_i) \mod \pi.$$

In view of Observation 8 the invariant is well defined, and if $\mathfrak{e}^*$ is integral it is actually well defined in $\mathbb{R}/2\pi\mathbb{Z}$. This invariant was first introduced when the authors were not fully aware of "the complex Ray–Singer torsion." The formula (26) is a more or less obvious expression of the phase of a holomorphic function in terms of its absolute value, the Ray–Singer torsion, as positive real-valued function. By Theorem 2 the invariant can be computed with combinatorial topology and by Section 7 quite explicitly in some cases. If the representations ρ_1, ρ_2 are unimodular then the coEuler structure is irrelevant. It is interesting to compare this invariant to the Atiyah–Patodi–Singer spectral flow; it is not the same but are related.

8.2. Novikov conjecture

Let X be a smooth vector field which satisfies (H), (L), (MS), (NCT). Suppose ω is a real-valued closed one form so that $\omega(X)_x < 0$, x not a rest point (Lyapunov form). Define the functions $I_{x,y}^X : \mathbb{R} \to \mathbb{Z}$ and $Z^X : \mathbb{R} \to \mathbb{Q}$ by

$$I_{x,y}^{X,\mathcal{O}}(R) := \sum_{\hat{\alpha}, \; \omega(\hat{\alpha}) < R} \mathbb{I}_{x,y}^{X,\mathcal{O}}(\hat{\alpha})$$

$$Z^X(R) := \sum_{\hat{\theta}, \; \omega(\hat{\theta}) < R} \mathbb{Z}^X(\hat{\theta}) \tag{27}$$

Part (a) of the following conjecture was formulated by Novikov for $X = \mathrm{grad}_g \omega$, ω a Morse closed one form when this vector field satisfies the above properties.

Conjecture 1.

a) *The function $I_{x,y}^{X,\mathcal{O}}(R)$ has exponential growth.*

b) *The function $Z^X(R)$ has exponential growth.*

Recall that a function $f : \mathbb{R} \to \mathbb{R}$ is said to have exponential growth iff there exists constants C_1, C_2 so that $|f(x)| < C_1 e^{C_2 x}$.

As a straightforward consequence of Proposition 7 we have

Theorem 4.

  a) *Suppose X satisfies* (H), (MS), (L) *and* (EG). *Then part* a) *of the conjecture above holds.*
  b) *Suppose M is V-acyclic for some V with $\dim(V) = 1$. Moreover, assume X satisfies* (H), (MS), (L), (NTC) *and* (EG). *Then part* b) *of the conjecture above holds.*

This result is proved in [10]; The V-acyclicity in part b) is not necessary if (EG) is replaced by an apparently stronger assumption (SEG). Prior to our work Pajitnov has considered for vector fields which satisfy (H), (L), (MS), (NCT) an additional property, condition ($\mathcal{CY}$), and has verified part (a) of this conjecture. He has also shown that the vector fields which satisfy (H), (L), (MS), (NCT) and ($\mathcal{CY}$) are actually C^0 dense in the space of vector fields which satisfy (H), (L), (MS), (NCT). It is shown in [10] that Pajitnov vector fields satisfy (EG), and in fact (SEG).

8.3. A question in dynamics

Let Γ be a finitely presented group, V a complex vector space and $\mathrm{Rep}(\Gamma; V)$ the variety of complex representations. Consider triples $\underline{a} := \{a, \epsilon_-, \epsilon_+\}$ where a is a conjugacy class of Γ and $\epsilon_\pm \in \{0, 1\}$. Define the rational function $\mathrm{let}_{\underline{a}} : \mathrm{Rep}(\Gamma; V) \to \mathbb{C}$ by

$$\mathrm{let}_{\underline{a}}(\rho) := \left(\det\left(\mathrm{id} - (-1)^{\epsilon_-} \rho(\alpha)^{-1} \right) \right)^{(-1)^{\epsilon_- + \epsilon_+}}$$

where $\alpha \in \Gamma$ is a representative of a.

Let (M, x_0) be a V-acyclic manifold and $\Gamma = \pi_1(M, x_0)$. Note that $[S^1, M]$ identifies with the conjugacy classes of Γ. Suppose X is a vector field satisfying (L) and (NCT). Every closed trajectory $\hat{\theta}$ gives rise to a conjugacy class $[\hat{\theta}] \in [S^1, M]$ and two signs $\epsilon_\pm(\hat{\theta})$. These signs are obtained from the differential of the return map in normal direction; $\epsilon_-(\hat{\theta})$ is the parity of the number of real eigenvalues larger than $+1$ and $\epsilon_-(\hat{\theta})$ is the parity of the number of real eigenvalues smaller than -1. For a simple closed trajectory, i.e., of period $p(\hat{\theta}) = 1$, let us consider the triple $\underline{\hat{\theta}} := \left([\hat{\theta}], \epsilon_-(\hat{\theta}), \epsilon_+(\hat{\theta}) \right)$. This gives a (at most countable) set of triples as in the previous paragraph.

Let $\xi \in H^1(M; \mathbb{R})$ be a Lyapunov cohomology class for X. Recall that for every R there are only finitely many closed trajectories $\hat{\theta}$ with $-\xi([\hat{\theta}]) \leq R$. Hence, we get a rational function $\zeta_R^{X, \xi} : \mathrm{Rep}(\Gamma; V) \to \mathbb{C}$

$$\zeta_R^{X, \xi} := \prod_{-\xi([\hat{\theta}]) \leq R} \mathrm{let}_{\underline{\hat{\theta}}}$$

where the product is over all triples $\hat{\underline{\theta}}$ associated to simple closed trajectories with $-\xi([\hat{\theta}]) \leq R$. It is easy to check that formally we have

$$\lim_{R \to \infty} \zeta_R^{X,\xi} = e^{P_X}.$$

It would be interesting to understand in what sense (if any) this can be made precise. We conjecture that there exists an open set with non-empty interior in each component of $\text{Rep}(\Gamma; V)$ on which we have true convergence. In fact there exist vector fields X where the sets of triples are finite in which case the conjecture is obviously true.

References

[1] M.F. Atiyah, V.K. Patodi and I.M. Singer, *Spectral asymmetry and Riemannian geometry. II,* Math. Proc. Cambridge Philos. Soc. **78**(1975), 405–432.

[2] J.M. Bismut and W. Zhang, *An extension of a theorem by Cheeger and Müller,* Astérisque **205**, Société Mathématique de France, 1992.

[3] R. Bott and L.W. Tu, *Differential forms in algebraic topology.* Graduate texts in Mathematics **82**. Springer Verlag, New York–Berlin, 1982.

[4] M. Braverman and T. Kappeler, *Refined analytic torsion,* preprint `math.DG/0505537`.

[5] D. Burghelea, *Removing Metric Anomalies from Ray–Singer Torsion,* Lett. Math. Phys. **47**(1999), 149–158.

[6] D. Burghelea, L. Friedlander and T. Kappeler, *Relative Torsion,* Commun. Contemp. Math. **3**(2001), 15–85.

[7] D. Burghelea and S. Haller, *Euler structures, the variety of representations and the Milnor–Turaev torsion,* Geom. and Topol. **10**(2006) 1185–1238.

[8] D. Burghelea and S. Haller, *Laplace transform, dynamics and spectral geometry (version 1),* to appear in Journal of Topology, Vol 1, 2008 preprint `math.DG/0405037`.

[9] D. Burghelea and S. Haller, *Dynamics, Laplace transform, and spectral geometry (version 2),* preprint `math.DG/0508216`.

[10] D. Burghelea and S. Haller, *Complex valued Ray–Singer torsion,* J. Funct. Anal. Vol 248 (2007) 27–78. preprint `math.DG/0604484`.

[11] D. Burghelea and S. Haller, *Complex valued Ray–Singer torsion II,* preprint `math.DG/0604484`.

[12] D. Burghelea and S. Haller, *Dynamical torsion, a non-commutative version of a theorem of Hutchings–Lee and Pajitnov,* in preparation.

[13] S. Dineen, *Complex Analysis in Locally Convex Spaces.* Mathematics Studies **57**, North-Holland, 1981.

[14] M. Farber, *Singularities of the analytic torsion,* J. Differential Geom. **41**(1995), 528–572.

[15] M. Farber and V. Turaev, *Poincaré–Reidemeister metric, Euler structures and torsion,* J. Reine Angew. Math. **520**(2000), 195–225.

[16] R. Hartshorne, *Algebraic Geometry.* Graduate Texts in Mathematics **52**. Springer Verlag, New York–Heidelberg, 1977.

[17] M. Hutchings, *Reidemeister torsion in generalized Morse Theory,* Forum Math. **14**(2002), 209–244.

[18] B. Jiang, *Estimation of the number of periodic orbits,* Pac. J. Math. **172**(1996), 151–185.

[19] A. Kriegl and P.W. Michor, *The convenient setting of global analysis.* Mathematical Surveys and Monographs **53**, American Mathematical Society, Providence, RI, 1997.

[20] J. Marcsik, *Analytic torsion and closed one forms.* Thesis, Ohio State University, 1995.

[21] V. Mathai and D. Quillen, *Superconnections, Thom Classes, and Equivariant differential forms,* Topology **25**(1986), 85–110.

[22] J. Milnor, *Whitehead torsion,* Bull. Amer. Math. Soc. **72**(1966), 358–426.

[23] W. Müller, private communication.

[24] L. Nicolaescu, *The Reidemeister Torsion of 3-manifolds.* De Gruyter Studies in Mathematics **30**, Walter de Gruyter & Co., Berlin, 2003.

[25] D. Quillen, *Determinants of Cauchy–Riemann operators on Riemann surfaces,* Functional Anal. Appl. **19**(1985), 31–34. Translation of Funktsional. Anal. i Prilozhen. **19**(1985), 37–41.

[26] D.B. Ray and I.M. Singer, *R-torsion and the Laplacian on Riemannian manifolds,* Adv. in Math. **7**(1971), 145–210.

[27] N. Steenrod, *The topology of fiber bundles.* Reprint of the 1957 edition. Princeton University Press, Princeton, NJ, 1999.

[28] G. Su and W. Zhang, *A Cheeger–Müller theorem for symmetric bilinear torsion,* preprint `math.DG/0610577`.

[29] V. Turaev, *Reidemeister torsion in Knot theory,* Uspekhi Mat. Nauk **41**(1986), 119–182.

[30] V. Turaev, *Euler structures, nonsingular vector fields, and Reidemeister-type torsions,* Math. USSR–Izv. **34**(1990), 627–662.

[31] V. Turaev, *Introduction to combinatorial torsion.* Notes taken by Felix Schlenk. Lectures in Mathematics ETH Zürich. Birkhäuser Verlag, Basel, 2001.

[32] V. Turaev, *Torsion of 3-dimensional manifolds.* Progress in Mathematics **208**. Birkhäuser Verlag, Basel–Boston–Berlin, 2002.

[33] G.W. Whitehead, *Elements of homotopy theory.* Graduate Texts in Mathematics **61**. Springer Verlag, New York–Berlin, 1978.

Dan Burghelea
Dept. of Mathematics, The Ohio State University,
231 West 18th Avenue, Columbus, OH 43210, USA.
e-mail: `burghele@mps.ohio-state.edu`

Stefan Haller
Department of Mathematics, University of Vienna,
Nordbergstrasse 15, A-1090, Vienna, Austria.
e-mail: `stefan.haller@univie.ac.at`

C^*-algebras and Elliptic Theory II

Trends in Mathematics, 67–86

The K-theory of Twisted Group Algebras

Siegfried Echterhoff

Abstract. We study the K-theory of twisted group algebras with the help of the Baum-Connes conjecture

Twisted group algebras for locally compact groups G first appeared in the study of projective unitary representations of G on Hilbert space H, i.e., continuous homomorphisms into the projective unitary group $\mathcal{P}U(H) = \mathcal{U}(H)/\mathbb{T}1$. Such representations appear naturally in quantum mechanics and in representation theory of locally compact groups (see [37]). On the other hand, some of the most basic examples in non-commutative geometry/topology, like the non-commutative n-tori A_Θ are realized as twisted group algebras of the integer group $\mathbb{Z}^n$.

In this article we want to discuss some applications of the Baum-Connes conjecture to the study of the K-theory of twisted group algebras. Some of the results reported here are easy consequences of the fundamental work of Kasparov, while others are outcomes of (or motivation for) recent joint work with some friends and/or collaborators, as there are Wolfgang Lück, Chris Phillips and Samuel Walters on a project on the structure of crossed products by irrational rotation algebras (see [21]), and Heath Emerson and Hyun Jeong Kim on a project on Poincaré duality (see [20]).

We should note that the K-theory of twisted group algebras in connection with the Baum-Connes conjecture plays also an important rôle in work of Mathai and coauthors (e.g., see [8, 9, 38, 39, 40, 41]). We will not discuss the very interesting results presented in those papers here and we therefore urge the reader to have his/her own look at them.

1. Twisted group algebras

In what follows let us always assume that G is a second countable locally compact group. A circle-valued 2-cocycle $\omega \in Z^2(G, \mathbb{T})$ (often called a *multiplier* on G) is a

---

This work was supported by the Deutsche Forschungsgemeinschaft (SFB 478).

Borel map $\omega : G \times G \to \mathbb{T}$ satisfying the equations

$$\omega(s, e) = \omega(e, s) = 1, \quad \text{and} \quad \omega(s, t)\omega(st, r) = \omega(s, tr)\omega(t, r)$$

for all $s, t, r \in G$. Two cocycles ω, ω' are cohomologous, if there exists a Borel map $f : G \to \mathbb{T}$ such that ω and ω' differ by the coboundary

$$\partial f(s, t) = f(s)f(t)\overline{f(st)}.$$

As usual, we write $H^2(G, \mathbb{T})$ for the group of all cohomolgy classes $[\omega]$ for $\omega \in Z^2(G, \mathbb{T})$. It is the second cohomology of G with Borel cochains and coefficient the trivial G-module $\mathbb{T}$ in the sense of C.C. Moore (see [42, 44]). If G is discrete, then $H^2(G, \mathbb{T})$ is just the ordinary second group-cohomology with coefficient $\mathbb{T}$.

If $\omega \in Z^2(G, \mathbb{T})$, then an ω-*representation* of G on a Hilbert-space H is a Borel map $V : G \to \mathcal{U}(H)$ (with respect to the strong operator topology on $\mathcal{U}(H)$) satisfying

$$V_s V_t = \omega(s, t) V_{st} \quad \text{for all } s, t \in G.$$

It is clear that composing V with the quotient map $\mathrm{Ad} : \mathcal{U}(H) \to \mathcal{P}U(H)$ gives a Borel homomorphism $s \mapsto \mathrm{Ad}(V_s)$ from G to $\mathcal{P}U(H)$, which is automatically continuous by [44, Proposition 5]. Conversely, given any continuous homomorphism $\beta : G \to \mathcal{P}U(H)$, we may choose a Borel section $c : \mathcal{P}U(H) \to \mathcal{U}(H)$ to obtain a Borel map $V := c \circ \beta : G \to \mathcal{U}(H)$, and a short computation using the fact that β is a homomorphism implies that there exists a unique cocycle $\omega \in Z^2(G, \mathbb{T})$ such that V is an ω-representation. Altering the section c results in multiplication of V by some Borel map $f : G \to \mathbb{T}$ and the cocycle ω by the coboundary $\partial(f)$, so that the cohomology class $[\omega]$ only depends on the given homomorphism β. Thus we see that the study of projective representations is equivalent to the study of ω-representations for cocycles $\omega \in Z^2(G, \mathbb{T})$, where it is enough to pick a representative for each cohomology class.

The regular ω-representation of G is the representation $L_\omega : G \to \mathcal{U}(L^2(G))$ given by the formula

$$(L_\omega(s)\xi)(t) := \omega(s, s^{-1}t)\xi(s^{-1}t) \quad \xi \in L^2(G).$$

Let $L^1(G, \omega)$ denote the ω-twisted L^1-algebra of G, i.e., the Banach space $L^1(G)$ (with respect to the Haar measure on G) equipped with multiplication and involution given by the formulas

$$f *_\omega g(t) = \int_G f(s)g(s^{-1}t)\omega(s, s^{-1}t)\, ds \quad \text{and} \quad f^*(s) = \overline{\omega(s, s^{-1})f(s^{-1})}.$$

If $V : G \to \mathcal{U}(H)$ is an ω-representation of G, then V extends to a $*$-homomorphism of $L^1(G, \omega)$ into $B(H)$ via the formula

$$V(f) := \int_G f(s)V(s)\, ds, \quad f \in L^1(G, \omega),$$

and every nondegenerate representation of $L^1(G, \omega)$ appears in this way. The full twisted group algebra $C^*(G, \omega)$ is defined as the enveloping C^*-algebra of $L^1(G, \omega)$ and the reduced twisted group algebra $C_r^*(G, \omega)$ is defined as the image of $C^*(G, \omega)$

under the regular ω-representation L_ω. Note that $C^*_{(r)}(G,\omega)$ depends, up to iso-morphism, only on the cohomology class $[\omega] \in H^2(G,\mathbb{T})$. Indeed, if $\omega' = \partial(\varphi)\omega$ for some Borel map $\varphi : G \to \mathbb{T}$, then a short computation shows that

$$\Phi : L^1(G,\omega') \to L^1(G,\omega); \Phi(f) = \varphi \cdot f \quad \text{(pointwise multiplication)}$$

is an isomorphism which extends also to isomorphisms of the C^*-completions. Note that the full and reduced twisted group algebras coincide if G is amenable.

Example 1.1. (1) Let $G = \mathbb{R}^2$. Then any real number $x \in \mathbb{R}$ determines a cocycle $\omega_x : \mathbb{R}^2 \times \mathbb{R}^2 \to \mathbb{T}$ via

$$\omega_x\big((s_1,t_1),(s_2,t_2)\big) = e^{\pi i x(s_1 t_2 - t_1 s_2)}.$$

One can show that $x \mapsto [\omega_x]$ gives an isomorphism $\mathbb{R} \cong H^2(\mathbb{R}^2,\mathbb{T})$. An ω_x-representation $V : \mathbb{R}^2 \to \mathcal{U}(H)$ is characterized by the Weyl-commutation relation

$$V_{(s,0)}V_{(0,t)} = e^{2\pi i x s t}V_{(0,t)}V_{(s,0)}. \tag{1.1}$$

A particular irreducible example (if $x \neq 0$) of such representation on $H = L^2(\mathbb{R})$ is given by

$$\big(W_{(s,0)}\xi\big)(r) = e^{2\pi i s r}\xi(r) \quad \text{and} \quad \big(W_{(0,t)}\xi\big)(r) = \xi(r - xt).$$

It follows from the Stone-von Neumann theorem (e.g., see [50, Corollary 10.4.1]) that, for $x \neq 0$, every representation satisfying the above relations is a multiple of W. But this implies that the integrated form of W induces a faithful representation of $C^*(\mathbb{R}^2,\omega_x)$ into $B(L^2(\mathbb{R}))$ via integration whenever $x \neq 0$. If $f \in L^1(\mathbb{R}^2)$, this representation takes the form

$$\big(W(f)\xi\big)(r) = \int_{\mathbb{R}^2} f(s,t)\big(W_{(s,t)}\xi\big)(r)\,d(s,t) = \int_{\mathbb{R}^2} f(s,t)e^{2\pi i s(r-\frac{x}{2}t)}\xi(r-xt)\,dsdt,$$

from which one can deduce (after performing some Fourier-transformation) that $C^*(\mathbb{R}^2,\omega_x)$ is isomorphic to the compact operators on $L^2(\mathbb{R})$ via W. Of course, if $x = 0$ we have $C^*(\mathbb{R}^2,\omega_0) = C^*(\mathbb{R}^2) \cong C_0(\widehat{\mathbb{R}^2})$.

(2) Let $\theta \in [0,1]$ and let $\omega_\theta : \mathbb{Z}^2 \times \mathbb{Z}^2 \to \mathbb{T}$ be the restriction of ω_θ, as defined in (1), to $\mathbb{Z}^2 \subseteq \mathbb{R}^2$. This gives an isomorphism $\mathbb{T} \to H^2(\mathbb{Z}^2,\mathbb{T})$, $z = e^{2\pi i\theta} \mapsto [\omega_\theta]$. If $(n,m) \mapsto V_{(n,m)}$ is any ω_θ-representation of $\mathbb{Z}^2$, we see that the generators $V_{(1,0)}$ and $V_{(0,1)}$ satisfy the commutation relation

$$V_{(1,0)}V_{(0,1)} = e^{2\pi i\theta}V_{(0,1)}V_{(1,0)}.$$

Thus we see that $C^*(\mathbb{Z}^2,\omega_\theta)$ is the universal C^*-algebra generated by two unitaries satisfying this relation, which is is precisely the non-commutative (if $\theta \neq 0$) two-torus A_θ. In a similar way, the higher-dimensional non-commutative n-tori are the twisted group algebras of $\mathbb{Z}^n$.

(3) Consider the standard action of $\mathrm{SL}(2,\mathbb{Z})$ on $\mathbb{Z}^2$ via matrix multiplication. Then the cocycle ω_θ on $\mathbb{Z}^2$ considered above can be extended to a cocycle of the semi-direct product $\mathbb{Z}^2 \rtimes \mathrm{SL}(2,\mathbb{Z})$ by the formula

$$\widetilde{\omega}_\theta\left(\left(\left(\begin{smallmatrix}n_1\\m_1\end{smallmatrix}\right),N_1\right),\left(\left(\begin{smallmatrix}n_2\\m_2\end{smallmatrix}\right),N_2\right)\right) = \omega_\theta\left(\left(\begin{smallmatrix}n_1\\m_1\end{smallmatrix}\right),N_1\cdot\left(\begin{smallmatrix}n_2\\m_2\end{smallmatrix}\right)\right). \tag{1.2}$$

These cocycles restrict to cocycles on $\mathbb{Z}^2 \rtimes H$ for any subgroup H of $\mathrm{SL}(2, \mathbb{Z})$ and the twisted group algebras $C^*_{(r)}(\mathbb{Z}^2 \rtimes H, \widetilde{\omega}_\theta)$ are isomorphic to the crossed products

$$C^*(\mathbb{Z}^2, \omega_\theta) \rtimes_{(r)} H = A_\theta \rtimes_{(r)} H,$$

where the action of $\mathrm{SL}(2, \mathbb{Z})$ on A_θ is given

$$\left(\alpha_N(f)\right)\left(\tbinom{n}{m}\right) = f\left(N^{-1}\left(\tbinom{n}{m}\right)\right), \quad f \in l^1(\mathbb{Z}^2, \omega_\theta)$$

(see [21] for details on this). In the special case of finite subgroups $F \subseteq \mathrm{SL}(2, \mathbb{Z})$, these crossed products have obtained considerable attention in the literature (e.g. see [6, 7, 24, 25, 26, 27, 34, 51, 52, 53]), since they can be considered as non-commutative analogues of the two-sphere: At least for irrational θ one can show that $A_\theta \rtimes F$ is Morita equivalent to the fixed-point algebra A_θ^F, which in the case $\theta = 0$ is isomorphic to $C(S^2)$ (this holds for all nontrivial *finite* $F \subseteq \mathrm{SL}(2, \mathbb{Z})$).

Up to stable isomorphism, the twisted group algebras can also be written as (full or reduced) crossed products $\mathcal{K}(H) \rtimes_{(r)} G$ of G by the compact operators $\mathcal{K}(H)$ on some separable Hilbert space H. To see this recall that the group of *-automorphism $\mathrm{Aut}(\mathcal{K}(H))$ is canonically isomorphic to $\mathcal{PU}(H)$ via $\mathrm{Ad}\, U(T) = UTU^*$ for $U \in \mathcal{U}(H)$. Hence, if we choose any ω-representation $V : G \to \mathcal{U}(H)$, we obtain an action $\alpha = \mathrm{Ad}\, V : G \to \mathrm{Aut}(\mathcal{K}(H))$ and it is then a very special case of [45, Theorem 3.4] that the full and reduced crossed products $\mathcal{K}(H) \rtimes_\alpha G$ and $\mathcal{K}(H) \rtimes_{\alpha, r} G$ are isomorphic to $C^*(G, \bar{\omega}) \otimes \mathcal{K}(H)$ and $C^*_r(G, \bar{\omega}) \otimes \mathcal{K}(H)$, respectively, where $\bar{\omega}$ denotes the complex conjugate (hence inverse) of ω. Indeed, both isomorphisms are given on the level of L^1-algebras by

$$\Phi : L^1(G, \bar{\omega}) \odot \mathcal{K}(H) \to L^1(G, \mathcal{K}(H)); \Phi(f \otimes k)(s) = f(s)kV_s^*.$$

Conversely, starting with any action $\beta : G \to \mathrm{Aut}(\mathcal{K}) = \mathcal{PU}(H)$ it follows from our previous discussion that there exist $\omega \in Z^2(G, \mathbb{T})$ and and an ω-representation $V : G \to \mathcal{U}(H)$ such that $\beta_s = \mathrm{Ad}\, V_s$ for all $s \in G$, and hence $\mathcal{K} \rtimes_{(r)} G \cong \mathcal{K} \otimes C^*_{(r)}(G, \bar{\omega})$. Note that two such actions β and β' are exterior (or Morita) equivalent if and only if $[\omega] = [\omega']$ in $H^2(G, \mathbb{T})$, where ω, ω' denote corresponding cocycles (e.g., see [16, 22] for more extensive discussions and the precise definition of these equivalence relations). Hence, up to equivalence, we may **(and will)** write $\mathcal{K}_\omega$ if we want to consider $\mathcal{K} = \mathcal{K}(H)$ equipped with a G-action corresponding to some $[\omega] \in H^2(G, \mathbb{T})$.

2. The Baum-Connes conjecture for twisted group algebras

The above described relation between twisted group algebras and crossed products by $\mathcal{K} = \mathcal{K}(H)$ make it easy to extend the formulation of the Baum-Connes conjecture for the K-theory of group C^*-algebras to the twisted case. In order to recall the conjecture (with coefficients) assume that G is a locally compact group and A is a G-algebra. Recall that a locally compact G-space Z is called *proper*, if the map $G \times Z \to Z \times Z; (s, x) \mapsto (s \cdot x, x)$ is proper in the sense that inverse images of compact sets are compact. Proper G-spaces behave very much like transformations by compact group actions, and indeed, it is shown in [1] that every proper G-space is locally induced from compact subgroups (the same result was obtained much later in [12] – unfortunately we were unaware about the existing result in [1] at that time). A proper G-space $\underline{E}G$ is called *universal* if for every other proper G-space X there exists a continuous G-map $F : X \to \underline{E}G$ which is unique up to G-equivariant homotopy. It is clear from the definition that $\underline{E}G$ is unique up to G-homotopy, and it is shown in [50] that a locally compact realization of $\underline{E}G$ always exists.

Following [4], the topological K-theory $K_*^{\text{top}}(G; A)$ of G with coefficient A is defined as

$$K_*^{\text{top}}(G; A) = \lim_X KK_*^G(C_0(X), A),$$

where X runs through the G-compact (i.e., $G\backslash X$ is compact) subsets of some realization of the universal proper G-space $\underline{E}G$ (see [4]) and $KK_*^G(C_0(X), A)$ denotes Kasparov's equivariant and bivariant K-functor (see [32, 33]).

If X is any G-compact proper G-space, then a cut-off function $c : X \to [0, \infty)$ is a continuous function with compact support such that $\int_G c(s^{-1}x)\,ds = 1$ for all $x \in X$. Then $p_X(s, x) = \sqrt{c(x)c(s^{-1}x)}$ is an idempotent in the convolution algebra $C_c(G \times X) \subseteq C_0(X) \rtimes_r G$ and hence determines a class $[p_X] \in K_0(C_0(X) \rtimes_r G)$ which, by an easy homotopy argument, does not depend on the particular choice of the cut-off function c. The Baum-Connes assembly map

$$\mu_A : K_*^{\text{top}}(G; A) \to K_*(A \rtimes_r G)$$

is then defined on the G-compact subsets $X \subseteq \underline{E}G$ via the composition

$$KK_*^G(C_0(X), A) \xrightarrow{\cdot \rtimes_r G} KK_*(C_0(X) \rtimes_r G, A \rtimes_r G) \xrightarrow{[p_X] \otimes \cdot} K_*(A \rtimes_r G),$$

where the first map is Kasparov's decent homomorphism (see [32, 33]), and the second map is given by taking Kasparov product over $C_0(X) \rtimes_r G$ with $[p_X] \in K_0(C_0(X) \rtimes_r G) = KK_0(\mathbb{C}, C_0(X) \rtimes_r G)$. Following [4], we say that G satisfies the Baum-Connes conjecture with coefficient A, if $\mu_A : K_*^{\text{top}}(G; A) \to K_*(A \rtimes_r G)$ is an isomorphism of abelian groups. We should mention the very deep result of Higson and Kasparov ([28]), which shows that every a-T-menable group (and, in particular, every amenable group) satisfies the Baum-Connes conjecture with arbitrary coefficients. On the other hand, the conjecture with coefficients apparently does **not hold** for all groups (see [29]).

Definition 2.1 (cf. [38, 9]). Let $\omega \in Z^2(G, \mathbb{T})$. Then the twisted topological K-theory of G with respect to ω is defined as the topological K-theory $K_*^{\text{top}}(G; \omega) := K_*^{\text{top}}(G; \mathcal{K}_{\bar\omega})$ of G with coefficient $\mathcal{K}_{\bar\omega}$, the compact operators with action corresponding to the inverse $\bar\omega$ of ω. The twisted assembly map for G with respect to ω is then given by the assembly map

$$\mu_\omega : K_*^{\text{top}}(G; \omega) = K_*^{\text{top}}(G; \mathcal{K}_{\bar\omega}) \xrightarrow{\mu_{\mathcal{K}_{\bar\omega}}} K_*(\mathcal{K}_{\bar\omega} \rtimes_r G) \cong K_*(C_r^*(G, \omega)).$$

We say that G satisfies the *twisted Baum-Connes conjecture*, if μ_ω is an isomorphism for all $[\omega] \in H^2(G, \mathbb{T})$.

There is an alternative but equivalent way to define the twisted assembly map via twisted equivariant KK-theory as introduced in [10] (see in particular [10, Definition 5.4]). But we wanted to avoid the general theory of twisted actions in this discussion, so we chose to work with the above approach, which, to our knowledge, was first used by Mathai in [38] (see also the papers [40, 41] by Marcolli and Mathai). Actually both approaches have been used in [13], where the following result is shown:

Theorem 2.2 ([13, Theorem 1.2]). *Suppose that G is a second countable group such that the quotient G/G_0 of G by the connected component G_0 satisfies the Baum-Connes conjecture with arbitrary coefficients (which is true if G/G_0 is a-T-menable by [28]). Then G satisfies the twisted Baum-Connes conjecture.*

The above result implies in particular that every almost connected locally compact group satisfies the twisted Baum-Connes conjecture, which in particular implies that these groups satisfy the ordinary Baum-Connes conjecture with trivial coefficient $\mathbb{C}$. This answered to the positive a long standing conjecture by Connes and Kasparov. The twisted Baum-Connes conjecture is actually very closely related to the conjecture with trivial coefficient $\mathbb{C}$. To see this, we first recall that every cocycle $\omega \in Z^2(G, \mathbb{T})$ determines a central extension

$$1 \to \mathbb{T} \to G_\omega \to G \to 1$$

G_ω of G by $\mathbb{T}$, where G_ω is the set $G \times \mathbb{T}$ with multiplication given by $(s, z)(t, w) = (st, \omega(s, t)zw)$ and equipped with the unique locally compact group topology which induces the product Borel structure on the set $G \times \mathbb{T}$ (this topology might not coincide with the product topology – see [36]). In this way we obtain the classification of all central extensions of G by $\mathbb{T}$ via $H^2(G, \mathbb{T})$. The following result is [13, Proposition 2.7]:[1]

Proposition 2.3. *Suppose that G is a second countable locally compact group. Then the following are equivalent:*

1. *G satisfies the twisted Baum-Connes conjecture.*
2. *For every central extension $1 \to \mathbb{T} \to \widetilde{G} \to G \to 1$, the group $\widetilde{G}$ satisfies the Baum-Connes conjecture with trivial coefficient $\mathbb{C}$.*

The above result shows in particular that the counterexamples to the Baum-Connes conjecture with coefficients, as constructed in [29], do not (so far) give counterexamples to the Baum-Connes conjecture for twisted group algebras, since they do not provide counterexamples for the conjecture with trivial coefficients.

---

[1]The assumption in [13, Proposition 2.7] that G/G_0 satisfies BC for arbitrary coefficients was nowhere used and is superfluous.

3. Closed subgroups of almost connected groups

In this section we recall some results on the topological K-theory group $K^{\mathrm{top}}_*(G; A)$, for any G-algebra A, in the special case where G can be embedded as a closed subgroup of some almost connected group L. Specializing to $A = \mathcal{K}_{\bar\omega}$ for $[\omega] \in H^2(G, \mathbb{T})$ will then give a description of the twisted topological K-theory of G. The results of this section are mostly due to Kasparov ([31, 32, 33]) and some of the results and methods presented here are also discussed in special cases in the work of Marcolli and Mathai [40, 41], but we hope that the presentation given below will give some additional insights.

So in what follows we shall always assume that G is a closed subgroup of the almost connected group L and we denote by C the maximal compact subgroup of L (which is unique up to conjugacy). It follows from work of Abels [2] that L/C is a universal proper L-space, which then implies that

$$K^{\mathrm{top}}_*(L; B) = KK^L_*(C_0(L/C), B)$$

for any L-algebra B. By a deep result of Kasparov (see [33, Theorem 5.8]), restricting actions from L to C gives a restriction isomorphism

$$KK^L_*(C_0(L/C), B) \overset{\mathrm{res}^L_C}{\cong} KK^C_*(C_0(L/C), B).$$

Moreover, using again a result of Abels [2], we know that L/C is C-equivariantly diffeomorphic to some finite-dimensional real vector space V equipped with a linear action of C. It follows from this that tensoring with $C_0(L/C)$ together with applying Kasparov's Bott-periodicity theorem ([31, Theorem 7]) to $L/C \times L/C \cong V \times V$, equipped with the diagonal C-action (which is spinor), gives a chain of isomorphisms

$$KK^C_*(C_0(L/C), B) \cong KK^C_*(C_0(L/C \times L/C), C_0(L/C) \otimes B)$$

$$\overset{\mathrm{Bott}}{\cong} KK^C_*(\mathbb{C}, C_0(L/C) \otimes B) = K^C_*(C_0(L/C) \otimes B)$$

(we refer to [10, Lemma 7.7] for more details on this chain of isomorphisms).

In many cases, the direct factor L/C in the above formulas can be removed by use of Kasparov's Bott-periodicity theorem. As mentioned above, we may identify L/C with some finite-dimensional real linear space V equipped with a linear C-action $\pi : C \to \mathrm{O}(V)$ (with respect to some choice of inner product on V). Assume further that V is even dimensional (which can be arranged by replacing L by $L \times \mathbb{R}$ if necessary – this gives the dimension shift of order $n = \dim(L/C)$ in Theorem 3.1 below). Let $\mathcal{C}l(V)$ denote the complex Clifford-algebra with respect to the given inner product. Since V is even dimensional, $\mathcal{C}l(V)$ is isomorphic to the full matrix algebra $M(k \times k, \mathbb{C})$, with $k = 2^{\frac{1}{2}\dim(V)}$. Then Kasparov's Bott periodicity theorem [31, Theorem 7] implies that $C_0(V)$ is KK^C-equivalent to $\mathcal{C}l(V)$ equipped with the action induced from the given action of C on V. If the action of C on V is **not** orientation preserving, i.e., $\pi(C) \nsubseteq \mathrm{SO}(V)$, then $\mathcal{C}l(V)$ has to be considered as a *graded* algebra, which makes things a bit unpleasant!

So assume from now on that $\pi(C) \subseteq SO(V)$ (which is automatic if L, and hence C, is connected). Then the grading can be ignored and we see that $C_0(V)$ becomes KK^C-equivalent to the ungraded full matrix algebra $M(k \times k, \mathbb{C}) \cong Cl(V)$. As explained in the previous section, the action of C on $M(k \times k, \mathbb{C}) = \mathcal{K}(\mathbb{C}^k)$ corresponds to a class $[\zeta_C] \in H^2(C, \mathbb{T})$, and hence we obtain an isomorphism

$$K_*^C(C_0(L/C) \otimes B) \cong K_*^C(\mathcal{K}_{\zeta_C} \otimes B) \cong K_*((\mathcal{K}_{\zeta_C} \otimes B) \rtimes C), \qquad (3.1)$$

where the second isomorphism is provided by the Green-Julg theorem. Notice that the class $[\zeta_C]$ (which is a class of order two) is the pull back of the class $[\zeta_V] \in H^2(SO(V), \mathbb{T})$ corresponding to the central extension

$$1 \to \mathbb{T} \to \mathrm{Spin}^c(V) \to SO(V) \to 1. \qquad (3.2)$$

which vanishes if and only if $V = L/C$ carries an L-equivariant spin^c-structure. In particular, if L/C is spin^c, then Kasparov's Bott-periodicity theorem gives an isomorphism

$$K_*^C(C_0(L/C) \otimes B) \cong K_*^C(B) \cong K_*(B \rtimes C).$$

For a more detailed discussion of these facts we refer to the discussion in [13, §7]. Summarizing, we obtain the following theorem:

Theorem 3.1 (Kasparov). *Suppose that L is an almost connected group and C is the maximal compact subgroup of L. Then, for every L-algebra B, there exists a canonical isomorphism*

$$K_*^{\mathrm{top}}(L; B) \cong K_*^C(C_0(L/C) \otimes B) \cong K_*((C_0(L/C) \otimes B) \rtimes C).$$

If the action of C on $V \cong L/C$ is orientation preserving, then

$$K_*^{\mathrm{top}}(L; B) \cong K_{*+n}^C(\mathcal{K}_{\zeta_C} \otimes B) \cong K_{*+n}((\mathcal{K}_{\zeta_C} \otimes B) \rtimes C)$$

with $n = \dim(L/C)$ and some class $[\zeta_C] \in H^2(C, \mathbb{T})$ of order two, which vanishes if and only if L/C carries an L-equivariant spin^c *structure.*

Let's now go back to the situation where we start with a closed subgroup G of L. If A is a G-algebra, then the induced C^*-algebra $\mathrm{Ind}_G^L A$ is defined as

$$\mathrm{Ind}_G^L A := \{ f \in C^b(L, A) : f(lg) = g^{-1}(f(l)) \text{ for all } g \in G \text{ and } l \in L,$$
$$\text{and } (l \mapsto \|f(l)\|) \in C_0(L/G) \}$$

equipped with the pointwise operations and the supremum-norm. $\mathrm{Ind}_G^L A$ is an L-algebra with action $(l \cdot f)(k) := f(l^{-1}k)$ for $f \in \mathrm{Ind}_G^L A$, $l, k \in L$. If $A = C_0(Y)$ for some locally compact G-space Y, then $\mathrm{Ind}_G^L A \cong C_0(L \times_G Y)$, where $L \times_G Y$ denotes the L-space induced from Y in the usual sense. Moreover, if the action of G on A is a restriction of an action of $\alpha : L \to \mathrm{Aut}(A)$, then $\mathrm{Ind}_G^L A$ is G-isomorphic to $C_0(L/G) \otimes A$ with respect to the diagonal action – the isomorphism given by $f \mapsto \Phi(f)$ with $\Phi(f)(l) = \alpha_l(f(l))$. It follows from [11, Theorem 2.2] that there is a natural induction isomorphism

$$\mathrm{I}_G^L : K_*^{\mathrm{top}}(G; A) \to K_*^{\mathrm{top}}(L, \mathrm{Ind}_G^L A) = KK_*^L(C_0(L/C), \mathrm{Ind}_G^L A).$$

To describe it, assume that $X \subseteq L/C = \underline{EG}$ is any G-compact set (by [11, Lemma 2.4] we may use L/C also as a model for $\underline{EG}$). Then we have a canonical map $F : L \times_G X \to L/C$ given by $F([l, x]) = lx$. The induction isomorphism I_G^L is then given on the level of $X \subseteq \underline{EG}$ by the composition

$$KK_*^G(C_0(X), A) \overset{i_G^L}{\to} KK_*^L(C_0(L \times_G X), \mathrm{Ind}_G^L A) \overset{F_*}{\to} KK_*^L(C_0(L/C), \mathrm{Ind}_G^L A),$$

where i_G^L denotes Kasparov's induction homomorphism (see [32] or [33, §3] for the construction). Hence, applying Kasparov's theorem above, we obtain

Proposition 3.2. *Suppose that G is a closed subgroup of the almost connected group L and let C denote the maximal compact subgroup of L. Then, for any G-algebra A, there is a canonical isomorphism*

$$K_*^{\mathrm{top}}(G; A) \cong K_*^C(C_0(L/C) \otimes \mathrm{Ind}_G^L A) \cong K_*\big((C_0(L/C) \otimes \mathrm{Ind}_G^L A) \rtimes C\big).$$

If the action of C on $V \cong L/C$ is orientation preserving, then

$$K_*^{\mathrm{top}}(G; A) \cong K_{*+n}^C(\mathcal{K}_{\zeta_C} \otimes \mathrm{Ind}_G^L A) \cong K_{*+n}\big((\mathcal{K}_{\zeta_C} \otimes \mathrm{Ind}_G^L A) \rtimes C\big),$$

with ζ_C as in Theorem 3.1 and $n = \dim(L/C)$. If L/C is L-equivariantly spin^c, then

$$K_*^{\mathrm{top}}(G; A) \cong K_{*+n}^C(\mathrm{Ind}_G^L A) \cong K_{*+n}(\mathrm{Ind}_G^L A \rtimes C).$$

Note that in general $\mathrm{Ind}_G^L A$ is the section algebra of a continuous C^*-algebra bundle over L/G with constant fibre A. If the quotient map $L \mapsto L/G$ has local continuous sections, which is automatic if L is a Lie-group by [46], then this bundle is locally trivial. Moreover, if the action of G on A is the restriction of some action of L on A, then we may replace $\mathrm{Ind}_G^L A$ by the algebra $C_0(L/G) \otimes A$ in the above proposition.

An interesting special case occurs when G has no compact subgroups. In that case the action of C on L/G is free, and there is a canonical Morita equivalence between the crossed product $(C_0(L/C) \otimes \mathrm{Ind}_G^L A) \rtimes C$ and the fixed-point algebra $(C_0(L/C) \otimes \mathrm{Ind}_G^L A)^C$ (and similarly with $C_0(L/C)$ replaced by $\mathcal{K}_{\zeta_C}$ in case of an orientation preserving action of C on L/C). This fixed point algebra is actually an algebra of C_0-sections of a continuous C^*-algebra bundle over $C \backslash (L/C \times L/G)$ with fibre A. In the spin^c-case, this reduces to a bundle over the classifying space $B(G) = C \backslash L/G$ of G, even if G is not discrete.

4. K-theory of twisted group algebras and twisted K-theory

We now want to use the results of the previous section to relate the K-theory of a twisted group algebra (or rather the twisted topological K-theory $K_*^{\mathrm{top}}(G; \omega)$) to equivariant twisted K-theory of certain spaces with respect to a compact group action.

For this we first have to recall the definition of twisted K-theory: Suppose that X is a locally compact space and that $\delta \in \check{H}^3(X, \mathbb{Z})$ is a third cohomology class. For a topological group G let $\underline{G}$ denote the sheaf of germs of G-valued functions on X. Then the short exact sequences of groups

$$1 \to \mathbb{T} \to U \to \mathcal{P}U \to 1 \quad \text{and} \quad 0 \to \mathbb{Z} \to \mathbb{R} \to \mathbb{T} \to 0 \qquad (4.1)$$

(with $U = U(H)$ for H the infinite-dimensional separable Hilbert space) induce long exact sequences in sheaf cohomology for the respective sheaf of germs of continuous functions. Since $\mathbb{R}$ and U are contractible (the latter by Kuiper's theorem), we have

$$\check{H}^l(X, \underline{U}) = \{0\} = \check{H}^n(X, \underline{\mathbb{R}^n})$$

for $l = 1, 2$ and $n \in \mathbb{N}$ [2]. Hence, the long exact sequences corresponding to (4.1) induce isomorphisms

$$\check{H}^1(X, \underline{\mathcal{P}U}) \cong \check{H}^2(X, \underline{\mathbb{T}}) \cong \check{H}^3(X, \mathbb{Z}).$$

Therefore we may regard δ as a class in $\check{H}^1(X, \underline{\mathcal{P}U})$. Since $\mathcal{P}U = \text{Aut}(\mathcal{K})$ with $\mathcal{K} = \mathcal{K}(H)$, the class δ determines (an isomorphism class of) a locally trivial bundle $\mathcal{A}$ over X with fibre $\mathcal{K}$ via the usual gluing process. We refer to [49] for a very detailed treatment of this and the general Dixmier-Douady classification of continuous-trace algebras.

Notations 4.1. Suppose that $\delta \in \check{H}^3(X, \mathbb{Z})$ is as above and let $\mathcal{K} \to \mathcal{A} \to X$ denote the corresponding locally trivial bundle with fibre $\mathcal{K}$. We denote by $C_0(X; \delta)$ the C^*-algebra of all continuous section of $\mathcal{A}$ which vanish at infinity on X. We call it the *continuous-trace algebra over X corresponding to the Dixmier-Douady class* $\delta \in \check{H}^3(X, \mathbb{Z})$.

If A is an arbitrary separable continuous-trace algebra with $\text{Prim}(A) = X$, as defined in [18, Chapter 4], then $A \otimes \mathcal{K}$ is isomorphic to some $C_0(X; \delta)$ as in the above definition. This gives the Dixmier-Douady classification of continuous-trace algebras first presented in [19]. Again, we refer to [49] for a modern treatment.

If C is a locally compact group acting on X, then we may consider all possible actions on algebras $C_0(X; \delta)$ which cover the given action on X. The C-equvariant *Brauer group* $\text{Br}_C(X)$ consists of all $X \rtimes C$-equivariant Morita equivalence classes of such actions. It is shown in [16] that $\text{Br}_C(X)$ is a group with multiplication given by taking diagonal actions on the balanced tensor product

$$C_0(X; \delta) \otimes_{C_0(X)} C_0(X; \delta') \cong C_0(X; \delta \cdot \delta').$$

In what follows, we shall denote elements in $\text{Br}_C(X)$ by $[C_0(X; \delta), \alpha]$, where $\alpha : C \to \text{Aut}(C_0(X; \delta))$ is the given action. The identity in $\text{Br}_C(X)$ is represented by $[C_0(X, \mathcal{K}), \text{id}]$. [3]

Definition 4.2 (cf. [3]). Suppose that X is a locally compact space and that C is a compact group. Let $[C_0(X, \delta), \alpha]$ be a class in $\text{Br}_C(X)$. Then the *twisted C-equivariant K-theory of X* with respect to this class is the group

$$K_*^C(C_0(X, \delta)) \cong K_*\big(C_0(X, \delta) \rtimes_\alpha C\big).$$

---

[2] Since U is not abelian, $\check{H}^l(X, \underline{U})$ is only defined for $l = 0, 1, 2$.
[3] Here we write $C_0(X, \mathcal{K})$ instead of $C_0(X; 1)$.

We refer to [3] for a detailed discussion on twisted K-theory. The connection between twisted topological K-theory $K_*^{\text{top}}(G;\omega)$ and twisted K-theory as defined above is given by

Proposition 4.3. *Suppose that G is a closed subgroup of the almost connected locally compact group L with maximal compact subgroup $C \subseteq L$. Then there is a homomorphism*

$$\mu : H^2(G, \mathbb{T}) \to \text{Br}_C(L/G); [\omega] \mapsto [C_0(L/G; \delta_\omega), \alpha_\omega]$$

such that

$$K_*^{\text{top}}(G;\omega) \cong K_*^C\big(C_0(L/C) \otimes C_0(L/G; \delta_\omega)\big).$$

If the action of C on $V \cong L/C$ is orientation preserving, the factor $C_0(L/C)$ can be replaced by $\mathcal{K}_{\zeta_C}$ (up to a dimension shift in K-theory by $n = \dim(L/C)$ and with ζ_C as in Theorem 3.1), and we have $K_*^{\text{top}}(G;\omega) \cong K_{*+n}^C(C_0(L/G; \delta_\omega))$ if L/G has a L-equivariant spin^c*-structure.*

Proof. By Proposition 3.2, a representative of the class $\mu(\omega) \in \text{Br}_C(L/G)$ is given by the algebra $\text{Ind}_G^L \mathcal{K}_{\bar{\omega}}$ with action α_ω of C given by left translation. That $[\omega] \to [\text{Ind}_G^L \mathcal{K}_{\bar{\omega}}, \alpha_\omega]$ defines a homomorphism from $H^2(G, \mathbb{T}) \to \text{Br}_C(L/G)$ follows from Lemma 4.6 below. $\qquad\square$

Remark 4.4. If G has no compact subgroups, which is equivalent to saying that $G \cap lCl^{-1} = \{e\}$ for all $l \in L$, then C acts freely on L/G, and hence the crossed product $\big(C_0(L/C) \otimes C_0(L/G; \delta_\omega)\big) \rtimes C$ will be Morita equivalent to a continuous trace algebra over $C\backslash(L/C \times L/G)$ (see [48]). If δ_ω^C denotes the corresponding Dixmier-Douady class, we get

$$K_*^{\text{top}}(G;\omega) \cong K_*(C_0(C\backslash(L/C \times L/G); \delta_\omega^C)).$$

As usual, we may simplify things if the action of C on $V \cong L/C$ is orientation preserving or spin^c – in both cases we get isomorphisms

$$K_*^{\text{top}}(G;\omega) \cong K_*(C_0(C\backslash L/G; \widetilde{\delta}_\omega^C))$$

for a suitable Dixmier-Douady class $\widetilde{\delta}_\omega^C \in \check{H}^3(C\backslash L/G, \mathbb{Z})$. The classes δ_ω^C and $\widetilde{\delta}_\omega^C$ can be explicitly computed from the class $\mu([\omega])$ via a general isomorphism $\text{Br}_C(X) \cong \check{H}^3(C\backslash X, \mathbb{Z})$ described in [16, 6.2], which holds whenever a locally compact group C acts freely and properly on a given space X.

In order to study the homomorphism $\mu : H^2(G, \mathbb{T}) \to \text{Br}_C(L/G)$ in the theorem we first recall that by the main result of [35] there is an isomorphism

$$\Theta : \text{Br}_G(C\backslash L) \cong \text{Br}_C(L/G),$$

where the action of G on L (and on $C\backslash L$) is given by right translation. It can be described as follows: Suppose that $[C_0(C\backslash L; \delta), \beta] \in \text{Br}_G(C\backslash L)$. Consider β as an action on the underlying bundle $\mathcal{B}$ rather than on the section algebra $C_0(C\backslash L; \delta)$ of sections of $\mathcal{B}$ (then β_g maps the fibre $\mathcal{K}_{C\cdot l}$ of $\mathcal{B}$ isomorphically to the fibre $\mathcal{K}_{C\cdot l\cdot g^{-1}}$).

Let us denote by $P : \mathcal{B} \to C \backslash L$ the bundle projection and let $p : L \to C \backslash L$ denote the quotient map. We construct a bundle over L/G with fibres $\mathcal{K}$ by defining

$$\mathcal{A} := \{(l, b) \in L \times \mathcal{B} : p(l) = P(b)\}/G$$

where G acts on $L \times \mathcal{B}$ via $g \cdot (l, b) = (lg^{-1}, \beta_g(b))$. The bundle projection $Q : \mathcal{A} \to L/G$ is given by $Q([l, b]) = q(l)$, where $q : L \to L/G$ denotes the quotient map and the action of C on $\mathcal{A}$ is given by

$$c[l, b] = [cl, b].$$

The class $\Theta\big([C_0(C \backslash L; \delta), \beta]\big)$ is then represented by the C^*-algebra of C_0-sections of $\mathcal{A}$ together with the action of C given by the above-defined action on $\mathcal{A}$. For later use, we also state the following fact shown in [35]:

Proposition 4.5 ([35, Theorem 1]). *Let*

$$[C_0(C \backslash L; \delta), \beta] \in \mathrm{Br}_G(C \backslash L) \quad \text{and let} \quad [C_0(L/G; \hat{\delta}), \alpha] \in \mathrm{Br}_C(L/G)$$

be the corresponding class with respect to $\Theta : \mathrm{Br}_G(C \backslash L) \overset{\cong}{\to} \mathrm{Br}_C(L/G)$ *as described above. Then* $C_0(C \backslash L; \delta) \rtimes_\beta G$ *is Morita equivalent to* $C_0(L/G, \hat{\delta}) \rtimes_\alpha C$.

On the other hand, we have an obvious homomorphism

$$\widetilde{\mu} : H^2(G, \mathbb{T}) \to \mathrm{Br}_G(C \backslash L)$$

which maps $[\omega]$ to the class $[C_0(C \backslash L, \mathcal{K}), \rho \otimes \beta_{\bar{\omega}}]$, where $\rho : G \to \mathrm{Aut}(C_0(C \backslash L))$ is the right translation action and where $\beta_{\bar{\omega}} : G \to \mathrm{Aut}(\mathcal{K})$ is an action corresponding to the inverse cocycle $\bar{\omega}$ as explained in Section 1. We then get:

Lemma 4.6. *The homomorphism* $\mu : H^2(G, \mathbb{T}) \to \mathrm{Br}_C(L/G)$ *of Proposition 4.3 is equal to the composition*

$$H^2(G, \mathbb{T}) \xrightarrow{\ \widetilde{\mu}\ } \mathrm{Br}_G(C \backslash L) \xrightarrow{\ \Theta\ } \mathrm{Br}_C(L/G).$$

Proof. Applying Proposition 3.2 to the G-algebra $\mathcal{K} = \mathcal{K}_{\bar{\omega}}$, we only have to check that $\mathrm{Ind}_G^L \mathcal{K}_{\bar{\omega}}$ is C-equivariantly isomorphic to the section algebra of the bundle $\mathcal{A}$ constructed out of $(C_0(C \backslash, \mathcal{K}), \rho \otimes \beta_{\bar{\omega}})$ as in the description of the map Θ given above. But the bundle $\mathcal{B}$ in that description is then the trivial bundle $(C \backslash L) \times \mathcal{K}$. The space $\{(l, b) : p(l) = P(b)\}$ is then easily identified with the trivial bundle $L \times \mathcal{K}$, and the C^*-algebra of C_0-sections of the quotient bundle $(L \times \mathcal{K})/G$ under the diagonal action $g(l, T) = (lg^{-1}, \beta_{\bar{\omega}}(g)(T))$ is then C-isomorphic to $\mathrm{Ind}_G^L \mathcal{K}$. $\qquad\square$

It follows from Proposition 4.3 that, whenever $[\omega]$ is in the kernel of μ, the twisted topological K-theory $K_*^{\mathrm{top}}(G; \omega)$ is isomorphic to the untwisted topological K-theory $K_*^{\mathrm{top}}(G)$ of G. Hence, if, in addition, G satisfies the twisted Baum-Connes conjecture, it will follow that the K-theory of the reduced twisted group algebra $C_r^*(G, \omega)$ is equal to the K-theory of the untwisted group algebra $C_r^*(G)$ of G.

This applies in particular to

Lemma 4.7. *Let G be a closed subgroup of the almost connected group L with maximal compact subgroup C and let $\mu : H^2(G, \mathbb{T}) \to \mathrm{Br}_C(L/G)$ be as in Proposition 4.3. Let*

$$\mathcal{R} = \{[\omega] \in H^2(G; \mathbb{T}) : \exists\, [\tilde{\omega}] \in H^2(L, \mathbb{T})$$

such that

$$[\omega] = [\tilde{\omega}|_{G \times G}] \quad and \quad [\tilde{\omega}|_{C \times C}] = [1]\}$$

Then $\mathcal{R} \subseteq \ker \mu$.

Proof. Since $[\omega]$ is the restriction of some $[\tilde{\omega}] \in H^2(L, \mathbb{T})$, the corresponding action of G on $\mathcal{K} = \mathcal{K}_{\tilde{\omega}}$ is the restriction of the action of L on $\mathcal{K}$ corresponding to $[\tilde{\omega}]$. Hence we have $\mathrm{Ind}_G^L \mathcal{K} \cong C_0(L/G, \mathcal{K})$ with respect to the diagonal action of L. Since the restriction of $\tilde{\omega}$ to C is trivial, the action of L on $\mathcal{K}$ also restricts to the trivial action of C on $\mathcal{K}$. $\qquad\square$

The above lemma actually applies to all examples presented in Example 1.1. In particular we get

Corollary 4.8. *Suppose that $H \subseteq \mathrm{SL}(2, \mathbb{Z})$ is any subgroup and let H act on the (rational or irrational) rotation algebra $A_\theta = C^*(\mathbb{Z}^2, \omega_\theta)$ as in Example 1.1. Then $K_*(A_\theta \rtimes_r H)$ is isomorphic to $K_*(C(\mathbb{T}^2) \rtimes_r H) = K_*(C_r^*(\mathbb{Z}^2 \rtimes H))$ for all $\theta \in [0, 1]$.*

Proof. In Example 1.1 we showed that $A \rtimes_r H \cong C_r^*(\mathbb{Z}^2 \rtimes H, \tilde{\omega}_\theta)$ with $\tilde{\omega}_\theta \in Z^2(\mathbb{Z}^2 \rtimes H, \mathbb{T})$ as given in (1.2). We also pointed out that ω_θ is the restriction of a cocycle, also denoted ω_θ, of $\mathbb{R}^2$, and the same formula as given in (1.2) will extend this cocycle to a cocycle $\tilde{\omega}_\theta$ on $\mathbb{R}^2 \rtimes \mathrm{SL}(2, \mathbb{R})$, which restricts to the same named cocycle on $\mathbb{Z}^2 \rtimes H \subseteq \mathbb{R}^2 \rtimes \mathrm{SL}(2, \mathbb{Z})$. Also, by construction, the cocycle $\tilde{\omega}_\theta$ restricts to the trivial cocycle on $\mathrm{SO}(2)$, the maximal compact subgroup of $\mathbb{R}^2 \rtimes \mathrm{SL}(2, \mathbb{R})$. Since $\mathbb{Z}^2 \rtimes \mathrm{SL}(2, \mathbb{Z})$ satisfies the Baum-Connes conjecture with coefficients, which follows from [14, Theorem 2.1] together with the fact that $\mathrm{SL}(2, \mathbb{Z})$ is a-T-menable, the same is true for $\mathbb{Z}^2 \rtimes H$ by [11, Theorem 2.5]. Hence, the result follows from Lemma 4.7 and the discussion preceding it. $\qquad\square$

Remark 4.9. The above result gives an easy answer to the problem of computing the K-theory groups of the crossed products $A_\theta \rtimes F$ for F a finite subgroup of $\mathrm{SL}(2, \mathbb{Z})$. Indeed, in this situation it is not very difficult to compute the groups $K_*(C(\mathbb{T}^2) \rtimes F)$ with the help of suitable six-term sequences. This solution to the problem was suggested to us by G. Skandalis at the C^*-algebra meeting in Oberwolfach in 2003. The K-theory groups have been computed before for all rational θ in the papers [25, 26, 27] and in case $F = \mathbb{Z}/2\mathbb{Z}$ for all θ in [34]. The case $F = \mathbb{Z}/4\mathbb{Z}$ was treated by Walters in [52] for a dense G_δ subset of the irrational numbers. I should also mention that the K_0-groups of $A_\theta \rtimes F$ have been computed very recently for all θ and all F by Polishchuk in [47] using completely different methods. In a later section we shall discuss a different method for computing the K-theory groups of $A_\theta \rtimes F$, which is developed in [21] and which also provides a very useful tool to compute explicit generators for $K_*(A_\theta \rtimes F)$. We should point out that the interest in computing the K-groups of these particular examples comes from the fact that this is part of a proof that $A_\theta \rtimes F$ for irrational θ is always an AF-algebra. This has been shown for $F = \mathbb{Z}/2\mathbb{Z}$ in [7] and it is shown for all other cases in [21].

Note that we actually have isomorphisms $K_*(A_\theta \rtimes H) \cong K_*(A_\theta \rtimes_r H)$ for all θ and all $H \subseteq \mathrm{SL}(2, \mathbb{Z})$, since every subgroup H of $\mathrm{SL}(2, \mathbb{Z})$ is K-amenable in the sense of Cuntz [17]. In particular, the above corollary holds also for the maximal crossed products.

5. Poincaré duality for twisted group algebras

In this section we want to reduce to the case where G is a **co-compact discrete subgroup** of the almost connected group L. In this case we have $\underline{EG} = L/C$ being G-compact, and therefore we get

$$K_*^{\mathrm{top}}(G; \omega) = KK_*^G(C_0(L/C), \mathcal{K}_{\bar\omega}),$$

since, as mentioned before, L/C is a model for $\underline{EG}$. Since the diagonal action of G on $\mathcal{K}_{\bar\omega} \otimes \mathcal{K}_\omega$ is given via the adjoint action of a tensor-product representation $V \otimes W$ with V an $\bar\omega$-representation and W an ω-representation, it follows that $V \otimes W$ is actually a homomorphism, which then implies that $\mathcal{K}_{\bar\omega} \otimes \mathcal{K}_\omega$ is G-equivariantly Morita equivalent to $\mathbb{C}$. This shows that the chain

$$KK_*^G(C_0(L/C), \mathcal{K}_{\bar\omega}) \xrightarrow{\otimes_{\mathbb{C}} 1_{\mathcal{K}_\omega}} KK^G(C_0(L/C) \otimes \mathcal{K}_\omega, \mathcal{K}_{\bar\omega} \otimes \mathcal{K}_\omega)$$
$$\xrightarrow{\cong} KK_*^G(C_0(L/C) \otimes K_\omega, \mathbb{C})$$

is an isomorphism with inverse given by tensoring with $\mathcal{K}_{\bar\omega}$. Since G is discrete, we get a natural isomorphism

$$KK_*^G(C_0(L/C) \otimes \mathcal{K}_\omega, \mathbb{C}) \cong KK_*(C_0(L/C, \mathcal{K}_\omega) \rtimes G), \mathbb{C})$$
$$= K^*(C_0(C_0(L/C, \mathcal{K}_\omega) \rtimes G).^4$$

So we may combine these maps to obtain an isomorphism

$$K_*^{\mathrm{top}}(G, \mathcal{K}_{\bar\omega}) \cong K^*(C_0(L/C, \mathcal{K}_\omega) \rtimes G).$$

On the other hand, if L/C carries an L-equivariant spin^c-structure, we obtain from Proposition 3.2 an isomorphism

$$K_*^{\mathrm{top}}(G; \omega) \cong K_*(C(L/G; \delta_\omega) \rtimes_\beta C),$$

---

[4] This follows from the correspondence between covariant representations of the system $(C_0(L/C, \mathcal{K}_{\bar\omega}), G)$ and representations of the crossed product $C_0(L/C, \mathcal{K}_{\bar\omega}) \rtimes G$, where the discreteness of G is needed to make sure that we can take the same generalized Fredholm operators in the KK-cycles.

where the class $[C(L/G; \delta_\omega), \beta] \in \mathrm{Br}_C(L/G)$ corresponds is the image of the system $[C_0(C \backslash L, \mathcal{K}_{\bar\omega}), \rho \otimes \alpha_{\bar\omega}] \in \mathrm{Br}_G(C \backslash L)$ under the isomorphism $\Theta : \mathrm{Br}_G(C \backslash L) \to \mathrm{Br}_C(L/G)$. By Proposition 4.5 we have a Morita equivalence

$$C(L/G; \delta_\omega) \rtimes C \sim_M C_0(C \backslash L, \mathcal{K}_{\bar\omega}) \rtimes G.$$

But it is straightforward to check that the transformation

$$\Phi : C_c(G \times C \backslash L, \mathcal{K}_{\bar\omega}) \to C_c(G \times L/C, \mathcal{K}_{\bar\omega}); \quad \Phi(f)(g, lC) = f(g, Cl^{-1})$$

extends to an isomorphism $C_0(C \backslash L, \mathcal{K}_{\bar\omega}) \rtimes G \cong C_0(L/C, \mathcal{K}_{\bar\omega}) \rtimes G$, where in the second crossed product we consider the **left** translation action of G on L/C. Combining all this, we arrive at:

Proposition 5.1. *Suppose that G is a discrete co-compact subgroup of the almost connected group L such that L/C, the quotient of L by the maximal compact subgroup C of L, has an L-invariant* spin^c-*structure. Then, for all $[\omega] \in H^2(G, \mathbb{T})$, there are isomorphisms*

$$K^*(C_0(L/C, \mathcal{K}_\omega) \rtimes G) \cong K_*^{\mathrm{top}}(G; \omega) \cong K_*(C_0(L/C, \mathcal{K}_{\bar\omega}) \rtimes G).$$

If G satisfies a strong version of the Baum-Connes conjecture, i.e., if the γ-element of G in the sense of [32] is the unit element in $KK^G(\mathbb{C}, \mathbb{C})$, then it follows from the spin^c condition in the above proposition that the Dirac and dual-Dirac elements constructed by Kasparov in [32, 33] induce a KK^G-equivalence between $C_0(L/C)$ and $\mathbb{C}$. But this implies that we also have KK-equivalences between $C_0(L/C, \mathcal{K}_\omega) \rtimes G$ and $\mathcal{K}_\omega \rtimes G$, and similar if ω is replaced by $\bar\omega$. Thus, from the above proposition we then obtain

Proposition 5.2. *Suppose that G is as in Proposition 5.1 such that in addition the γ-element of G in the sense of [32] is the identity in $KK^G(\mathbb{C}, \mathbb{C})$. Then there is an isomorphism $K_*(C^*(G, \omega)) \cong K^*(C^*(G, \bar\omega))$.*

The above propositions should be interpreted as a weak version of K-theory Poincaré duality for the (dual) pair of C^*-algebras

$$\big(C_0(L/C, \mathcal{K}_\omega) \rtimes G, C_0(L/C, \mathcal{K}_{\bar\omega}) \rtimes G\big) \quad \text{and} \quad (C^*(G, \bar\omega), C^*(G, \omega)),$$

respectively. In the case $G = \mathbb{Z}^2$ this comprises to the well-studied Poincaré duality between the rotation algebra A_θ and its opposite algebra $A_{-\theta}$ (e.g., see [15]). Of course, we arrive at the isomorphisms of Propositions 5.1 and 5.2 only after a quite long chain of isomorphisms, which certainly leads to some confusion if one really wants to work with it. Therefore, Heath Emerson, Hyun Jeong Kim and the author took the above observations as a starting point for an extensive study of Poincaré dualities related to group actions on manifolds and to twisted group algebras ([20]). One outcome (among others) of this investigation is the construction of fundamental classes

$$\Delta \in K_*(C^*(G, \omega) \otimes C^*(G, \bar\omega)) \quad \text{and} \quad \widehat{\Delta} \in K^*(C^*(G, \bar\omega) \otimes C^*(G, \omega))$$

which induce a KK-theoretic Poincaré duality for the pair $(C^*(G, \bar\omega), C^*(G, \omega))$ under somewhat weaker conditions as the ones given in Proposition 5.2 above. For example, this result applies to all cocompact lattices in $SU(n, 1)$ and in (suitable double covers of) $SO(n, 1)$, for the fundamental groups Γ_g of Riemannian surfaces of genus g, and to all finitely generated torsion free nilpotent groups.

In a more general setting we obtain KK-theoretic Poincaré dualities for pairs $(C_0(X, \mathcal{K}_\omega) \rtimes G, C_0(X, \mathcal{K}_{\bar\omega}) \rtimes G)$ whenever G is a discrete group acting isometrically and co-compactly on on the spin^c-manifold X.

6. Homotopy invariance of twisted topological K-theory

The methods used in [21] for the computation of the groups $K_*(A_\theta \rtimes F)$ (compare with Remark 4.9 above) depend on the homotopy invariance of twisted topological K-theory, which we shall briefly discuss below.

Definition 6.1. Let G be a second countable locally compact group. Consider $C([0, 1], \mathbb{T})$, equipped with the topology of uniform convergence, as a trivial G-module. Then two classes $[\omega_0], [\omega_1] \in H^2(G, \mathbb{T})$ are called *homotopic*, if there exists a Borel cocycle $\Omega \in Z^2(G, C([0, 1], \mathbb{T}))$ such that $[\omega_0] = [ev_0(\Omega)]$ and $[\omega_1] = [ev_1(\Omega)]$ in $H^2(G, \mathbb{T})$, where $ev_t : Z^2(G, C([0, 1], \mathbb{T})) \to Z^2(G, \mathbb{T})$ denotes evaluation at t for each $t \in [0, 1]$

If $\Omega \in Z^2(G, C([0, 1], \mathbb{T}))$ is a homotopy, then it follows from [30, Proposition 3.1] that $\bar\Omega$ determines an action α of G on $C([0, 1], \mathcal{K})$ as follows: for each $t \in [0, 1]$ let $\bar\omega_t = ev_t(\bar\Omega)$ and let $L_{\bar\omega_t} : G \to \mathcal{U}(L^2(G))$ denote the regular $\bar\omega_t$-representation of G (see §1 for the definition). Let $\mathcal{K} = \mathcal{K}(L^2(G))$. Then α is given by the formula

$$(\alpha(g)(f))(t) = \operatorname{Ad} L_{\bar\omega_t}(g)(f(t)) \quad \text{for } f \in C([0, 1], \mathcal{K}), \ g \in G \text{ and } t \in [0, 1].$$

Clearly, the evaluations of α at $t = 0$ and $t = 1$ give actions corresponding to $[\bar\omega_0]$ and $[\bar\omega_1]$. Depending on results of [14] we show in [21]:

Theorem 6.2. *Suppose that $[\omega_0], [\omega_1] \in H^2(G, \mathbb{T})$ are homotopic. Then the twisted topological K-theories $K_*^{\mathrm{top}}(G; \omega_0)$ and $K_*^{\mathrm{top}}(G; \omega_1)$ are isomorphic. To be more precise: If $\Omega \in H^2(G, C([0, 1], \mathbb{T}))$ is any homotopy and if $\alpha : G \to \operatorname{Aut}(C([0, 1], \mathcal{K}))$ is the action corresponding to $\bar\Omega$ as above, then the evaluation maps induce isomorphisms*

$$ev_{t,*} : K_*^{\mathrm{top}}(G; C([0, 1], \mathcal{K})) \to K_*^{\mathrm{top}}(G; \mathcal{K}_{\bar\omega_t}) = K_*^{\mathrm{top}}(G; \omega_t)$$

for all $t \in [0, 1]$.

Of course, if G satisfies the twisted Baum-Connes conjecture, it follows from this result that $K_*(C_r^*(G, \omega_0)) \cong K_*(C_r^*(G, \omega_1))$ whenever ω_0 is homotopic to ω_1.

Example 6.3. Suppose that $[\omega] \in H^2(G, \mathbb{T})$ is the image of some class $c \in H^2(G, \mathbb{R})$ under the exponential map $\exp : \mathbb{R} \to \mathbb{T}; r \mapsto \exp(2\pi i r)$ (we then say that $[\omega]$ is a *real class* in $H^2(G, \mathbb{T})$). Then

$$\Omega(g, h)(t) := \exp(2\pi i t c(g, h)), \quad g, h \in G, t \in [0, 1]$$

is a homotopy $\Omega \in Z^2(G, C([0,1], \mathbb{T}))$ between $[\omega]$ and $[1]$. Thus, it follows immediately that $K_*^{\mathrm{top}}(G; \omega)$ is isomorphic to $K_*^{\mathrm{top}}(G)$ for every real cocycle $\omega \in Z^2(G, \mathbb{T})$. Since the cocycles considered in Example 1.1 are all real, this gives another way to see that $K_*(A_\theta \rtimes_r H) \cong K_*(C(\mathbb{T}^2) \rtimes_r H)$ for every subgroup H of $\mathrm{SL}(2, \mathbb{Z})$.

If $\Omega \in Z^2(G, C([0,1], \mathbb{T}))$ is a homotopy of cocycles, then one can actually form a "bundle" of twisted group algebras $C_{(r)}^*(G, \Omega) := C([0,1]) \rtimes_{\Omega,(r)} G$ over $[0,1]$ with fibres the twisted group algebras $C_{(r)}^*(G, \omega_t)$, $t \in [0,1]$. We refer to [23] for a general discussion on the construction and structure of such bundles. Let us just mention here that it follows from the Packer-Raeburn stabilization trick ([45, Theorem 3.4]) that there is an isomorphism

$$C_{(r)}^*(G, \Omega) \otimes \mathcal{K} \cong C([0,1], \mathcal{K}) \rtimes_{\alpha,(r)} G$$

with α as in Theorem 6.2. Combining this with Theorem 6.2 we get

Corollary 6.4 ([21, Corollary 1.10]). *Suppose that G satisfies the twisted Baum-Connes conjecture for homotopies, i.e., G satisfies BC with coefficient $C([0,1], \mathcal{K})$ for any action $\alpha : G \to \mathrm{Aut}(C([0,1], \mathcal{K}))$ as in Theorem 6.2.[5] Then the evaluation maps*

$$\mathrm{ev}_{t,*} : K_*(C_r^*(G, \Omega)) \to K_*(C_r^*(G, \omega_t))$$

are isomorphisms for all $t \in [0,1]$.

It is this result which is used in [21] to compute explicit generators for $K_*(A_\theta \rtimes F) = K_*(C^*(\mathbb{Z}^2 \rtimes F, \widetilde{\omega}_\theta)$ with F a finite subgroup of $\mathrm{SL}(2, \mathbb{Z})$ as in Example 1.1. The basic idea is quite easy: Since $\widetilde{\omega}_\theta$ is real, there is a canonical homotopy between $\widetilde{\omega}_\theta$ and the trivial cocycle as explained above. Now compute explicit generators for the untwisted case $K_*(C^*(\mathbb{Z}^2 \rtimes F))$ and simply guess how these generators extend to K-theory classes of the bundle $C^*(\mathbb{Z}^2 \rtimes F, \Omega)$. Corollary 6.4 then implies that these classes restrict to generators in all other fibres.

Using the above-described computations together with some deep results from C^*-algebra classification theory, we obtain as the main result of [21] the following

Theorem 6.5 (Echterhoff, Lück, Phillips, Walters). *Let F be any of the finite subgroups $\mathbb{Z}_2, \mathbb{Z}_3, \mathbb{Z}_4, \mathbb{Z}_6 \subseteq \mathrm{SL}_2(\mathbb{Z})$ (which are unique up to conjugacy) and let $\theta \in \mathbb{R} \setminus \mathbb{Q}$. Then the crossed product $A_\theta \rtimes_\alpha F$ is an AF algebra. For all $\theta \in \mathbb{R}$ we have*

$$K_0(A_\theta \rtimes \mathbb{Z}_2) \cong \mathbb{Z}^6, \ K_0(A_\theta \rtimes \mathbb{Z}_3) \cong \mathbb{Z}^8, \ K_0(A_\theta \rtimes \mathbb{Z}_4) \cong \mathbb{Z}^9 \ and \ K_0(A_\theta \rtimes \mathbb{Z}_6) \cong \mathbb{Z}^{10},$$

and the image of the canonical normalized trace on $A_\theta \rtimes_\alpha F$ (which is the unique trace if θ is irrational) is equal to $\frac{1}{k}(\mathbb{Z} + \theta\mathbb{Z})$ if $F = \mathbb{Z}_k$ for $k = 2, 3, 4, 6$. As a consequence, for all $\theta, \theta' \in \mathbb{R} \setminus \mathbb{Q}$, $A_\theta \rtimes_\alpha \mathbb{Z}_k$ is isomorphic to $A_{\theta'} \rtimes_\alpha \mathbb{Z}_l$ if and only if $k = l$ and $\theta' = \pm\theta \mod \mathbb{Z}$.

---

[5]By considering constant homotopies one can then deduce that G satisfies the twisted Baum-Connes conjecture – the converse holds if G is exact and has a γ-element in the sense of Kasparov by [13, Proposition 3.1].

Note that this theorem generalizes the results of Bratteli and Kishimote in [7] (which treats the case $F = \mathbb{Z}/2\mathbb{Z}$) and of Walters in [53] (which treats the case $F = \mathbb{Z}/4\mathbb{Z}$ for a dense G_δ-subset of $[0,1] \setminus \mathbb{Q}$). But our methods differ substantially from the methods used in those papers. Along the same lines, a similar result is obtained in [21] for $A_\Theta \rtimes \mathbb{Z}/2\mathbb{Z}$, the crossed product of a simple higher-dimensional noncommutative torus by the flip action.

References

[1] H. Abels. *A universal proper G-space*. Math. Z. **159** (1978), no. 2, 143–158.

[2] H. Abels. *Parallelizability of proper actions, global K-slices and maximal compact subgroups*. Math. Ann. 212 (1974/75), 1–19.

[3] M. Atiyah, G. Segal. *Twisted K-theory*. Ukr. Math. Bull. **1** (2004), no. 3, 291–334.

[4] P. Baum, A. Connes and N. Higson. *Classifying space for proper actions and K-theory of group C^*-algebras*, Contemporary Mathematics, **167**, 241–291 (1994).

[5] L. Baggett and A. Kleppner. *Multiplier representations of abelian groups*. J. Functional Analysis 14 (1973), 299–324.

[6] O. Bratteli, G.A. Elliott, D.E. Evans, and A. Kishimoto. *Noncommutative spheres. I*, International J. Math. **2**(1991), 139–166.

[7] O. Bratteli and A. Kishimoto. *Noncommutative spheres. III. Irrational rotations*, Commun. Math. Physics **147**(1992), 605–624.

[8] J. Brodzki, V. Mathai, J. Rosenberg, and R.J. Szabo. *D-Branes, RR-fields and duality on noncommutative manifolds*. Preprint (arXiv: hep-th/0607020 v1). To appear in Journal of Physics Conference Series.

[9] A.L. Carey, K.C. Hannabuss, V. Mathai, P. McCann. *Quantum Hall effect on the hyperbolic plane*. Comm. Math. Phys. **190** (1998), 629–673.

[10] J. Chabert and S. Echterhoff. *Twisted equivariant KK-theory and the Baum-Connes conjecture for group extensions*. K-Theory **23**, 157–200 (2001).

[11] J. Chabert and S. Echterhoff. *Permanence properties of the Baum-Connes conjecture*. Doc. Math. **6**, 127–183 (2001).

[12] J. Chabert, S. Echterhoff, R. Meyer. *Deux remarques sur l'application de Baum-Connes*. C. R. Acad. Sci. Paris Sér. I Math. **332** (2001), no. 7, 607–610.

[13] J. Chabert, S. Echterhoff, R. Nest. *The Connes-Kasparov conjecture for almost connected groups and for linear p-adic groups*. Publ. Math. Inst. Hautes tudes Sci. **97** (2003), 239–278.

[14] J. Chabert, S. Echterhoff, H. Oyono-Oyono. *Going-down functors, the Künneth formula, and the Baum-Connes conjecture*. Geom. Funct. Anal. **14** (2004), 491–528.

[15] A. Connes. *Gravity coupled with matter and the foundation of non-commutative geometry*. Comm. Math. Phys. **182** (1996), no. 1, 155–176.

[16] D. Crocker, A. Kumjian, I. Raeburn, D.P. Williams. *An equivariant Brauer group and actions of groups on C^*-algebras*. J. Funct. Anal. **146** (1997), 151–184.

[17] J. Cuntz. *K-theoretic amenability for discrete groups*. J. Reine Angew. Math. **344** (1983), 180–195.

[18] J. Dixmier. *C*-algebras*. North-Holland Mathematical Library, Vol. 15. North-Holland Publishing Co., Amsterdam-New York-Oxford, 1977.

[19] J. Dixmier, A. Douady. *Champs continus d'espaces hilbertiens et de C*-algèbres*. Bull. Soc. Math. France **91** (1963) 227–284.

[20] S. Echterhoff, H. Emerson, H.J. Kim. *KK-theoretic duality for proper twisted actions*. To appear in Math. Annalen.

[21] S. Echterhoff, W. Lück, C. Phillips, S. Walters. *The structure of crossed products of irrational rotation algebras by finite subgroups of* SL(2, ℤ). Preprint.

[22] S. Echterhoff, D.P. Williams. *Locally inner actions on $C_0(X)$-algebras*. J. Operator Theory **45** (2001), 131–160.

[23] S. Echterhoff, D.P. Williams. *Central twisted transformation groups and group C*-algebras of central group extensions*. Indiana Univ. Math. J. **51** (2002), no. 6, 1277–1304.

[24] C. Farsi and N. Watling. *Fixed point subalgebras of the rotation algebra*, C. R. Math. Rep. Acad. Sci. Canada **14**(1991), 75–80; corrigendum **14**(1991), 234.

[25] C. Farsi and N. Watling. *Quartic algebras*, Canad. J. Math. **44** (1992), 1167–1191.

[26] C. Farsi and N. Watling. *Cubic algebras*, J. Operator Theory **30** (1993), 243–266.

[27] C. Farsi and N. Watling. *Elliptic algebras*, J. Funct. Anal. **118** (1993), 1–21.

[28] N. Higson and G. Kasparov. *E-theory and KK-theory for groups which act properly and isometrically on Hilbert space*, Invent. Math. **144**, 23–74, (2001).

[29] N. Higson, V. Lafforgue and G. Skandalis. *Counterexamples to the Baum-Connes conjecture*. Geom. Funct. Anal. **12** (2002), no. 2, 330–354.

[30] S. Hurder, D. Olesen, I. Raeburn, J. Rosenberg. *The Connes spectrum for actions of abelian groups on continuous-trace algebras*. Ergodic Theory Dynam. Systems **6** (1986), 541–560.

[31] G. Kasparov. *The operator K-functor and extensions of C*-algebras*. Izv. Akad. Nauk SSSR Ser. Mat. **44** (1980), 571–636.

[32] G. Kasparov. *K-theory, group C*-algebras, and higher signatures (conspectus)*. In: Novikov conjectures, index theorems and rigidity, Vol. 1 (Oberwolfach, 1993), 101–146, London Math. Soc. Lecture Note Ser., 226, Cambridge Univ. Press, Cambridge, 1995.

[33] G. Kasparov. *Equivariant KK-theory and the Novikov conjecture*, Invent. Math. **91**, 147–201 (1988).

[34] A. Kumjian. *On the K-theory of the symmetrized non-commutative torus*, C. R. Math. Rep. Acad. Sci. Canada **12** (1990), 87–89.

[35] A. Kumjian, I. Raeburn, D.P. Williams. *The equivariant Brauer groups of commuting free and proper actions are isomorphic*. Proc. Amer. Math. Soc. **124** (1996), no. 3, 809–817.

[36] G.W. Mackey. *Les ensembles boréliens et les extensions des groupes*. J. Math. Pures Appl. (9) **36** (1957), 171–178.

[37] G.W. Mackey. *Unitary representations of group extensions. I*. Acta Math. **99** (1958) 265–311.

[38] V. Mathai. *K-theory of twisted group C^*-algebras and positive scalar curvature.* Tel Aviv Topology Conference: Rothenberg Festschrift, Contemp. Math. **231** (1998), 203–225.

[39] V. Mathai. *The Novikov conjecture for low degree cohomology classes.* Geom. Dedicata **99** (2003), 1–15.

[40] M. Marcolli, V. Mathai. *Twisted index theory on good orbifolds. I. Noncommutative Bloch theory.* Commun. Contemp. Math. **1** (1999), 553–587.

[41] M. Marcolli, V. Mathai. *Twisted index theory on good orbifolds. II. Fractional quantum numbers.* Comm. Math. Phys. **217** (2001), 55–87.

[42] C.C. Moore. *Extensions and low dimensional cohomology theory of locally compact groups. I.* Trans. Amer. Math. Soc. **113** (1964), 40–63.

[43] C.C. Moore. *Extensions and low dimensional cohomology theory of locally compact groups. II.* Trans. Amer. Math. Soc. **113** (1964), 64–86.

[44] C.C. Moore. *Group extensions and cohomology for locally compact groups. III.* Trans. Amer. Math. Soc. **221** (1976), 1–33.

[45] J.A. Packer and I. Raeburn. *Twisted crossed products of C^*-algebras.* Math. Proc. Cambridge Philos. Soc. **106** (1989), no. 2, 293–311.

[46] R.S. Palais. *On the existence of slices for actions of non-compact Lie groups.* Ann. of Math. (2) **73** (1961), 295–323.

[47] A. Polishchuk. *Holomorphic bundles on 2-dimensional noncommutative toric orbifolds*, Noncommutative geometry and number theory, 341–359, Aspects Math., E37, Vieweg, Wiesbaden, 2006.

[48] I. Raeburn, D.P. Williams. *Pull-backs of C^*-algebras and crossed products by certain diagonal actions.* Trans. Amer. Math. Soc. **287** (1985), no. 2, 755–777.

[49] I. Raeburn, D.P. Williams. Morita equivalence and continuous-trace C^*-algebras. Mathematical Surveys and Monographs, 60. American Mathematical Society, Providence, RI, 1998.

[50] I.E. Segal and R.A. Kunze. Integrals and Operators. 2nd ed. Grundlehren der Mathematischen Wissenschaften 228. Springer Verlag. Berlin, Heidelberg, New York 1978.

[51] S.G. Walters. *Chern characters of Fourier modules*, Canad. J. Math. **52** (2000), 633–672.

[52] S.G. Walters. *K-theory of non-commutative spheres arising from the Fourier automorphism*, Canadian J. Math. **53** (2001), 631–672.

[53] S.G. Walters. *The AF structure of non commutative toroidal $\mathbf{Z}/4\mathbf{Z}$ orbifolds*, J. reine angew. Math. **568** (2004), 139–196.

Siegfried Echterhoff
Westfälische Wilhelms-Universität Münster
Mathematisches Institut
Einsteinstr. 62 D-48149 Münster
Germany
e-mail: `echters@math.uni-muenster.de`

C*-algebras and Elliptic Theory II
Trends in Mathematics, 87–101
© 2008 Birkhäuser Verlag Basel/Switzerland

Twisted Burnside Theorem for Two-step Torsion-free Nilpotent Groups

Alexander Fel'shtyn, Fedor Indukaev and Evgenij Troitsky

Abstract. It is proved that the Reidemeister number of any automorphism of any finitely generated torsion-free two-step nilpotent group coincides with the number of fixed points of the corresponding homeomorphism of the finite-dimensional part of the dual space (of equivalence classes of unitary representations) provided that at least one of these numbers is finite. An important example of the discrete Heisenberg group is studied in detail.

Mathematics Subject Classification (2000). 20C; 20E45; 22D10; 22D25; 22D30; 37C25; 43A30; 46L.

Keywords. Reidemeister number, twisted conjugacy classes, Burnside theorem, two-step nilpotent group, Heisenberg group.

Contents

The second and third authors are partially supported by RFFI Grant 05-01-00923 and Grant "Universities of Russia".

1. Introduction

Definition 1.1. Let G be a countable discrete group and $\phi : G \to G$ an endomorphism. Two elements $x, x' \in G$ are said to be ϕ-*conjugate* or *twisted conjugate* iff there exists a $g \in G$ such that

$$x' = gx\phi(g^{-1}).$$

We write $\{x\}_\phi$ for the ϕ-*conjugacy* or *twisted conjugacy* class of the element $x \in G$. The number of ϕ-conjugacy classes is called the *Reidemeister number* of the endomorphism ϕ and is denoted by $R(\phi)$. If ϕ is the identity map, then the ϕ-conjugacy classes are the ordinary conjugacy classes in G.

If G is a finite group, then the classical Burnside theorem (see, e.g., [13, p. 140]) says that the number of classes of irreducible representations is equal to the number of conjugacy classes of elements of G. Let $\widehat{G}$ be the *unitary dual* of G, i.e., the set of equivalence classes of unitary irreducible representations of G.

Therefore, by Burnside's theorem, if ϕ is the identity automorphism of any finite group G, then $R(\phi) = \# \operatorname{Fix}(\widehat{\phi})$.

At present, one of the main achievements in the field is the following result.

Theorem 1.2 ([6]). *Let G be a finitely generated discrete group of type* I, *ϕ an endomorphism of G, $R(\phi)$ the number of ϕ-conjugacy classes, and $S(\phi) = \# \operatorname{Fix}(\widehat{\phi})$ the number of the $\widehat{\phi}$-invariant equivalence classes of irreducible unitary representations. If one of the numbers $R(\phi)$ and $S(\phi)$ is finite, then it is equal to the other.*

The research is motivated not only by a natural desire to extend the classical Burnside theorem to the case of infinite groups and twisted conjugacy classes but also by dynamical applications. Namely, a natural identification of the Reidemeister number with the number of fixed points has very interesting consequences in Dynamics and Number Theory (see [6]).

On the other hand, one can introduce the number $R_*(\phi)$ of "Reidemeister classes related to twisted invariant functions on G in the Fourier–Stieltjes algebra $B(G)$" or, more precisely, the dimension of the space of twisted invariant functions on G that can be extended to bounded functionals on the group C^*-algebra $C^*(G)$. Let $S_*(\phi)$ be the sum of codimensions of subspaces of the form $L_I \subset C^*(G)/I$, where L_I is spanned by elements $a - L_g a L_{\phi(g^{-1})}$ and I ranges over the Glimm spectrum of G, i.e., over the complete regularization of $\widehat{G}$. We refer to $S_*(\phi)$ as the number of generalized fixed points of $\widehat{\phi}$ on the Glimm spectrum of G.

Theorem 1.3 (weak twisted Burnside theorem, [15]). *The number $R_*(\phi)$ is equal to the number $S_*(\phi)$ of generalized fixed points of $\widehat{\phi}$ on the Glimm spectrum of G if one of the numbers $R_*(\phi)$ and $S_*(\phi)$ is finite.*

This result enables one to obtain a strong form of the twisted Burnside theorem $R(\phi) = S(\phi)$ in a number of cases.

The interest to the twisted conjugacy relations originates, in particular, from the Nielsen–Reidemeister fixed point theory (see, e.g., [12, 4]), from Selberg theory (see, e.g., [14, 1]), and from Algebraic Geometry (see, e.g., [10]).

The congruences give necessary conditions for the realization problem for Reidemeister numbers in topological dynamics.

Note that, as it is known, the Reidemeister number of an endomorphism of a finitely generated Abelian group is finite iff 1 is not in the spectrum of the restriction of this endomorphism to the free part of the group (see, e.g., [12]), and the Reidemeister number is infinite for any automorphism of a nonelementary Gromov hyperbolic group [7].

The main results of the present paper are as follows.

- The Reidemeister number of any automorphism of any finitely generated torsion-free two-step nilpotent group coincides with the number of fixed points of the corresponding homeomorphism of the finite-dimensional part of the unitary dual if one of the numbers is finite.
- For the discrete Heisenberg group and any even number $2N$, an automorphism ϕ with $R(\phi) = 2N$ is constructed.
- For $N = 1$, the related fixed points are found explicitly.

Acknowledgment

The present research is a part of the joint research programm of A. Fel'shtyn and E. Troitsky in Max-Planck-Institut für Mathematik (MPI) in Bonn. We would like to thank the MPI for its kind support and hospitality during the completion of the most part of this work.

The authors are grateful to V. Balantsev, M. B. Bekka, R. Hill, V. Manuilov, A. Mishchenko, A. Shtern, L. Vainerman, and A. Vershik for helpful discussions, and to the referee for kind suggestions.

The results of Sections 2 and 3 were obtained by A. Fel'shtyn and E. Troitsky, the results of Sections 4 and 5 were obtained by E. Troitsky (they had appeared in [5] in more detail and for a much more general case), and the results of Sections 6 and 7 were obtained by F. Indukaev.

2. Preliminary considerations

Lemma 2.1. *Let G be Abelian. The twisted conjugacy class H of e is a subgroup, and the other are cosets of the form gH.*

Proof. The first statement follows from the equalities
$$h\phi(h^{-1})g\phi(g^{-1}) = gh\phi((gh)^{-1}, \quad (h\phi(h^{-1}))^{-1} = \phi(h)h^{-1} = h^{-1}\phi(h).$$

For the second statement, let $a \sim b$, i.e., $b = ha\phi(h^{-1})$. In this case,
$$gb = gha\phi(h^{-1}) = h(ga)\phi(h^{-1}), \qquad gb \sim ga. \qquad \square$$

Lemma 2.2 ([4, 12]). *An automorphism $\phi : \mathbb{Z}^k \to \mathbb{Z}^k$ with $R(\phi) < \infty$ has a unique fixed point, namely, the identity element.*

Denote by $\tau_g : G \to G$ the automorphism $\tau_g(\widetilde{g}) = g\widetilde{g}\phi(g^{-1})$, $g \in G$ and preserve this notation for the restriction of t_g to any normal subgroup.

Lemma 2.3. $\{g\}_\phi k = \{g\,k\}_{\tau_{k^{-1}}\circ\phi}.$

Proof. Let $g' = f\,g\,\phi(f^{-1})$ be ϕ-conjugate to g. Then

$$g'\,k = f\,g\,\phi(f^{-1})\,k = f\,g\,k\,k^{-1}\,\phi(f^{-1})\,k = f\,(g\,k)\,(\tau_{k^{-1}} \circ \phi)(f^{-1}).$$

Conversely, if g' is $\tau_{k^{-1}} \circ \phi$-conjugate to g, then

$$g'\,k^{-1} = f\,g\,(\tau_{k^{-1}} \circ \phi)(f^{-1})k^{-1} = f\,g\,k^{-1}\,\phi(f^{-1}).$$

Hence, a shift takes ϕ-conjugacy classes onto classes related to another automorphism. $\square$

3. Extensions and Reidemeister classes

Consider a group extension respecting a homomorphism ϕ:

$$
\begin{array}{ccccccccc}
0 & \longrightarrow & H & \overset{i}{\longrightarrow} & G & \overset{p}{\longrightarrow} & G/H & \longrightarrow & 0 \\
& & \phi' \downarrow & & \phi \downarrow & & \overline{\phi} \downarrow & & \\
0 & \longrightarrow & H & \overset{i}{\longrightarrow} & G & \overset{p}{\longrightarrow} & G/H & \longrightarrow & 0,
\end{array}
\tag{1}
$$

where H is a normal subgroup of G. The following argument has partial intersection with [8, 9].

First of all, note that the Reidemeister classes of ϕ in G are taken epimorphically onto classes of $\overline{\phi}$ in G/H. Indeed,

$$p(\widetilde{g})p(g)\overline{\phi}(p(\widetilde{g}^{-1})) = p(\widetilde{g}g\phi(\widetilde{g}^{-1})).
\tag{2}$$

Let $R(\phi) < \infty$. Then the previous remark implies $R(\overline{\phi}) < \infty$. Consider a class $K = \{h\}_{\tau_g\phi'}$, where $\tau_g(h) := ghg^{-1}$, $g \in G$, $h \in H$. The corresponding equivalence relation is

$$h \sim \widetilde{h}hg\phi'(\widetilde{h}^{-1})g^{-1}.
\tag{3}$$

Since H is normal, the automorphism $\tau_g : H \to H$ is well defined. Denote by K the image iK as well. By (3), the shifted set Kg is a subset of Hg characterized by

$$hg \sim \widetilde{h}(hg)\phi'(\widetilde{h}^{-1}).
\tag{4}$$

Hence, Kg is a subset of $\{hg\}_\phi \cap Hg$, and the partition $Hg = \cup(\{h\}_{\tau_g\phi'})g$ is a subpartition of $Hg = \cup(Hg \cap \{hg\}_\phi)$.

Lemma 3.1 ([8]). *Let $\# \operatorname{Fix}(\tau_z \overline{\phi}) = 1$ for some representative z of some class $\{y\}_{\overline{\phi}}$. Let $\{z_\alpha\}$ be the full collection of representatives of this kind, and let g_α be some elements of G such that $p(g_\alpha) = z_\alpha$. If $R(\phi) < \infty$, then*

$$R(\phi) = \sum_\alpha R(\tau_{g_\alpha} \phi').$$

4. Torsion-free two-step nilpotent groups

A torsion-free finitely generated two-step nilpotent group G is an extension of the form (1) with $H \cong \mathbb{Z}^m$ and $G/H \cong \mathbb{Z}^k$, where H is the center of G, and hence the extension respects any homomorphism ϕ.

Theorem 4.1. *Let ϕ be an automorphism of a torsion-free finitely generated two-step nilpotent group G and let $R(\phi) < \infty$. Then all ϕ-class functions are coefficients of finite-dimensional representations of G.*

Proof. By Lemma 2.2 and Lemma 3.1, one has $R(\phi') < \infty$. Taking a quotient by the subgroup $H_1 = \{e\}_{\phi'}$ (see Lemma 2.1), one obtains a quotient group $G_1 = G/H_1$ with a bijection of ϕ-conjugacy classes under the projection. This means that it suffices to prove the statement for G_1. For G_1, we have the following ϕ-invariant extension:

$$H/H_1 = A \to G_1 \to \mathbb{Z}^k,$$

where the Abelian group A is finite.

By Lemma 2.3, the number of different sets among the shifts of a ϕ-conjugacy class is less than or equal to $R(\phi) \times M$, where M is the number of pairwise distinct automorphisms $\tau_g : G_1 \to G_1$. Since G_1 is finitely generated, A is finite, and G_1/A is Abelian, it follows that $M < \infty$. Hence, the common stabilizer of classes has a finite index. Let S be a characteristic subgroup of finite index inside this stabilizer. Then the characteristic function of ϕ-conjugacy class is the inverse image (under the projection) of a function on the finite group G_1/S. This function is a coefficient of a finite-dimensional representation $\rho : G_1/S \to \operatorname{End} V$. Therefore, the characteristic function of the original set is the corresponding coefficient of the representation

$$G \to G/H = G_1 \to G_1/S \xrightarrow{\rho} \operatorname{End} V.$$

$\square$

5. Twisted Burnside theorem

Definition 5.1. Denote by $\widehat{G}_f$ the subset of the unitary dual $\widehat{G}$ related to finite-dimensional representations.

Theorem 5.2. *Let G be a torsion-free finitely generated two-step nilpotent group and let ϕ be its automorphism. Denote by $S_f(\phi)$ the number of fixed points of $\widehat{\phi}_f$ on $\widehat{G}_f$. Then $R(\phi) = S_f(\phi)$ if one of these two numbers is finite.*

Proof. Let us start from the following observation. Let Σ be the universal compact group associated with G and $\alpha : G \to \Sigma$ the canonical morphism (see, e.g., [3, Sect. 16.1]). Then $\widehat{G}_f = \widehat{\Sigma}$ [3, 16.1.3]. The coefficients of (finite-dimensional) nonequivalent irreducible representations of Σ are linear independent as functions on Σ by the Peter–Weyl theorem. Hence, the corresponding functions on G are linearly independent.

It is sufficient to verify the following three statements:

1) If $R(\phi) < \infty$, than each ϕ-class function is a finite linear combination of twisted-invariant functionals that are coefficients of points of Fix $\widehat{\phi}_f$.
2) If $\rho \in \mathrm{Fix}\,\widehat{\phi}_f$, there exists one and only one (up to scaling) twisted invariant functional on $\rho(C^*(G))$ ($\rho(C^*(G))$ is a finite-dimensional full matrix algebra).
3) For different ρ, the corresponding ϕ-class functions are linearly independent. This follows from the remark at the beginning of the proof.

Note that Theorem 4.1 implies, in particular, that the ϕ-central functions (for ϕ with $R(\phi) < \infty$) are functionals not only on $L^1(G)$ but also on $C^*(G)$, i.e., they belong to the Fourier–Stieltjes algebra $B(G)$.

Assertion 1) follows from Theorem 4.1. Indeed, the twisted action takes any functional related to some representation to another functional related to the same representation. Since these functionals are linearly independent, this component of the linear combination has to be twisted-invariant. For any $\rho \in \widehat{G}_f$, any functional has the form $a \mapsto \mathrm{Tr}(ba)$ for some fixed b. The twisted invariance implies the twisted invariance of b (evident details can be found in [6, Sect. 3]). Hence, b intertwines ρ and $\rho \circ \phi$ and $\rho \in \mathrm{Fix}(\widehat{\phi}_f)$. The uniqueness of an intertwining operator (up to scaling) implies 2). $\square$

6. An infinite series of automorphisms of the discrete Heisenberg group

In the present section we describe the automorphisms of the discrete Heisenberg group which is two-step nilpotent. After this, for any $N \in \mathbb{N}$, we shall present an automorphism (of the group) whose Reidemeister number is equal to $2N$.

Definition 6.1. The *discrete Heisenberg group*, which we denote by H, is defined as the following semidirect product of $\mathbb{Z}^2$ by $\mathbb{Z}$:

$$H = \mathbb{Z}^2 \rtimes \mathbb{Z}, \qquad \widetilde{\alpha} : \mathbb{Z} \to Aut(\mathbb{Z}^2);$$

$$s \mapsto \alpha^s, \quad \alpha = (*, *)\begin{pmatrix} 1 & 1 \\ 0 & 1 \end{pmatrix}.$$

Thus, the group consists of triples of integers $((m, k), s)$ with the following multiplication law:

$$((m, k), s) * ((m', k'), s') = ((m, k) + \alpha^s(m', k')) = ((m + m', k + k' + sm'), s + s');$$

In particular,

$$((m, k), 0) * ((0, 0), s) = ((m, k), s);$$

The inverse element can be found by the formula:

$$((m, k), s)^{-1} = ((-m, sm - k), -s).$$

The group H can also be regarded as the group of integral 3×3-matrices of the form

$$\begin{pmatrix} 1 & s & k \\ 0 & 1 & m \\ 0 & 0 & 1 \end{pmatrix}$$

with respect to the matrix multiplication.

The group H has 3 generators,

$$a = ((1,0),0); \quad b = ((0,1),0); \quad c = ((0,0),1),$$

and the relations are

$$[a, b] = e; \quad [b, c] = e; \quad [c, a] = b, \tag{5}$$

where $[\cdot, \cdot]$ stands for the commutator and $e = ((0,0),0)$ for the group identity.

One can immediately see that

$$((m, k), s) = a^m b^k c^s, \tag{6}$$

$$c^\alpha a^\beta = a^\beta c^\alpha b^{\alpha\beta}. \tag{7}$$

It is also clear that the commutator subgroup of H is the infinite cyclic subgroup generated by $b = ((0,1),0)$. Hence, H is two-step nilpotent.

Let ϕ be an automorphism of H. Any automorphism takes the commutator subgroup isomorphically onto itself. Hence, $\phi(b) = b^\alpha, \alpha = \pm 1$. Let ϕ act on the generators as follows:

$$\begin{cases} \phi(a) = \phi((1,0),0) = ((p,q),r) = a^p b^q c^r; \\ \phi(b) = \phi((0,1),0) = ((0,\alpha),0) = b^\alpha, \quad \alpha = \pm 1; \\ \phi(c) = \phi((0,0),1) = ((u,v),w) = a^u b^v c^w; \end{cases} \tag{8}$$

Then

$$\phi((m, k), s) = \phi(a^m b^k c^s) = (\phi(a))^m (\phi(b))^k (\phi(c))^s = (a^p b^q c^r)^m \cdot (b^\alpha)^k \cdot (a^u b^v c^w)^s$$

for an arbitrary element $((m, k), s) \in H$. Consider the first factor,

$$(a^p b^q c^r)^m = \underbrace{(a^p b^q c^r)(a^p b^q c^r) \cdots (a^p b^q c^r)}_{m}.$$

Recall that b commutes with any element of H. Represent the last product in the form $a^\alpha b^\beta c^\gamma$ by moving all factors a^p to the left. The transposition of a^p with b^q gives nothing new. However, if a^p is transposed with c^r, then we must add a factor b^{pr} according to (7). This transposition occurs $1 + 2 + \ldots + (m - 1) = \frac{m(m-1)}{2}$ times when moving the factors of the form a^p to the left. Thus,

$$(a^p b^q c^r)^m = a^{mp} b^{mq + pr\frac{m(m-1)}{2}} c^{mr}.$$

Analogously,

$$(a^u b^v c^w)^s = a^{su} b^{sv + uw\frac{s(s-1)}{2}} c^{sw}.$$

Hence,

$$\phi((m,k),s) = a^{mp} \cdot b^{mq+pr\frac{m(m-1)}{2}} \cdot c^{mr} \cdot b^{\alpha k} \cdot a^{su} \cdot b^{sv+uw\frac{s(s-1)}{2}} \cdot c^{sw}$$

$$= a^{mp+su} b^{mq+sv+\alpha k+pr\frac{m(m-1)}{2}+uw\frac{s(s-1)}{2}+mrsu} c^{mr+sw}.$$

That is, the automorphisms of the discrete Heisenberg group must satisfy the condition

$$\phi\colon ((m,k),s)$$

$$\mapsto ((mp+su, mq+sv+\alpha k+pr\frac{m(m-1)}{2}+uw\frac{s(s-1)}{2}+mrsu), mr+sw),$$

$$p,q,r,u,v,w \in \mathbb{Z}, \alpha = \pm 1. \quad (9)$$

Lemma 6.2. *A mapping of the form* (9) *is an automorphism of H iff $wp - ru = \alpha$.*

Proof. Let us prove that the condition $wp - ru = \alpha$ is necessary and sufficient for ϕ of the form (9) to be an endomorphism.

To do this, it suffices to show that $\phi(a)$, $\phi(b)$, and $\phi(c)$ satisfy the same relations as a, b, c.

The relations $[\phi(a), \phi(b)] = e$ and $[\phi(c), \phi(b)] = e$ obviously hold by (8). It remains to verify the following relation:

$$\phi(ca) = \phi(bac)$$

$$\Leftrightarrow \quad ((u,v),w)((p,q),r) = ((0,\alpha),0)((p,q),r)((u,v),w)$$

$$\Leftrightarrow \quad ((u+p,v+q+wp),w+r) = ((p+u,q+v+\alpha+ru),r+w)$$

$$\Leftrightarrow \quad v+q+wp = q+v+\alpha+ru \quad \Leftrightarrow \quad wp-ru = \alpha.$$

Thus, the condition $wp - ru = \alpha$ is necessary and sufficient for any ϕ of the form (9) to preserve the group operation.

Let us now show that the same condition is sufficient for the bijectivity of ϕ. Indeed, consider an equation $\phi((m,k),s) = (x_1, x_2, x_3)$ for unknowns $((m,k),s)$. It can be represented by the set of equations

$$\begin{cases} mp+su = x_1; \\ mr+sw = x_3; \\ \alpha k + F(m,r) = x_2. \end{cases}$$

Since $wp - ru = \alpha = \pm 1$, this set of equations has a unique solution for all x_1, x_2, x_3. Hence, ϕ is a bijection, and we are done. $\square$

Lemma 6.3. *For each $N \in \mathbb{N}$, there is an automorphism ϕ of H such that $R(\phi) = 2N$.*

Proof. An arbitrary automorphism ϕ of H is given by formula (9) with parameters $p,q,r,u,v,w \in \mathbb{Z}$ such that $wp - ru = \alpha = \pm 1$. Let N be a positive integer. Our aim is to construct an automorphism ϕ with $R(\phi) = 2N$. Take the following values of parameters:

$$q = v = 0, p = 1 - N, u = N, r = 1, w = -1.$$

Then $\alpha = wp - ru = -1$, and one has

$$\phi((m,k),s) =$$
$$\left(\left(m(1-N) + sN, -k + \frac{m(m-1)}{2}(1-N) - \frac{s(s-1)}{2}N + msN\right), m - s\right).$$

Write

$$Q_0(m,s) = \frac{m(m-1)}{2}(1-N) + \frac{s(s-1)}{2}N + msN.$$

Then

$$\phi((m,k),s) = \left((m(1-N) + sN, -k + Q_0(m,s)), m - s\right).$$

Let us describe the class of ϕ-conjugacy of an arbitrary element $h = ((m,k),s)$ of the group H. Let $g = ((g_1, g_2), g_3)$. Then $g^{-1} = ((-g_1, g_1 g_3 - g_2), -g_3)$ and

$$\phi(g^{-1}) = \left((-g_1(1-N) - g_3 N, g_2 - g_1 g_3 + Q_0(-g_1, -g_3)), g_3 - g_1\right).$$

Writing $-g_1 g_3 + Q_0(-g_1, -g_3) =: Q_1(g_1, g_3)$, we obtain

$$g h \phi(g^{-1}) = ((g_1, g_2), g_3) \cdot ((m,k),s) \cdot ((-g_1(1-N) - g_3 N, g_2 + Q_1(g_1, g_3)), g_3 - g_1)$$
$$= ((g_1 + m, g_2 + k + g_3 m), g_3 + s) \cdot ((-g_1(1-N) - g_3 N, g_2 + Q_1(g_1, g_3)), g_3 - g_1)$$
$$= ((g_1 + m - g_1 + g_1 N - g_3 N, 2g_2 + k + Q_2(g_1, g_3, s, m)), s + 2g_3 - g_1)$$
$$= ((m + (g_1 - g_3)N, k + 2g_2 + Q_2(g_1, g_3, s, m)), s + 2g_3 - g_1),$$

where $Q_2(g_1, g_3, s, m) = g_3 m + Q_1(g_1, g_3) - g_3 g_1(1-N) - sg_1(1-N) - g_3^2 N - sg_3 N$. Thus, the class of ϕ-conjugacy of $h = ((m,k),s)$ is given by:

$$\{h\}_\phi = \{g h \phi(g^{-1}) \mid h \in H\}$$
$$= \{((m + (g_1 - g_3)N, k + 2g_2 + Q_2(g_1, g_3, s, m)), s + 2g_3 - g_1) \mid g_1, g_2, g_3 \in \mathbb{Z}\}.$$

One can make the following change of variables:

$$\begin{cases} g_1 - g_3 =: f_1; \\ g_2 =: f_2; \\ 2g_3 - g_1 =: f_3. \end{cases}$$

This change is invertible because its determinant is equal to -1, and one has

$$\{((m,k),s)\}_\phi = \{((m + f_1 N, k + 2f_2 + Q_3(f_1, f_3, s, m)), s + f_3) \mid f_1, f_2, f_3 \in \mathbb{Z}\}. \tag{10}$$

Hence, if the m-components of two elements do not coincide modulo N, then they are not equivalent (i.e., not ϕ-conjugate). Let

$$H_r = \{((m,k),s) \mid m \equiv r \mod N\}; \quad r = \overline{0, \ldots, N-1}.$$

Then elements of H_i are not equivalent to element of H_j if $i \neq j$. It is also obvious that the sets H_r form a partition of H.

Let us choose an r and consider equivalence classes of the elements $((r,0),0)$ and $((r,1),0)$.

$$\{((r,0),0)\}_\phi = \{((r+f_1N, 2f_2 + Q_3(f_1,f_3,r,0)), f_3) \mid f_1, f_2, f_3 \in \mathbb{Z}\};$$
$$\{((r,1),0)\}_\phi = \{((r+f_1N, 2f_2 + 1 + Q_3(f_1,f_3,r,0)), f_3) \mid f_1, f_2, f_3 \in \mathbb{Z}\}. \quad (11)$$

Let us take an arbitrary element $((m_0,k_0), s_0)$, $\quad m_0 \equiv r \mod N$ of H_r and show that it belongs to one of these two classes.

Indeed, one can choose in (11) some f_1 and f_3 such that $r + f_1 N = m_0$, $f_3 = s_0$. Thus, one fixes the value of $Q_3(f_1, f_3, r, 0)$. According to the parity of this value, it is possible to choose f_2 in such a way that either $2f_2 + Q_3(f_1,f_3,s,m) = k_0$ or $2f_2 + 1 + Q_3(f_1,f_3,s,m) = k_0$. We have $((m_0,k_0),s_0) \sim ((r,0),0)$ in the first case and $((m_0,k_0),s_0) \sim ((r,1),0)$ in the other case.

Thus, the group H is divided into N subsets H_r such that elements of different subsets are nonequivalent and each H_r is subdivided into exactly two classes of equivalence. Hence, $R(\phi) = 2N$. $\qquad \square$

7. Fixed representations determined explicitly

In this section we describe finite-dimensional irreducible unitary representations of the discrete Heisenberg group by using the notion of induced representation. After this, we find the fixed representations for a special automorphism of the Reidemeister number 2.

The dual object for $\mathbb{Z}^2$ is the torus $\mathbb{T}^2$. A pair $\chi = (\xi, \eta) \in \mathbb{T}^2$ corresponds to the character $(m,k) \mapsto e^{2\pi i(m\xi + k\eta)}$. The torus is a right G-space for the action

$$\chi h(m,k) = \chi(h * ((m,k),0) * h^{-1}).$$

The action of $((m,k),s)$ is defined by the formula

$$(\xi, \eta) \mapsto (\xi, \eta)\begin{pmatrix} 1 & 0 \\ s & 1 \end{pmatrix} = (\xi + s\eta, \eta).$$

We need the following facts.

Theorem 7.1 (Glimm). [13, Sect. 9.1] *The following properties are equivalent for any complete separable metric H-space X:*

1. *Each H-orbit in X is locally closed.*
2. *The quotient space X/H is a T_0-space.*
3. *There exists a countable family of H-invariant Borel subsets in X separating any two H-orbits.*
4. *Each H-ergodic Borel measure on X is supported by one of these orbits.*

Theorem 7.2 (Mackey). [13, p. 197] *If N is a closed commutative normal subgroup of a locally compact group H and an action of H on $\widehat{N}$ satisfies the properties of the Glimm theorem, then any irreducible representation ρ of H has the form $\mathrm{Ind}(H, Y, \beta)$, where Y is the stabilizer of some point $\chi \in \widehat{N}$ and the restriction of β on N is a scalar one and is a multiple of the character χ.*

Conversely, [2, Theorem 5, p. 509] implies that, if β is an irreducible representation of Y whose restriction to N is a multiple of some character, then $\mathrm{Ind}(H, Y, \beta)$ is irreducible.

Theorem 7.3. [11, Theorem 3.2, Chapter II] *Let G be a locally compact group with countable base and let G act transitively on a locally compact Hausdorff space Γ. Let γ be an arbitrary point of Γ with the stabilizer H. Then H is closed, and the map $gH \mapsto g\gamma$ is a homeomorphism of G/H onto Γ.*

Hence, for these G and Γ, each orbit O is homeomorphic to the quotient space of G by the stabilizer of any point of this orbit.

Let us describe the finite-dimensional irreducible representations of H. According to Mackey's theorem and to what has been said above, all these representations are induced from the stabilizer of some point $\chi = (\xi, \eta) \in \mathbb{T}^2$ such that its orbit $\widehat{O}_\chi$ is finite. Let this orbit consist of p points for some χ. The stabilizer of χ is

$$Y = \{((m, k), ps) \mid m, k, s \in \mathbb{Z}\} = \mathbb{Z}^2 \times (p\mathbb{Z}).$$

By [2, Lemma 3, p. 508], the multiplication by the character of χ is a bijection between the collections of irreducible unitary representations $p\mathbb{Z}$ and $Y = \mathbb{Z}^2 \times (p\mathbb{Z})$.

Hence, any p-dimensional irreducible unitary representation of H ($P < \infty$) can be obtained as a result of the following procedure: 1) choose a point $\chi = (\xi, \eta) \in \mathbb{T}^2$ with the orbit of cardinality p; 2) choose an irreducible representation α of the subgroup $p\mathbb{Z} = \{((0, 0), ps) \mid s \in \mathbb{Z}\}$; 3) multiply this representation by the character of χ and obtain a representation β of the corresponding subgroup Y; 4) form the representation ρ of H induced by β.

Choose an arbitrary $\alpha \in [0, 1)$ and, consequently, an irreducible representation π of the subgroup $(p\mathbb{Z})$:

$$\pi((0, 0), ps) = e^{2\pi i s\alpha}.$$

The multiplication by the character $\chi = (\xi, \eta)$ defines the following representation β of the subgroup Y:

$$\beta((m, k), ps) = \chi(m, k)e^{2\pi i s\alpha} = e^{2\pi i(m\xi + k\eta + s\alpha)};$$

Now we must construct the representation ρ of H induced by β. To this end, recall the realization of induced representations in the space of L^2-functions on $X = Y \backslash H$ (see, e.g., [13, pp. 188–190]) for a discrete group.

Let H be a discrete group, Y its subgroup, β a unitary representation of Y in a Hilbert space V, $X = Y \backslash H$ the corresponding right homogeneous space. Let us choose a map $s \colon X \to G$ such that $s(Hg) \in Hg$. Then the induced representation ρ in the space $L^2(X, V)$ is defined by the following formula:

$$[\rho(h)f](x) = A(h, x)f(xh),$$

where the operator-valued function $A(h, x)$ is defined by

$$A(h, x) = \beta(y),$$

where $y \in Y$ is given by the rule

$$s(x)h = ys(xh).$$

In our case, β is 1-dimensional and $A(h, x)$ is a complex-valued function. One has $Y((m,k),s) = Y((0,0), s \bmod p)$. Hence,

$$Y \backslash H = \{Y((0,0),0),\ Y((0,0),1),\ \dots,\ Y((0,0),p-1)\} =: \{x_0, x_1, \dots, x_{p-1}\}.$$

Choose $s : X \to G$ in such a way that

$$s \colon Y((m,k),s) \longmapsto ((0,0), s \bmod p).$$

Now we must find an $y \in Y$ (for given $x \in X = Y \backslash H$ and $h \in H$) in accordance with $s(x)h = ys(xh)$. Let $h = ((m,k),s)$, $x = Y((0,0),j)$. Then $s(x) = ((0,0),j)$, and

$$s(x)h = ((0,0),j)((m,k),s) = ((m,k+jm), s+j);$$

$$s(xh) = s(Y((0,0),j)((m,k),s))$$
$$= s(Y(m,k+jm), s+j) = ((0,0),(s+j) \bmod p).$$

Let the desired y be $y = ((y_k, y_m), y_s)$. Then

$$ys(xh) = ((y_k, y_m), y_s)((0,0),(s+j) \bmod p) = ((y_{m,k}), y_s + (s+j) \bmod p).$$

Hence,

$$y_m = m, \quad y_k = k + jm, \quad y_s = s + j - (s+j) \bmod p.$$

Write $[l]_p := l - l \bmod p$. In this case, $y = ((m, k+jm), [s+j]_p)$ and

$$A(h,x) = \beta(y) = \beta((m, k+jm), [s+j]_p) = e^{2\pi i (m\xi + (k+jm)\eta + \frac{[s+j]_p}{p})}$$
$$= e^{2\pi i (m\xi + (k+jm)\eta + [\frac{s+j}{p}])},$$

where $[r]$ is the entire part of r. Finally, the induced representation ρ in $L^2(X) = L^2(\{x_0, x_1, \dots, x_{p-1}\})$ is given by

$$[\rho(h)f](x) = A(h,x)f(xh) = e^{2\pi i (m\xi + (k+jm)\eta + [\frac{s+j}{p}])}f(xh);$$

$$[\rho((m,k),s)f](x_j) = e^{2\pi i (m\xi + (k+jm)\eta + [\frac{s+j}{p}])}f(x_{(j+s) \bmod p}); \quad j = \overline{0, p-1}.$$

Let us choose the base $\epsilon_0, \epsilon_1, \dots, \epsilon_{p-1}$ in $L^2(X) = L^2(\{x_0, x_1, \dots, x_{p-1}\})$, where ϵ_j is the indicator of a point $x_j \in X$. With respect to this base, our representation is defined by

$$\rho((m,k),s) \colon \epsilon_j \mapsto e^{2\pi i (m\xi + (k+jm)\eta + [\frac{s+j}{p}]\alpha)}\epsilon_{(j-s) \bmod p}; \quad j = \overline{0, p-1}. \qquad (12)$$

Thus, all finite-dimensional irreducible unitary representations of H are of the form (12) for some $\xi, \eta, \alpha \in [0,1)$, and the orbit of $\chi = (\xi, \eta) \in \mathbb{T}^2$ consists of p points. The action of $((m,k),s)$ on $\mathbb{T}^2$ is given by

$$(\xi, \eta) \mapsto (\xi + s\eta, \eta).$$

Hence, the orbit of this action has cardinality p iff η is a reduced fraction with denominator p.

Let us find the character χ_ρ of the representation (12). The matrix of the element $\rho((m,k),s)$ is diagonal for $s \equiv 0 \mod p$ and has zeros on the diagonal for other s. Hence, for s non divisible by p one has $\chi_\rho((m,k),s) = 0$. So,

$$\chi_\rho((m,k),s) = \delta^0_{s \bmod p} \sum_{j=0}^{p-1} \exp\left(2\pi i\left(m\xi + k\eta + jm\eta + \left[\frac{s+j}{p}\right]\alpha\right)\right).$$

For s divisible by p and $j \in \overline{0,p-1}$ one has $[\frac{s+j}{p}] = \frac{s}{p}$. Let us transform the expression:

$$\chi_\rho((m,k),s) = \delta^0_{s \bmod p} \exp\left(2\pi i\left(m\xi + k\eta + \frac{s}{p}\alpha\right)\right) \sum_{j=0}^{p-1} e^{2\pi i m\eta j}.$$

To calculate the above sum of p elements in terms of a geometric progression, let us observe that $e^{2\pi i m\eta} = 1$ iff $m\eta \in \mathbb{Z}$, i.e., $m \equiv 0 \mod p$. Hence,

$$\sum_{j=0}^{p-1} e^{2\pi i m\eta j} = \begin{cases} p, & \text{if } m \equiv 0 \mod p, \\ \frac{\exp(2\pi i m\eta p)-1}{\exp(2\pi i m\eta)-1} & \text{if } m \not\equiv 0 \mod p. \end{cases}$$

However, $\eta p \in \mathbb{Z}$. Hence, $\exp(2\pi i m\eta p) - 1 = 0$. As a result, we obtain

$$\chi_\rho((m,k),s) = \begin{cases} p \cdot e^{2\pi i(m\xi + k\eta + \frac{s}{p}\alpha)}, & \text{if } s \equiv 0 \mod p \text{ and } m \equiv 0 \mod p, \\ 0, & \text{otherwise.} \end{cases}$$

Consider the following automorphism ϕ of G:

$$\phi((m,k),s) = \left(\left(s+m, -k + \frac{m(m-1)}{2} + sm\right), m\right).$$

After an identification of the representation in the present section with that in the previous one, we can readily see that $R(\phi) = 2$. Let us find implicitly the character of the representation $\rho\phi$.,

One has:

$$\rho\phi((m,k),s)\epsilon_j = \rho((s+m, -k + \frac{m(m-1)}{2} + sm), m)\epsilon_j$$

$$= \exp 2\pi i\left((s+m)\xi + \left(-k + \frac{m(m-1)}{2} + sm + js + jm\right)\eta + \left[\frac{m+j}{p}\right]\alpha\right)$$

$$\times \epsilon_{(j-m) \bmod p}.$$

As in the above calculations, we can see that $\chi_{\rho\phi}$ vanishes at the elements with $m \not\equiv 0 \mod p$, and then we can use the fact that $[\frac{m+j}{p}] = \frac{m}{p}$ if $m \equiv 0 \mod p$ and $j \in \overline{1,p}$,

$$\chi_{\rho\phi}((m,k),s) =$$

$$\delta^0_{m \bmod p} \exp 2\pi i\left((s+m)\xi + \left(-k + \frac{m(m-1)}{2} + sm\right)\eta + \frac{m}{p}\alpha\right) \sum_{j=0}^{p-1} e^{2\pi i(s+m)\eta j},$$

and

$$\sum_{j=0}^{p-1} e^{2\pi i(s+m)\eta j} = \begin{cases} p, & \text{if } (s+m)\eta \in \mathbb{Z}, \text{ i.e., if } (s+m) \equiv 0 \bmod p, \\ 0, & \text{if } (s+m)\eta \notin \mathbb{Z} \end{cases}$$

Hence,

$$\chi_{\rho\phi}((m,k),s) = \begin{cases} p \cdot e^{2\pi i((s+m)\xi + (-k + \frac{m(m-1)}{2} + sm)\eta + \frac{m}{p}\alpha)}, & \text{if } s \text{ and } m \equiv 0 \bmod p, \\ 0, & \text{otherwise.} \end{cases}$$

Let us find the finite-dimensional fixed points of $\widehat{\phi}$. One of them is the trivial 1-dimensional representation.

To find the other fixed point, we write out the coincidence condition for the characters:

$$e^{2\pi i((s+m)\xi + (-k + \frac{m(m-1)}{2} + sm)\eta + \frac{m}{p}\alpha)} = e^{2\pi i(m\xi + k\eta + \frac{s}{p}\alpha)} \text{ for } s, m \equiv 0 \bmod p.$$

Let $p = 2$. Then $\eta = \frac{1}{2}$, and therefore

$$e^{2\pi i((s+m)\xi + \frac{1}{2}(-k + \frac{m(m-1)}{2} + sm) + \frac{m}{2}\alpha)} = e^{2\pi i(m\xi + \frac{1}{2}k + \frac{s}{2}\alpha)} \text{ for even } s, m.$$

Setting $s =: 2t, m =: 2q$, we obtain:

$$e^{2\pi i((2t+2q)\xi + \frac{1}{2}(-k + q(2q-1) + 4tq) + q\alpha)} = e^{2\pi i(2q\xi + \frac{1}{2}k + t\alpha)} \text{ for any } s, m \in \mathbb{Z}.$$

This is equivalent to

$$\left((2t+2q)\xi + \frac{1}{2}(-k + q(2q-1) + 4tq) + q\alpha\right) - \left(2q\xi + \frac{1}{2}k + t\alpha\right) \in \mathbb{Z} \text{ for any } s, m \in \mathbb{Z}.$$

After reducing and cancelling entire summands $q^2, 2tq, -k$, we obtain

$$2t\xi - \frac{q}{2} + q\alpha - t\alpha \in \mathbb{Z} \text{ for any } s, m \in \mathbb{Z}$$

$$\Leftrightarrow t(2\xi - \alpha) + q(\alpha - \frac{1}{2}) \in \mathbb{Z}, \forall s, m \in \mathbb{Z}.$$

This relation is evidently satisfied for $\alpha = \frac{1}{2}$ and $\xi = \frac{1}{4}$.

Thus, the fixed class representation is two-dimensional. It is defined by (12) with $\alpha = \frac{1}{2}, \xi = \frac{1}{4}$, and $\eta = \frac{1}{2}$, i.e.,

$$\rho_2((m,k),s) \colon \epsilon_j \mapsto e^{2\pi i(\frac{m}{4} + \frac{k+jm}{2} + \frac{1}{2}[\frac{s+j}{2}])} \epsilon_{(j-s) \bmod 2}; \quad j = 0, 1.$$

References

[1] J. Arthur and L. Clozel, *Simple algebras, base change, and the advanced theory of the trace formula*, Princeton University Press, Princeton, NJ, 1989. MR **90m**:22041

[2] A.O. Barut and R. Rączka, *Theory of group representations and applications*, second ed., World Scientific Publishing Co., Singapore, 1986. MR **88c**:22013

[3] J. Dixmier, *C*-algebras*, North-Holland, Amsterdam, 1982.

[4] A. Fel'shtyn, *Dynamical zeta functions, Nielsen theory and Reidemeister torsion*, Mem. Amer. Math. Soc. **147** (2000), no. 699, xii+146. MR **2001a**:37031

[5] A. Fel'shtyn and E. Troitsky, *Twisted Burnside theorem*, Preprint 46, Max-Planck-Institut für Mathematik, 2005, math.GR/0606179, accepted in *Crelle's journal*.

[6] _____, *A twisted Burnside theorem for countable groups and Reidemeister numbers*, Noncommutative Geometry and Number Theory (C. Consani and M. Marcolli, eds.), Vieweg, Braunschweig, 2006, pp. 141–154 (Preprint MPIM2004–65, math.RT/0606155).

[7] A.L. Fel'shtyn, *The Reidemeister number of any automorphism of a Gromov hyperbolic group is infinite*, Zap. Nauchn. Sem. S.-Peterburg. Otdel. Mat. Inst. Steklov. (POMI) **279** (2001), no. 6 (Geom. i Topol.), 229–240, 250. MR **2002e:**20081

[8] D. Gonçalves and P. Wong, *Twisted conjugacy classes in exponential growth groups*, Bull. London Math. Soc. **35** (2003), no. 2, 261–268. MR **2003j:**20054

[9] Daciberg L. Gonçalves, *The coincidence Reidemeister classes of maps on nilmanifolds*, Topol. Methods Nonlinear Anal. **12** (1998), no. 2, 375–386. MR **MR1701269 (2000d:**55004)

[10] A. Grothendieck, *Formules de Nielsen-Wecken et de Lefschetz en géométrie algébrique*, Séminaire de Géométrie Algébrique du Bois-Marie 1965-66. SGA 5, Lecture Notes in Math., vol. 569, Springer-Verlag, Berlin, 1977, pp. 407–441.

[11] Sigurður Helgason, *Differential geometry and symmetric spaces*, Pure and Applied Mathematics, Vol. XII, Academic Press, New York, 1962. MR MR0145455 (26 #2986)

[12] B. Jiang, *Lectures on Nielsen fixed point theory*, Contemp. Math., vol. 14, Amer. Math. Soc., Providence, RI, 1983.

[13] A.A. Kirillov, *Elements of the theory of representations*, Springer-Verlag, Berlin Heidelberg New York, 1976.

[14] Salahoddin Shokranian, *The Selberg-Arthur trace formula*, Lecture Notes in Mathematics, vol. 1503, Springer-Verlag, Berlin, 1992, Based on lectures by James Arthur. MR **MR1176101 (93j:**11029)

[15] E. Troitsky, *Noncommutative Riesz theorem and weak Burnside type theorem on twisted conjugacy*, Funct. Anal. Pril. **40** (2006), no. 2, 44–54, In Russian, English translation: *Funct. Anal. Appl.* **40** (2006), No. 2, 117–125 (Preprint 86 (2004), Max-Planck-Institut für Mathematik, math.OA/0606191).

Alexander Fel'shtyn
Instytut Matematyki, Uniwersytet Szczecinski, ul. Wielkopolska 15, 70-451 Szczecin, Poland
e-mail: felshtyn@mpim-bonn.mpg.de

Fedor Indukaev
Dept. of Mech. and Math., Moscow State University, 119992 GSP-2 Moscow, Russia
e-mail: indukaev@mail.ru

Evgenij Troitsky
Dept. of Mech. and Math., Moscow State University, 119992 GSP-2 Moscow, Russia
e-mail: troitsky@mech.math.msu.su
URL: http://mech.math.msu.su/~troitsky

C^*-algebras and Elliptic Theory II
Trends in Mathematics, 103–121

Ihara Zeta Functions for Periodic Simple Graphs

Daniele Guido, Tommaso Isola and Michel L. Lapidus

Abstract. The definition and main properties of the Ihara zeta function for graphs are reviewed, focusing mainly on the case of periodic simple graphs. Moreover, we give a new proof of the associated determinant formula, based on the treatment developed by Stark and Terras for finite graphs.

Mathematics Subject Classification (2000). 05C25; 05C38; 46Lxx; 11M41.

Keywords. Periodic graphs, Ihara zeta function, analytic determinant, determinant formula, functional equations.

1. Introduction

The zeta functions associated to finite graphs by Ihara [17], Hashimoto [12, 13], Bass [2] and others, combine features of Riemann's zeta function, Artin L-functions, and Selberg's zeta function, and may be viewed as analogues of the Dedekind zeta functions of a number field. They are defined by an Euler product and have an analytic continuation to a meromorphic function satisfying a functional equation. They can be expressed as the determinant of a perturbation of the graph Laplacian and, for Ramanujan graphs, satisfy the Riemann hypothesis [26].

The first attempt in this context to study infinite graphs was made by Grigorchuk and Żuk [9], who considered graphs obtained as a suitable limit of a sequence of finite graphs. They proved that their definition does not depend on the approximating sequence in case of Cayley graphs of finitely generated residually finite groups, and, more generally, in case of graphs obtained as Schreier graphs of a pair (G, H) of a finitely generated group G and a separable subgroup H.

The first and second authors were partially supported by MIUR, GNAMPA and by the European Network "Quantum Spaces – Noncommutative Geometry" HPRN-CT-2002-00280. The third author was partially supported by the National Science Foundation, the Academic Senate of the University of California, and GNAMPA.

The definition of the zeta function was extended to (countable) periodic graphs by Clair and Mokhtari-Sharghi in [4], where the determinant formula has been proved. They deduce this result as a specialization of the treatment of group actions on trees (the so-called theory of tree lattices, as developed by Bass, Lubotzky and others, see [3]).

The purpose of this work is to give a more direct proof of that result, for the case of periodic simple graphs with a free action. We hope that our treatment, being quite elementary, could be useful for someone seeking an introduction to the subject. In a sequel to this paper [11], we shall prove that for periodic amenable graphs, the Ihara zeta function can be approximated by the zeta functions of a suitable sequence of finite graphs, thereby answering in the affirmative a question raised by Grigorchuk and Żuk in [9].

In order to provide a self-contained approach to the subject, we start by recalling the definition and some properties of the zeta function for finite graphs. Then, after having introduced some preliminary notions, we define in Section 4 the analogue of the Ihara zeta function, and show that it is a holomorphic function, while, in Section 6, we prove a corresponding determinant formula. The latter requires some care, because it involves the definition and properties of a determinant for bounded operators (acting on an infinite-dimensional Hilbert space and) belonging to a von Neumann algebra with a finite trace. This question is addressed in Section 5. In the final section, we establish several functional equations.

In closing this introduction, we note that the operator-algebraic techniques used here are introduced by the authors in [10] in order to study the Ihara zeta functions attached to a new class of infinite graphs, called self-similar fractal graphs.

2. Zeta function for finite graphs

The Ihara zeta function is defined by means of equivalence classes of prime cycles. Therefore, we need to introduce some terminology from graph theory, following [24, 26] with some modifications.

A *graph* $X = (VX, EX)$ consists of a collection VX of objects, called *vertices*, and a collection EX of objects called (oriented) *edges*, together with two maps $e \in EX \mapsto (o(e), t(e)) \in VX \times VX$ and $e \in EX \mapsto \overline{e} \in EX$, satisfying the following conditions: $\overline{\overline{e}} = e$, $o(\overline{e}) = t(e)$, $\forall e \in EX$. The vertex $o(e)$ is called the *origin* of e, while $t(e)$ is called the *terminus* of e. The couple $\{e, \overline{e}\}$ is called a *geometric edge*. A graph is called *simple* if $EX \subset \{(u, v) \in VX \times VX : u \neq v\}$, $o(u, v) = u$, $t(u, v) = v$, $\overline{(u, v)} = (v, u)$; therefore, the set of geometric edges can be identified with a set of unordered pairs of distinct vertices. Observe that in the literature what we have called graph is also called a multigraph, while a simple graph is also called a graph. We will only deal with simple graphs. The edge $e = \{u, v\}$ is said to join the vertices u, v, while u and v are said to be *adjacent*, which is denoted $u \sim v$. A *path* (of length m) in X from $v_0 \in VX$ to $v_m \in VX$, is $(v_0, \ldots, v_m)$, where $v_i \in VX$ and $v_{i+1} \sim v_i$, for $i = 0, \ldots, m - 1$. In the following, we denote by $|C|$ the length of a path C. A path is *closed* if $v_m = v_0$. A graph is *connected* if there is a path between any pair of distinct vertices.

Definition 2.1 (Proper closed Paths).
  (i) A path in X has *backtracking* if $v_{i-1} = v_{i+1}$, for some $i \in \{1, \dots, m-1\}$. A
       path with no backtracking is also called *proper*. Denote by $\mathcal{C}$ the set of proper
       closed paths.
 (ii) A proper closed path $C = (v_0, \dots, v_m = v_0)$ has a *tail* if there is $k \in$
       $\{1, \dots, [m/2] - 1\}$ s.t. $v_j = v_{m-j}$, for $j = 1, \dots, k$. Denote by $\mathcal{C}^{\text{tail}}$ the set
       of proper closed paths with tail, and by $\mathcal{C}^{\text{notail}}$ the set of proper tail-less
       closed paths, also called *reduced* closed paths. Observe that $\mathcal{C} = \mathcal{C}^{\text{tail}} \cup \mathcal{C}^{\text{notail}}$,
       $\mathcal{C}^{\text{tail}} \cap \mathcal{C}^{\text{notail}} = \emptyset$.
(iii) A reduced closed path is *primitive* if it is not obtained by going $n \geq 2$ times
       around some other closed path.

Example 2.2. Some examples of non reduced closed paths are shown in Figure 1.

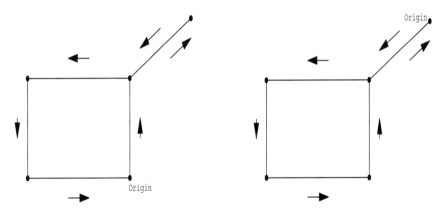

FIGURE 1. Backtracking path   Path with tail   .

We also need an equivalence relation for closed paths

Definition 2.3 (Cycles). We say that two closed paths $C = (v_0, \dots, v_m = v_0)$,
$D = (w_0, \dots, w_m = w_0)$ are *equivalent*, and write $C \sim_o D$, if there is k s.t.
$w_j = v_{j+k}$, for all j, where the addition is taken mod m, that is, the origin of D is
shifted k steps w.r.t. the origin of C. The equivalence class of C is denoted $[C]_o$.
An equivalence class is also called a *cycle*. Therefore, a closed path is just a cycle
with a specified origin.

Denote by $\mathcal{R}$ the set of reduced cycles, and by $\mathcal{P} \subset \mathcal{R}$ the subset of primitive
reduced cycles, also called *prime* cycles.

Then Ihara [17] defined the zeta function of a finite graph, that is, a graph
$X = (VX, EX)$ with VX and EX finite sets, as

Definition 2.4 (Zeta function).

$$Z_X(u) := \prod_{C \in \mathcal{P}} (1 - u^{|C|})^{-1}, \qquad u \in \mathbb{C}.$$

Ihara also proved the main result of this theory, though in the particular case of regular graphs; subsequently, through the efforts of Sunada [29], Hashimoto [12, 13] and Bass [2], that result was proved in full generality. Nowadays, there exist many different proofs of Theorem 2.5, *e.g.*, [26, 7, 18]. To state it, we need to introduce some more notation. Let us denote by $A = [A(v, w)]$, $v, w \in VX$, the adjacency matrix of X, that is,

$$A(v, w) = \begin{cases} 1 & \{v, w\} \in EX \\ 0 & \text{otherwise.} \end{cases}$$

Let $Q := \text{diag}(\deg(v_1) - 1, \deg(v_2) - 1, \ldots)$, where $\deg(v)$ is the number of vertices adjacent to v, and $\Delta(u) := I - Au + Qu^2$, $u \in \mathbb{C}$, a deformation of the usual Laplacian on the graph, which is $\Delta(1) = (Q + I) - A$. Then, with $d := \max_{v \in VX} \deg(v)$, and $\chi(X) = |VX| - |EX|$, the Euler characteristic of X, we get

Theorem 2.5 (Determinant formula). [17, 29, 12, 13, 2]

$$\frac{1}{Z_X(u)} = (1 - u^2)^{-\chi(X)} \det(\Delta(u)), \quad \text{for } |u| < \frac{1}{d-1}.$$

Example 2.6. We can compute the zeta function of the example shown in Figure 2 by using the determinant formula. We obtain $Z_X(u)^{-1} = (1 - u^2)^2(1 - u)(1 - 2u)(1 + u + 2u^2)^3$.

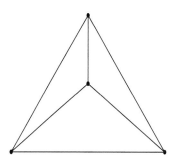

FIGURE 2. A graph

The zeta function has been used to establish some properties of the graphs. For example

Theorem 2.7. [13, 14, 2, 23, 18] *Let X be a finite graph, $r = |EX| - |VX| + 1$ the rank of the fundamental group $\pi_1(X, x_0)$. Then r is the order of the pole of $Z_X(u)$*

at $u = 1$. If $r > 1$

$$\lim_{u \to 1^-} Z_X(u)(1-u)^r = -\frac{1}{2^r(r-1)\kappa_X},$$

where κ_X is the number of spanning trees in X.

Theorem 2.8. [15, 16] *Let X be a finite graph, R_X be the radius of the greatest circle of convergence of Z_X. Denote by π_n the number of prime cycles which have length n. If $g.c.d.\{|C| : C \in \mathcal{P}\} = 1$, then*

$$\pi_n \sim \frac{R_X^{-n}}{n}, \quad n \to \infty.$$

Theorem 2.9. [17, 19, 26] *Let X be a finite graph which is $(q+1)$-regular, i.e., $\deg(v) = q + 1$ for all $v \in VX$. Then the following are equivalent*

(i) (RH) $\begin{cases} Z_X(q^{-s})^{-1} = 0 \\ \Re s \in (0,1) \end{cases} \implies \Re s = \frac{1}{2}$.

(ii) *X is a Ramanujan graph, i.e., $\lambda \in \sigma(A), |\lambda| < q + 1 \implies |\lambda| \le 2\sqrt{q}$.*

More results on the Ihara zeta function are contained in [25, 27, 28] and in various papers by Mizuno and Sato. In closing this section, we mention a generalization of the Ihara zeta function recently introduced by Bartholdi [1] and studied by Mizuno and Sato (see [20] and references therein).

3. Periodic simple graphs

Let $X = (VX, EX)$ be a simple graph, which we assume to be (countable and) with bounded degree, *i.e.*, the degree of the vertices is uniformly bounded. Let Γ be a countable discrete subgroup of automorphisms of X, which acts freely on X (*i.e.*, any $\gamma \in \Gamma$, $\gamma \ne id$ doesn't have fixed points), and with finite quotient $B := X/\Gamma$. Denote by $\mathcal{F} \subset VX$ a set of representatives for VX/Γ, the vertices of the quotient graph B. Let us define a unitary representation of Γ on $\ell^2(VX)$ by $(\lambda(\gamma)f)(x) := f(\gamma^{-1}x)$, for $\gamma \in \Gamma$, $f \in \ell^2(VX)$, $x \in V(X)$. Then the von Neumann algebra $\mathcal{N}(X, \Gamma) := \{\lambda(\gamma) : \gamma \in \Gamma\}'$ of bounded operators on $\ell^2(VX)$ commuting with the action of Γ inherits a trace given by $Tr_\Gamma(T) = \sum_{x \in \mathcal{F}} T(x,x)$, for $T \in \mathcal{N}(X, \Gamma)$.

Let us denote by A the adjacency matrix of X. Then (by [21], [22]) $\|A\| \le d := \sup_{v \in VX} \deg(v) < \infty$, and it is easy to see that $A \in \mathcal{N}(X, \Gamma)$.

For any $m \in \mathbb{N}$, let us denote by $A_m(x,y)$ the number of proper paths in X, of length m, with initial vertex x and terminal vertex y, for $x, y \in VX$. Then $A_1 = A$. Let $A_0 := I$ and $Q := \operatorname{diag}(\deg(v_1) - 1, \deg(v_2) - 1, \ldots)$. Then

Lemma 3.1.
(i) $A_2 = A^2 - Q - I \in \mathcal{N}(X, \Gamma)$,
(ii) *for $m \ge 3$, $A_m = A_{m-1}A - A_{m-2}Q \in \mathcal{N}(X, \Gamma)$,*
(iii) *let $\alpha := \frac{d + \sqrt{d^2 + 4d}}{2}$; then $\|A_m\| \le \alpha^m$, for $m \ge 0$.*

Proof. (i) if $x = y$ then $A_2(x, x) = 0$ because there are no proper closed paths of length 2 starting at x, whereas $A^2(x, x) = \deg(x) = (Q + I)(x, x)$, so that $A_2(x, x) = A^2(x, x) - (Q + I)(x, x)$. If $x \neq y$, then $A^2(x, y)$ is the number of paths of length 2 (necessarily proper) from x to y, so $A_2(x, y) = A^2(x, y) = A^2(x, y) - (Q + I)(x, y)$.

(ii) for $x, y \in VX$, the sum $\sum_{z \in VX} A_{m-1}(x, z) A(z, y)$ counts the proper paths of length m from x to y, plus additional paths formed of a proper path of length $m - 2$ from x to y followed by a path of length 2 from y to z and back; since the path from x to y and then to z is a proper path of length $m - 1$ (one of those counted by $A_{m-1}(x, z)$), z can only be one of the $\deg(y) - 1 = Q(y, y)$ vertices adjacent to y, the last one being on the proper path from x to y. Therefore $\sum_{z \in VX} A_{m-1}(x, z) A(z, y) = A_m(x, y) + A_{m-2}(x, y) Q(y, y)$, and the statement follows.

(iii) We have $\|A_1\| = \|A\| \leq d$, $\|A_2\| \leq d^2 + d$, and $\|A_m\| \leq d(\|A_{m-1}\| + \|A_{m-2}\|)$, from which the claim follows by induction. $\qquad\square$

Denote by $\mathcal{C}_m$ the subset of $\mathcal{C}$ consisting of the proper closed paths of length m, and attach a similar meaning to $\mathcal{C}_m^{tail}$, $\mathcal{C}_m^{notail}$, $\mathcal{R}_m$ and $\mathcal{P}_m$.

Lemma 3.2. *Denote by* $t_m := \sum_{x \in \mathcal{F}} |\{C \in \mathcal{C}_m^{tail} : C \text{ starts at } x\}|$, *where* $|\cdot|$ *denotes the cardinality of a set. Then*

(i) $t_1 = t_2 = 0$, *and, for* $m \geq 3$, $t_m = Tr_\Gamma((Q - I) A_{m-2}) + t_{m-2}$,

(ii) $t_m = Tr_\Gamma \left((Q - I) \sum_{j=1}^{[\frac{m-1}{2}]} A_{m-2j} \right)$.

Proof. (i) Indeed, we have

$$t_m = \sum_{x \in \mathcal{F}} |\{C \in \mathcal{C}_m^{tail} : C \text{ starts at } x\}|$$

$$= \sum_{x \in \mathcal{F}} \sum_{y \sim x} |\{C \in \mathcal{C}_m^{tail} : C \text{ starts at } x \text{ goes to } y \text{ at first step}\}|$$

$$= \sum_{y \in \mathcal{F}} \sum_{x \sim y} |\{C \in \mathcal{C}_m^{tail} : C \text{ starts at } x \text{ goes to } y \text{ at first step}\}|,$$

where the last equality follows from the fact that the cardinalities above are Γ-invariant, and we can choose $\gamma \in \Gamma$ for which the second vertex y of γC is in $\mathcal{F}$. A path C in the last set goes from x to y, then over a closed path D of length $m - 2$, and then back to x. There are two kinds of closed paths D at y: those with tails and those without. If D has no tail, then there are $Q(y, y) + 1$ possibilities for x to be adjacent to y, but x cannot be on D (otherwise, C would have backtracking), which leaves $Q(y, y) - 1$ possibilities. If D has a tail, x cannot be on D (otherwise,

C would have backtracking), which leaves $Q(y, y)$ possibilities. Therefore, we get

$$\sum_{x \sim y} |\{C \in \mathcal{C}_m^{\text{tail}} : C \text{ starts at } x \text{ goes to } y \text{ at first step}\}|$$

$$= (Q(y, y) - 1) \cdot |\{D \in \mathcal{C}_{m-2}^{\text{notail}} : D \text{ starts at } y\}|$$
$$+ Q(y, y) \cdot |\{D \in \mathcal{C}_{m-2}^{\text{tail}} : D \text{ starts at } y\}|$$
$$= (Q(y, y) - 1) \cdot |\{D \in \mathcal{C}_{m-2} : D \text{ starts at } y\}|$$
$$+ |\{D \in \mathcal{C}_{m-2}^{\text{tail}} : D \text{ starts at } y\}|,$$

so that

$$t_m = \sum_{y \in \mathcal{F}} (Q(y, y) - 1) \cdot |\{D \in \mathcal{C}_{m-2} : D \text{ starts at } y\}|$$

$$+ \sum_{y \in \mathcal{F}} |\{D \in \mathcal{C}_{m-2}^{\text{tail}} : D \text{ starts at } y\}|$$

$$= \sum_{y \in \mathcal{F}} (Q(y, y) - 1) A_{m-2}(y, y) + t_{m-2}$$

$$= Tr_{\Gamma}((Q - I) A_{m-2}) + t_{m-2}.$$

(ii) Follows from (i). $\qquad\square$

We need to introduce an equivalence relation between reduced cycles.

Definition 3.3 (Equivalence relation between reduced cycles). Given C, $D \in \mathcal{R}$, we say that C and D are Γ-*equivalent*, and write $C \sim_{\Gamma} D$, if there is an isomorphism $\gamma \in \Gamma$ s.t. $D = \gamma(C)$. We denote by $[\mathcal{R}]_{\Gamma}$ the set of Γ-equivalence classes of reduced cycles, and analogously for the subset $\mathcal{P}$.

For the purposes of the next result, for any closed path $D = (v_0, \ldots, v_m = v_0)$, we also denote v_j by $v_j(D)$.

Let us now assume that C is a prime cycle of length m. Then the *stabilizer* of C in Γ is the subgroup $\Gamma_C = \{\gamma \in \Gamma : \gamma(C) = C\}$ or, equivalently, $\gamma \in \Gamma_C$ if there exists $p(\gamma) \in \mathbb{Z}_m$ s.t., for any choice of the origin of C, $v_j(\gamma C) = v_{j-p}(C)$, for any j. Let us observe that $p(\gamma)$ is a group homomorphism from Γ_C to $\mathbb{Z}_m$, which is injective because Γ acts freely. As a consequence, $|\Gamma_C|$ divides m.

Definition 3.4. Let $C \in \mathcal{P}$ and define $\nu(C) := \dfrac{|C|}{|\Gamma_C|}$. If $C = D^k \in \mathcal{R}$, where $D \in \mathcal{P}$, define $\nu(C) = \nu(D)$. Observe that $\nu(C)$ only depends on $[C]_{\Gamma} \in [\mathcal{R}]_{\Gamma}$.

Lemma 3.5. *Let us set* $N_m := \sum_{[C]_{\Gamma} \in [\mathcal{R}_m]_{\Gamma}} \nu(C)$. *Then*
  (i) $N_m = Tr_{\Gamma}(A_m) - t_m$,
  (ii) $N_m \leq d(d-1)^{m-1} |\mathcal{F}|$.

Proof. (i) Let us assume that $[C]_{\Gamma}$ is an equivalence class of prime cycles in $[\mathcal{P}_m]_{\Gamma}$, and consider the set U of all primitive closed paths with the origin in $\mathcal{F}$ and representing $[C]_{\Gamma}$. If C is such a representative, any other representative can be obtained in this way: choose $k \in \mathbb{Z}_m$, let $\gamma(k)$ be the (unique) element in Γ for which $\gamma(k)v_k(C) \in \mathcal{F}$, and define C_k as

$$v_j(C_k) = \gamma(k)v_{j+k}(C), \; j \in \mathbb{Z}_m.$$

If we want to count the elements of U, we should know how many of the elements C_k above coincide with C. For this to happen, γ should clearly be in the stabilizer of the cycle $[C]_o$. Conversely, for any $\gamma \in \Gamma_C$, there exists $p = p(\gamma) \in \mathbb{Z}_m$ such that $\gamma v_j(C) = v_{j-p}(C)$, therefore $\gamma = \gamma(p)$. As a consequence, $v_j(C_{p(\gamma)}) = \gamma(p)v_{j+p}(C) = v_j(C)$, so that $C_{p(\gamma)} = C$. We have proved that the cardinality of U is equal to $\nu(C)$. The proof for a non-prime cycle is analogous. Therefore,

$$
\begin{aligned}
N_m &= \sum_{[C]_\Gamma \in [\mathcal{R}_m]_\Gamma} |\{D \in \mathcal{C}_m^{notail} : [D]_o \sim_\Gamma C, v_0(D) \in \mathcal{F}\}| \\
&= |\{C \in \mathcal{C}_m^{notail}, v_0(C) \in \mathcal{F}\}| \\
&= |\{C \in \mathcal{C}_m, v_0(C) \in \mathcal{F}\}| - |\{C \in \mathcal{C}_m^{tail}, v_0(C) \in \mathcal{F}\}| \\
&= Tr_\Gamma(A_m) - t_m.
\end{aligned}
\tag{3.1}
$$

(ii) Follows from (3.1). □

4. The zeta function

In this section, we define the Ihara zeta function for a periodic graph, and prove that it is a holomorphic function.

Definition 4.1 (Zeta function). We let

$$Z_{X,\Gamma}(u) := \prod_{[C]_\Gamma \in [\mathcal{P}]_\Gamma} (1 - u^{|C|})^{-\frac{1}{|\Gamma_C|}},$$

for all $u \in \mathbb{C}$ sufficiently small so that the infinite product converges.

Lemma 4.2.
  (i) $Z(u) := \prod_{[C] \in [\mathcal{P}]_\Gamma} (1 - u^{|C|})^{-\frac{1}{|\Gamma_C|}}$, *defines a holomorphic function in* $\{u \in \mathbb{C} : |u| < \frac{1}{d-1}\}$,
  (ii) $u \frac{Z'(u)}{Z(u)} = \sum_{m=1}^{\infty} N_m u^m$, *for* $|u| < \frac{1}{d-1}$,
  (iii) $Z(u) = \exp\left(\sum_{m=1}^{\infty} \frac{N_m}{m} u^m\right)$, *for* $|u| < \frac{1}{d-1}$.

Proof. Let us observe that, for $|u| < \frac{1}{d-1}$,

$$
\begin{aligned}
\sum_{m=1}^{\infty} N_m u^m &= \sum_{[C]_\Gamma \in [\mathcal{R}]_\Gamma} \nu(C) u^{|C|} \\
&= \sum_{m=1}^{\infty} \sum_{[C]_\Gamma \in [\mathcal{P}]_\Gamma} \frac{|C|}{|\Gamma_C|} u^{|C^m|}
\end{aligned}
$$

$$= \sum_{[C]_\Gamma \in [\mathcal{P}]_\Gamma} \frac{1}{|\Gamma_C|} \sum_{m=1}^{\infty} |C| u^{|C|m}$$

$$= \sum_{[C]_\Gamma \in [\mathcal{P}]_\Gamma} \frac{1}{|\Gamma_C|} u \frac{d}{du} \sum_{m=1}^{\infty} \frac{u^{|C|m}}{m}$$

$$= - \sum_{[C]_\Gamma \in [\mathcal{P}]_\Gamma} \frac{1}{|\Gamma_C|} u \frac{d}{du} \log(1 - u^{|C|})$$

$$= u \frac{d}{du} \log Z(u),$$

where, in the last equality we used uniform convergence on compact subsets of $\left\{ u \in \mathbb{C} : |u| < \frac{1}{d-1} \right\}$. The rest of the proof is clear. $\qquad \square$

Example 4.3. Some examples of cycles with different stabilizers are shown in Figure 3. They refer to the standard lattice graph $X = \mathbb{Z}^2$ endowed with the action of the group Γ which is generated by the rotation by $\frac{\pi}{2}$ around the point P and the translations by elements $(m, n) \in \mathbb{Z}^2$ acting as $(m, n)(v_1, v_2) := (v_1 + 2m, v_2 + 2n)$, for $v = (v_1, v_2) \in VX = \mathbb{Z}^2$.

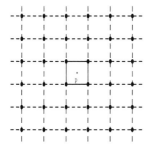 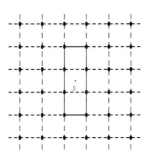

FIGURE 3. A cycle with $|\Gamma_C| = 4$.   A cycle with $|\Gamma_C| = 2$

The interested reader can find the computation of the Ihara zeta function for several periodic simple graphs in [9, 4, 5, 6].

5. An analytic determinant for von Neumann algebras with a finite trace

In this section, we define a determinant for a suitable class of not necessarily normal operators in a von Neumann algebra with a finite trace. The results obtained are used in Section 6 to prove a determinant formula for the zeta function.

In a celebrated paper [8], Fuglede and Kadison defined a positive-valued determinant for finite factors (*i.e.*, von Neumann algebras with trivial center and finite trace). Such a determinant is defined on all invertible elements and enjoys the main properties of a determinant function, but it is positive-valued. Indeed, for an invertible operator A with polar decomposition $A = UH$, where U is a unitary operator and $H := \sqrt{A^*A}$ is a positive self-adjoint operator, the Fuglede–Kadison determinant is defined by

$$\det(A) = \exp \circ \tau \circ \log H,$$

where $\log H$ may be defined via the functional calculus.

For the purposes of the present paper, we need a determinant which is an analytic function. As we shall see, this can be achieved, but corresponds to a restriction of the domain of the determinant function and implies the loss of some important properties. Let $(\mathcal{A}, \tau)$ be a von Neumann algebra endowed with a finite trace. Then, a natural way to obtain an analytic function is to define, for $A \in \mathcal{A}$, $\det_\tau(A) = \exp \circ \tau \circ \log A$, where

$$\log(A) := \frac{1}{2\pi i} \int_\Gamma \log \lambda (\lambda - A)^{-1} d\lambda,$$

and Γ is the boundary of a connected, simply connected region Ω containing the spectrum of A. Clearly, once the branch of the logarithm is chosen, the integral above does not depend on Γ, provided Γ is given as above.

Then a naïve way of defining det is to allow all elements A for which there exists an Ω as above, and a branch of the logarithm whose domain contains Ω. Indeed the following holds.

Lemma 5.1. *Let A, Ω, Γ be as above, and φ, ψ two branches of the logarithm such that both domains contain Ω. Then*

$$\exp \circ \tau \circ \varphi(A) = \exp \circ \tau \circ \psi(A).$$

Proof. The function $\varphi(\lambda) - \psi(\lambda)$ is continuous and everywhere defined on Γ. Since it takes its values in $2\pi i \mathbb{Z}$, it should be constant on Γ. Therefore

$$\exp \circ \tau \circ \varphi(A) = \exp \circ \tau \left(\frac{1}{2\pi i} \int_\Gamma 2\pi i n_0 (\lambda - A)^{-1} d\lambda \right) \exp \circ \tau \circ \psi(A)$$

$$= \exp \circ \tau \circ \psi(A). \qquad \square$$

The problem with the previous definition is its dependence on the choice of Ω. Indeed, it is easy to see that when $A = \begin{pmatrix} 1 & 0 \\ 0 & i \end{pmatrix}$ and we choose Ω containing $\{e^{i\vartheta}, \vartheta \in [0, \pi/2]\}$ and any suitable branch of the logarithm, we get $det(A) = e^{i\pi/4}$, by using the normalized trace on 2×2 matrices. On the other hand, if we choose Ω containing $\{e^{i\vartheta}, \vartheta \in [\pi/2, 2\pi]\}$ and a corresponding branch of the logarithm, we get $det(A) = e^{5i\pi/4}$. Therefore, we make the following choice.

Definition 5.2. Let $(\mathcal{A}, \tau)$ be a von Neumann algebra endowed with a finite trace, and consider the subset $\mathcal{A}_0 = \{A \in \mathcal{A} : 0 \notin \text{conv}\, \sigma(A)\}$, where $\sigma(A)$ denotes the spectrum of A. For any $A \in \mathcal{A}_0$ we set

$$\det_\tau(A) = \exp \circ \tau \circ \left(\frac{1}{2\pi i} \int_\Gamma \log \lambda (\lambda - A)^{-1} d\lambda \right),$$

where Γ is the boundary of a connected, simply connected region Ω containing $\operatorname{conv} \sigma(A)$, and log is a branch of the logarithm whose domain contains Ω.

Corollary 5.3. *The determinant function defined above is well defined and analytic on $\mathcal{A}_0$.*

We collect some properties of our determinant in the following result.

Proposition 5.4. *Let $(\mathcal{A}, \tau)$ be a von Neumann algebra endowed with a finite trace, $A \in \mathcal{A}_0$. Then*
  (i) $\det_\tau(zA) = z^{\tau(I)} \det_\tau(A)$, *for any $z \in \mathbb{C} \setminus \{0\}$,*
  (ii) *if A is normal, and $A = UH$ is its polar decomposition,*

$$\det_\tau(A) = \det_\tau(U) \det_\tau(H),$$

  (iii) *if A is positive, $\det_\tau(A) = \det(A)$, where the latter is the Fuglede-Kadison determinant.*

Proof. (i) If the half-line $\{\rho e^{i\vartheta_0} \in \mathbb{C} : \rho > 0\}$ does not intersect $\operatorname{conv} \sigma(A)$, then the half-line $\{\rho e^{i(\vartheta_0 + t)} \in \mathbb{C} : \rho > 0\}$ does not intersect $\operatorname{conv} \sigma(zA)$, where $z = re^{it}$. If log is the branch of the logarithm defined on the complement of the real negative half-line, then $\varphi(x) = i(\vartheta_0 - \pi) + \log(e^{-i(\vartheta_0 - \pi)}x)$ is suitable for defining $\det_\tau(A)$, while $\psi(x) = i(\vartheta_0 + t - \pi) + \log(e^{-i(\vartheta_0 + t - \pi)}x)$ is suitable for defining $\det_\tau(zA)$. Moreover, if Γ is the boundary of a connected, simply connected region Ω containing $\operatorname{conv} \sigma(A)$, then $z\Gamma$ is the boundary of a connected, simply connected region $z\Omega$ containing $\operatorname{conv} \sigma(zA)$. Therefore,

$$\det_\tau(zA) = \exp \circ \tau \left(\frac{1}{2\pi i} \int_{z\Gamma} \psi(\lambda)(\lambda - zA)^{-1} d\lambda \right)$$

$$= \exp \circ \tau \left(\frac{1}{2\pi i} \int_\Gamma (i(\vartheta_0 + t - \pi) + \log(e^{-i(\vartheta_0 + t - \pi)}re^{it}\mu))(\mu - A)^{-1} d\mu \right)$$

$$= \exp \circ \tau \left((\log r + it)I + \frac{1}{2\pi i} \int_\Gamma \varphi(\mu)(\mu - A)^{-1} d\mu \right)$$

$$= z^{\tau(I)} \det_\tau(A).$$

(ii) When $A = UH$ is normal, $U = \int_{[0,2\pi]} e^{i\vartheta} \, du(\vartheta)$, $H = \int_{[0,\infty)} r \, dh(r)$, then $A = \int_{[0,\infty) \times [0,2\pi]} re^{i\vartheta} \, d(h(r) \otimes u(\vartheta))$. The property $0 \notin \operatorname{conv} \sigma(A)$ is equivalent to the fact that the support of the measure $d(h(r) \otimes u(\vartheta))$ is compactly contained in some open half-plane

$$\{\rho e^{i\vartheta} : \rho > 0, \vartheta \in (\vartheta_0 - \pi/2, \vartheta_0 + \pi/2)\},$$

or, equivalently, that the support of the measure $dh(r)$ is compactly contained in $(0, \infty)$, and the support of the measure $du(\vartheta)$ is compactly contained in $(\vartheta_0 - \pi/2, \vartheta_0 + \pi/2)$. Therefore $A \in \mathcal{A}_0$ is equivalent to $U, H \in \mathcal{A}_0$. Then

$$\log A = \int_{[0,\infty) \times (\vartheta_0 - \pi/2, \vartheta_0 + \pi/2)} (\log r + i\vartheta)\, d(h(r) \otimes u(\vartheta)),$$

which implies that

$$\det_\tau(A) = \exp \circ \tau \left(\int_0^\infty \log r\, dh(r) + \int_{\vartheta_0 - \pi/2}^{\vartheta_0 + \pi/2} i\vartheta\, du(\vartheta) \right) = \det_\tau(U) \cdot \det_\tau(H).$$

(iii) Follows by the above argument. □

Remark 5.5. We note that the above-defined determinant function strongly violates the product property $\det_\tau(AB) = \det_\tau(A) \det_\tau(B)$. Indeed, the fact that $A, B \in \mathcal{A}_0$ does not imply $AB \in \mathcal{A}_0$, as is seen, e.g., by taking $A = B = \begin{pmatrix} 1 & 0 \\ 0 & i \end{pmatrix}$. Moreover, even if $A, B, AB \in \mathcal{A}_0$ and A and B commute, the product property may be violated, as is shown by choosing $A = B = \begin{pmatrix} 1 & 0 \\ 0 & e^{3i\pi/4} \end{pmatrix}$, and using the normalized trace on 2×2 matrices.

6. The determinant formula

In this section, we prove the main result in the theory of Ihara zeta functions, which says that Z is the reciprocal of a holomorphic function, which, up to a factor, is the determinant of a deformed Laplacian on the graph. We first need some technical results. Let us recall that $d := \sup_{v \in VX} \deg(v)$, and $\alpha := \frac{d + \sqrt{d^2 + 4d}}{2}$.

Lemma 6.1.

(i) $\left(\sum_{m \geq 0} A_m u^m \right)(I - Au + Qu^2) = (1 - u^2)I$, *for* $|u| < \frac{1}{\alpha}$,

(ii) $\left(\sum_{m \geq 0} \left(\sum_{k=0}^{[m/2]} A_{m-2k} \right) u^m \right)(I - Au + Qu^2) = I$, *for* $|u| < \frac{1}{\alpha}$.

Proof. (i) From Lemma 3.1 we obtain

$$\left(\sum_{m \geq 0} A_m u^m \right)(I - Au + Qu^2) = \sum_{m \geq 0} A_m u^m - \sum_{m \geq 0} \left(A_m A u^{m+1} - A_m Q u^{m+2} \right)$$

$$= \sum_{m \geq 0} A_m u^m - A_0 A u - A_1 A u^2 + A_0 Q u^2$$

$$- \sum_{m \geq 3} (A_{m-1} A - A_{m-2} Q)\, u^m$$

$$= \sum_{m \geq 0} A_m u^m - Au - A^2 u^2 + Q u^2 - \sum_{m \geq 3} A_m u^m$$

$$= I + Au + A_2 u^2 - Au - A^2 u^2 + Q u^2$$

$$= (1 - u^2)I.$$

(ii)
$$I = (1 - u^2)^{-1} \left(\sum_{m \geq 0} A_m u^m \right) (I - Au + Qu^2)$$

$$= \left(\sum_{m \geq 0} A_m u^m \right) \left(\sum_{j=0}^{\infty} u^{2j} \right) (I - Au + Qu^2)$$

$$= \left(\sum_{k \geq 0} \sum_{j=0}^{\infty} A_k u^{k+2j} \right) (I - Au + Qu^2)$$

$$= \left(\sum_{m \geq 0} \left(\sum_{j=0}^{[m/2]} A_{m-2j} \right) u^m \right) (I - Au + Qu^2). \qquad \square$$

Lemma 6.2. *Denote by* $B_m := A_m - (Q - I) \sum_{k=1}^{[m/2]} A_{m-2k} \in \mathcal{N}(X, \Gamma)$, *for* $m \geq 0$. *Then*

(i) $B_0 = I$, $B_1 = A$,

(ii) $B_m = QA_m - (Q - I) \sum_{k=0}^{[m/2]} A_{m-2k}$,

(iii) $Tr_\Gamma B_m = \begin{cases} N_m - Tr_\Gamma(Q - I) & m \text{ even} \\ N_m & m \text{ odd,} \end{cases}$

(iv) $\sum_{m \geq 1} B_m u^m = \left(Au - 2Qu^2 \right) \left(I - Au + Qu^2 \right)^{-1}$, *for* $|u| < \dfrac{1}{\alpha}$.

Proof. (i), (ii) follow from computations involving bounded operators.

(iii) It follows from Lemma 3.2 (ii) that, if m is odd,
$$Tr_\Gamma B_m = Tr_\Gamma(A_m) - t_m = N_m,$$

whereas, if m is even,
$$Tr_\Gamma B_m = Tr_\Gamma(A_m) - t_m - Tr_\Gamma((Q - I)A_0) = N_m - Tr_\Gamma(Q - I).$$

(iv) $\left(\sum_{m \geq 0} B_m u^m \right) (I - Au + Qu^2)$

$$= \left(Q \sum_{m \geq 0} A_m u^m - (Q - I) \sum_{m \geq 0} \sum_{j=0}^{[m/2]} A_{m-2j} u^m \right) (I - Au + Qu^2)$$

$$= Q(1 - u^2)I - (Q - I) \left(\sum_{m \geq 0} \sum_{j=0}^{[m/2]} A_{m-2j} u^m \right) (I - Au + Qu^2)$$

$$= (1 - u^2)Q - (Q - I) = I - u^2 Q,$$

where the second equality follows by Lemma 6.1 (i) and the third equality follows by Lemma 6.1 (ii). Since $B_0 = I$, we get

$$\left(\sum_{m\geq 1} B_m u^m\right)(I - Au + Qu^2) = I - u^2 Q - B_0(I - Au + Qu^2)$$

$$= Au - 2Qu^2. \qquad \square$$

Lemma 6.3. *Let $f : u \in B_\varepsilon \equiv \{u \in \mathbb{C} : |u| < \varepsilon\} \mapsto f(u) \in \mathcal{N}(X, \Gamma)$, be a C^1-function, $f(0) = 0$, and $\|f(u)\| < 1$, for all $u \in B_\varepsilon$. Then*

$$Tr_\Gamma\left(-\frac{d}{du}\log(I - f(u))\right) = Tr_\Gamma\left(f'(u)(I - f(u))^{-1}\right).$$

Proof. To begin with, $-\log(I - f(u)) = \sum_{n\geq 1}\frac{1}{n}f(u)^n$, converges in operator norm, uniformly on compact subsets of B_ε. Moreover,

$$\frac{d}{du}f(u)^n = \sum_{j=0}^{n-1} f(u)^j f'(u) f(u)^{n-j-1}.$$

Therefore, $-\frac{d}{du}\log(I - f(u)) = \sum_{n\geq 1}\frac{1}{n}\sum_{j=0}^{n-1} f(u)^j f'(u) f(u)^{n-j-1}$, so that

$$Tr_\Gamma\left(-\frac{d}{du}\log(I - f(u))\right) = \sum_{n\geq 1}\frac{1}{n}\sum_{j=0}^{n-1} Tr_\Gamma\left(f(u)^j f'(u) f(u)^{n-j-1}\right)$$

$$= \sum_{n\geq 1} Tr_\Gamma(f(u)^{n-1} f'(u))$$

$$= Tr_\Gamma\left(\sum_{n\geq 0} f(u)^n f'(u)\right)$$

$$= Tr_\Gamma(f'(u)(I - f(u))^{-1}),$$

where we have used the fact that Tr_Γ is norm continuous. $\qquad \square$

Corollary 6.4.

$$Tr_\Gamma\left(\sum_{m\geq 1} B_m u^m\right) = Tr_\Gamma\left(-u\frac{d}{du}\log(I - Au + Qu^2)\right), \quad |u| < \frac{1}{\alpha}.$$

Proof. It follows from Lemma 6.2 (iv) that

$$Tr_\Gamma\left(\sum_{m\geq 1} B_m u^m\right) = Tr_\Gamma((Au - 2Qu^2)(I - Au + Qu^2)^{-1})$$

$$= Tr_\Gamma\left(-u\frac{d}{du}\log(I - Au + Qu^2)\right),$$

where the last equality follows from the previous lemma applied with $f(u) := Au - Qu^2$. $\qquad \square$

Observe that for the L^2-Euler characteristic of X we have

$$\chi^{(2)}(X) := -\frac{1}{2}Tr_\Gamma(Q - I) = |V(B)| - |E(B)| = \chi(B),$$

where $\chi(B)$ is the Euler characteristic of the quotient graph $B = X/\Gamma$.

Theorem 6.5 (Determinant formula).

$$\frac{1}{Z_{X,\Gamma}(u)} = (1 - u^2)^{-\chi(B)} \det\nolimits_\Gamma(I - Au + Qu^2), \quad for \ |u| < \frac{1}{\alpha}.$$

Proof.
$$Tr_\Gamma\left(\sum_{m\geq 1} B_m u^m\right) = \sum_{m\geq 1} Tr_\Gamma(B_m)u^m$$

$$= \sum_{m\geq 1} N_m u^m - \sum_{k\geq 1} Tr_\Gamma(Q - I)u^{2k}$$

$$= \sum_{m\geq 1} N_m u^m - Tr_\Gamma(Q - I)\frac{u^2}{1 - u^2},$$

where the second equality follows by Lemma 6.2 (iii). Therefore,

$$u\frac{d}{du}\log Z_{X,\Gamma}(u) = \sum_{m\geq 1} N_m u^m$$

$$= Tr_\Gamma\left(-u\frac{d}{du}\log(I - Au + Qu^2)\right) - \frac{u}{2}\frac{d}{du}\log(1 - u^2)Tr_\Gamma(Q - I)$$

so that, dividing by u and integrating from $u = 0$ to u, we get

$$\log Z_{X,\Gamma}(u) = -Tr_\Gamma\left(\log(I - Au + Qu^2)\right) - \frac{1}{2}Tr_\Gamma(Q - I)\log(1 - u^2),$$

which implies that, for $|u| < \frac{1}{\alpha}$, we have

$$\frac{1}{Z_{X,\Gamma}(u)} = (1 - u^2)^{\frac{1}{2}Tr_\Gamma(Q-I)} \cdot \exp Tr_\Gamma \log(I - Au + Qu^2). \qquad \square$$

7. Functional equations

In this final section, we obtain several functional equations for the Ihara zeta functions of $(q + 1)$−regular graphs, *i.e.*, graphs with $\deg(v) = q + 1$, for any $v \in VX$. The various functional equations correspond to different ways of completing the zeta functions, as is done in [26] for finite graphs.

Lemma 7.1. *Let X be a $(q + 1)$-regular graph and $\Delta(u) := (1 + qu^2)I - uA$. Then*
  (i) $\chi^{(2)}(X) = \chi(B) = |V(B)|(1 - q)/2 \in \mathbb{Z}$,
  (ii) $Z_{X,\Gamma}(u) = (1 - u^2)^{\chi(B)} \det\nolimits_\Gamma(\Delta(u))^{-1}$, *for $|u| < \frac{1}{q}$,*

(iii) *by using the determinant formula in (ii), $Z_{X,\Gamma}$ can be extended to a function holomorphic at least in the open set*

$$\Omega := \mathbb{R}^2 \setminus \left(\left\{ (x,y) \in \mathbb{R}^2 : x^2 + y^2 = \frac{1}{q} \right\} \cup \left\{ (x,0) \in \mathbb{R}^2 : \frac{1}{q} \le |x| \le 1 \right\} \right).$$

See Figure 4.

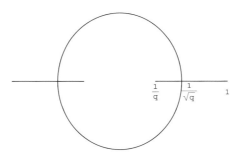

FIGURE 4. The open set Ω

(iv) $\det_{\Gamma}\left(\Delta(\frac{1}{qu}) \right) = (qu^2)^{-|VB|} \det_{\Gamma}(\Delta(u))$, *for $u \in \Omega \setminus \{0\}$.*

Proof. (i) This follows by a simple computation.

(ii) This follows from (i).

(iii) Let us observe that

$$\sigma(\Delta(u)) = \left\{ 1 + qu^2 - u\lambda : \lambda \in \sigma(A) \right\} \subset \left\{ 1 + qu^2 - u\lambda : \lambda \in [-d, d] \right\}.$$

It follows that $0 \notin \operatorname{conv} \sigma(\Delta(u))$ at least for $u \in \mathbb{C}$ such that $1 + qu^2 - u\lambda \ne 0$ for $\lambda \in [-d, d]$, that is for $u = 0$ or $\frac{1+qu^2}{u} \notin [-d, d]$, or equivalently, at least for $u \in \Omega$. The rest of the proof follows from Corollary 5.3.

    (iv) This follows by Proposition 5.4 (i) and the fact that $Tr_{\Gamma}(I_V) = |VB|$. $\qquad\square$

Proposition 7.2 (Functional equations). *Let X be $(q+1)$-regular. Then, for all $u \in \Omega$, we have*

  (i) $\Lambda_{X,\Gamma}(u) := (1-u^2)^{-\chi(B)}(1-u^2)^{|VB|/2}(1-q^2u^2)^{|VB|/2} Z_{X,\Gamma}(u) = -\Lambda_{X,\Gamma}\left(\frac{1}{qu} \right)$,

  (ii) $\xi_{X,\Gamma}(u) := (1-u^2)^{-\chi(B)}(1-u)^{|VB|}(1-qu)^{|VB|} Z_{X,\Gamma}(u) = \xi_{X,\Gamma}\left(\frac{1}{qu} \right)$,

  (iii) $\Xi_{X,\Gamma}(u) := (1-u^2)^{-\chi(B)}(1+qu^2)^{|VB|} Z_{X,\Gamma}(u) = \Xi_{X,\Gamma}\left(\frac{1}{qu} \right)$.

Proof.

(i) $\Lambda_X(u) = (1-u^2)^{|VB|/2}(1-q^2u^2)^{|VB|/2}\det_\Gamma(\Delta(u))^{-1}$

$$= u^{|VB|}\left(\frac{q^2}{q^2u^2}-1\right)^{|VB|/2}(qu)^{|VB|}\left(\frac{1}{q^2u^2}-1\right)^{|VB|/2}\frac{1}{(qu^2)^{|VB|}}\det_\Gamma\left(\Delta(\frac{1}{qu})\right)^{-1}$$

$$= -\Lambda_X\left(\frac{1}{qu}\right).$$

(ii) $\xi_X(u) = (1-u)^{|VB|}(1-qu)^{|VB|}\det_\Gamma(\Delta(u))^{-1}$

$$= u^{|VB|}\left(\frac{q}{qu}-1\right)^{|VB|}(qu)^{|VB|}\left(\frac{1}{qu}-1\right)^{|VB|}\frac{1}{(qu^2)^{|VB|}}\det_\Gamma\left(\Delta(\frac{1}{qu})\right)^{-1}$$

$$= \xi_X\left(\frac{1}{qu}\right).$$

(iii) $\Xi_X(u) = (1+qu^2)^{|VB|}\det_\Gamma(\Delta(u))^{-1}$

$$= (qu^2)^{|VB|}\left(\frac{q}{q^2u^2}+1\right)^{|VB|}\frac{1}{(qu^2)^{|VB|}}\det_\Gamma\left(\Delta(\frac{1}{qu})\right)^{-1}$$

$$= \Xi_X\left(\frac{1}{qu}\right). \qquad \square$$

Acknowledgment

The second and third named authors would like to thank respectively the University of California, Riverside, and the University of Roma "Tor Vergata" for their hospitality at different stages of the preparation of this paper.

References

[1] L. Bartholdi. *Counting paths in graphs*, Enseign. Math. **45** (1999), 83–131.

[2] H. Bass. *The Ihara-Selberg zeta function of a tree lattice*, Internat. J. Math. **3** (1992), 717–797.

[3] H. Bass, A. Lubotzky. *Tree lattices*, Progress in Math. **176**, Birkhäuser, Boston, 2001.

[4] B. Clair, S. Mokhtari-Sharghi. *Zeta functions of discrete groups acting on trees*, J. Algebra **237** (2001), 591–620.

[5] B. Clair, S. Mokhtari-Sharghi. *Convergence of zeta functions of graphs*, Proc. Amer. Math. Soc. **130** (2002), 1881–1886.

[6] B. Clair. *Zeta functions of graphs with $\mathbb{Z}$ actions*, preprint, 2006, arXiv:math.NT/0607689.

[7] D. Foata, D. Zeilberger. *A combinatorial proof of Bass's evaluations of the Ihara-Selberg zeta function for graphs*, Trans. Amer. Math. Soc. **351** (1999), 2257–2274.

[8] B. Fuglede, R.V. Kadison. *Determinant theory in finite factors*, Ann. Math. **55** (1952), 520–530.

[9] R.I. Grigorchuk, A. Żuk. *The Ihara zeta function of infinite graphs, the KNS spectral measure and integrable maps*, in: "Random Walks and Geometry", Proc. Workshop (Vienna, 2001), V.A. Kaimanovich *et al.*, eds., de Gruyter, Berlin, 2004, pp. 141–180.

[10] D. Guido, T. Isola, M.L. Lapidus. *A trace on fractal graphs and the Ihara zeta function*, to appear in Trans. Amer. Math. Soc. , arXiv:math.OA/0608060.

[11] D. Guido, T. Isola, M.L. Lapidus. *Ihara's zeta function for periodic graphs and its approximation in the amenable case*, preprint, 2006, arXiv:math.OA/0608229.

[12] K. Hashimoto, A. Hori. *Selberg-Ihara's zeta function for p-adic discrete groups*, in: "Automorphic Forms and Geometry of Arithmetic Varieties", Adv. Stud. Pure Math. **15**, Academic Press, Boston, MA, 1989, pp. 171–210.

[13] K. Hashimoto. *Zeta functions of finite graphs and representations of p-adic groups*, in: "Automorphic Forms and Geometry of Arithmetic Varieties", Adv. Stud. Pure Math. **15**, Academic Press, Boston, MA, 1989, pp. 211–280.

[14] K. Hashimoto. *On zeta and L-functions of finite graphs*, Internat. J. Math. **1** (1990), 381–396.

[15] K. Hashimoto. *Artin type L-functions and the density theorem for prime cycles on finite graphs*, Internat. J. Math. **3** (1992), 809–826.

[16] M.D. Horton, H.M. Stark, A.A. Terras. *What are zeta functions of graphs and what are they good for?*, Quantum graphs and their applications, 173–189, Contemp. Math., 415, Amer. Math. Soc., Providence, RI, 2006.

[17] Y. Ihara. *On discrete subgroups of the two by two projective linear group over $\mathfrak{p}$-adic fields*, J. Math. Soc. Japan **18** (1966), 219–235.

[18] M. Kotani, T. Sunada. *Zeta functions of finite graphs*, J. Math. Sci. Univ. Tokyo **7** (2000), 7–25.

[19] A. Lubotzky. *Discrete groups, expanding graphs and invariant measures*, Progress in Math. **125**, Birkhäuser, Basel, 1994.

[20] H. Mizuno, I. Sato. *Bartholdi zeta functions of some graphs*, Discrete Math. **206** (2006), 220–230.

[21] B. Mohar. *The spectrum of an infinite graph*, Linear Algebra Appl. **48** (1982), 245–256.

[22] B. Mohar, W. Woess. *A survey on spectra of infinite graphs*, Bull. London Math. Soc. **21** (1989), 209–234.

[23] S. Northshield. *A note on the zeta function of a graph*, J. Combin. Theory Series B **74** (1998), 408–410.

[24] J.-P. Serre. *Trees*, Springer-Verlag, New York, 1980.

[25] J.-P. Serre. *Répartition asymptotique des valeurs propres de l'opérateur de Hecke T_p*, J. Amer. Math. Soc. **10** (1997), 75–102.

[26] H.M. Stark, A.A. Terras. *Zeta functions of finite graphs and coverings*, Adv. Math. **121** (1996), 126–165.

[27] H.M. Stark, A.A. Terras. *Zeta functions of finite graphs and coverings. II*, Adv. Math. **154** (2000), 132–195.

[28] H.M. Stark, A.A. Terras. *Zeta functions of finite graphs and coverings. III*, Adv. Math. **208** (2007), 467–489.

[29] T. Sunada. *L-functions in geometry and applications*, Springer Lecture Notes in Math. **1201**, 1986, pp. 266–284.

Daniele Guido
Dipartimento di Matematica
Università di Roma "Tor Vergata"
I–00133 Roma, Italy.
e-mail: guido@mat.uniroma2.it

Tommaso Isola
Dipartimento di Matematica
Università di Roma "Tor Vergata"
I–00133 Roma, Italy.
e-mail: isola@mat.uniroma2.it

Michel L. Lapidus
Department of Mathematics
University of California, Riverside
CA 92521-0135, USA.
e-mail: lapidus@math.ucr.edu

C^*-algebras and Elliptic Theory II

Trends in Mathematics, 123–144

Adiabatic Limits and the Spectrum of the Laplacian on Foliated Manifolds

Yuri A. Kordyukov and Andrey A. Yakovlev

Abstract. We present some recent results on the behavior of the spectrum of the differential form Laplacian on a Riemannian foliated manifold when the metric on the ambient manifold is blown up in directions normal to the leaves (in the adiabatic limit).

Contents

Introduction

Let $(M, \mathcal{F})$ be a closed foliated manifold, $\dim M = n$, $\dim \mathcal{F} = p$, $p + q = n$, endowed with a Riemannian metric g. Then we have a decomposition of the tangent bundle to M into a direct sum $TM = F \oplus H$, where $F = T\mathcal{F}$ is the tangent bundle to $\mathcal{F}$ and $H = F^\perp$ is the orthogonal complement of F, and the corresponding

Supported by the Russian Foundation of Basic Research (grant no. 06-01-00208).

decomposition of the metric: $g = g_F + g_H$. Define a one-parameter family g_h of Riemannian metrics on M by

$$g_h = g_F + h^{-2}g_H, \quad h > 0. \tag{1}$$

By the adiabatic limit, we will mean the asymptotic behavior of Riemannian manifolds (M, g_h) as $h \to 0$.

In this form, the notion of the adiabatic limit was introduced by Witten [38] in the study of the global anomaly. He considered a family of Dirac operators acting along the fibers of a Riemannian fiber bundle over the circle and gave an argument relating the holonomy of the determinant line bundle of this family to the adiabatic limit of the eta invariant of the Dirac operator on the total space. Witten's result was proved rigorously in [6], [7] and [10], and extended to general Riemannian bundles in [5] and [13]. This study gave rise to the development of adiabatic limit technique for analyzing the behavior of certain spectral invariants under degeneration that has many applications in the local index theory (see, for instance, [8]).

New properties of adiabatic limits were discovered by Mazzeo and Melrose [30]. They showed that in the case of a fibration, a Taylor series analysis of so called small eigenvalues in the adiabatic limit and the corresponding eigenforms leads directly to a spectral sequence, which is isomorphic to the Leray spectral sequence. This result was used in [13], and further developed in [16], where the very general setting of any pair of complementary distributions is considered. Nevertheless, the most interesting results of [16] are only proved for foliations satisfying very restrictive conditions. The ideas from [30] and [16] were also applied in the case of the contact-adiabatic (or sub-Riemannian) limit in [18, 36].

In this paper, we will discuss extensions of the results mentioned above to the adiabatic limits on foliated manifolds. For any $h > 0$, we will consider the Laplace operator Δ_h on differential forms defined by the metric g_h. It is a self-adjoint, elliptic, differential operator with the positive, scalar principal symbol in the Hilbert space $L^2(M, \Lambda T^*M, g_h)$ of square integrable differential forms on M, endowed with the inner product induced by g_h, which has discrete spectrum. Denote by

$$0 \le \lambda_0(h) \le \lambda_1(h) \le \lambda_2(h) \le \cdots$$

the spectrum of Δ_h, taking multiplicities into account.

In Section 1, we discuss the asymptotic behavior as $h \to 0$ of the trace of $f(\Delta_h)$:

$$\operatorname{tr} f(\Delta_h) = \sum_{i=0}^{+\infty} f(\lambda_i(h)),$$

for any sufficiently nice function f, say, for $f \in S(\mathbb{R})$. The results given in this section should be viewed as a very first step in extending the adiabatic limit technique to analyze the behavior of spectral invariants to the case of foliations.

In Section 2 we study "branches" of eigenvalues $\lambda_i(h)$ that are convergent to zero as $h \to 0$ (the "small" eigenvalues) and the corresponding eigenspaces and discuss the differentiable spectral sequence of the foliation, which is a direct generalization of the Leray spectral sequence, and its Hodge theoretic description.

We will consider two basic classes of foliations – Riemannian foliations and one-dimensional foliations defined by the orbits of invariant flows on Riemannian Heisenberg manifolds.

We remark that the adiabatic limit is, up to scaling, an example of collapsing (in general, without bounded curvature) in the sense of [9]. For a discussion of the behavior of the spectrum of the differential form Laplacian on a compact Riemannian manifold under collapse, we refer, for instance, to [4, 11, 17, 22, 27] and references therein.

We are grateful to J. Álvarez López for useful discussions. We also thank the referee for suggestions to improve the paper.

1. Adiabatic limits and eigenvalue distribution

Let $(M, \mathcal{F})$ be a closed foliated manifold, endowed with a Riemannian metric g. In this section, we will discuss the asymptotic behavior of the trace of $f(\Delta_h)$ in the adiabatic limit.

1.1. Riemannian foliations

For Riemannian foliations, the problem was studied in [25]. Recall (see, for instance, [33, 34, 31, 32]) that a foliation $\mathcal{F}$ is called Riemannian, if there exists a Riemannian metric g on M such that the induced metric g_τ on the normal bundle $\tau = TM/F$ is holonomy invariant, or, equivalently, in any foliated chart $\phi : U \to I^p \times I^q$ with local coordinates (x, y), the restriction g_H of g to $H = F^\perp$ is written in the form

$$g_H = \sum_{\alpha, \beta = 1}^{q} g_{\alpha\beta}(y) \theta^\alpha \theta^\beta,$$

where $\theta^\alpha \in H^*$ is the 1-form, corresponding to the form dy^α under the isomorphism $H^* \cong T^*\mathbb{R}^q$, and $g_{\alpha\beta}(y)$ depend only on the transverse variables $y \in \mathbb{R}^q$. Such a Riemannian metric is called bundle-like.

It turns out that the adiabatic spectral limit on a Riemannian foliation can be considered as a semiclassical spectral problem for a Schrödinger operator on the leaf space $M/\mathcal{F}$, and the resulting asymptotic formula for the trace of $f(\Delta_h)$ can be written in the form of the semiclassical Weyl formula for a Schrödinger operator on a compact Riemannian manifold, if we replace the classical objects, entering to this formula by their noncommutative analogues. This observation provides a very natural interpretation of the asymptotic formula for the trace of $f(\Delta_h)$ (see Theorem 1.1 below).

First, we transfer the operators Δ_h to the fixed Hilbert space

$$L^2\Omega = L^2(M, \Lambda T^*M, g),$$

using an isomorphism Θ_h from $L^2(M, \Lambda T^* M, g_h)$ to $L^2 \Omega$ defined as follows. With respect to a bigrading on $\Lambda T^* M$ given by

$$\Lambda^k T^* M = \bigoplus_{i=0}^{k} \Lambda^{i,k-i} T^* M, \quad \Lambda^{i,j} T^* M = \Lambda^i H^* \otimes \Lambda^j F^*,$$

we have, for $u \in L^2(M, \Lambda^{i,j} T^* M, g_h)$,

$$\Theta_h u = h^i u. \tag{2}$$

The operator Δ_h in $L^2(M, \Lambda T^* M, g_h)$ corresponds under the isometry Θ_h to the operator $L_h = \Theta_h \Delta_h \Theta_h^{-1}$ in $L^2 \Omega$.

With respect to the above bigrading of $\Lambda T^* M$, the de Rham differential d can be written as

$$d = d_F + d_H + \theta,$$

where

1. $d_F = d_{0,1} : C^\infty(M, \Lambda^{i,j} T^* M) \to C^\infty(M, \Lambda^{i,j+1} T^* M)$ is the tangential de Rham differential, which is a first-order tangentially elliptic operator, independent of the choice of g;
2. $d_H = d_{1,0} : C^\infty(M, \Lambda^{i,j} T^* M) \to C^\infty(M, \Lambda^{i+1,j} T^* M)$ is the transversal de Rham differential, which is a first-order transversally elliptic operator;
3. $\theta = d_{2,-1} : C^\infty(M, \Lambda^{i,j} T^* M) \to C^\infty(M, \Lambda^{i+2,j-1} T^* M)$ is a zeroth-order differential operator.

One can show that

$$d_h = \Theta_h d \Theta_h^{-1} = d_F + h d_H + h^2 \theta,$$

and the adjoint of d_h in $L^2 \Omega$ is

$$\delta_h = \Theta_h \delta \Theta_h^{-1} = \delta_F + h \delta_H + h^2 \theta^*.$$

Therefore, one has

$$L_h = d_h \delta_h + \delta_h d_h$$
$$= \Delta_F + h^2 \Delta_H + h^4 \Delta_\theta + h K_1 + h^2 K_2 + h^3 K_3,$$

where $\Delta_F = d_F d_F^* + d_F^* d_F$ is the tangential Laplacian, $\Delta_H = d_H d_H^* + d_H^* d_H$ is the transverse Laplacian, $\Delta_\theta = \theta \theta^* + \theta^* \theta$ and $K_2 = d_F \theta^* + \theta^* d_F + \delta_F \theta + \theta \delta_F$ are of zeroth order, and $K_1 = d_F \delta_H + \delta_H d_F + \delta_F d_H + d_H \delta_F$ and $K_3 = d_H \theta^* + \theta^* d_H + \delta_H \theta + \theta \delta_H$ are first-order differential operators.

Suppose that $\mathcal{F}$ is a Riemannian foliation and g is a bundle-like metric. The key observation is that, in this case, the transverse principal symbol of the operator δ_H is holonomy invariant, and, therefore, the first order-differential operator K_1 is a *leafwise* differential operator. Using this fact, one can show that the leading term in the asymptotic expansion of the trace of $f(\Delta_h)$ or, equivalently, of the trace of $f(L_h)$ as $h \to 0$ coincides with the leading term in the asymptotic expansion of the trace of $f(\bar{L}_h)$ as $h \to 0$, where

$$\bar{L}_h = \Delta_F + h^2 \Delta_H.$$

More precisely, we have the following estimates (with some $C_1, C_2 > 0$):

$$|\operatorname{tr} f(L_h)| < C_1 h^{-q}, \quad |\operatorname{tr} f(L_h) - \operatorname{tr} f(\bar{L}_h)| < C_2 h^{1-q}, \quad 0 < h \leq 1,$$

where we recall that q denotes the codimension of $\mathcal{F}$.

We observe that the operator $\bar{L}_h$ has the form of a Schrödinger operator on the leaf space $M/\mathcal{F}$, where Δ_H plays the role of the Laplace operator, and Δ_F the role of the operator-valued potential on $M/\mathcal{F}$.

Recall that, for a Schrödinger operator H_h on a compact Riemannian manifold X, $\dim X = n$, with a matrix-valued potential $V \in C^\infty(X, \mathcal{L}(E))$, where E is a finite-dimensional Euclidean space and $V(x)^* = V(x)$:

$$H_h = -h^2 \Delta + V(x), \quad x \in X,$$

the corresponding asymptotic formula (the semiclassical Weyl formula) has the following form:

$$\operatorname{tr} f(H_h) = (2\pi)^{-n} h^{-n} \int_{T^*X} \operatorname{Tr} f(p(x, \xi))\, dx\, d\xi + o(h^{-n}), \quad h \to 0+, \quad (3)$$

where $p \in C^\infty(T^*X, \mathcal{L}(E))$ is the principal h-symbol of H_h:

$$p(x, \xi) = |\xi|^2 + V(x), \quad (x, \xi) \in T^*X.$$

Now we demonstrate how the asymptotic formula for the trace of $f(\Delta_h)$ in the adiabatic limit can be written in a similar form, using noncommutative geometry. (For the basic information on noncommutative geometry of foliations, we refer the reader to [26] and references therein.)

Let G be the holonomy groupoid of $\mathcal{F}$. Let us briefly recall its definition. Denote by $\sim_h$ the equivalence relation on the set of piecewise smooth leafwise paths $\gamma : [0, 1] \to M$, setting $\gamma_1 \sim_h \gamma_2$ if γ_1 and γ_2 have the same initial and final points and the same holonomy maps. The holonomy groupoid G is the set of $\sim_h$ equivalence classes of leafwise paths. G is equipped with the source and the range maps $s, r : G \to M$ defined by $s(\gamma) = \gamma(0)$ and $r(\gamma) = \gamma(1)$. Recall also that, for any $x \in M$, the set $G^x = \{\gamma \in G : r(\gamma) = x\}$ is the covering of the leaf L_x through the point x, associated with the holonomy group of the leaf. We will identify any $x \in M$ with the element of G given by the constant path $\gamma(t) = x, t \in [0, 1]$.

Let λ_L denote the Riemannian volume form on a leaf L given by the induced metric, and $\lambda^x, x \in M$, denote the lift of λ_{L_x} via the holonomy covering map $s : G^x \to L_x$.

Denote by $\pi : N^*\mathcal{F} \to M$ the conormal bundle to $\mathcal{F}$ and by $\mathcal{F}_N$ the linearized foliation in $N^*\mathcal{F}$ (cf., for instance, [32, 26]). Recall that, for any $\gamma \in G, s(\gamma) = x, r(\gamma) = y$, the codifferential of the corresponding holonomy map defines a linear map $dh_\gamma^* : N_y^*\mathcal{F} \to N_x^*\mathcal{F}$. Then the leaf of the foliation $\mathcal{F}_N$ through $\nu \in N^*\mathcal{F}$ is the set of all $dh_\gamma^*(\nu) \in N^*\mathcal{F}$, where $\gamma \in G, r(\gamma) = \pi(\nu)$.

The holonomy groupoid $G_{\mathcal{F}_N}$ of the linearized foliation $\mathcal{F}_N$ can be described as the set of all $(\gamma, \nu) \in G \times N^*\mathcal{F}$ such that $r(\gamma) = \pi(\nu)$. The source map $s_N : G_{\mathcal{F}_N} \to N^*\mathcal{F}$ and the range map $r_N : G_{\mathcal{F}_N} \to N^*\mathcal{F}$ are defined as $s_N(\gamma, \nu) = dh_\gamma^*(\nu)$ and $r_N(\gamma, \nu) = \nu$. We have a map $\pi_G : G_{\mathcal{F}_N} \to G$ given by $\pi_G(\gamma, \nu) = \gamma$.

Denote by $\mathcal{L}(\pi^*\Lambda T^*M)$ the vector bundle on $G_{\mathcal{F}_N}$, whose fiber at a point $(\gamma, \nu) \in G_{\mathcal{F}_N}$ is the space of linear maps

$$(\pi^*\Lambda T^*M)_{s_N(\gamma,\nu)} \to (\pi^*\Lambda T^*M)_{r_N(\gamma,\nu)}.$$

There is a standard way (due to Connes [12]) to introduce the structure of involutive algebra on the space $C_c^\infty(G_{\mathcal{F}_N}, \mathcal{L}(\pi^*\Lambda T^*M))$ of smooth, compactly supported sections of $\mathcal{L}(\pi^*\Lambda T^*M)$. For any $\nu \in N^*\mathcal{F}$, this algebra has a natural representation R_ν in the Hilbert space $L^2(G^\nu_{\mathcal{F}_N}, s_N^*(\pi^*\Lambda T^*M))$ that determines its embedding to the C^*-algebra of all bounded operators in $L^2(G^\nu_{\mathcal{F}_N}, s_N^*(\pi^*\Lambda T^*M))$. Taking the closure of the image of this embedding, we get a C^*-algebra, called the twisted foliation C^*-algebra and denoted by $C^*(N^*\mathcal{F}, \mathcal{F}_N, \pi^*\Lambda T^*M)$. The leaf space $N^*\mathcal{F}/\mathcal{F}_N$ can be informally considered as the cotangent bundle to $M/\mathcal{F}$, and the algebra $C^*(N^*\mathcal{F}, \mathcal{F}_N, \pi^*\Lambda T^*M)$ can be viewed as a noncommutative analogue of the algebra of continuous vector-valued differential forms on this singular space.

Let $g_N \in C^\infty(N^*\mathcal{F})$ be the fiberwise Riemannian metric on $N^*\mathcal{F}$ induced by the metric on M. The principal h-symbol of Δ_h is a tangentially elliptic operator in $C^\infty(N^*\mathcal{F}, \pi^*\Lambda T^*M)$ given by

$$\sigma_h(\Delta_h) = \Delta_{\mathcal{F}_N} + g_N,$$

where $\Delta_{\mathcal{F}_N}$ is the lift of the tangential Laplacian Δ_F to a tangentially elliptic (relative to $\mathcal{F}_N$) operator in $C^\infty(N^*\mathcal{F}, \pi^*\Lambda T^*M)$, and g_N denotes the multiplication operator in $C^\infty(N^*\mathcal{F}, \pi^*\Lambda T^*M)$ by the function g_N. (Observe that g_N coincides with the transversal principal symbol of Δ_H.) Consider $\sigma_h(\Delta_h)$ as a family of elliptic operators along the leaves of the foliation $\mathcal{F}_N$ and lift these operators to the holonomy coverings of the leaves. For any $\nu \in N^*\mathcal{F}$, we get a formally self-adjoint uniformly elliptic operator $\sigma_h(\Delta_h)_\nu$ in $C^\infty(G^\nu_{\mathcal{F}_N}, s_N^*(\pi^*\Lambda T^*M))$, which essentially self-adjoint in the Hilbert space $L^2(G^\nu_{\mathcal{F}_N}, s_N^*(\pi^*\Lambda T^*M))$. For any $f \in S(\mathbb{R})$, the family $\{f(\sigma_h(\Delta_h)_\nu), \nu \in N^*\mathcal{F}\}$ defines an element $f(\sigma_h(\Delta_h))$ of the C^*-algebra $C^*(N^*\mathcal{F}, \mathcal{F}_N, \pi^*\Lambda T^*M)$.

The foliation $\mathcal{F}_N$ has a natural transverse symplectic structure, which can be described as follows. Consider a foliated chart $\varkappa : U \subset M \to I^p \times I^q$ on M with coordinates $(x, y) \in I^p \times I^q$ (I is the open interval $(0, 1)$) such that the restriction of $\mathcal{F}$ to U is given by the sets $y = \text{const}$. One has the corresponding coordinate chart in T^*M with coordinates denoted by $(x, y, \xi, \eta) \in I^p \times I^q \times \mathbb{R}^p \times \mathbb{R}^q$. In these coordinates, the restriction of the conormal bundle $N^*\mathcal{F}$ to U is given by the equation $\xi = 0$. So we have a coordinate chart $\varkappa_n : U_1 \subset N^*\mathcal{F} \longrightarrow I^p \times I^q \times \mathbb{R}^q$ on $N^*\mathcal{F}$ with the coordinates $(x, y, \eta) \in I^p \times I^q \times \mathbb{R}^q$. Indeed, the coordinate chart $\varkappa_n$ is a foliated coordinate chart for $\mathcal{F}_N$, and the restriction of $\mathcal{F}_N$ to U_1 is given by the level sets $y = \text{const}, \eta = \text{const}$. The transverse symplectic structure for $\mathcal{F}_N$ is given by the transverse two-form $\sum_j dy_j \wedge d\eta_j$.

The corresponding canonical transverse Liouville measure $dy\, d\eta$ is holonomy invariant and, by noncommutative integration theory [12], defines the trace $\text{tr}_{\mathcal{F}_N}$ on the C^*-algebra $C^*(N^*\mathcal{F}, \mathcal{F}_N, \pi^*\Lambda T^*M)$. Combining the Riemannian volume forms λ_L and the transverse Liouville measure, we get a volume form $d\nu$ on $N^*\mathcal{F}$. For any $k \in C_c^\infty(G_{\mathcal{F}_N}, \mathcal{L}(\pi^*\Lambda T^*M))$, its trace is given by the formula

$$\mathrm{tr}_{\mathcal{F}_N}(k) = \int_{N^*\mathcal{F}} k(\nu)d\nu.$$

The trace $\mathrm{tr}_{\mathcal{F}_N}$ is a noncommutative analogue of the integral over the leaf space $N^*\mathcal{F}/\mathcal{F}_N$ with respect to the transverse Liouville measure. One can show that the value of this trace on $f(\sigma_h(\Delta_h))$ is finite.

Replacing in the formula (3) the integration over T^*X and the matrix trace Tr by the trace $\mathrm{tr}_{\mathcal{F}_N}$ and the principal h-symbol p by $\sigma_h(\Delta_h)$, we obtain the correct formula for $\mathrm{tr}\, f(\Delta_h)$ in the adiabatic limit.

Theorem 1.1 ([25]). *For any $f \in S(\mathbb{R})$, the asymptotic formula holds:*

$$\mathrm{tr}\, f(\Delta_h) = (2\pi)^{-q}h^{-q}\, \mathrm{tr}_{\mathcal{F}_N}\, f(\sigma_h(\Delta_h)) + o(h^{-q}), \quad h \to 0. \tag{4}$$

The formula (4) can be rewritten in terms of the spectral data of leafwise Laplace operators. We will formulate the corresponding result for the spectrum distribution function

$$N_h(\lambda) = \sharp\{i : \lambda_i(h) \leq \lambda\}.$$

Restricting the tangential Laplace operator Δ_F to the leaves of the foliation $\mathcal{F}$ and lifting the restrictions to the holonomy coverings of leaves, we get the the Laplacian Δ_x acting in $C_c^\infty(G^x, s^*\Lambda T^*M)$. Using the assumption that $\mathcal{F}$ is Riemannian, it can be checked that, for any $x \in M$, Δ_x is formally self-adjoint in $L^2(G^x, s^*\Lambda T^*M)$, that, in turn, implies its essential self-adjointness in this Hilbert space (with initial domain $C_c^\infty(G^x, s^*\Lambda T^*M)$). For each $\lambda \in \mathbb{R}$, let $E_x(\lambda)$ be the spectral projection of Δ_x, corresponding to the semi-axis $(-\infty, \lambda]$. The Schwartz kernels of the operators $E_x(\lambda)$ define a leafwise smooth section e_λ of the bundle $\mathcal{L}(\Lambda T^*M)$ over G.

We introduce the spectrum distribution function $N_{\mathcal{F}}(\lambda)$ of the operator Δ_F by the formula

$$N_{\mathcal{F}}(\lambda) = \int_M \mathrm{Tr}\, e_\lambda(x)\, dx, \quad \lambda \in \mathbb{R},$$

where dx denotes the Riemannian volume form on M. By [24], for any $\lambda \in \mathbb{R}$, the function $\mathrm{Tr}\, e_\lambda$ is a bounded measurable function on M, therefore, the spectrum distribution function $N_{\mathcal{F}}(\lambda)$ is well defined and takes finite values.

As above, one can show that the family $\{E_x(\lambda) : x \in M\}$ defines an element $E(\lambda)$ of the twisted von Neumann foliation algebra $W^*(G, \Lambda T^*M)$, the holonomy invariant transverse Riemannian volume form for $\mathcal{F}$ defines a trace $\mathrm{tr}_{\mathcal{F}}$ on $W^*(G, \Lambda T^*M)$, and the right-hand side of the last formula can be interpreted as the value of this trace on $E(\lambda)$.

Theorem 1.2 ([25]). *Let $(M, \mathcal{F})$ be a Riemannian foliation, equipped with a bundle-like Riemannian metric g. Then the asymptotic formula for $N_h(\lambda)$ has the following form:*

$$N_h(\lambda) = h^{-q} \frac{(4\pi)^{-q/2}}{\Gamma((q/2)+1)} \int_{-\infty}^{\lambda} (\lambda - \tau)^{q/2} d_\tau N_\mathcal{F}(\tau) + o(h^{-q}), \quad h \to 0.$$

1.2. A linear foliation on the 2-torus

In this section, we consider the simplest example of the situation studied in the previous section, namely, the example of a linear foliation on the 2-torus. So consider the two-dimensional torus $\mathbb{T}^2 = \mathbb{R}^2/\mathbb{Z}^2$ with the coordinates $(x, y) \in \mathbb{R}^2$, taken modulo integer translations, and the Euclidean metric g on $\mathbb{T}^2$:

$$g = dx^2 + dy^2.$$

Let $\widetilde{X}$ be the vector field on $\mathbb{R}^2$ given by

$$\widetilde{X} = \frac{\partial}{\partial x} + \alpha \frac{\partial}{\partial y},$$

where $\alpha \in \mathbb{R}$. Since $\widetilde{X}$ is translation invariant, it determines a vector field X on $\mathbb{T}^2$. The orbits of X define a one-dimensional foliation $\mathcal{F}$ on $\mathbb{T}^2$. The leaves of $\mathcal{F}$ are the images of the parallel lines $\widetilde{L}_{(x_0,y_0)} = \{(x_0 + t, y_0 + t\alpha) : t \in \mathbb{R}\}$, parameterized by $(x_0, y_0) \in \mathbb{R}^2$, under the projection $\mathbb{R}^2 \to \mathbb{T}^2$.

In the case when α is rational, all leaves of $\mathcal{F}$ are closed and are circles, and $\mathcal{F}$ is given by the fibers of a fibration of $\mathbb{T}^2$ over $\mathbb{S}^1$. In the case when α is irrational, all leaves of $\mathcal{F}$ are everywhere dense in $\mathbb{T}^2$.

The one-parameter family g_h of Riemannian metrics on $\mathbb{T}^2$ defined by (1) is given by

$$g_h = \frac{1 + h^{-2}\alpha^2}{1 + \alpha^2} dx^2 + 2\alpha \frac{1 - h^{-2}}{1 + \alpha^2} dx dy + \frac{\alpha^2 + h^{-2}}{1 + \alpha^2} dy^2.$$

The Laplace operator (on functions) defined by g_h has the form $\Delta_h = \Delta_F + h^2 \Delta_H$, where

$$\Delta_F = -\frac{1}{1 + \alpha^2}\left(\frac{\partial}{\partial x} + \alpha \frac{\partial}{\partial y}\right)^2, \quad \Delta_H = -\frac{h^2}{1 + \alpha^2}\left(-\alpha \frac{\partial}{\partial x} + \frac{\partial}{\partial y}\right)^2$$

are the tangential and the transverse Laplace operators respectively.

The operator Δ_h has a complete orthogonal system of eigenfunctions

$$u_{kl}(x, y) = e^{2\pi i(kx + ly)}, \quad (x, y) \in \mathbb{T}^2,$$

with the corresponding eigenvalues

$$\lambda_{kl}(h) = (2\pi)^2 \left(\frac{1}{1 + \alpha^2}(k + \alpha l)^2 + \frac{h^2}{1 + \alpha^2}(-\alpha k + l)^2\right), \quad (k, l) \in \mathbb{Z}^2. \quad (5)$$

The eigenvalue distribution function of Δ_h has the form

$$N_h = \#\{(k, l) \in \mathbb{Z}^2 : (2\pi)^2 \left(\frac{1}{1 + \alpha^2}(k + \alpha l)^2 + \frac{h^2}{1 + \alpha^2}(-\alpha k + l)^2\right) < \lambda\}.$$

Thus we come to the following problem of number theory:

Problem 1.3. Find the asymptotic for $h \to 0$ of the number of integer points in the ellipse

$$\{(\xi, \eta) \in \mathbb{R}^2 : (2\pi)^2 \left(\frac{1}{1+\alpha^2}(\xi + \alpha\eta)^2 + \frac{h^2}{1+\alpha^2}(-\alpha\xi + \eta)^2 \right) < \lambda \}.$$

In the case when α is rational, this problem can be easily solved by elementary methods of analysis. In the case when α is irrational, such an elementary solution seems to be unknown, and, in order to solve the problem, the connection of this problem with the spectral theory of the Laplace operator and with adiabatic limits plays an important role.

Theorem 1.4 ([40]). *The following asymptotic formula for the spectrum distribution function $N_h(\lambda)$ of the operator Δ_h for a fixed $\lambda \in \mathbb{R}$ holds:*
1 *For $\alpha \notin \mathbb{Q}$,*

$$N_h(\lambda) = \frac{1}{4\pi} h^{-1}\lambda + o(h^{-1}), \quad h \to 0. \tag{6}$$

2. *For $\alpha \in \mathbb{Q}$ of the form $\alpha = \frac{p}{q}$, where p and q are coprime,*

$$N_h(\lambda) = h^{-1} \sum_{\substack{k\in\mathbb{Z} \\ |k| < \frac{\sqrt{\lambda}}{2\pi}\sqrt{p^2+q^2}}} \frac{1}{\pi\sqrt{p^2+q^2}} \left(\lambda - \frac{4\pi^2}{p^2+q^2}k^2\right)^{1/2} + o(h^{-1}), \quad h \to 0. \tag{7}$$

Remark 1. The asymptotic formulas (6) and (7) of Theorem 1.4 look quite different. Nevertheless, it can be shown that

$$\lim_{\substack{p\to+\infty \\ q\to+\infty}} \sum_{\substack{k\in\mathbb{Z} \\ |k| < \frac{\sqrt{\lambda}}{2\pi}\sqrt{p^2+q^2}}} \frac{1}{\pi\sqrt{p^2+q^2}} \left(\lambda - \frac{4\pi^2}{p^2+q^2}k^2\right)^{1/2} = \frac{1}{4\pi}\lambda.$$

Indeed, one can write

$$\sum_k \frac{1}{\pi\sqrt{p^2+q^2}} \left(\lambda - \frac{4\pi^2}{p^2+q^2}k^2\right)^{1/2} = \sum_k g(\xi_k)\Delta\xi_k,$$

where

$$g(\xi) = \frac{1}{2\pi^2}(\lambda - \xi^2)^{1/2}, \quad \xi_k = \frac{2\pi}{\sqrt{p^2+q^2}}k.$$

This immediately implies that

$$\lim_{\substack{p\to+\infty \\ q\to+\infty}} \sum_{k\in\mathbb{Z}|k| < \frac{\sqrt{\lambda}}{2\pi}\sqrt{p^2+q^2}} \frac{1}{\pi\sqrt{p^2+q^2}} \left(\lambda - \frac{4\pi^2}{p^2+q^2}k^2\right)^{1/2} = \frac{1}{2\pi^2}\int_{-\sqrt{\lambda}}^{\sqrt{\lambda}}(\lambda - \xi^2)^{1/2}\,d\xi$$

$$= \frac{1}{4\pi}\lambda.$$

We now show how to derive the asymptotic formulae of Theorem 1.4 from Theorem 1.2 (see [40] for more details).

Case 1: $\alpha \notin \mathbb{Q}$. In this case $G = \mathbb{T}^2 \times \mathbb{R}$. The source and the range maps $s, r : G \to \mathbb{T}^2$ are defined for any $\gamma = (x, y, t) \in G$ by $s(\gamma) = (x - t, y - \alpha t)$ and $r(\gamma) = (x, y)$. For any $(x, y) \in \mathbb{T}^2$, the set $G^{(x,y)}$ coincides with the leaf $L_{(x,y)}$ through (x, y) and is diffeomorphic to $\mathbb{R}$:

$$L_{(x,y)} = \{(x - t, y - \alpha t) : t \in \mathbb{R}\}.$$

The Riemannian volume form $\lambda^{(x,y)}$ on $L_{(x,y)}$ equals $\sqrt{1 + \alpha^2}\, dt$. Finally, the restriction of the operator $\Delta_{\mathcal{F}}$ to each leaf $L_{(x,y)}$ coincides with the operator

$$A = -\frac{1}{1 + \alpha^2} \frac{d^2}{dt^2},$$

acting in the space $L^2(\mathbb{R}, \sqrt{1 + \alpha^2}\, dt)$.

Using the Fourier transform, one can easily compute the Schwartz kernel $E_\lambda(t, t_1)$ of the spectral projection $\chi_\lambda(A)$ of the operator A, corresponding to the semi-axis $(-\infty, \lambda]$ (relative to the volume form $\sqrt{1 + \alpha^2}\, dt$):

$$E_\lambda(t, t_1) = \frac{1}{2\pi\sqrt{1 + \alpha^2}} \int_{\mathbb{R}} e^{i(t - t_1)\xi} \chi_\lambda\left(\frac{|\xi|^2}{1 + \alpha^2}\right) d\xi.$$

Then, for any $\gamma = (x, y, t) \in G = \mathbb{T}^2 \times \mathbb{R}$, we have

$$e_\lambda(\gamma) = E_\lambda(0, t) = \frac{1}{2\pi\sqrt{1 + \alpha^2}} \int_{\mathbb{R}} e^{-it\xi} \chi_\lambda\left(\frac{|\xi|^2}{1 + \alpha^2}\right) d\xi.$$

The restriction of e_λ to $\mathbb{T}^2$ is given by

$$e_\lambda(x, y) = E_\lambda(0, 0) = \frac{1}{2\pi\sqrt{1 + \alpha^2}} \int_{\mathbb{R}} \chi_\lambda\left(\frac{|\xi|^2}{1 + \alpha^2}\right) d\xi = \frac{1}{\pi}\sqrt{\lambda}, \quad \lambda > 0.$$

We get that the spectrum distribution function $N_{\mathcal{F}}(\lambda)$ of the operator $\Delta_{\mathcal{F}}$ has the form:

$$N_{\mathcal{F}}(\lambda) = \int_{\mathbb{T}^2} e_\lambda(x, y)\, dx\, dy = \frac{1}{\pi}\sqrt{\lambda}, \quad \lambda > 0.$$

By Theorem 1.2, we obtain

$$N_h(\lambda) = h^{-1}\frac{1}{\pi} \int_{-\infty}^{\lambda} (\lambda - \tau)^{1/2} d_\tau N_{\mathcal{F}}(\tau) + o(h^{-1})$$

$$= \frac{1}{4\pi} h^{-1}\lambda + o(h^{-1}), \quad h \to 0.$$

Case 2: $\alpha \in \mathbb{Q}$ of the form $\alpha = \frac{p}{q}$, where p and q are coprime. In this case, the holonomy groupoid is $\mathbb{T}^2 \times (\mathbb{R}/q\mathbb{Z})$. The leaf $L_{(x,y)}$ through any (x, y) is the circle $\{(x + t, y + \alpha t) : t \in \mathbb{R}/q\mathbb{Z}\}$ of length $l = \sqrt{p^2 + q^2}$. The restriction of the operator $\Delta_{\mathcal{F}}$ to each $L_{(x,y)}$ coincides with the operator

$$A = -\frac{1}{1 + \alpha^2} \frac{d^2}{dt^2},$$

acting in the space $L^2(\mathbb{R}/q\mathbb{Z}, \sqrt{1 + \alpha^2}\, dt)$.

Using the Fourier transform, it is easy to see that the kernel of the spectral projection $\chi_\lambda(A)$ in $L^2(\mathbb{R}/q\mathbb{Z}, \sqrt{1+\alpha^2}\,dt)$ is given by the formula

$$E_\lambda(t, t_1) = \frac{1}{\sqrt{p^2+q^2}} \sum_{\substack{k \in \mathbb{Z} \\ |k| < \frac{\sqrt{\lambda}}{2\pi}\sqrt{p^2+q^2}}} e^{\frac{2\pi i}{q}k(t-t_1)}.$$

For any $\gamma = (x, y, t) \in G = \mathbb{T}^2 \times (\mathbb{R}/q\mathbb{Z})$, we have

$$e_\lambda(\gamma) = E_\lambda(0, t) = \frac{1}{\sqrt{p^2+q^2}} \sum_{\substack{k \in \mathbb{Z} \\ |k| < \frac{\sqrt{\lambda}}{2\pi}\sqrt{p^2+q^2}}} e^{-\frac{2\pi i}{q}kt}.$$

We get that the spectrum distribution function $N_{\mathcal{F}}(\lambda)$ of Δ_F is of the form:

$$N_{\mathcal{F}}(\lambda) = \int_{\mathbb{T}^2} e_\lambda(x, y)\,dx\,dy = \frac{1}{\sqrt{p^2+q^2}} \#\{k \in \mathbb{Z} : |k| < \frac{\sqrt{\lambda}}{2\pi}\sqrt{p^2+q^2}\}.$$

By Theorem 1.2, we obtain for $h \to 0$

$$N_h(\lambda) = h^{-1}\frac{1}{\pi}\int_{-\infty}^\lambda (\lambda - \tau)^{1/2} d_\tau N_{\mathcal{F}}(\tau) + o(h^{-1})$$

$$= h^{-1}\frac{1}{\pi\sqrt{p^2+q^2}} \sum_{\substack{k \in \mathbb{Z} \\ |k| < \frac{\sqrt{\lambda}}{2\pi}\sqrt{p^2+q^2}}} (\lambda - \frac{4\pi^2}{p^2+q^2}k^2)^{1/2} + o(h^{-1}).$$

1.3. Riemannian Heisenberg manifolds

In this section we consider the adiabatic limits associated with one-dimensional foliations given by the orbits of invariant flows on Riemannian Heisenberg manifolds. These foliations are examples of non-Riemannian foliations.

Recall that the real three-dimensional Heisenberg group H is the Lie subgroup of $\mathrm{GL}(3, \mathbb{R})$ consisting of all matrices of the form

$$\gamma(x, y, z) = \begin{bmatrix} 1 & x & z \\ 0 & 1 & y \\ 0 & 0 & 1 \end{bmatrix}, \quad x, y, z \in \mathbb{R}.$$

Its Lie algebra $\mathfrak{h}$ is the Lie subalgebra of $gl(3, \mathbb{R})$ consisting of all matrices of the form

$$X(x, y, z) = \begin{bmatrix} 0 & x & z \\ 0 & 0 & y \\ 0 & 0 & 0 \end{bmatrix}, \quad x, y, z \in \mathbb{R}.$$

A Riemannian Heisenberg manifold M is defined to be a pair $(\Gamma\backslash H, g)$, where $\Gamma = \{\gamma(x, y, z) : x, y, z \in \mathbb{Z}\}$ is a uniform discrete subgroup of H and g is a Riemannian metric on $\Gamma\backslash H$ whose lift to H is left H-invariant.

It is easy to see that g is uniquely determined by the value of its lift to H at the identity $\gamma(0, 0, 0)$, that is, by a symmetric positive definite 3×3-matrix.

In the following, we will assume that the metric g corresponds to a 3×3-matrix of the form

$$\begin{pmatrix} h_{11} & h_{12} & 0 \\ h_{12} & h_{22} & 0 \\ 0 & 0 & g_{33} \end{pmatrix}. \tag{8}$$

The lift of g to H is given by the formula

$$g(\gamma(x,y,z)) = h_{11}dx^2 + 2h_{12}dx\,dy + h_{22}dy^2 + g_{33}(dz - x\,dy)^2,$$

$$(x,y,z) \in \mathbb{R}^3.$$

The corresponding Laplace operator has the form

$$\Delta = -\left\{ \frac{1}{h_{11}h_{22} - h_{12}^2} \left[h_{22}\frac{\partial^2}{\partial x^2} - h_{12}\left[\frac{\partial}{\partial x}\left(\frac{\partial}{\partial y} + x\frac{\partial}{\partial z}\right)\right.\right.\right.$$
$$\left.\left.\left. + \left(\frac{\partial}{\partial y} + x\frac{\partial}{\partial z}\right)\frac{\partial}{\partial x}\right] + h_{11}\left(\frac{\partial}{\partial y} + x\frac{\partial}{\partial z}\right)^2\right] + \frac{1}{g_{33}}\frac{\partial^2}{\partial z^2}\right\}.$$

Theorem 1.5 ([19]). *The spectrum of the Laplace operator Δ on functions on M (with multiplicities) has the form*

$$\operatorname{spec}\Delta = \Sigma_1 \cup \Sigma_2,$$

where

$$\Sigma_1 = \left\{\lambda(a,b) = 4\pi^2\frac{h_{22}a^2 - 2h_{12}ab + h_{11}b^2}{h_{11}h_{22} - h_{12}^2} : a,b \in \mathbb{Z}\right\},$$

$$\Sigma_2 = \left\{\mu(c,k) = \frac{4\pi^2c^2}{g_{33}} + \frac{2\pi c(2k+1)}{\sqrt{h_{11}h_{22} - h_{12}^2}} \text{ with mult. } 2c :\right.$$

$$\left. c \in \mathbb{Z}^+, \quad k \in \mathbb{Z}^+ \cup \{0\}\right\}.$$

Remark 2. As shown in [19], for an arbitrary left H-invariant metric g on H, there exists a left H-invariant metric g_1, which corresponds to a 3×3-matrix of the form (8), such that Riemannian Heisenberg manifolds $(\Gamma\backslash H, g)$ and $(\Gamma\backslash H, g_1)$ are isometric. Therefore, Theorem 1.5 provides a solution of the problem of calculation of the spectrum of the Laplace operator on functions for an arbitrary Riemannian Heisenberg manifold.

Now we assume that the metric g on M corresponds to a 3×3-matrix of the form

$$\begin{pmatrix} h_{11} & 0 & 0 \\ 0 & h_{22} & 0 \\ 0 & 0 & g_{33} \end{pmatrix}.$$

In this case, one can write down explicitly all the eigenfunctions of the corresponding Laplace operator on functions. This fact plays an important role in the proof of the following theorem.

Theorem 1.6 ([41]). *The spectrum of the Laplace operator* Δ *on differential one forms on* M *(with multiplicities) has the form*

$$\operatorname{spec}\Delta = \Sigma_1 \cup \Sigma_2 \cup \Sigma_3,$$

where

$$\Sigma_1 = \{\lambda_{\pm}(a,b) = 4\pi^2\left(\frac{a^2}{h_{11}} + \frac{b^2}{h_{22}}\right)$$

$$+ \frac{\frac{g_{33}}{h_{11}h_{22}} \pm \sqrt{\frac{g_{33}^2}{h_{11}^2 h_{22}^2} + 16\pi^2 \frac{g_{33}}{h_{11}h_{22}}\left(\frac{a^2}{h_{11}} + \frac{b^2}{h_{22}}\right)}}{2} \quad \textit{with mult. } 2 : a, b \in \mathbb{Z}\},$$

$$\Sigma_2 = \{\mu(c,k) = \frac{4\pi^2 c^2}{g_{33}} + \frac{2\pi c(2k+1)}{\sqrt{h_{11}h_{22}}} \quad \textit{with mult. } 2c :$$

$$c \in \mathbb{Z}^+, k \in \mathbb{Z}^+ \cup \{0\}\}.$$

$$\Sigma_3 = \{\mu_{\pm}(c,k) = \frac{4\pi^2 c^2}{g_{33}} + \frac{2\pi c(2k+1)}{\sqrt{h_{11}h_{22}}}$$

$$+ \frac{\frac{g_{33}}{h_{11}h_{22}} \pm \sqrt{\left(\frac{4\pi c}{\sqrt{h_{11}h_{22}}} + \frac{g_{33}}{h_{11}h_{22}}\right)^2 + 8k\frac{2\pi c g_{33}}{(\sqrt{h_{11}h_{22}})^3}}}{2} \quad \textit{with mult. } 2c :$$

$$c \in \mathbb{Z}^+, k \in \mathbb{Z}^+ \cup \{0\}\}.$$

We refer the reader to [4] for a similar calculation of the spectrum of the Dirac operator on Riemannian Heisenberg manifolds.

Let $\alpha \in \mathbb{R}$. Consider the left-invariant vector field on H associated with

$$X(1,\alpha,0) = \begin{bmatrix} 0 & 1 & 0 \\ 0 & 0 & \alpha \\ 0 & 0 & 0 \end{bmatrix} \in \mathfrak{h}.$$

Since $X(1,\alpha,0)$ is a left-invariant vector field, it determines a vector field on $M = \Gamma\backslash H$. The orbits of this vector field define a one-dimensional foliation $\mathcal{F}$ on M. The leaf through a point $\Gamma\gamma(x,y,z) \in M$ is described as

$$L_{\Gamma\gamma(x,y,z)} = \{\Gamma\gamma(x+t, y+\alpha t, z + \alpha tx + \frac{\alpha t^2}{2}) \in \Gamma\backslash H : t \in \mathbb{R}\}.$$

We assume that g corresponds to the identity 3×3-matrix. Consider the adiabatic limit associated with the Riemannian Heisenberg manifold $(\Gamma\backslash H, g)$ and the one-dimensional foliation $\mathcal{F}$. The Riemannian metric g_h on $\Gamma\backslash H$ defined by (1) corresponds to the matrix

$$\begin{pmatrix} \frac{1+h^{-2}\alpha^2}{1+\alpha^2} & \alpha\frac{1-h^{-2}}{1+\alpha^2} & 0 \\ \alpha\frac{1-h^{-2}}{1+\alpha^2} & \frac{\alpha^2 + h^{-2}}{1+\alpha^2} & 0 \\ 0 & 0 & h^{-2} \end{pmatrix}, \quad h > 0.$$

The corresponding Laplacian (on functions) on the group H has the form:

$$\Delta_h = -\frac{1}{1+\alpha^2}\left[\left(\frac{\partial}{\partial x} + \alpha\left(\frac{\partial}{\partial y} + x\frac{\partial}{\partial z}\right)\right)^2 + h^2\left(-\alpha\frac{\partial}{\partial x} + \frac{\partial}{\partial y} + x\frac{\partial}{\partial z}\right)^2\right] - h^2\frac{\partial^2}{\partial z^2}.$$

Using an explicit computation of the heat kernel on the Heisenberg group, one can show the following asymptotic formula.

Theorem 1.7 ([39]). *For any $t > 0$, we have as $h \to 0$*

$$\operatorname{tr} e^{-t\Delta_h} = \frac{h^{-2}}{4\pi}\int_{-\infty}^{+\infty}\frac{\eta}{\sinh(t\eta)}e^{-t\eta^2}\,d\eta + o(h^{-2}). \tag{9}$$

Remark 3. The formula (9) looks quite different from what we have in the case of a Riemannian foliation. For instance, if $\mathcal{F}$ is a one-dimensional Riemannian foliation on a three-dimensional closed Riemannian manifold M given by the orbits of an non-singular isometric flow such that the set of closed orbits has measure zero, then, by Theorem 1.1 (or, equivalently, by Theorem 1.2), the asymptotic formula for the trace of the heat operator $e^{-t\Delta_h}$ in the adiabatic limit has the following form: for any $t > 0$,

$$\begin{aligned}
\operatorname{tr} e^{-t\Delta_h} &= \frac{h^{-2}}{4\pi t}\int_{-\infty}^{+\infty}e^{-t\eta^2}\,d\eta + o(h^{-2}) \\
&= \frac{h^{-2}}{4\sqrt{\pi t^3}} + o(h^{-2}), \quad h \to 0.
\end{aligned}$$

So, in comparison with the case of Riemannian foliations, the formula (9) contains an additional factor $\frac{t\eta}{\sinh(t\eta)}$ related with the distortion of the transverse part of the Riemannian metric along the orbits of the flow.

Remark 4. It would be quite interesting to write the formula (9) in a form similar to the formula (4).

2. Adiabatic limits and differentiable spectral sequence

In this section, we will discuss the problem of "small eigenvalues" of the Laplace operator in the adiabatic limit and its relation with the differentiable spectral sequence of the foliation. We will start with some background information on the differentiable spectral sequence.

2.1. Preliminaries on the differentiable spectral sequence

As usual, let $\mathcal{F}$ be a codimension q foliation on a closed manifold M. The differentiable spectral sequence (E_k, d_k) of $\mathcal{F}$ is a direct generalization of (the differentiable version of) the Leray spectral sequence for fibrations, which converges to the de Rham cohomology of M.

Denote by Ω the space of smooth differential forms and by Ω^r the space of smooth differential r-forms on M. Similar to the bundle case, the *differentiable spectral sequence* (E_k, d_k) of $\mathcal{F}$ is defined by the decreasing filtration by differential subspaces

$$\Omega = \Omega_0 \supset \Omega_1 \supset \cdots \supset \Omega_q \supset \Omega_{q+1} = 0 \, ,$$

where the space Ω_k^r of r-forms of filtration degree $\geq k$ consists of all $\omega \in \Omega^r$ such that $i_X \omega = 0$ for all $X = X_1 \wedge \cdots \wedge X_{r-k+1}$, where the X_i are vector fields tangent to the leaves. Roughly speaking, ω in Ω_k^r iff it is of degree $\geq k$ transversely to the leaves.

Recall that the induced spectral sequence (E_k, d_k) is defined in the following standard way (see, for instance, [28]):

$$Z_k^{u,v} = \Omega_u^{u+v} \cap d^{-1}\left(\Omega_{u+k}^{u+v+1}\right) \, , \quad Z_\infty^{u,v} = \Omega_u^{u+v} \cap \ker d \, ,$$

$$B_k^{u,v} = \Omega_u^{u+v} \cap d\left(\Omega_{u-k}^{u+v-1}\right) \, , \quad B_\infty^{u,v} = \Omega_u^{u+v} \cap \operatorname{Im} d \, ,$$

$$E_k^{u,v} = \frac{Z_k^{u,v}}{Z_{k-1}^{u+1,v-1} + B_{k-1}^{u,v}} \, , \quad E_\infty^{u,v} = \frac{Z_\infty^{u,v}}{Z_\infty^{u+1,v-1} + B_\infty^{u,v}} \, .$$

We assume $B_{-1}^{u,v} = 0$, so $E_0^{u,v} = \Omega_u^{u+v}/\Omega_{u+1}^{u+v}$. Each homomorphism $d_k : E_k^{u,v} \to E_k^{u+k,v-k+1}$ is canonically induced by d.

The terms $E_1^{0,*}$ and $E_2^{*,0}$ are respectively called *leafwise cohomology* and *basic cohomology*, and $E_2^{*,p}$ is isomorphic to the *transverse cohomology* [20] (also called *Haefliger cohomology*).

The C^∞ topology of Ω induces a topological vector space structure on each term E_k such that d_k is continuous. A subtle problem here is that E_k may not be Hausdorff [20]. So it makes sense to consider the subcomplex given by the closure of the trivial subspace, $\bar{0}_k \subset E_k$, as well as the quotient complex $\widehat{E}_k = E_k/\bar{0}_k$, whose differential operator will be also denoted by d_k.

2.2. Riemannian foliations

For a Riemannian foliation $\mathcal{F}$, each term E_k of the differentiable spectral sequence (E_k, d_k) is Hausdorff of finite dimension if $k \geq 2$, and $H(\bar{0}_1) = 0$. So $E_k \cong \widehat{E}_k$ for $k \geq 2$. The proof of this result given in [29] uses the structure theorem for Riemannian foliations due to Molino [31, 32] to reduce the problem to transitive foliations, and, for transitive foliations, it uses a construction of a parametrix for the de Rham complex given by Sarkaria [37]. Moreover, it turns out that, for $k \geq 2$, the terms E_k are homotopy invariants of Riemannian foliations [3]. (This result generalizes a previous work showing the topological invariance of the basic cohomology [15].)

Now return to adiabatic limits. So let g be a Riemannian metric on M and g_h be the one-parameter family of metrics defined by (1). Denote by Δ_h^r the Laplace operator on differential r-forms on M defined by g_h, and by

$$0 \leq \lambda_0^r(h) \leq \lambda_1^r(h) \leq \lambda_2^r(h) \leq \cdots$$

its spectrum (with multiplicities). It is well known that the eigenvalues of the Laplacian on differential forms vary continuously under continuous perturbations of the metric, and thus the "branches" of eigenvalues $\lambda_i^r(h)$ depend continuously on $h > 0$. In this section, we shall only consider the "branches" $\lambda_i^r(h)$ that are convergent to zero as $h \to 0$; roughly speaking, the "small" eigenvalues. The asymptotics as $h \to 0$ of these metric invariants are related to the differential invariant $\widehat{E}_1^r$ and the homotopy invariants E_k^r, $k \geq 2$, as follows.

Theorem 2.1 ([1]). *With the above notation, for Riemannian foliations on closed Riemannian manifolds we have*

$$\dim \widehat{E}_1^r = \sharp \left\{ i \mid \lambda_i^r(h) = O\left(h^2\right) \quad as \quad h \to 0 \right\},$$
$$\dim E_k^r = \sharp \left\{ i \mid \lambda_i^r(h) = O\left(h^{2k}\right) \quad as \quad h \to 0 \right\}, \quad k \geq 2.$$

We refer to [23] for a particular form of this theorem in the case of Riemannian flows.

As a part of the proof of Theorem 2.1 and also because of its own interest, the asymptotics of eigenforms of Δ_h corresponding to "small" eigenvalues were also studied. This study was begun in [30] for the case of Riemannian bundles, and continued in [16] for general complementary distributions.

Here we formulate the results obtained in [1] for the case of Riemannian foliations. Recall that Θ_h is an isomorphism of Hilbert spaces, which moves our setting to the fixed Hilbert space $L^2\Omega$ (see (2)). The "rescaled Laplacian" $L_h = \Theta_h \Delta_h \Theta_h^{-1}$ has the same spectrum as Δ_h, and eigenspaces of Δ_h are transformed into eigenspaces of L_h by Θ_h. It turns out that eigenspaces of L_h corresponding to "small" eigenvalues are convergent as $h \to 0$ when the metric g is bundle-like, and the limit is given by a nested sequence of bigraded subspaces,

$$\Omega \supset \mathcal{H}_1 \supset \mathcal{H}_2 \supset \mathcal{H}_3 \supset \cdots \supset \mathcal{H}_\infty.$$

The definition of $\mathcal{H}_1, \mathcal{H}_2$ was given in [2] as a Hodge theoretic approach to (E_1, d_1) and (E_2, d_2), which is based on the study of leafwise heat flow. The space $\mathcal{H}_1$ is defined as the space of smooth leafwise harmonic forms:

$$\mathcal{H}_1 = \{\omega \in \Omega : \Delta_F \omega = 0\}.$$

As shown in [2], the orthogonal projection in $L^2\Omega$ on the kernel of Δ_F in $L^2\Omega$ restricts to smooth differential forms, yielding an operator $\Pi : \Omega \to \mathcal{H}_1$. We define the operator d_1 on $\mathcal{H}_1$ as $d_1 = \Pi d_H$. The adjoint of d_1 in $\mathcal{H}_1$ equals $\delta_1 = \Pi \delta_H$. Finally, we take $\Delta_1 = d_1 \delta_1 + \delta_1 d_1$ on $\mathcal{H}_1$ and put

$$\mathcal{H}_2 = \ker \Delta_1.$$

The other spaces $\mathcal{H}_k$ are defined in [1] as an extension of this Hodge theoretic approach to the whole spectral sequence (E_k, d_k). In particular,

$$\mathcal{H}_1 \cong \widehat{E}_1, \quad \mathcal{H}_k \cong E_k, \quad k = 2, 3, \ldots, \infty,$$

as bigraded topological vector spaces. Thus this sequence stabilizes (that is, $\mathcal{H}_k = \mathcal{H}_\infty$ for k large enough) because the differentiable spectral sequence is convergent in a finite number of steps. The convergence of eigenforms corresponding to "small" eigenvalues is precisely stated in the following result, where $L^2\mathcal{H}_1$ denotes the closure of $\mathcal{H}_1$ in $L^2\Omega$.

Theorem 2.2. *For any Riemannian foliation on a closed manifold with a bundle-like metric, let ω_i be a sequence in Ω^r such that $\|\omega_i\| = 1$ and*

$$\langle L_{h_i}\omega_i, \omega_i \rangle \in o\left(h_i^{2(k-1)}\right)$$

for some fixed integer $k \geq 1$ and some sequence $h_i \to 0$. Then some subsequence of the ω_i is strongly convergent, and its limit is in $L^2\mathcal{H}_1^r$ if $k = 1$, and in $\mathcal{H}_k^r$ if $k \geq 2$.

To simplify notation let $m_1^r = \dim \widehat{E}_1^r$, and let $m_k^r = \dim E_k^r$ for each $k = 2, 3, \ldots, \infty$. Thus Theorem 2.1 establishes $\lambda_i^r(h) = O\left(h^{2k}\right)$ for $i \leq m_k^r$, yielding $\lambda_i^r(h) \equiv 0$ for i large enough. For every $h > 0$, consider the nested sequence of graded subspaces

$$\Omega \supset \mathcal{H}_1(h) \supset \mathcal{H}_2(h) \supset \mathcal{H}_3(h) \supset \cdots \supset \mathcal{H}_\infty(h),$$

where $\mathcal{H}_k^r(h)$ is the space generated by the eigenforms of Δ_h corresponding to eigenvalues $\lambda_i^r(h)$ with $i \leq m_k^r$; in particular, we have $\mathcal{H}_k(h) = \mathcal{H}_\infty(h) = \ker \Delta_h$ for k large enough. Set also $\mathcal{H}_k(0) = \mathcal{H}_k$. We have $\dim \mathcal{H}_k^r(h) = m_k^r$ for all $h > 0$, so the following result is a sharpening of Theorem 2.1.

Corollary 2.3. *For any Riemannian foliation on a closed manifold with a bundle-like metric and $k = 2, 3, \ldots, \infty$, the assignment $h \mapsto \mathcal{H}_k^r(h)$ defines a continuous map from $[0, \infty)$ to the space of finite-dimensional linear subspaces of $L^2\Omega^r$ for all $r \geq 0$. If $\dim \widehat{E}_1^r < \infty$, then this also holds for $k = 1$.*

By the standard perturbation theory, the map $h \mapsto \mathcal{H}_k^r(h)$ is, clearly, C^∞ on $(0, \infty)$ for any Riemannian foliation on a closed manifold with a bundle-like metric, $k = 2, 3, \ldots, \infty$ and $r \geq 0$. As shown in [30], this map is C^∞ up to $h = 0$, if the foliation is given by the fibers of a Riemannian fibration. In the next section, we will see an example of a Riemannian foliation and a bundle-like metric such that the map $h \mapsto \mathcal{H}_k^r(h)$ is not C^∞ at $h = 0$.

2.3. A linear foliation on the 2-torus

In this section, we consider the simplest example of the situation studied in the previous section, namely – the example of a linear foliation on the 2-torus. So, as in Section 1.2, consider the two-dimensional torus $\mathbb{T}^2 = \mathbb{R}^2/\mathbb{Z}^2$ with the coordinates (x, y), the one-dimensional foliation $\mathcal{F}$ defined by the orbits of the vector field $\widetilde{X} = \frac{\partial}{\partial x} + \alpha \frac{\partial}{\partial y}$, where $\alpha \in \mathbb{R}$, and the Euclidean metric $g = dx^2 + dy^2$ on $\mathbb{T}^2$. The eigenvalues of the corresponding Laplace operator Δ_h (counted with multiplicities) are described as follows:

$$\mathrm{spec}\,\Delta_h^0 = \mathrm{spec}\,\Delta_h^2 = \{\lambda_{kl}(h) : (k, l) \in \mathbb{Z}^2\},$$

$$\mathrm{spec}\,\Delta_h^1 = \{\lambda_{k_1 l_1}(h) + \lambda_{k_2 l_2}(h) : (k_1, l_1) \in \mathbb{Z}^2, (k_2, l_2) \in \mathbb{Z}^2\},$$

where $\lambda_{kl}(h)$ are given by (5). So, for $\alpha \notin \mathbb{Q}$, small eigenvalues appear only if $(k, l) = (0, 0)$ and $(k_1, l_1) = (k_2, l_2) = (0, 0)$ and have the form

$$\lambda_0^0(h) = \lambda_0^2(h) = 0, \quad \lambda_0^1(h) = \lambda_1^1(h) = 0. \tag{10}$$

For $\alpha \in \mathbb{Q}$ of the form $\alpha = \frac{p}{q}$, where p and q are coprime, small eigenvalues appear only if $(k,l) = t(p,q), t \in \mathbb{Z}$, and $(k_1, l_1) = t_1(p,q), (k_2, l_2) = t_2(p,q), t_1, t_2 \in \mathbb{Z}$. So there are infinitely many different branches of eigenvalues λ_h with $\lambda_h = O(h^2)$ as $h \to 0$, and all the branches of eigenvalues λ_h with $\lambda_h = O(h^4)$ as $h \to 0$ are given by (10).

Now let us turn to the differential spectral sequence. By a straightforward computation, one can show that

$$E_2^{u,v} = E_\infty^{u,v} = \mathbb{R}, \quad u = 0, 1, \quad v = 0, 1,$$

that agrees with the above description of small eigenvalues.

The case of E_1 is more interesting. First of all, it depends on whether α is rational or not. For $\alpha \in \mathbb{Q}$, $\mathcal{F}$ is given by the fibers of a trivial fibration $\mathbb{T}^2 \to S^1$, and, therefore, for any $u = 0, 1$ and $v = 0, 1$, we have

$$E_1^{u,v} = \widehat{E}_1^{u,v} = \Omega^u(S^1) \otimes H^v(S^1) = C^\infty(S^1).$$

For $\alpha \notin \mathbb{Q}$, we have

$$\widehat{E}_1^{u,v} = \mathbb{R}, \quad u = 0, 1, \quad v = 0, 1.$$

The description of E_1 is more complicated and depends on the diophantine properties of α. Recall that $\alpha \notin \mathbb{Q}$ is called diophantine, if there exist $c > 0$ and $d > 1$ such that, for any $p \in \mathbb{Z} \setminus \{0\}$ and $q \in \mathbb{Z} \setminus \{0\}$, we have

$$|q\alpha - p| > \frac{c}{|q|^d}.$$

Otherwise, α is called Liouville. It is easy to see that $E_1^{1,0} = E_1^{0,0}$ and $E_1^{1,1} = E_1^{0,1}$. As shown in [21] and [35], we have

- $E_1^{0,0} = \mathbb{R}$;
- $E_1^{0,1} = \mathbb{R}$ if α is diophantine and $E_1^{0,1}$ is infinite-dimensional if α is Liouville.

So when α is a Liouville's number, $\bar{0}_1^{0,1} = \bar{0}_1^{1,1} \neq 0$. As a direct consequence of this fact and [1, Theorem D], we obtain that, when α is a Liouville's number, there exists a bundle-like metric on $\mathbb{T}^2$ such that the associated map $h \mapsto \mathcal{H}_\infty^1(h)$, which is continuous on $[0, \infty)$ by Corollary 2.3 and C^∞ on $(0, \infty)$, is not C^∞ at $h = 0$.

2.4. Riemannian Heisenberg manifolds

In this section, we discuss similar problems for adiabatic limits associated with the Riemannian Heisenberg manifold $(\Gamma \backslash H, g)$ and the one-dimensional foliation $\mathcal{F}$ introduced in Section 1.3 (see [41]). We assume that g corresponds to the identity matrix, and the one-dimensional foliation $\mathcal{F}$ is defined by the vector field $X(1, 0, 0)$.

The corresponding Riemannian metric g_h on $\Gamma \backslash H$ defined by (1) is given by the matrix

$$\begin{pmatrix} 1 & 0 & 0 \\ 0 & h^{-2} & 0 \\ 0 & 0 & h^{-2} \end{pmatrix}, \quad h > 0.$$

By Theorem 1.5, it follows that the spectrum of the Laplacian Δ_h on 0- and 3- forms on M (with multiplicities) is described as

$$\operatorname{spec}\Delta_h^0 = \operatorname{spec}\Delta_h^3 = \Sigma_{1,h} \cup \Sigma_{2,h},$$

where

$$\Sigma_{1,h} = \{\lambda_h(a,b) = 4\pi^2(a^2 + h^2b^2) : a,b \in \mathbb{Z}\},$$
$$\Sigma_{2,h} = \{\mu_h(c,k) = 2\pi c(2k+1)h + 4\pi^2c^2h^2 \text{ with mult. } 2c,$$
$$c \in \mathbb{Z}^+, k \in \mathbb{Z}^+ \cup \{0\}\}.$$

First, note that, for any $a \in \mathbb{Z} \setminus \{0\}$ and $b \in \mathbb{Z} \setminus \{0\}$,

$$\lambda_h(a,b) > 4\pi^2h^2, \quad h > 0,$$

and for any $c \in \mathbb{Z}^+$ and $k \in \mathbb{Z}^+ \cup \{0\}\}$

$$\mu_h(c,k) > 4\pi^2h^2, \quad h > 0.$$

Therefore, for any $h > 0$,

$$\lambda_0^0(h) = \lambda_0^3(h) = 0, \quad \lambda_1^0(h) = \lambda_1^3(h) > 4\pi^2h^2.$$

Next, we see that, for any $b \in \mathbb{Z} \setminus \{0\}$,

$$\lambda_h(0,b) = 4\pi^2b^2h^2 = O(h^2), \quad h \to 0.$$

Since we have infinitely many different branches of eigenvalues λ_h with $\lambda_h = O(h^2)$ as $h \to 0$, we conclude that, for any $i > 0$,

$$\lambda_i^0(h) = \lambda_i^3(h) = O(h^2), \quad h \to 0.$$

By Theorem 1.6, the spectrum of the Laplace operator Δ_h on one and two forms on M (with multiplicities) has the form

$$\operatorname{spec}\Delta_h^1 = \operatorname{spec}\Delta_h^2 = \Sigma_{1,h} \cup \Sigma_{2,h} \cup \Sigma_{3,h},$$

where

$$\Sigma_{1,h} = \{\lambda_{h,\pm}(a,b) = 4\pi^2\left(a^2 + h^2b^2\right) + \frac{1 \pm \sqrt{1 + 16\pi^2(a^2 + h^2b^2)}}{2}$$
$$\text{with mult. } 2 : a,b \in \mathbb{Z}\},$$
$$\Sigma_{2,h} = \{\mu_h(c,k) = 4\pi^2c^2h^2 + 2\pi c(2k+1)h$$
$$\text{with mult. } 2c : c \in \mathbb{Z}^+, k \in \mathbb{Z}^+ \cup \{0\}\},$$
$$\Sigma_{3,h} = \{\mu_{h,\pm}(c,k) = 4\pi^2c^2h^2 + 2\pi c(2k+1)h + \frac{1 \pm \sqrt{(4\pi ch + 1)^2 + 16k\pi ch}}{2}$$
$$\text{with mult. } 2c : c \in \mathbb{Z}^+, k \in \mathbb{Z}^+ \cup \{0\}\}.$$

Observe that, for any $b \in \mathbb{Z} \setminus \{0\}$,

$$\lambda_{h,-}(0,b) = 4\pi^2b^2h^2 + \frac{1 - \sqrt{1 + 16\pi^2b^2h^2}}{2} = 16\pi^4b^4h^4 + O(h^4), \quad h \to 0,$$

and, for any $\lambda \in \operatorname{spec} \Delta_h \setminus \{0\}$,

$$\lambda > Ch^4, \quad h > 0,$$

with some constant $C > 0$. Therefore, we have

$$\lambda_0^1(h) = \lambda_0^2(h) = \lambda_1^1(h) = \lambda_1^2(h) = 0, \quad h > 0,$$

and, by the above argument, we obtain, for any $i > 1$

$$\lambda_i^1(h) = \lambda_i^2(h) = O(h^4), \quad h \to 0.$$

We now turn to the differentiable spectral sequence. By a straightforward computation, one can show that $E_3 = E_\infty$, all the terms $\widehat{E}_1^r$ are infinite-dimensional and, for the basic cohomology, we have

$$E_2^{0,0} = \mathbb{R}, \quad E_2^{1,0} = \mathbb{R}, \quad E_2^{2,0} = C^\infty(S^1).$$

So we get that, in this case, for $r = 0$ and $r = 3$,

$$\dim \widehat{E}_1^r = \sharp \left\{ i \mid \lambda_i^r(h) = O\left(h^2\right) \quad \text{as} \quad h \to 0 \right\} = \infty \,,$$

$$\dim E_k^r = \sharp \left\{ i \mid \lambda_i^r(h) = O\left(h^{2k}\right) \quad \text{as} \quad h \to 0 \right\} = 1 \,, \quad k \geq 2 \,.$$

and, for $r = 1$ and $r = 2$

$$\dim \widehat{E}_1^r = \sharp \left\{ i \mid \lambda_i^r(h) = O\left(h^2\right) \quad \text{as} \quad h \to 0 \right\} = \infty \,,$$

$$\dim E_2^r = \sharp \left\{ i \mid \lambda_i^r(h) = O\left(h^4\right) \quad \text{as} \quad h \to 0 \right\} = \infty \,,$$

$$\dim E_k^r = \sharp \left\{ i \mid \lambda_i^r(h) = O\left(h^{2k}\right) \quad \text{as} \quad h \to 0 \right\} = 2 \,, \quad k \geq 3 \,.$$

A more precise information can be obtained from the consideration of the corresponding eigenspaces that will be discussed elsewhere.

Remark 5. It is quite possible that both the asymptotic formula of Theorem 1.7 and the investigation of small eigenvalues given in this section can be extended to the differential form Laplace operator on an arbitrary Riemannian Heisenberg manifold. Nevertheless, we believe that many essentially new features of adiabatic limits on Riemannian Heisenberg manifolds can be already seen in the particular cases, which were considered in this paper, and we don't expect anything rather different in the general case.

References

[1] J. Álvarez López, Yu.A. Kordyukov, *Adiabatic limits and spectral sequences for Riemannian foliations.* Geom. Funct. Anal. **10** (2000), 977–1027.

[2] J. Álvarez López, Yu.A. Kordyukov, *Long time behavior of leafwise heat flow for Riemannian foliations.* Compositio Math. **125** (2001), 129–153.

[3] J. Álvarez López, X.M. Masa, *Morphisms of pseudogroups and foliation maps.* In: "Foliations 2005": Proceedings of the International Conference, Lódz, Poland, 13–24 June 2005, pp. 1–19, World Sci. Publ., Singapore, 2006.

[4] B. Ammann, Ch. Bär, *The Dirac operator on nilmanifolds and collapsing circle bundles.* Ann. Global Anal. Geom. **16** (1998), 221–253.

[5] J.M. Bismut, J. Cheeger, *η-invariants and their adiabatic limits.* J. Amer. Math. Soc. **2** (1989), 33–70.

[6] J.M. Bismut, D.S. Freed, *The analysis of elliptic families, I. Metrics and connections on determinant bundles.* Comm. Math. Phys. **106** (1986), 159–176.

[7] J.M. Bismut, D.S. Freed, *The analysis of elliptic families, II. Dirac operators, eta invariants and the holonomy theorem.* Comm. Math. Phys. **107** (1986), 103–163.

[8] J.M. Bismut, *Local index theory and higher analytic torsion.* In: Proceedings of the International Congress of Mathematicians, Vol. I (Berlin, 1998). Doc. Math. 1998, Extra Vol. I, 143–162

[9] J. Cheeger, M. Gromov, *Collapsing Riemannian manifolds while keeping their curvature bounded. I.* J. Differential Geom. **23** (1986), 309–346.

[10] J. Cheeger, *η-invariants, the adiabatic approximation and conical singularities. I. The adiabatic approximation.* J. Differential Geom. **26** (1987), 175–221.

[11] B. Colbois, G. Courtois, *Petites valeurs propres et classe d'Euler des S^1-fibres.* Ann. Sci. Ecole Norm. Sup. (4) **33** (2000), 611–645.

[12] A. Connes, *Sur la théorie non commutative de l'intégration.* In *Algèbres d'opérateurs (Sém., Les Plans-sur-Bex, 1978),* Lecture Notes in Math. Vol. 725, pp. 19–143. Springer, Berlin, Heidelberg, New York, 1979.

[13] X. Dai, *Adiabatic limits, non-multiplicity of signature and the Leray spectral sequence.* J. Amer. Math. Soc. **4** (1991), 265–231.

[14] X. Dai, *APS boundary conditions, eta invariants and adiabatic limits.* Trans. Amer. Math. Soc. **354** (2002), 107–122.

[15] A. El Kacimi-Alaoui, M. Nicolau, *On the topological invariance of the basic cohomology.* Math. Ann. **295** (1993), 627–634.

[16] R. Forman, *Spectral sequences and adiabatic limits.* Comm. Math. Phys. **168** (1995), 57–116.

[17] K. Fukaya, *Collapsing of Riemannian manifolds and eigenvalues of Laplace operator.* Invent. Math. **87** (1987), 517–547.

[18] Z. Ge, *Adiabatic limits and Rumin's complex.* C. R. Acad. Sci. Paris. **320** (1995), 699–702.

[19] C.S. Gordon, E.N. Wilson, *The spectrum of the Laplacian on Riemannian Heisenberg manifolds.* Michigan Math. J. **33** (1986), 253–271.

[20] A. Haefliger, *Some remarks on foliations with minimal leaves.* J. Differential Geom. **15** (1980), 269–284.

[21] J.L. Heitsch, *A cohomology for foliated manifolds.* Comment. Math. Helv. **50** (1975), 197–218.

[22] P. Jammes, *Sur le spectre des fibres en tore qui s'effondrent.* Manuscripta Math. **110** (2003), 13–31.

[23] P. Jammes, *Effondrement, spectre et prorrétés diophantiennes des flots riemanniens.* Preprint math.DG/0505417, 2005.

[24] Yu.A. Kordyukov. *Functional calculus for tangentially elliptic operators on foliated manifolds.* In *Analysis and Geometry in Foliated Manifolds, Proceedings of the VII International Colloquium on Differential Geometry, Santiago de Compostela, 1994,* 113–136, Singapore, 1995. World Scientific.

[25] Yu.A. Kordyukov, *Adiabatic limits and spectral geometry of foliations*. Math. Ann. **313** (1999), 763–783.

[26] Yu.A. Kordyukov, Noncommutative geometry of foliations, Preprint math.DG/ 0504095, 2005.

[27] J. Lott, *Collapsing and the differential form Laplacian: the case of a smooth limit space*. Duke Math. J. **114** (2002), 267–306.

[28] J. McCleary, *User's Guide to Spectral Sequences*, volume 12 of *Mathematics Lecture Series*. Publish or Perish Inc., Wilmington, Del., 1985.

[29] X. Masa, *Duality and minimality in Riemannian foliations*. Comment. Math. Helv. **67** (1992), 17–27.

[30] R.R. Mazzeo, R.B. Melrose, *The adiabatic limit, Hodge cohomology and Leray's spectral sequence for a fibration*. J. Differential Geom. **31** (1990), 185–213.

[31] P. Molino, *Géométrie globale des feuilletages riemanniens*. Nederl. Akad. Wetensch. Indag. Math. **44** (1982), 45–76.

[32] P. Molino, *Riemannian foliations*, Progress in Mathematics. Vol. 73. Birkhäuser Boston Inc., Boston, MA, 1988.

[33] B.L. Reinhart, *Foliated manifolds with bundle-like metrics*. Ann. of Math. (2) **69** (1959), 119–132.

[34] B.L. Reinhart, *Differential geometry of foliations*, Springer-Verlag, Berlin, 1983.

[35] C. Roger, *Méthodes Homotopiques et Cohomologiques en Théorie de Feuilletages*. Université de Paris XI, Paris, 1976.

[36] M. Rumin, *Sub-Riemannian limit of the differential form spectrum of contact manifolds*. Geom. Func. Anal. **10** (2000), 407–452.

[37] K.S. Sarkaria, *A finiteness theorem for foliated manifolds*. J. Math. Soc. Japan **30** (1978), 687–696.

[38] E. Witten, *Global gravitational anomalies*. Comm. Math. Phys. **100** (1985), 197–229.

[39] A.A. Yakovlev, *Adiabatic limits on Riemannian Heisenberg manifolds*. Preprint, 2007, to appear in Mat. sb.

[40] A.A. Yakovlev, *The spectrum of the Laplace-Beltrami operator on the two-dimensional torus in adiabatic limit*. Preprint math.DG/0612695, 2006.

[41] A.A. Yakovlev, *Adiabatic limits and spectral sequences for one-dimensional foliations on Riemannian Heisenberg manifolds*, in preparation.

Yuri A. Kordyukov
Institute of Mathematics, Russian Academy of Sciences
112 Chernyshevsky str.
450077 Ufa, Russia
e-mail: `yurikor@matem.anrb.ru`

Andrey A. Yakovlev
Department of Mathematics, Ufa State Aviation Technical University
12 K. Marx str.
450000 Ufa, Russia
e-mail: `yakovlevandrey@yandex.ru`

C^*-algebras and Elliptic Theory II

Trends in Mathematics, 145–147

© 2008 Birkhäuser Verlag Basel/Switzerland

On the Non-standard Podleś Spheres

Ulrich Krähmer

Abstract. It was shown in [1, 5] that the C*-completion of Podleś' generic quantum spheres $A_{q\rho}$ [4] is independent of the parameter ρ. In the present note we provide a proof that this is not true for the $A_{q\rho}$ themselves which remained a conjecture in [1]. As a byproduct we obtain that $\mathrm{Aut}(A_{q\rho}) = \mathbb{C}^\times$.

1. Introduction

The quantum spheres of Podleś [4] constitute a family of algebras $A_{q\rho}$, $q \in \mathbb{C}^\times = \mathbb{C} \setminus \{0\}$ not a root of unity, $\rho \in \mathbb{C} \cup \{\infty\}$, that can be considered as deformations of the complex coordinate ring of the real affine variety $S^2 \subset \mathbb{R}^3$. They can be embedded as left coideal subalgebras into the standard quantized coordinate ring $\mathbb{C}_q[SL(2)]$ and become in this way the paradigmatic examples of homogeneous spaces of quantum groups. If $q \in \mathbb{R}$ and $\rho \in \mathbb{R} \cup \{\infty\}$, then $A_{q\rho}$ are *-subalgebras of the 'compact real form' of $\mathbb{C}_q[SL(2)]$. See, e.g., [3] for details and more information.

It was shown in [1, 5] that the C*-completion of these *-algebras does not depend on ρ, but it remained a conjecture that this is not the case for the $A_{q\rho}$ themselves. The present contribution gives a proof of this fact, see Theorem 3.1 below.

2. The algebras $A_{q\rho}$ and some of their properties

Let $q \in \mathbb{C}^\times$ be not a root of unity and $\rho \in \mathbb{C}$. Define $A_{q\rho}$ as the unital associative algebra with generators x_{-1}, x_0, x_1 and relations

$$x_0 x_{\pm 1} = q^{\pm 2} x_{\pm 1} x_0, \quad x_{\mp 1} x_{\pm 1} = q^{\pm 2} x_0^2 + (1 + q^{\pm 2})\rho x_0 - 1. \qquad (2.1)$$

Analogously one defines $A_{q\infty}$ by the relations

$$x_0 x_{\pm 1} = q^{\pm 2} x_{\pm 1} x_0, \quad x_{\mp 1} x_{\pm 1} = q^{\pm 2} x_0^2 + (1 + q^{\pm 2}) x_0. \qquad (2.2)$$

This work was supported by the EU Marie Curie postdoctoral fellowship EIF 515144. It is a pleasure to thank the authors of [1] for pointing out to me this problem and for all the other discussions we had.

The defining relations imply (see [3], p. 125 for the details) that the elements

$$e_{ij} := \begin{cases} x_0^i x_1^j & j \geq 0 \\ x_0^i x_{-1}^{-j} & j < 0. \end{cases}, i \in \mathbb{N}_0, j \in \mathbb{Z}$$

form a vector space basis of $A_{q\rho}$. It is immediate that $A_{q\rho}$ is $\mathbb{Z}$-graded,

$$A_{q\rho} = \bigoplus_{j \in \mathbb{Z}} A^j, \quad A^j := \mathrm{span}\{e_{ij} \mid i \in \mathbb{N}_0\} = \{f \in A_{q\rho} \mid x_0 f = q^{2j} f x_0\}.$$

We denote by I the ideal generated by x_0 and by $\pi : A_{q\rho} \to A_{q\rho}/I$ the canonical projection. Using the basis $\{e_{ij}\}$ one sees that $I = x_0 A_{q\rho} = A_{q\rho} x_0$.

Proposition 2.1. *$A_{q\rho}$ is an integral domain and any invertible element is a scalar.*

Proof. $A_{q\rho}$ can be embedded into the quantized coordinate ring $\mathbb{C}_q[SL(2)]$ ([3], Proposition 4.31) which has these properties ([2], 9.1.9 and 9.1.14). □

Besides this we will need the well-known and easily verified fact that the following is a complete list of the characters of $A_{q\rho}$:

$$\rho \neq \infty, \pm i: \quad \chi_\lambda(x_0) = 0, \chi_\lambda(x_{\pm 1}) = \lambda^{\pm 1}, \quad \lambda \in \mathbb{C}^\times,$$
$$\rho = \pm i: \quad \chi_\lambda(x_0) = 0, \chi_\lambda(x_{\pm 1}) = \lambda^{\pm 1}, \quad \lambda \in \mathbb{C}^\times,$$
$$\chi'(x_{\pm 1}) = 0, \chi'(x_0) = \mp i,$$
$$\rho = \infty: \quad \chi_\lambda^\pm(x_{\pm 1}) = \chi_\lambda^\pm(x_0) = 0, \chi_\lambda^\pm(x_{\mp 1}) = \lambda, \quad \lambda \in \mathbb{C}.$$

We denote by $J \subset A_{q\rho}$ the intersection of the kernels of all characters. For $\rho \neq \infty, \pm i$ an element $x = \sum_{ij} \xi_{ij} e_{ij} \in A_{q\rho}$, $\xi_{ij} \in \mathbb{C}$, is mapped by χ_λ to $f(\lambda)$, where f is the Laurent polynomial $f(z) = \sum_{j \in \mathbb{Z}} \xi_{0j} z^j$. Thus $\chi_\lambda(x) = 0$ for all $\lambda \in \mathbb{C}^\times$ iff $f = 0$. Hence $J = I$. The same is true for $\rho = \infty$ as one checks similarly. For $\rho = \pm i$ one obtains the smaller ideal $I \cap \ker \chi'$.

3. The algebra $A_{q\rho}$ depends on ρ

The aim of this note is to prove the following fact that was conjectured in [1]:

Theorem 3.1. *The algebras $A_{q\rho}, A_{q\rho'}$ are isomorphic iff $\rho' = \pm\rho$ $(-\infty = \infty)$.*

Proof. We first note that $A_{q\infty}$ cannot be isomorphic to $A_{q\rho}$ with $\rho \neq \infty$: Otherwise $A_{q\infty}/J$ would be isomorphic to $A_{q\rho}/J$. The first algebra is isomorphic to $\mathbb{C}[z] \oplus \mathbb{C}[z]$ with $\pi(x_{\pm 1})$ as generators. This follows from adding $x_0 = 0$ to (2.2). For $\rho \neq \infty, \pm i$ the algebra $A_{q\rho}/J$ is instead isomorphic to $\mathbb{C}[z, z^{-1}]$ with $z^{\pm 1}$ corresponding to $\pm\pi(x_{\pm 1})$. For $\rho = \pm i$ we have $J = I \cap \ker \chi' \subset I$, and $A_{q\pm i}/I$ is as before isomorphic to $\mathbb{C}[z, z^{-1}]$. That is, $\mathbb{C}[z, z^{-1}]$ is a quotient algebra of $A_{q\pm i}/J$, hence the latter can also not be isomorphic to $A_{q\infty}/J = \mathbb{C}[z] \oplus \mathbb{C}[z]$.

Suppose now that $\psi : A_{q\rho'} \to A_{q\rho}$ is an isomorphism with $\rho, \rho' \neq \infty$. We denote by $X_i \in A_{q\rho}$ the images of the generators of $A_{q\rho'}$ under ψ.

Since X_i generate $A_{q\rho}$, $\pi(X_i)$ generate $\pi(A_{q\rho}) = \mathbb{C}[z, z^{-1}]$. This algebra is a commutative integral domain, so $\pi(X_0)\pi(X_{\pm 1}) = q^{\pm 2}\pi(X_{\pm 1})\pi(X_0)$ implies that either $\pi(X_0)$ or both $\pi(X_{\pm 1})$ vanish. But $\mathbb{C}[z, z^{-1}]$ cannot be generated by a single element, so $\pi(X_0) = 0$. Hence $X_0 = \lambda_0 x_0$ for some $\lambda_0 \in A_{q\rho}$. Repeating the whole argumentation with the roles of x_i and X_i interchanged one gets $x_0 = \mu_0 X_0$, that is, $X_0 = \mu_0 \lambda_0 X_0$ for some $\mu_0 \in A_{q\rho}$. Proposition 2.1 now implies $\lambda_0 = \mu_0^{-1} \in \mathbb{C}^\times$.

Therefore $x_0 X_{\pm 1} = q^{\pm 2} X_{\pm 1} x_0$. Hence $X_{\pm 1} \in A^{\pm 1}$, so $X_{\pm 1} = P_\pm(x_0) x_{\pm 1}$ for some polynomials $P_\pm \in \mathbb{C}[z]$. Inserting this into (2.1) one sees that both $P_\pm$ must be of degree zero. So $X_i = \lambda_i x_i$ for three non-zero constants λ_i. Inserting this again into the relations (2.1) we get

$$q^{\pm 2} \lambda_0^2 x_0^2 + (1 + q^{\pm 2}) \rho' \lambda_0 x_0 - 1 = \lambda_1 \lambda_{-1} (q^{\pm 2} x_0^2 + (1 + q^{\pm 2}) \rho x_0 - 1),$$

which is equivalent to

$$\lambda_0 = \pm 1, \quad \rho' = \pm \rho, \quad \lambda_1 \lambda_{-1} = 1.$$

If conversely $\rho' = -\rho$, then it is immediate that the assignment $x_{-1}, x_0, x_1 \mapsto x_{-1}, -x_0, x_1$ extends to an isomorphism $A_{q\rho'} \to A_{q\rho}$. $\qquad \square$

Note that we have proven en passant (only for $\rho \neq \infty$, but $\rho = \infty$ is treated analogously):

Corollary 3.2. *For $\rho \neq 0$, the map $\lambda \mapsto \sigma_\lambda$, $\sigma_\lambda(x_i) = \lambda^i x_i$ is an isomorphism $\mathbb{C}^\times \to \mathrm{Aut}(A_{q\rho})$. For $\rho = 0$, it is an embedding, and $\mathrm{Aut}(A_{q0})$ is a semidirect product of its image and the subgroup $\mathbb{Z}_2$ generated by the automorphism which fixes $x_{\pm 1}$ and maps x_0 to $-x_0$.*

References

[1] P.M. Hajac, R. Matthes, W. Szymański: Quantum Real Projective Space, Disc and Sphere. Algebr. Represent. Theory 6 No. 2 (2003), 169–192

[2] A. Joseph: Quantum Groups and Their Primitive Ideals. Springer, 1995

[3] A.U. Klimyk, K. Schmüdgen: Quantum Groups and Their Representations. Springer, 1997

[4] P. Podleś: Quantum Spheres. Lett. Math. Phys. 14 (1987), 193–202

[5] A. Sheu: Quantization of the Poisson $SU(2)$ and its Poisson Homogeneous Space – The 2-Sphere. Comm. Math. Phys. 155 (1991), 217–232

Ulrich Krähmer
Mathematics Department
University of Glasgow
University Gardens
Glasgow G12 8QW
United Kingdom
e-mail: ukraehmer@maths.gla.ac.uk

C^*-algebras and Elliptic Theory II
Trends in Mathematics, 149–181
© 2008 Birkhäuser Verlag Basel/Switzerland

Boundaries, Eta Invariant and the Determinant Bundle

Richard Melrose and Frédéric Rochon

Abstract. Cobordism invariance shows that the index, in K-theory, of a family of pseudodifferential operators on the boundary of a fibration vanishes if the symbol family extends to be elliptic across the whole fibration. For Dirac operators with spectral boundary condition, Dai and Freed [5] gave an explicit version of this at the level of the determinant bundle. Their result, that the eta invariant of the interior family trivializes the determinant bundle of the boundary family, is extended here to the wider context of pseudodifferential families of cusp type.

Mathematics Subject Classification (2000). Primary 58J52; Secondary 58J28.

Keywords. Eta invariant, determinant line bundle.

Introduction

For a fibration of compact manifolds $M \longrightarrow B$ where the fibre is a compact manifold with boundary, the cobordism of the index can be interpreted as the vanishing of the (suspended) family index for the boundary

$$\mathrm{ind}_{\mathrm{AS}} : K_c^1(T^*(\partial M/B)) \longrightarrow \mathrm{K}^1(B) \tag{1}$$

on the image of the restriction map

$$K_c(T^*(M/B)) \longrightarrow K_c(\mathbb{R} \times T^*(\partial M/B)) = K_c^1(T^*(\partial M/B)). \tag{2}$$

This was realized analytically in [9] in terms of cusp pseudodifferential operators, namely that any elliptic family of cusp pseudodifferential operators can be perturbed by a family of order $-\infty$ to be invertible; this is described as the universal case in [1]. For the even version of (1) and in the special case of Dirac operators, Dai and Freed in [5], showed that the τ (i.e., exponentiated η) invariant of a

The first author acknowledges the support of the National Science Foundation under grant DMS0408993,the second author acknowledges support of the Fonds québécois sur la nature et les technologies and NSERC while part of this work was conducted.

self-adjoint Dirac operator on an odd-dimensional compact oriented manifold with
boundary, with augmented Atiyah-Patodi-Singer boundary condition, defines an
element of the inverse determinant line for the boundary Dirac operator. Here
we give a full pseudodifferential version of this, showing that the τ invariant for
a suspended (hence 'odd') family of elliptic cusp pseudodifferential operators, P,
trivializes the determinant bundle for the indicial family $I(P)$

$$\tau = \exp(i\pi\eta(P)) : \operatorname{Det}(I(P)) \longrightarrow \mathbb{C}^*, \qquad (3)$$

which in this case is a doubly-suspended family of elliptic pseudodifferential operators; the relation to the Dirac case is discussed in detail.

This paper depends substantially on [10] where the determinant on $2n$ times
suspended smoothing families is discussed. This determinant in the doubly suspended case is used to define the determinant bundle for any doubly-suspended
elliptic family of pseudodifferential operators on a fibration (without boundary).
As in the unsuspended case (see Bismut and Freed [4]), the first Chern class of the
determinant bundle is the 2-form part of the Chern character of the index bundle.
The realization of the eta invariant for singly suspended invertible families in [7]
is extended here to the case of invertible families of suspended cusp operators. In
the Dirac case this is shown to reduce to the eta invariant of Atiyah, Patodi and
Singer for a self-adjoint Dirac operator with augmented APS boundary condition.

In the main body of the paper the consideration of a self-adjoint Dirac operator, $\eth$, is replaced by that of the suspended family, generalizing $\eth + it$, where t
is the suspension variable. This effectively replaces the self-adjoint Fredholm operators, as a classifying space for odd K-theory, by the loop group of the small
unitary group (see [2], p. 81). One advantage of using suspended operators in this
way is that the regularization techniques of [7] can be applied to define the eta
invariant as an extension of the index. In order to discuss self-adjoint (cusp) pseudodifferential operators using this suspension approach, it is necessary to consider
somewhat less regular (product-type) families, generalizing $A + it$, so we show how
to extend the analysis to this larger setting.

As in [10], we introduce the various determinant bundles in a direct global
form, as associated bundles to principal bundles (of invertible perturbations) instead of using the original spectral definition of Quillen [12]. In this way, the fact
that the τ invariant gives a trivialization of the determinant bundle follows rather
directly from the log-multiplicative property

$$\eta(A * B) = \eta(A) + \eta(B)$$

of the eta invariant.

The paper is organized as follows. In Section 1, we review the main properties
of cusp operators. In Section 2, we consider a conceptually simpler situation which
can be thought as an 'even' counterpart of our result. In Section 3, we present the
determinant bundle as an associated bundle to a principal bundle; this definition is
extended to family of $2n$-suspended elliptic operators in Section 4. This allows us
in Section 5 to rederive a well-known consequence of the cobordism invariance of

the index at the level of determinant bundles using the contractibility result of [9]. In Section 6, we introduce the notion of cusp suspended ∗-algebra, which is used in Section 7 to lift the determinant from the boundary. This lifted determinant is defined using the eta invariant for invertible suspended cusp operators introduced in Section 8. In Section 9, we prove the trivialization result and in Section 10 we relate it to the result of Dai and Freed [5] for Dirac operators. Finally, in Section 11, these results are extended to include the case of a self-adjoint family of elliptic cusp pseudodifferential operators. This involves the use of product-type suspended operators, which are discussed in the Appendix.

1. Cusp pseudodifferential operators

This section is intended to be a quick summary of the main properties of cusp pseudodifferential operators and ellipticity. We refer to [6], [8] and [9] for more details.

Let Z be a compact manifold with non-empty boundary ∂Z. Let $x \in \mathcal{C}^\infty(Z)$ be a defining function for the boundary, that is, $x \geq 0$ everywhere on Z,

$$\partial Z = \{z \in Z; x(z) = 0\}$$

and $dx(z) \neq 0$ for all $z \in \partial Z$. Such a choice of boundary defining function determines a cusp structure on the manifold Z, which is an identification of the normal bundle of the boundary ∂Z in Z with $\partial Z \times L$ for a 1-dimensional real vector space L. If E and F are complex vector bundles on Z, then $\Psi_{cu}^m(Z; E, F)$ denotes the space of cusp pseudodifferential operators acting from $\mathcal{C}^\infty(Z; E)$ to $\mathcal{C}^\infty(Z; F)$ associated to the choice of cusp structure. Different choices lead to different algebras of cusp pseudodifferential operators, but all are isomorphic. We therefore generally ignore the particular choice of cusp structure.

A *cusp vector field* $V \in \mathcal{C}^\infty(Z, TZ)$ is a vector field such that $Vx \in x^2\mathcal{C}^\infty(Z)$ for any defining function consistent with the chosen cusp structure. We denote by $\mathcal{V}_{cu}(Z)$ the Lie algebra of such vector fields. The *cusp tangent bundle* ^{cu}TZ is the smooth vector bundle on Z such that $\mathcal{V}_{cu} = \mathcal{C}^\infty(Z; {}^{cu}TZ)$; it is isomorphic to TZ as a vector bundle, but not naturally so.

Let $^{cu}S^*Z = ({}^{cu}T^*Z \setminus 0)/\mathbb{R}^+$ be the *cusp cosphere bundle* and let R^m be the trivial complex line bundle on $^{cu}S^*Z$ with sections given by functions over $^{cu}T^*Z \setminus 0$ which are positively homogeneous of degree m.

Proposition 1.1. *For each* $m \in \mathbb{Z}$, *there is a* symbol map *giving a short exact sequence*

$$\Psi_{cu}^{m-1}(Z; E, F) \longrightarrow \Psi_{cu}^m(Z; E, F) \xrightarrow{\sigma_m} \mathcal{C}^\infty({}^{cu}S^*Z; \hom(E, F) \otimes R^m). \quad (1.1)$$

Then $A \in \Psi_{cu}^m(Z; E, F)$ is said to be *elliptic* if its symbol is invertible. In this context, ellipticity is not a sufficient condition for an operator of order 0 to be Fredholm on L^2.

More generally, one can consider the space of full symbols of order m

$$\mathcal{S}_{\mathrm{cu}}^m(Z; E, F) = \rho^{-m} \mathcal{C}^\infty(\overline{{}^{\mathrm{cu}}T^*Z}; \hom(E, F)),$$

where ρ is a defining function for the boundary (at infinity) in the radial compactification of ${}^{\mathrm{cu}}T^*Z$. After choosing appropriate metrics and connections, one can define a quantization map following standard constructions

$$q : \mathcal{S}_{\mathrm{cu}}^m(Z; E, F) \longrightarrow \Psi_{\mathrm{cu}}^m(Z; E, F) \tag{1.2}$$

which induces an isomorphism of vector spaces

$$\mathcal{S}_{\mathrm{cu}}^m(Z; E, F)/\mathcal{S}_{\mathrm{cu}}^{-\infty}(Z; E, F) \cong \Psi_{\mathrm{cu}}^m(Z; E, F)/\Psi_{\mathrm{cu}}^{-\infty}(Z; E, F). \tag{1.3}$$

If Y is a compact manifold without boundary and E is a complex vector bundle over Y, there is a naturally defined algebra of suspended pseudodifferential operators, which is denoted here $\Psi_{\mathrm{sus}}^*(Y; E)$. For a detailed discussion of this algebra (and the associated modules of operators between bundles) see [7]. An element $A \in \Psi_{\mathrm{sus}}^m(Y; E)$ is a one-parameter family of pseudodifferential operators in $\Psi^m(Y; E)$ in which the parameter enters symbolically. A suspended pseudodifferential operator is associated to each cusp pseudodifferential operator by 'freezing coefficients at the boundary.' Given $A \in \Psi_{\mathrm{cu}}^m(Z; E, F)$, for each $u \in \mathcal{C}^\infty(Z; E)$, $Au|_{\partial Z} \in \mathcal{C}^\infty(\partial Z; F)$ depends only on $u|_{\partial Z} \in \mathcal{C}^\infty(\partial Z; E)$. The resulting operator $A_\partial : \mathcal{C}^\infty(\partial Z; E) \longrightarrow \mathcal{C}^\infty(\partial Z; F)$ is an element of $\Psi^m(\partial Z; E, F)$. More generally, if $\tau \in \mathbb{R}$ then

$$\Psi_{\mathrm{cu}}^m(Z; E, F) \ni A \longmapsto e^{i\frac{\tau}{x}} A e^{-i\frac{\tau}{x}} \in \Psi_{\mathrm{cu}}^m(Z; E, F) \text{ and}$$

$$I(A, \tau) = (e^{i\frac{\tau}{x}} A e^{-i\frac{\tau}{x}})_\partial \in \Psi_{\mathrm{sus}}^m(\partial Z; E, F) \tag{1.4}$$

is the *indicial family* of A.

Proposition 1.2. *The* indicial homomorphism *gives a short exact sequence,*

$$x\Psi_{\mathrm{cu}}^m(Z; E, F) \longrightarrow \Psi_{\mathrm{cu}}^m(Z; E, F) \xrightarrow{I} \Psi_{\mathrm{sus}}^m(\partial Z; E, F).$$

There is a power series expansion for operators $A \in \Psi_{\mathrm{cu}}^m(Z; E, F)$ at the boundary of which $I(A)$ is the first term. Namely, if x is a boundary defining function consistent with the chosen cusp structure there is a choice of product decomposition near the boundary consistent with x and a choice of identifications of E and F with their restrictions to the boundary. Given such a choice the 'asymptotically translation-invariant' elements of $\Psi_{\mathrm{cu}}^m(Z; E, F)$ are well defined by

$$[x^2 D_x, A] \in x^\infty \Psi_{\mathrm{cu}}^m(Z; E, F) \tag{1.5}$$

where D_x acts through the product decomposition. In fact

$$\{A \in \Psi_{\mathrm{cu}}^m(Z; E, F); \ (1.5) \text{ holds}\} / x^\infty \Psi_{\mathrm{cu}}^m(Z; E, F) \xrightarrow{I} \Psi_{\mathrm{sus}}^m(\partial Z; E, F) \tag{1.6}$$

is an isomorphism. Applying Proposition 1.2 repeatedly and using this observation, *any* element of $\Psi_{\mathrm{cu}}^m(Z; E, F)$ then has a power series expansion

$$A \sim \sum_{j=0}^{\infty} x^j A_j, \ A_j \in \Psi_{\mathrm{cu}}^m(Z; E, F), \ [x^2 D_x, A_j] \in x^\infty \Psi_{\mathrm{cu}}^m(Z; E, F) \qquad (1.7)$$

which determines it modulo $x^\infty \Psi_{\mathrm{cu}}^m(Z; E, F)$. Setting $I_j(A) = I(A_j)$ this gives a short exact sequence

$$x^\infty \Psi_{\mathrm{cu}}^m(Z; E, F) \longrightarrow \Psi_{\mathrm{cu}}^m(Z; E, F) \overset{I_*}{\longrightarrow} \Psi_{\mathrm{sus}}^m(\partial Z; E, F)[[x]],$$
$$I_*(A) = \sum_{j=0}^{\infty} x^j I_j(A) \qquad (1.8)$$

which is multiplicative provided the image modules are given the induced product

$$I_*(A) * I_* B = \sum_{j=0}^{\infty} \frac{(ix^2)^j}{j!} (D_\tau^j I_*(A))(D_x^j I_*(B)). \qquad (1.9)$$

This is equivalent to a star product although not immediately in the appropriate form because of the asymmetry inherent in (1.8); forcing the latter to be symmetric by iteratively commuting $x^{j/2}$ to the right induces an explicit star product in x^2. In contrast to Proposition 1.2, the sequence (1.8) does depend on the choice of product structure, on manifold and bundles, and the choice of the defining function.

A cusp pseudodifferential operator $A \in \Psi_{\mathrm{cu}}^m(Z; E, F)$ is said to be *fully elliptic* if it is elliptic and if its indicial family $I(A)$ is invertible in $\Psi_{\mathrm{sus}}^*(\partial Z; E, F)$; this is equivalent to the invertibility of $I(A, \tau)$ for each τ and to $I_*(A)$ with respect to the star product.

Proposition 1.3. *A cusp pseudodifferential operator is Fredholm acting on the natural cusp Sobolev spaces if and only if it is fully elliptic.*

For bundles on a compact manifold without boundary, let $G_{\mathrm{sus}}^m(Y; E, F) \subset \Psi_{\mathrm{sus}}^m(Y; E, F)$ denote the subset of elliptic and invertible elements. The η invariant of Atiyah, Patodi and Singer, after reinterpretation, is extended in [7] to a map

$$\eta : G_{\mathrm{sus}}^m(Y; E, F) \longrightarrow \mathbb{C},$$
$$\eta(AB) = \eta(A) + \eta(B), \ A \in G_{\mathrm{sus}}^m(Y; F, G), \ B \in G_{\mathrm{sus}}^{m'}(Y; E, F). \qquad (1.10)$$

In [8] an index theorem for fully elliptic fibred cusp operators is obtained, as a generalization of the Atiyah-Patodi-Singer index theorem.

Theorem 1.4 ([8]). *Let $P \in \Psi_{\mathrm{cu}}^m(X; E, F)$ be a fully elliptic operator, then the index of P is given by the formula*

$$\mathrm{ind}(P) = \overline{\mathrm{AS}}(P) - \frac{1}{2} \eta(I(P)) \qquad (1.11)$$

where $\overline{\mathrm{AS}}$ is a regularized integral involving only a finite number of terms in the full symbol expansion of P, $I(P) \in \Psi_{\mathrm{sus}}^m(\partial X; E)$ is the indicial family of P and η is the functional (1.10) introduced in [7].

Note that the ellipticity condition on the symbol of P implies that E and F are isomorphic as bundles over the boundary, since $\sigma_m(P)$ restricted to the inward-pointing normal gives such an isomorphism. Thus one can freely assume that E and F are identified near the boundary.

In the case of a Dirac operator arising from a product structure near the boundary with invertible boundary Dirac operator and spectral boundary condition, the theorem applies by adding a cylindrical end on which the Dirac operator extends to be translation-invariant, with the indicial family becoming the spectral family for the boundary Dirac operator (for pure imaginary values of the spectral parameter). The formula (1.11) then reduces to the Atiyah-Patodi-Singer index theorem.

The result (1.11) is not really in final form, since the integral $\overline{\mathrm{AS}}(P)$ is not given explicitly nor interpreted in any topological sense. However, since it is symbolic, $\overline{\mathrm{AS}}(P)$ makes sense if P is only elliptic, without assuming the invertibility of the indicial family. It therefore defines a smooth function

$$\overline{\mathrm{AS}} : \mathrm{Ell}^m_{\mathrm{cu}}(X; E, F) \longrightarrow \mathbb{C} \tag{1.12}$$

for each m. We show in Theorem 2.3 below that this function is a log-determinant for the indicial family.

Cusp operators of order $-\infty$ are in general not compact, so in particular not of trace class. Nevertheless, it is possible to define a regularized trace which will be substantially used in this paper.

Proposition 1.5. *For $A \in \Psi^{-n-1}_{\mathrm{cu}}(Z)$, $n = \dim(Z)$ and $z \in \mathbb{C}$, the function $z \mapsto \mathrm{Tr}(x^z A)$ is holomorphic for $\mathrm{Re}\, z > 1$ and has a meromorphic extension to the whole complex plane with at most simple poles at $1 - \mathbb{N}_0$, $\mathbb{N}_0 = \{0, 1, 2, \ldots\}$.*

For $A \in \Psi^{-n-1}_{\mathrm{cu}}(Z)$, the *boundary residue trace* of A, denoted $\mathrm{Tr}_{R,\partial}(A)$, is the residue at $z = 0$ of the meromorphic function $z \mapsto \mathrm{Tr}(x^z A)$. In terms of the expansion (1.7)

$$\mathrm{Tr}_{R,\partial}(A) = \frac{1}{2\pi} \int_{\mathbb{R}} \mathrm{Tr}(I_1(A, \tau)) d\tau. \tag{1.13}$$

The *regularized trace* is defined to be

$$\overline{\mathrm{Tr}}(A) = \lim_{z \to 0} \left(\mathrm{Tr}(x^z A) - \frac{\mathrm{Tr}_{R,\partial}(A)}{z} \right), \quad \text{for } A \in \Psi^{-n-1}_{\mathrm{cu}}(Z).$$

For $A \in x^2 \Psi^{-n-1}_{\mathrm{cu}}(Z)$ this reduces to the usual trace but in general it is not a trace, since it does not vanish on all commutators. Rather, there is a *trace-defect formula*

$$\overline{\mathrm{Tr}}([A, B]) = \frac{1}{2\pi i} \int_{\mathbb{R}} \mathrm{Tr}\left(I(A, \tau) \frac{\partial}{\partial \tau} I(B, \tau) \right) d\tau,$$

$$A \in \Psi^m_{\mathrm{cu}}(Z),\ B \in \Psi^{m'}_{\mathrm{cu}}(Z),\ m + m' \leq -n - 1. \tag{1.14}$$

The sign of this formula is correct provided we use (1.4) to define the indicial family. Notice that there is a (harmless) sign mistake in the trace-defect formula of [9], where a different convention for the indicial family is used.

2. Logarithm of the determinant

As a prelude to the discussion of the determinant bundle, we will consider the conceptually simpler situation of the principal $\mathbb{Z}$-bundle corresponding to the 1-dimensional part of the odd index. We first recall the generalization of the notion of principal bundle introduced in [10].

Definition 2.1. Let G be a smooth group (possibly infinite-dimensional), then a smooth fibration $\mathcal{G} \longrightarrow B$ over a compact manifold B with typical fibre G is called a **bundle of groups** with model G if its structure group is contained in $\mathrm{Aut}(G)$, the group of smooth automorphisms of G.

Definition 2.2. Let $\phi : \mathcal{G} \longrightarrow B$ be a bundle of groups with model G, then a (right) **principal $\mathcal{G}$-bundle** is a smooth fibration $\pi : \mathcal{P} \longrightarrow B$ with typical fibre G together with a smooth fibrewise (right) group action

$$h : \mathcal{P} \times_B \mathcal{G} \ni (p, g) \longmapsto p \cdot g \in \mathcal{P}$$

which is free and transitive, where

$$\mathcal{P} \times_B \mathcal{G} = \{(p, g) \in \mathcal{P} \times \mathcal{G}; \quad \pi(p) = \phi(g)\}.$$

In particular, a principal G-bundle $\pi : \mathcal{P} \longrightarrow B$ is automatically a principal $\mathcal{G}$-bundle where $\mathcal{G}$ is the trivial bundle of groups

$$\mathcal{G} = G \times B \longrightarrow B$$

given by the projection on the right factor. In that sense, definition 2.2 is a generalization of the notion of a principal bundle.

Notice also that given a bundle of groups $\mathcal{G} \longrightarrow B$, then $\mathcal{G}$ itself is a principal $\mathcal{G}$-bundle. It is the **trivial principal $\mathcal{G}$-bundle**. More generally, we say that a principal $\mathcal{G}$-bundle $\mathcal{P} \longrightarrow B$ is **trivial** if there exists a diffeomorphism $\Psi : \mathcal{P} \longrightarrow \mathcal{G}$ which preserves the fibrewise group action:

$$\Psi(h(p, g)) = \Psi(p)g, \quad \forall (p, g) \in \mathcal{P} \times_B \mathcal{G}.$$

In this section, the type of principal $\mathcal{G}$-bundle of interest arises by considering an elliptic family $Q \in \Psi^m_{\mathrm{sus}}(M/B; E, F)$ of suspended operators over a fibration

$$Y \overline{} M \qquad\qquad (2.1)$$
$$\Big\downarrow \phi$$
$$B$$

of compact manifolds without boundary (not necessarily bounding a fibration with boundary). Namely, it is given by the smooth fibration $\mathcal{Q} \longrightarrow B$, with fibre at $b \in B$

$$\mathcal{Q}_b =$$
$$\left\{ Q_b + R_b; R_b \in \Psi_{\mathrm{sus}}^{-\infty}(Y_b, E_b, F_b); \; \exists \, (Q_b + R_b)^{-1} \in \Psi_{\mathrm{sus}}^{-m}(Y_b; F_b, E_b) \right\}, \quad (2.2)$$

the set of all invertible perturbations of Q_b. The fibre is non-empty and is a principal space for the action of the once-suspended smoothing group

$$G_{\mathrm{sus}}^{-\infty}(Y_b; E_b) = \left\{ \mathrm{Id} + A; A \in \Psi_{\mathrm{sus}}^{-\infty}(Y; E_b), \; (\mathrm{Id} + A)^{-1} \in \mathrm{Id} + \Psi_{\mathrm{sus}}^{-\infty}(Y; E_b) \right\} \quad (2.3)$$

acting on the right. Thus, $\mathcal{Q}$ is a principal $G_{\mathrm{sus}}^{-\infty}(M/B; E)$-bundle with respect to the bundle of groups $G_{\mathrm{sus}}^{-\infty}(M/B; E) \longrightarrow B$ with fibre at $b \in B$ given by (2.3).

The structure group at each point is a classifying space for even K-theory and carries an index homomorphism

$$\mathrm{ind} : G_{\mathrm{sus}}^{-\infty}(Y_b; E_b) \longrightarrow \mathbb{Z}, \quad \mathrm{ind}(\mathrm{Id} + A) = \frac{1}{2\pi i} \int_{\mathbb{R}} \mathrm{Tr} \left(\frac{dA(t)}{dt} (\mathrm{Id} + A(t))^{-1} \right) dt \quad (2.4)$$

labelling the components, i.e., giving the 0-dimensional cohomology. For a suspended elliptic family this induces an integral 1-class on B; namely the first Chern class of the odd index bundle of the family. This can be seen in terms of the induced principal $\mathbb{Z}$-bundle $\mathcal{Q}_{\mathbb{Z}}$ associated to $\mathcal{Q}$

$$\mathcal{Q}_{\mathbb{Z}} = \mathcal{Q} \times \mathbb{Z}/\sim, \quad (Ag, m - \mathrm{ind}(g)) \sim (A, m), \quad \forall g \in G_{\mathrm{sus}}^{-\infty}(Y; E_b). \quad (2.5)$$

Since $\mathbb{C}^* = \mathbb{C} \setminus \{0\}$ is a classifying space for $\mathbb{Z}$, such bundles are classified up to equivalence by the integral 1-cohomology of the base.

More explicitly, any principal $\mathbb{Z}$-bundle $\phi : P \longrightarrow B$ admits a 'connection' in the sense of a map $h : P \longrightarrow \mathbb{C}$ such that $h(mp) = h(p) + m$ for the action of $m \in \mathbb{Z}$. Then the integral 1-class of the principal $\mathbb{Z}$-bundle P is given by the map

$$e^{2\pi i h} : B \longrightarrow \mathbb{C}^* \quad (2.6)$$

or the cohomology class of dh seen as a 1-form on B. The triviality of the principal $\mathbb{Z}$-bundle is equivalent to the vanishing of the integral 1-class, that is, to the existence of a function $f : B \longrightarrow \mathbb{C}$ such that $h - \phi^* f$ is locally constant.

Moreover, restricted to the 'residual' subgroup $G_{\mathrm{sus}}^{-\infty}(Y; E)$, the eta functional of (1.10) reduces to twice the index

$$\eta \big|_{G_{\mathrm{sus}}^{-\infty}(Y;E)} = 2 \, \mathrm{ind} \, . \quad (2.7)$$

In case the fibration is the boundary of a fibration of compact manifolds with boundary, as in [9], and the suspended family is the indicial family of an elliptic family of cusp pseudodifferential operators then we know that the whole odd index of the indicial family vanishes in odd K-theory. In particular the first Chern class vanishes and the associated principal $\mathbb{Z}$-bundle is trivial.

Theorem 2.3. *The eta invariant defines a connection $\frac{1}{2}\eta(A)$ on the principal $\mathbb{Z}$-bundle in (2.5) (so the first odd Chern class is $\frac{1}{2}d\eta$) and in the case of the indicial operators of a family of elliptic cusp operators, the Atiyah-Singer term in the index formula (1.11) is a log-determinant for the indicial family, so trivializing the $\mathbb{Z}$-bundle.*

Proof. By (2.7), the function on $\mathcal{Q} \times \mathbb{Z}$

$$h(A, m) = \frac{1}{2}\eta(A) + m \tag{2.8}$$

descends to $\mathcal{Q}_{\mathbb{Z}}$ and defines a connection on it. Thus the map

$$\tau = \exp(i\pi\eta) : B \longrightarrow \mathbb{C}^* \tag{2.9}$$

gives the classifying 1-class, the first odd Chern class in $H^1(B, \mathbb{Z})$ of the index bundle. In general this class is not trivial, but when $Q = I(Q_{\text{cu}})$ is the indicial family of a family of fully elliptic cusp operators Q_{cu}, the Atiyah-Singer term $\overline{\text{AS}}(Q_{\text{cu}})$ is a well-defined smooth function which does not depend on the choice of the indicial family modulo $G_{\text{sus}}^{-\infty}(Y; E)$. From formula (1.11)

$$h - \overline{\text{AS}}(Q_{\text{cu}}) = -\overline{\text{AS}}(Q_{\text{cu}}) + \frac{1}{2}\eta(A) + m = -\text{ind}(A_{\text{cu}}, b) + m \tag{2.10}$$

is locally constant. This shows that the Atiyah-Singer term explicitly trivializes the principal $\mathbb{Z}$-bundle $\mathcal{Q}_{\mathbb{Z}}$.   □

3. The determinant line bundle

Consider a fibration of closed manifolds as in (2.1) and let E and F be complex vector bundles on M. Let $P \in \Psi^m(M/B; E, F)$ be a smooth family of elliptic pseudodifferential operators acting on the fibres. If the numerical index of the family vanishes, then one can, for each $b \in B$, find $Q_b \in \Psi^{-\infty}(Y_b; E_b, F_b)$ such that $P_b + Q_b$ is invertible. The families index, which is an element of the even K-theory of the base $K^0(B)$ (see [3] for a definition), is the obstruction to the existence of a smooth family of such perturbations. This obstruction can be realized as the non-triviality of the bundle with fibre

$$\mathcal{P}_b = \left\{ P_b + Q_b; \ Q_b \in \Psi^{-\infty}(Y_b, E_b, F_b), \ \exists \ (P_b + Q_b)^{-1} \in \Psi^{-k}(Y_b; F_b, E_b) \right\}. \tag{3.1}$$

As in the odd case discussed above, the fibre is non-trivial (here because the numerical index is assumed to vanish) and is a bundle of principal G-spaces for the groups

$$G^{-\infty}(Y_b; E) = \left\{ \text{Id} + Q; Q \in \Psi^{-\infty}(Y_b; E), \ \exists \ (\text{Id} + Q)^{-1} \in \Psi^0(Y_b; E) \right\} \tag{3.2}$$

acting on the right. Thus, $\mathcal{P} \longrightarrow B$ is a principal $G^{-\infty}(M/B; E)$-bundle for the bundle of groups $G^{-\infty}(M/B; E) \longrightarrow B$ with fibre at $b \in B$ given by (3.2).

The Fredholm determinant

$$\det : \text{Id} + \Psi^{-\infty}(X; W) \longrightarrow \mathbb{C}$$

is well defined for any compact manifold X and vector bundle W. It is multiplicative

$$\det(AB) = \det(A)\det(B)$$

and is non-vanishing precisely on the group $G^{-\infty}(X;W)$. Explicitly, it may be defined by

$$\det(B) = \exp\left(\int_{[0,1]} \gamma^* \operatorname{Tr}(A^{-1}dA)\right) \tag{3.3}$$

where $\gamma : [0,1] \longrightarrow G^{-\infty}(X;W)$ is any smooth path with $\gamma(0) = \operatorname{Id}$ and $\gamma(1) = B$. Such a path exists since $G^{-\infty}(X;W)$ is connected and the result does not depend on the choice of γ in view of the integrality of the 1-form $\frac{1}{2\pi i}\operatorname{Tr}(A^{-1}dA)$ (which gives the index for the loop group).

Definition 3.1. If $P \in \Psi^m(M/B;E,F)$ is a family of elliptic pseudodifferential operators with vanishing numerical index and $\mathcal{P} \longrightarrow B$ is the bundle given by (3.1), then the determinant line bundle $\operatorname{Det}(P) \longrightarrow B$ of P is the associated line bundle given by

$$\operatorname{Det}(P) = \mathcal{P} \times_{G^{-\infty}(M/B;E)} \mathbb{C} \tag{3.4}$$

where $G^{-\infty}(Y_b;F_b)$ acts on $\mathbb{C}$ via the determinant; thus, $\operatorname{Det}(P)$ is the space $\mathcal{P} \times \mathbb{C}$ with the equivalence relation

$$(A,c) \sim (Ag^{-1},\det(g)c)$$

for $A \in \mathcal{P}_b$, $g \in G^{-\infty}(Y_b;F_b)$, $b \in B$ and $c \in \mathbb{C}$.

As discussed in [10], this definition is equivalent to the original spectral definition due to Quillen [12].

   If $P \in \Psi^m(M/B;E,F)$ is a general elliptic family, with possibly non-vanishing numerical index, it is possible to give a similar definition but depending on some additional choices. Assuming for definiteness that the numerical index is $l \geq 0$ one can choose a trivial l-dimensional subbundle $K \subset \mathcal{C}^\infty(M/B;E)$ as a bundle over B, a Hermitian inner product on E and a volume form on B. Then the fibre in (3.1) may be replaced by

$$\mathcal{P}_{b,K} = \left\{P_b + Q_b;\ Q_b \in \Psi^{-\infty}(Y_b,E_b,F_b),\ \ker(P_b+Q_b) = K_b\right\}. \tag{3.5}$$

This fibre is non-empty and for each such choice of Q_b there is a unique element $L_b \in \Psi^{-m}(Y_b;F_b,E_b)$ which is a left inverse of P_b+Q_b with range $K_b^\perp$ at each point of B. The action of the bundle of groups $G^{-\infty}(M/B;F)$ on the left makes this into a (left) principal $G^{-\infty}(M/B;F)$-bundle. Then the fibre of the determinant bundle may be taken to be

$$\operatorname{Det}(P)_{b,K} = \mathcal{P}_{b,K} \times \mathbb{C}/\sim,\ \ (A,c) \sim (BA,\det(B)c). \tag{3.6}$$

In case the numerical index is negative there is a similar construction intermediate between the two cases.

4. The $2n$-suspended determinant bundle

As described in [10], it is possible to extend the notion of determinant, and hence that of the determinant line bundle, to suspended pseudodifferential operators with an even number of parameters.

Let $L \in \Psi^m_{s(2n)}(M/B; E, F)$ be an elliptic family of $(2n)$-suspended pseudodifferential operators. Ellipticity (in view of the symbolic dependence on the parameters) implies that such a family is invertible near infinity in $\mathbb{R}^{2n}$. Thus the families index is well defined as an element of the compactly supported K-theory $K_c(\mathbb{R}^{2n}) = \mathbb{Z}$. By Bott periodicity this index may be identified with the numerical index of a family where the parameters are quantized, see the discussion in [10]. Even assuming the vanishing of this numerical index, to get an explicitly defined determinant bundle, as above, we need to introduce a formal parameter ϵ.

Let $\Psi^m_{s(2n)}(Y; F)[[\epsilon]]$ denote the space of formal power series in ϵ with coefficients in $\Psi^m_{s(2n)}(Y; F)$. For $A \in \Psi^m_{s(2n)}(Y; F)[[\epsilon]]$ and $B \in \Psi^{m'}_{s(2n)}(Y; F)[[\epsilon]]$, consider the *-*product* $A * B \in \Psi^{m+m'}_{s(2n)}(Y; F)[[\epsilon]]$ given by

$$
\begin{aligned}
A * B(u) &= \left(\sum_{\mu=0}^{\infty} a_\mu \epsilon^\mu\right) * \left(\sum_{\nu=0}^{\infty} b_\nu \epsilon^\nu\right) \\
&= \sum_{\mu=0}^{\infty}\sum_{\nu=0}^{\infty} \epsilon^{\mu+\nu} \left(\sum_{p=0}^{\infty} \frac{i^p \epsilon^p}{2^p p!} \omega(D_v, D_w)^p A(v) B(w)\Big|_{v=w=u}\right)
\end{aligned}
\tag{4.1}
$$

where $u, v, w \in \mathbb{R}^{2n}$ and ω is the standard symplectic form on $\mathbb{R}^{2n}$, $\omega(v, w) = v^T J w$ with

$$
J = \begin{pmatrix} 0 & -\operatorname{Id}_n \\ \operatorname{Id}_n & 0 \end{pmatrix}.
\tag{4.2}
$$

That (4.1) is an associative product follows from its identification with the usual 'Moyal product' arising as the symbolic product for pseudodifferential operators on $\mathbb{R}^n$.

Definition 4.1. The module $\Psi^m_{s(2n)}(Y; E, F)[[\epsilon]]$ with *-product as in (4.1) will be denoted $\Psi^m_{s\star(2n)}(Y; E, F)[[\epsilon]]$ and the quotient by the ideal $\epsilon^{n+1}\Psi^m_{s(2n)}(Y; E, F)[[\epsilon]]$, $n = \dim(Y)$, by $\Psi^m_{s\star(2n)}(Y; E, F)$.

The quotient here corresponds formally to setting

$$
\epsilon^{n+1} = 0.
\tag{4.3}
$$

Proposition 4.2. (*Essentially from* [10]) *The group*

$$
G^{-\infty}_{s\star(2n)}(Y; F) = \big\{ \operatorname{Id} + S; S \in \Psi^{-\infty}_{s\star(2n)}(Y; F),
$$
$$
\exists\, (\operatorname{Id} + S)^{-1} \in \Psi^0_{s\star(2n)}(Y; F)\big\}, \tag{4.4}
$$

*with composition given by the *-product, admits a determinant homomorphism*

$$
\det : G^{-\infty}_{s\star(2n)}(Y; F) \longrightarrow \mathbb{C}, \ \det(A * B) = \det(A)\det(B), \tag{4.5}
$$

given by

$$\det(B) = \exp\left(\int_{[0,1]} \gamma^* \alpha_n\right) \tag{4.6}$$

where α_n is the coefficient of ϵ^n in the 1-form $\mathrm{Tr}(A^{-1} * dA)$ and $\gamma : [0,1] \longrightarrow G^{-\infty}_{s\star(2n)}(Y;F)$ is any smooth path with $\gamma(0) = \mathrm{Id}$ and $\gamma(1) = B$.

Proof. In [10] the determinant is defined via (4.6) for the full formal power series algebra with $*$-product. Since the 1-form α_n only depends on the term of order n in the formal power series, and this term for a product only depends on the first n terms of the factors, we can work in the quotient and (4.5) follows. $\qquad\square$

For the group $G^{-\infty}_{s\star(2)}(X;E)$, the form α_2 can be computed explicitly.

Proposition 4.3. *On $G^{-\infty}_{s\star(2)}(X;E)$*

$$\alpha_2 = i\pi d\mu(a) - \frac{1}{4\pi i}\int_{\mathbb{R}^2} \mathrm{Tr}\left((a_0^{-1}\frac{\partial a_0}{\partial t})(a_0^{-1}\frac{\partial a_0}{\partial \tau})a_0^{-1}da_0\right.$$
$$\left. - (a_0^{-1}\frac{\partial a_0}{\partial \tau})(a_0^{-1}\frac{\partial a_0}{\partial t})a_0^{-1}da_0\right)dtd\tau, \tag{4.7}$$

where

$$\mu(a) = \frac{1}{2\pi^2 i}\int_{\mathbb{R}^2} \mathrm{Tr}(a_0^{-1}a_1)dtd\tau. \tag{4.8}$$

Proof. For $a = a_0 + \epsilon a_1 \in G^{-\infty}_{s\star(2)}(X;E)$, the inverse a^{-1} of a with respect to the $*$-product is

$$a^{-1} = a_0^{-1} - \epsilon(a_0^{-1}a_1a_0^{-1} - \frac{i}{2}\{a_0^{-1}, a_0\}a_0^{-1}), \tag{4.9}$$

where a_0^{-1} is the inverse of a_0 in $G^{-\infty}_{s(2)}(X;E)$ and

$$\{a,b\} = D_t a D_\tau b - D_\tau a D_t b = \partial_\tau a \partial_t b - \partial_t a \partial_\tau b$$

is the Poisson Bracket. Hence,

$$\mathrm{Tr}(a^{-1} * da) = \mathrm{Tr}\left(a_0^{-1}da_0 + \epsilon(-\frac{i}{2}\{a_0^{-1}, da_0\} - a_0^{-1}a_1a_0^{-1}da_0\right.$$
$$\left. + \frac{i}{2}\{a_0^{-1}, a_0\}a_0^{-1}da_0 + a_0^{-1}da_1)\right)$$
$$= \frac{1}{2\pi}\int_{\mathbb{R}^2}\left(\mathrm{Tr}(a_0^{-1}da_0) + \epsilon\,\mathrm{Tr}(-\frac{i}{2}\{a_0^{-1}, da_0\}\right.$$
$$\left. - a_0^{-1}a_1a_0^{-1}da_0 + \frac{i}{2}\{a_0^{-1}, a_0\}a_0^{-1}da_0 + a_0^{-1}da_1)\right)dtd\tau. \tag{4.10}$$

So

$$\alpha_2(a) = \frac{1}{2\pi} \int_{\mathbb{R}^2} \mathrm{Tr} \left(-\frac{i}{2}\{a_0^{-1}, da_0\} - a_0^{-1}a_1a_0^{-1}da_0 \right.$$

$$\left. + \frac{i}{2}\{a_0^{-1}, a_0\}a_0^{-1}da_0 + a_0^{-1}da_1 \right) dt d\tau . \quad (4.11)$$

On the right-hand side of (4.11), the first term vanishes since it is the integral of the trace of a Poisson bracket. Indeed, integrating by parts one of the terms with respect to t and τ,

$$\int_{\mathbb{R}^2} \mathrm{Tr}(\{a_0^{-1}, da_0\})dt d\tau = \int_{\mathbb{R}^2} \mathrm{Tr}(D_t a_0^{-1} D_\tau(da_0) - D_\tau a_0^{-1} D_t(da_0))dt d\tau$$

$$= \int_{\mathbb{R}^2} \mathrm{Tr}(D_t a_0^{-1} D_\tau(da_0) + D_t D_\tau a_0^{-1}(da_0))dt d\tau \quad (4.12)$$

$$= \int_{\mathbb{R}^2} \mathrm{Tr}(D_t a_0^{-1} D_\tau(da_0) - D_t a_0^{-1} D_\tau(da_0))dt d\tau$$

$$= 0.$$

Hence,

$$\alpha_2(a) = \frac{1}{2\pi} \int_{\mathbb{R}^2} \mathrm{Tr}(\frac{i}{2}\{a_0^{-1}, a_0\}a_0^{-1}da_0 + a_0^{-1}da_1 - a_0^{-1}a_1a_0^{-1}da_0)dt d\tau. \quad (4.13)$$

The last two terms on the right combine to give $i\pi d\mu(a)$. Writing out the Poisson bracket in terms of t and τ, gives (4.7). □

Proposition 4.4. *Integration of the 1-form α_2 gives an isomorphism*

$$\phi : \pi_1 \left(G_{s\star(2)}^{-\infty}(X; E) \right) \ni f \longmapsto \frac{1}{2\pi i} \int_{\mathbb{S}^1} f^*\alpha_2 \in \mathbb{Z}. \quad (4.14)$$

Proof. By the previous proposition and Stokes' theorem,

$$\int_{\mathbb{S}^1} f^*\alpha_2 = \frac{1}{12\pi i} \int_{\mathbb{S}^3} g^*(\mathrm{Tr}((a^{-1}da)^3))$$

$$= 2\pi i \int_{\mathbb{S}^3} g^*\beta_2^{\mathrm{odd}}, \ \forall \ f : \mathbb{S}^1 \longrightarrow G_{s(2)}^{-\infty}(X; E), \quad (4.15)$$

where the map g is defined by

$$g : \mathbb{S}^3 \ni (s, \tau, t) \longmapsto f(s)(t, \tau) \in G^{-\infty}(X; E), \quad (4.16)$$

and $\beta_2^{\mathrm{odd}} = \frac{1}{6(2\pi i)^2} \mathrm{Tr}((a^{-1}da)^3)$. Our convention is that the orientation on $\mathbb{R}^2$ is given by the symplectic form $\omega = d\tau \wedge dt$. The 3-form β_2^{odd} on $G^{-\infty}(X; E)$ is such that

$$\lambda : \pi_3(G^{-\infty}(X; E)) \ni h \longmapsto \int_{\mathbb{S}^3} h^*\beta_2^{\mathrm{odd}} \in \mathbb{Z} \quad (4.17)$$

is an isomorphism (see [9]) where

$$G^{-\infty}(X; E) = \{\mathrm{Id} + S; S \in \Psi^{-\infty}(X; E), ((\mathrm{Id} + S)^{-1} \in \Psi^0(X; E)\} . \quad (4.18)$$

Up to homotopy, $G_{s(2)}^{-\infty}(X;E) \cong [\mathbb{S}^2, G^{-\infty}(X;E)]$, so the map $f \mapsto g$ is an iso-morphism $\pi_1(G_{s(2)}^{-\infty}(X;E)) \cong \pi_3(G^{-\infty}(X;E))$. Hence the proposition follows from (4.15) and (4.17). $\qquad\square$

We may identify

$$\Psi_{s(2n)}^m(M/B;E,F) \subset \Psi_{s\star(2n)}^m(M/B;E,F) \tag{4.19}$$

as the subspace of elements independent of ϵ. For an elliptic family L with vanishing numerical index one can then consider in the same way as above the (non-empty) principal $G_{s\star(2n)}^{-\infty}(Y_b;F_b)$ spaces

$$\mathcal{L}_b = \{L_b + S_b; S_b \in \Psi_{s\star(2n)}^{-\infty}(Y_b;E_b,F_b)$$

$$\exists\,(L_b + S_b)^{-1} \in \Psi_{s\star(2n)}^{-m}(Y_b;F_b,E_b)\} \tag{4.20}$$

forming a smooth infinite-dimensional bundle over B.

Definition 4.5. For an elliptic family $L \in \Psi_{s\star(2n)}^m(M/B;E,F)$ with vanishing numerical index, the determinant line bundle is given by

$$\mathrm{Det}(L) = \mathcal{L} \times_{G_{s\star(2n)}^{-\infty}(M/B;E)} \mathbb{C} \tag{4.21}$$

where each fibre of $G_{s\star(2n)}^{-\infty}(M/B;E)$ acts on $\mathbb{C}$ via the determinant of Proposition 4.2.

5. Cobordism invariance of the index

Suppose that the fibration (2.1) arises as the boundary of a fibration where the fibre is a compact manifold with boundary:

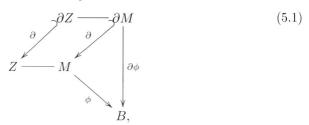

$$\tag{5.1}$$

so Z and M are compact manifolds with boundary. Let E and F be complex vector bundles over the manifold M. Suspending the short exact sequence of Proposition 1.2 one arrives at the short exact sequence

$$x\Psi_{cs(k)}^m(Z;E,F) \longrightarrow \Psi_{cs(k)}^m(Z;E,F) \overset{I}{\longrightarrow} \Psi_{s(k+1)}^m(\partial Z;E,F), \ k \in \mathbb{N}. \tag{5.2}$$

Theorem 5.1. *Let $L \in \Psi_{s(2n)}^m(\partial M/B;E,F)$ be an elliptic family of $2n$-suspended pseudodifferential operators and suppose that the fibration arises as the boundary of a fibration as in (5.1) and that L is the indicial family $L = I(P)$ of an elliptic* *family $P \in \Psi^m_{cs(2n-1)}(M/B; E, F)$ of $(2n-1)$-suspended cusp pseudodifferential operators, then the index bundle of (4.20) is trivial.*

Proof. Given $b \in B$, we claim that P_b can be perturbed by

$$Q_b \in \Psi^{-\infty}_{cs(2n-1)}(M_b; E_b, F_b)$$

to become invertible. Indeed, we may think of P_b as a family of cusp operators on $\mathbb{R}^{2n-1}$. To this family we can associate the bundle $\mathcal{I}_b$ over $\mathbb{R}^{2n-1}$ of invertible perturbations by elements in $\Psi^{-\infty}_{cu}(M_b; E_b, F_b)$. This bundle is well defined in the sense that invertible perturbations exist for all $t \in \mathbb{R}^{2n-1}$ by Theorem 5.2 of [9]. The ellipticity of P_b ensures that there exists $R > 0$ such that $P_b(t)$ is invertible for $|t| \geq R$. By the contractibility result of [9], there exists an invertible section $P_b(t) + Q_b(t)$ of $\mathcal{I}_b$ such that $Q_b(t) = 0$ for $|t| > R$. In particular Q_b is an element of $\Psi^{-\infty}_{cs(2n-1)}(M_b; E_b, F_b)$, and so $P_b + Q_b$ is the desired invertible perturbation.

It follows that there exists $S_b \in \Psi^{-\infty}_{s(2n)}(\partial Z_b; E_b, F_b)$ such that $I(P_b) = L_b$ is invertible. This could also have been seen directly using K-theory and the cobordism invariance of the index. In any case, this shows that the family P gives rise to a bundle $\mathcal{P}_{cs(2n-1)}$ on the manifold B with fibre at $b \in B$

$$\mathcal{P}_{cs(2n-1),b} = \{P_b + Q_b; \ Q_b \in \Psi^{-\infty}_{cs(2n-1)}(Z_b; E_b, F_b),$$
$$(P_b + Q_b)^{-1} \in \Psi^{-k}_{cs(2n-1)}(Z_b; F_b, E_b)\}. \quad (5.3)$$

If we consider the bundle of groups $G^{-\infty}_{cs(2n-1)}(M/B; E) \longrightarrow B$ with fibre at $b \in B$

$$G^{-\infty}_{cs(2n-1)}(Z_b; E_b) = \{\mathrm{Id} + Q_b; \ Q_b \in \Psi^{-\infty}_{cs(2n-1)}(Z_b; E_b),$$
$$(\mathrm{Id} + Q_b)^{-1} \in \Psi^0_{cs(2n-1)}(Z_b; E_b)\}, \quad (5.4)$$

then $\mathcal{P}_{cs(2n-1)}$ may be thought as a principal $G^{-\infty}_{cs(2n-1)}(M/B; E)$-bundle, where the group $G^{-\infty}_{cs(2n-1)}(Z_b; E_b)$ acts on the right in the obvious way. From [9] it follows $G^{-\infty}_{cs(2n-1)}(Z_b; F_b)$, is weakly contractible. Hence, $\mathcal{P}_{cs(2n-1)}$ has a global section defined over B, so is trivial as a principal $G^{-\infty}_{cs(2n-1)}(M/B; E)$-bundle. Taking the indicial family of this global section gives a global section of the bundle $\mathcal{L}$ which is therefore trivial as a principal $G^{-\infty}_{s\star(2n)}(M/B; E)$-bundle. $\qquad \square$

As an immediate consequence, the determinant bundle of a $2n$-suspended family which arises as the indicial family of elliptic cusp operators is necessarily trivial. Indeed, it is an associated bundle to the index bundle, which is trivial in that case. In the case of a twice-suspended family we will give an explicit trivialization in terms of the extended τ invariant of the elliptic cusp family. To do so we first need to define the η invariant in this context. As for the determinant of a suspended family discussed in [10] and in §4 above, the extended η invariant is only defined on the $\star$-extended operators which we discuss first.

6. Suspended cusp $\star$-algebra

On a compact manifold Z with boundary, consider, for a given boundary defining function x, the space of formal power series

$$\mathcal{A}^m(Z; E) = \sum_{j=0}^{\infty} \varepsilon^j x^j \Psi_{\mathrm{cs}}^m(Z; E) \tag{6.1}$$

in which the coefficients have increasing order of vanishing at the boundary. The exterior derivations D_t (differentiation with respect to the suspending parameter) and $D_{\log x}$ can be combined to give an exterior derivative $D = (D_t, D_{\log x})$-valued in $\mathbb{R}^2$. Here, the derivation $D_{\log x}$ is defined to be

$$D_{\log x} A = \frac{d}{dz} x^z A x^{-z}\big|_{z=0}.$$

for $A \in \Psi_{\mathrm{cs}}^m(Z; E)$. It is such that (cf. [9] where a different convention is used for the indicial family)

$$I(D_{\log x} A) = 0, \quad I(\frac{1}{x} D_{\log x} A)) = D_\tau I(A)$$

where τ is the suspension variable for the indicial family. Combining this with the symplectic form on $\mathbb{R}^2$ gives a star product

$$A * B = \sum_{j,k,p} \varepsilon^{j+k+p} \frac{i^p}{2^p p!} \omega(D_A, D_B)^p A_j B_k,$$
$$A = \sum_j \varepsilon^j A_j, \quad B = \sum_k \varepsilon^k B_k. \tag{6.2}$$

Here, the differential operator $\omega(D_A, D_B)^p$ is first to be expanded out, with D_A being D acting on A and D_B being D acting on B and then the product is taken in $\Psi_{\mathrm{cs}}^m(Z; E)$. Note that

$$D_{\log x} : x^p \Psi_{\mathrm{cs}}^m(Z; E) \longrightarrow x^{p+1} \Psi_{\mathrm{cs}}^{m-1}(Z; E)$$

so the series in (6.2) does lie in the space (6.1). The same formal argument as in the usual case shows that this is an associative product. We take the quotient by the ideal spanned by $(\varepsilon x)^2$ and denote the resulting algebra $\Psi_{\mathrm{cs}\star}^m(Z; E)$. Its elements are sums $A + \varepsilon A'$, $A' \in x\Psi_{\mathrm{cs}}^m(Z; E)$ and the product is just

$$(A + \varepsilon A') * (B + \varepsilon B') = AB + \varepsilon(AB' + A'B)$$

$$- \frac{i\varepsilon}{2} (D_t A D_{\log x} B - D_{\log x} A D_t B) \mod \varepsilon^2 x^2. \tag{6.3}$$

The minus sign comes from our definition of the symplectic form (4.2). The boundary asymptotic expansion (1.8), now for suspended operators, extends to the power series to give a map into triangular, doubly-suspended, double power series

$$I_* : \mathcal{A}^m(Z; E) \longrightarrow \Psi_{s(2)}^m(\partial Z; E)[[\varepsilon x, x]]. \tag{6.4}$$

To relate this more directly to the earlier discussion of star products on the suspended algebras we take the quotient by the ideal generated by x^2 giving a map

$$\tilde{I} : \Psi^m_{\mathrm{cs}\star}(Z; E) \longrightarrow \left\{ a_0 + xe + \varepsilon x a_1, \ a_0, \ e, a_1 \in \Psi^m_{\mathrm{s}(2)}(\partial Z; E) \right\}. \tag{6.5}$$

The surjectivity of the indicial map shows that this map too is surjective and so induces a product on the image.

Proposition 6.1. *The surjective map $\tilde{I}$ in (6.5) is an algebra homomorphism for the product generated by*

$$a_0 \tilde{\star} b_0 = a_0 b_0 - \varepsilon x \frac{i}{2} \left(D_t a_0 D_\tau b_0 - D_\tau a_0 D_t b_0 \right), \ a_0, \ b_0 \in \Psi^m_{\mathrm{s}(2)}(\partial Z; E) \tag{6.6}$$

extending formally over the parameters εx, x to the range in (6.5).

Proof. First observe that in terms of the expansions (1.7) for $A \in \Psi^m_{\mathrm{cs}}(Z; E)$ and $B \in \Psi^{m'}_{\mathrm{cs}}(Z; E)$ at the boundary

$$I_*(AB) = A_0 B_0 + x(A_0 F + E B_0) + O(x^2),$$
$$A = A_0 + xE + \mathcal{O}(x^2), \quad B = B_0 + xF + \mathcal{O}(x^2).$$

It follows that for

$$A = A_0 + xE + \varepsilon x A_1, \ B = B_0 + xF + \varepsilon x B_1$$

the image of the product is

$$\tilde{I}(A * B) = \tilde{I}(A)\tilde{I}(B) - \varepsilon x \frac{i}{2}(D_t a_0 D_\tau b_0 - D_\tau a_0 D_t b_0), \tag{6.7}$$

where $a_0 = I(A_0)$ and $b_0 = I(B_0)$. This is precisely what is claimed. $\qquad\square$

For any manifold without boundary Y we will denote by $\Psi^m_{\mathrm{s}\tilde{\star}(2)}(Y; E)$ the corresponding algebra with the product coming from (6.7) so that (6.5) becomes the homomorphism of algebras

$$\tilde{I} : \Psi^m_{\mathrm{cs}\star}(Z; E) \longrightarrow \Psi^m_{\mathrm{s}\tilde{\star}(2)}(\partial Z; E). \tag{6.8}$$

As the notation indicates, this algebra is closely related to $\Psi^m_{\mathrm{s}\star(2)}(Y; E)$ discussed in §4. Namely, by identifying the parameter as $\epsilon = \varepsilon x$ the latter may be identified with the quotient by the ideal

$$x\Psi^m_{\mathrm{s}(2)} \longrightarrow \Psi^m_{\mathrm{s}\tilde{\star}(2)}(Y; E) \xrightarrow{\epsilon = \varepsilon x} \Psi^m_{\mathrm{s}\star(2)}(Y; E). \tag{6.9}$$

Similarly, for the invertible elements of order zero,

$$\tilde{G}^0_{\mathrm{s}(2)}(Y; E) \longrightarrow G^0_{\mathrm{s}\tilde{\star}(2)}(Y; E) \xrightarrow{\epsilon = \varepsilon x} G^0_{\mathrm{s}\star(2)}(Y; E) \tag{6.10}$$

is exact, where

$$\tilde{G}^0_{\mathrm{s}(2)}(Y; E) = \{\mathrm{Id} + Q \in G^0_{\mathrm{s}\tilde{\star}(2)}(Y; E); Q \in x\Psi^0_{\mathrm{s}(2)}(Y; E)\}.$$

7. Lifting the determinant from the boundary

As a special case of (6.10) the groups of order $-\infty$ perturbations of the identity are related in the same way:

$$
\widetilde{G}_{s(2)}^{-\infty}(Y;E) \longrightarrow G_{s\star(2)}^{-\infty}(Y;E)
$$

$$
= \left\{ \mathrm{Id} + A_0 + xE + \varepsilon x A_1, \ A_0, \ E, \ A_1 \in \Psi_{s(2)}^{-\infty}(Y;E); \mathrm{Id} + A_0 \in G_{s(2)}^0(Y;E) \right\}
$$

$$
\longrightarrow G_{s\star(2)}^{-\infty}(Y;E), \quad (7.1)
$$

where

$$
\widetilde{G}_{s(2)}^{-\infty}(Y;E) = \{ \mathrm{Id} + Q \in G_{s(2)}^{-\infty}(Y;E); Q \in x\Psi_{s(2)}^{-\infty}(Y;E) \}.
$$

The determinant defined on the quotient group lifts to a homomorphism on the larger group with essentially the same properties. In fact, it can be defined directly as

$$
\det(b) = \exp\left(\int_0^1 \gamma^* \tilde{\alpha}_2 \right), \ b \in G_{s\star(2)}^{-\infty}(Y;E) \quad (7.2)
$$

where $\tilde{\alpha}_2$ is the coefficient of εx in the expansion of $a^{-1}\tilde{\star}da$ and γ is a curve from the identity to b. Since the normal subgroup in (7.1) is affine, the larger group is contractible to the smaller. Certainly the pull-back of $\tilde{\alpha}_2$ to the subgroup is α_2, with ϵ replaced by εx, so Proposition 4.4 holds for the larger group as well. Indeed a minor extension of the computations in the proof of Proposition 4.3 shows that at $a = (a_0 + xe_2 + \varepsilon x a_1) \in G_{s\star(2)}^{-\infty}(Y;E)$

$$
a^{-1} = a_0^{-1} - x\left(a_0^{-1}ea_0^{-1}\right) + \varepsilon x \left(a_0^{-1}a_1 a_0^{-1} + \frac{i}{2}\{a_0^{-1}, a_0\}a_0^{-1} \right)
$$
$$
\implies \tilde{\alpha}_2 = \alpha_2 \quad (7.3)
$$

in terms of formula (4.7).

Since the group

$$
G_{cs\star}^{-\infty}(Z;E) = \left\{ \mathrm{Id} + A, \ A \in \Psi_{cs\star}^{-\infty}(Z;E); \ \exists \ (\mathrm{Id}+A)^{-1} \in \mathrm{Id} + \Psi_{cs\star}^{-\infty}(Z;E) \right\} \quad (7.4)
$$

is homotopic to its principal part, and hence is contractible, the lift of $d\log\det$ under $\tilde{I}$ must be exact; we compute an explicit formula for the lift of the determinant.

Theorem 7.1. *On* $G_{cs\star}^{-\infty}(Z;F)$,

$$
\det(\tilde{I}(A)) = e^{i\pi \eta_{cu}(A)} \quad (7.5)
$$

where

$$
\eta_{cu}(A) = \frac{1}{2\pi i} \int_{\mathbb{R}} \overline{\mathrm{Tr}} \left(A_0^{-1}\frac{\partial A_0}{\partial t} + \frac{\partial A_0}{\partial t} A_0^{-1} \right) dt + \mu(\tilde{I}(A)), \quad (7.6)
$$

with μ *defined in Proposition 4.3.*

Proof. We proceed to compute $d\eta_{\mathrm{cu}}$,

$$d(\eta_{\mathrm{cu}} - \mu)(A) = \frac{1}{2\pi i} \int_{\mathbb{R}} \overline{\mathrm{Tr}}\Big(-(A_0^{-1}dA_0)(A_0^{-1}\frac{\partial A_0}{\partial t}) + A_0^{-1}\frac{\partial dA_0}{\partial t}$$
$$+ \frac{\partial dA_0}{\partial t}A_0^{-1} - (\frac{\partial A_0}{\partial t}A_0^{-1})(dA_0 A_0^{-1})\Big)dt. \quad (7.7)$$

Integrating by parts in the second and third terms gives

$$d(\eta_{\mathrm{cu}} - \mu)(A) = \frac{1}{2\pi i} \int_{\mathbb{R}} \overline{\mathrm{Tr}}\left([A_0^{-1}\frac{\partial A_0}{\partial t}, A_0^{-1}dA_0] - [\frac{\partial A_0}{\partial t}A_0^{-1}, dA_0 A_0^{-1}]\right)dt. \quad (7.8)$$

Using the trace-defect formula, this becomes

$$d(\eta_{\mathrm{cu}} - \mu)(A) = -\frac{1}{4\pi^2} \int_{\mathbb{R}^2} \mathrm{Tr}\left(a_0^{-1}\frac{\partial a_0}{\partial t}\frac{\partial}{\partial \tau}(a_0^{-1}da_0)\right.$$
$$\left. - \frac{\partial a_0}{\partial t}a_0^{-1}\frac{\partial}{\partial \tau}(da_0 a_0^{-1})\right)dt d\tau, \quad (7.9)$$

where $a_0 = I(A_0)$. Expanding out the derivative with respect to τ and simplifying

$$d(\eta_{\mathrm{cu}} - \mu)(A) = \frac{1}{4\pi^2} \int_{\mathbb{R}^2} \mathrm{Tr}\left((a_0^{-1}\frac{\partial a_0}{\partial t})(a_0^{-1}\frac{\partial a_0}{\partial \tau})a_0^{-1}da_0\right.$$
$$\left. - (a_0^{-1}\frac{\partial a_0}{\partial \tau})(a_0^{-1}\frac{\partial a_0}{\partial t})a_0^{-1}da_0\right)dt d\tau, \quad (7.10)$$

which shows that $i\pi d\eta_{\mathrm{cu}}(A) = d\log \det(\tilde{I}(A))$. $\qquad \square$

Now, consider the subgroup

$$G^{-\infty}_{\mathrm{cs}\star, \tilde{I}=\mathrm{Id}}(Z; E) \subset G^{-\infty}_{\mathrm{cs}\star}(Z; E) \quad (7.11)$$

consisting of elements of the form $\mathrm{Id} + Q$ with $Q \in \Psi^{-\infty}_{\mathrm{cs}}(Z; F)$ and $\tilde{I}(Q) = 0$. In particular

$$\mathrm{Id} + Q \in G^{-\infty}_{\mathrm{cs}\star, \tilde{I}=\mathrm{Id}}(Z; E) \implies I(Q_0) = 0. \quad (7.12)$$

Proposition 7.2. *In the commutative diagram*

$$\begin{array}{ccccc}
G^{-\infty}_{\mathrm{cs}\star, \tilde{I}=\mathrm{Id}}(Z; F) & \longrightarrow & G^{-\infty}_{\mathrm{cs}\star}(Z; F) & \xrightarrow{\tilde{I}} & G^{-\infty}_{\mathrm{s}\hat{\star}(2)}(\partial Z; F) \\
\downarrow{\scriptstyle \mathrm{ind}} & & \downarrow{\scriptstyle \frac{1}{2}\eta_{\mathrm{cu}}} & & \downarrow{\scriptstyle \det} \\
\mathbb{Z} & \xrightarrow{\hspace{1cm}} & \mathbb{C} & \xrightarrow{\exp(2\pi i \cdot)} & \mathbb{C}^*
\end{array} \quad (7.13)$$

the top row is an even-odd classifying sequence for K-theory.

Proof. We already know the contractibility of the central group, and the end groups are contractible to their principal parts, which are classifying for even and odd K-theory respectively. For $A = \mathrm{Id} + Q \in G^{-\infty}_{\mathrm{cs}\star, \tilde{I}=\mathrm{Id}}(Z; E)$, $Q = Q_0 + \epsilon Q_1$, and the condition $\tilde{I}(Q) = 0$ reduces to $Q_0, Q_1 \in x^2 \Psi_{cs}^{-\infty}(Z; F)$ so are all of trace class. Then

$$\frac{1}{2}\eta_{cu}(A) = \frac{1}{2\pi i} \int_{\mathbb{R}} \text{Tr}(A_0^{-1} \frac{\partial A_0}{\partial t}) dt \tag{7.14}$$

which is the formula for the odd index (2.4). $\hspace{2cm} \square$

8. The extended η invariant

Next we show that the cusp η-invariant defined in (7.6) can be extended to a function on the elliptic invertible elements of $\Psi_{cs\star}^m(Z; E)$. To do so the boundary-regularized trace $\overline{\text{Tr}}(A)$, defined on operators of order $-\dim(Z) - 1$, is replaced by a fully regularized trace functional on $\Psi_{cs}^{\mathbb{Z}}(Z; E)$ following the same approach as in [7].

For $m \in \mathbb{Z}$ arbitrary and $A \in \Psi_{cs}^m(Z; E)$,

$$\frac{d^p A(t)}{dt^p} \in \Psi_{cs}^{m-p}(Z; E), \tag{8.1}$$

so the function

$$h_p(t) = \overline{\text{Tr}}\left(\frac{d^p A(t)}{dt^p}\right) \in \mathcal{C}^\infty(\mathbb{R}), \tag{8.2}$$

is well defined for $p > m + \dim(Z) + 1$. Since the regularization in the trace functional is in the normal variable to the boundary, $h_p(t)$ has, as in the boundaryless case, a complete asymptotic expansion as $t \longrightarrow \pm\infty$,

$$h_p(t) \sim \sum_{l \geq 0} h_{p,l}^\pm |t|^{m-p+\dim(Z)-l}. \tag{8.3}$$

So

$$g_p(t) = \int_{-t}^t \int_0^{t_p} \cdots \int_0^{t_1} h_p(r) dr dt_1 \ldots dt_p \tag{8.4}$$

also has an asymptotic expansion as $t \to \infty$,

$$g_p(t) \sim \sum_{j \geq 0} g_{p,j} t^{m+1+\dim(Z)-j} + g_p'(t) + g_p''(t) \log t, \tag{8.5}$$

where $g_p'(t)$ and $g_p''(t)$ are polynomials of degree at most p. Increasing p to $p + 1$ involves an additional derivative in (8.2) and an additional integral in (8.4). This changes the integrand of the final integral in (8.4) by a polynomial so $g_{p+1}(t) - g_p(t)$ is a polynomial without constant term. This justifies

Definition 8.1. The *doubly regularized trace* is the continuous linear map

$$\overline{\overline{\text{Tr}}} : \Psi_{cs}^{\mathbb{Z}}(Z; E) \longrightarrow \mathbb{C} \tag{8.6}$$

given by the coefficient of t^0 in the expansion (8.5).

When $m < -1 - \dim(Z)$, this reduces to the integral of the boundary-regularized trace

$$\overline{\overline{\mathrm{Tr}}}(A) = \int_{\mathbb{R}} \overline{\mathrm{Tr}}(A(t))dt. \tag{8.7}$$

In general, the doubly regularized trace does not vanish on commutators. However, it does vanish on commutators where one factor vanishes to high order at the boundary so the associated trace-defect can only involve boundary terms.

The trace-defect formula involves a similar regularization of the trace functional on the boundary for doubly suspended operators. So for a vector bundle over a compact manifold without boundary consider

$$\int_{\mathbb{R}^2} \mathrm{Tr}(b)dtd\tau, \ b \in \Psi_{\mathrm{s}(2)}^m(Y;E), \ m < -\dim(Y) - 2. \tag{8.8}$$

For general $b \in \Psi_{\mathrm{s}(2)}^{\mathbb{Z}}(Y;E)$ set

$$\tilde{h}_p(t) = \int_{\mathbb{R}} \mathrm{Tr}\left(\frac{\partial^p b(t,\tau)}{\partial t^p}\right) d\tau, \ p > m + \dim(Y) + 2. \tag{8.9}$$

As $t \to \pm\infty$, there is again a complete asymptotic expansion

$$\tilde{h}_p(t) \sim \sum_{l \geq 0} h_{p,l}^{\pm} |t|^{m+1+\dim(Y)-p-l} \tag{8.10}$$

so

$$\tilde{g}_p(t) = \int_{-t}^{t} \int_0^{t_p} \cdots \int_0^{t_1} \tilde{h}_p(r)drdt_1 \ldots dt_p \tag{8.11}$$

has an asymptotic expansion as $t \to +\infty$

$$\tilde{g}_p(t) \sim \sum_{j \geq 0} \tilde{g}_{p,j} t^{m+2+\dim(Y)-j} + \tilde{g}_p'(t) + \tilde{g}_p''(t) \log t, \tag{8.12}$$

where $\tilde{g}_p'(t)$ and $\tilde{g}_p''(t)$ are polynomials of degree at most p.

Proposition 8.2. *For $a \in \Psi_{\mathrm{s}(2)}^{\mathbb{Z}}(Y)$ the regularized trace $\overline{\mathrm{Tr}}_{\mathrm{s}(2)}(a)$, defined as the coefficient of t^0 in the expansion (8.12) is a well-defined trace functional, reducing to*

$$\overline{\mathrm{Tr}}_{\mathrm{s}(2)}(a) = \int_{\mathbb{R}^2} \mathrm{Tr}(a)dtd\tau, \ when \ m < -\dim(Y) - 2 \tag{8.13}$$

and it satisfies

$$\overline{\mathrm{Tr}}_{\mathrm{s}(2)}\left(\frac{\partial a}{\partial \tau}\right) = 0. \tag{8.14}$$

Proof. That $\overline{\mathrm{Tr}}_{\mathrm{s}(2)}(a)$ is well defined follows from the discussion above. That it vanishes on commutators follows from the same arguments as in [7]. Namely, the derivatives of a commutator, $\frac{d^p}{dt^p}[A,B]$, are themselves commutators and the sums of the orders of the operators decreases as p increases. Thus, for large p and for a commutator, the function $\tilde{h}_p(t)$ vanishes. The identity (8.14) follows similarly. $\square$

Proposition 8.3 (Trace-defect formula). *For A, $B \in \Psi^{\mathbb{Z}}_{\mathrm{cs}}(Z)$,*

$$\overline{\overline{\mathrm{Tr}}}([A, B]) = \frac{1}{2\pi i} \overline{\mathrm{Tr}}_{s(2)} \left(I(A, \tau) \frac{\partial I(B, \tau)}{\partial \tau} \right) = -\frac{1}{2\pi i} \overline{\mathrm{Tr}}_{s(2)} \left(I(B, \tau) \frac{\partial I(A, \tau)}{\partial \tau} \right).$$

$$(8.15)$$

Proof. For $p \in \mathbb{N}$ large enough, we can apply the trace-defect formula (1.14) to get

$$\overline{\mathrm{Tr}} \left(\frac{\partial^p}{\partial t^p} [A, B] \right) = \frac{1}{2\pi i} \int_{\mathbb{R}} \mathrm{Tr} \left(\frac{\partial^p}{\partial t^p} \left(I(A, \tau) \frac{\partial}{\partial \tau} I(B, \tau) \right) \right) d\tau,$$

from which the result follows. $\qquad\square$

Using the regularized trace functional, μ may be extended from $G^{-\infty}_{s\tilde{\star}(2)}(Y; E, F)$ to $G^m_{s\tilde{\star}(2)}(Y; E)$ by setting

$$\mu(a) = \frac{1}{2\pi^2 i} \overline{\mathrm{Tr}}_{s(2)}(a_0^{-1} a_1), \ \ a = (a_0 + xe + \varepsilon x a_1) \in G^m_{s\tilde{\star}(2)}(Y; E). \qquad (8.16)$$

Proposition 8.4. *For $A = A_0 + \varepsilon x A_1 \in G^m_{\mathrm{cs}\star}(Z; E, F)$, the set of invertible elements of $\Psi^m_{\mathrm{cs}\star}(Z; E, F)$,*

$$\eta_{\mathrm{cu}}(A) := \frac{1}{2\pi i} \overline{\overline{\mathrm{Tr}}} \left(A_0^{-1} \frac{\partial A_0}{\partial t} + \frac{\partial A_0}{\partial t} A_0^{-1} \right) + \mu(\tilde{I}(A)), \qquad (8.17)$$

is log-multiplicative under composition

$$\eta_{\mathrm{cu}}(A * B) = \eta_{\mathrm{cu}}(A) + \eta_{\mathrm{cu}}(B), \ \forall \ A \in G^m_{\mathrm{cs}\star}(Z; E, F), \ B \in G^{m'}_{\mathrm{cs}\star}(Z; F, G). \quad (8.18)$$

Proof. If $a = \tilde{I}(A, \tau)$ and $b = \tilde{I}(B, \tau)$ denote the associated boundary operators, a straightforward calculation shows that

$$\mu(a * b) = \mu(a) + \mu(b) - \frac{1}{4\pi^2} \overline{\mathrm{Tr}}_{s(2)}(b_0^{-1} a_0^{-1} \{a_0, b_0\}). \qquad (8.19)$$

On the other hand,

$$\overline{\overline{\mathrm{Tr}}} \left((A_0 B_0)^{-1} \frac{\partial (A_0 B_0)}{\partial t} + \frac{\partial (A_0 B_0)}{\partial t} (A_0 B_0)^{-1} \right)$$

$$= \overline{\overline{\mathrm{Tr}}} \left(A_0^{-1} \frac{\partial A_0}{\partial t} + \frac{\partial A_0}{\partial t} A_0^{-1} \right) + \overline{\overline{\mathrm{Tr}}} \left(B_0^{-1} \frac{\partial B_0}{\partial \tau} + \frac{\partial B_0}{\partial \tau} B_0^{-1} \right) + \alpha, \quad (8.20)$$

where

$$\alpha = \overline{\overline{\mathrm{Tr}}} \left([B_0^{-1} A_0^{-1} \frac{\partial A_0}{\partial t}, B_0] + [A_0, \frac{\partial B_0}{\partial t} B_0^{-1} A_0^{-1}] \right). \qquad (8.21)$$

Using the trace-defect formula (8.15),

$$\alpha = \frac{1}{2\pi i} \overline{\mathrm{Tr}}_{s(2)} \left(b_0^{-1} a_0^{-1} \frac{\partial a_0}{\partial t} \frac{\partial b_0}{\partial \tau} - \frac{\partial a_0}{\partial \tau} \left(\frac{\partial b_0}{\partial t} b_0^{-1} a_0^{-1} \right) \right)$$

$$= -\frac{1}{2\pi i} \overline{\mathrm{Tr}}_{s(2)} \left(b_0^{-1} a_0^{-1} \{a_0, b_0\} \right). \qquad (8.22)$$

Combining (8.19), (8.20) and (8.22) gives (8.18). $\qquad\square$

9. Trivialization of the determinant bundle

In §4 the determinant bundle is defined for a family of elliptic, doubly-suspended, pseudodifferential operators on the fibres of a fibration of compact manifolds without boundary. When the family arises as the indicial family of a family of once-suspended elliptic cusp pseudodifferential operators on the fibres of fibration, (5.1), the determinant bundle is necessarily trivial, following the discussion in §5, as a bundle associated to a trivial bundle.

Theorem 9.1. *If $P \in \Psi_{cs}^m(M/B; E, F)$ is an elliptic family of once-suspended cusp pseudodifferential operators and $\mathcal{P}$ is the bundle of invertible perturbations by elements of $\Psi_{cs\star}^{-\infty}(M/B; E, F)$ then the τ invariant*

$$\tau = \exp(i\pi\eta_{cu}) : \mathcal{P} \longrightarrow \mathbb{C}^* \qquad (9.1)$$

descends to a non-vanishing linear function on the determinant bundle of the indicial family

$$\tau : \mathrm{Det}(I(P)) \longrightarrow \mathbb{C}. \qquad (9.2)$$

Proof. As discussed in §5, the bundle $\mathcal{P}$ has non-empty fibres and is a principal $G_{cs\star}^{-\infty}(M/B; E)$-bundle. The cusp eta invariant, defined by (8.17) is a well-defined function

$$\eta_{cu} : \mathcal{P} \longrightarrow \mathbb{C}. \qquad (9.3)$$

Moreover, under the action of the normal subgroup $G_{cs\star,\tilde{I}=\mathrm{Id}}^{-\infty}(M/B; E)$ in (7.13) it follows that the exponential, τ, of η_{cu} in (9.1) is constant. Thus

$$\tau : \mathcal{P}' \longrightarrow \mathbb{C}^* \qquad (9.4)$$

is well defined where $\mathcal{P}' = \mathcal{P}/G_{cs\star,\tilde{I}=\mathrm{Id}}^{-\infty}(M/B; E)$ is the quotient bundle with fibres which are principal spaces for the action of the quotient group $G_{s\tilde{\star}(2)}^{-\infty}(\partial; E)$ in (2.9). In fact $\tilde{I}$ identifies the fibres of $\mathcal{P}'$ with the bundle $\mathcal{L}$ of invertible perturbations by $\Psi_{s\tilde{\star}(2)}^{-\infty}(\partial Z; E, F)$ of the indicial family of the original family P, so

$$\tau : \mathcal{L} \longrightarrow \mathbb{C}^*. \qquad (9.5)$$

Now, the additivity of the cusp η invariant in (8.19) and the identification in (7.13) of τ with the determinant on the structure group shows that τ transforms precisely as a linear function on $\mathrm{Det}(I(P))$:

$$\tau : \mathrm{Det}(I(P)) \longrightarrow \mathbb{C}. \qquad \square$$

10. Dirac families

In this section, we show that Theorem 9.1 can be interpreted as a generalization of a theorem of Dai and Freed in [5] for Dirac operators defined on odd-dimensional Riemannian manifolds with boundary. This essentially amounts to two things. First, that the eta functional defined in(8.17) corresponds to the usual eta invariant in the Dirac case, which is established in Proposition 10.2 below. Since we are only defining this eta functional for invertible operators, $e^{i\pi\eta}$ really corresponds to the τ functional which trivializes the inverse determinant line bundle in [5]. Secondly, that in the Dirac case, the determinant bundle $\det(I(P))$ is isomorphic to the determinant line bundle of the associated family of boundary Dirac operators, which is the content of Proposition 10.3 below.

As a first step, let us recall the usual definition of the eta function on a manifold with boundary (see [11]). Let X be a Riemannian manifold with nonempty boundary $\partial X = Y$. Near the boundary, suppose that the Riemannian metric is of product type, so there is a neighborhood $Y \times [0, 1) \subset X$ of the boundary in which the metric takes the form

$$g = du^2 + h_Y \tag{10.1}$$

where $u \in [0, 1)$ is the coordinate normal to the boundary and h_Y is the pullback of a metric on Y via the projection $Y \times [0, 1) \longrightarrow Y$. Let S be a Hermitian vector bundle over X and let $D : \mathcal{C}^\infty(X, S) \longrightarrow \mathcal{C}^\infty(X, S)$ be a first-order elliptic differential operator on X which is formally selfadjoint with respect to the inner product defined by the fibre metric of S and the metric on X. In the neighborhood $Y \times [0, 1) \subset X$ of the boundary described above, assume that the operator D takes the form

$$D = \gamma \left(\frac{\partial}{\partial u} + A \right) \tag{10.2}$$

where $\gamma : S|_Y \longrightarrow S|_Y$ is a bundle isomorphism and $A : \mathcal{C}^\infty(Y, S|_Y) \longrightarrow \mathcal{C}^\infty(Y, S|_Y)$ is a first-order elliptic operator on Y such that

$$\gamma^2 = -\operatorname{Id}, \ \gamma^* = -\gamma, \ A\gamma = -\gamma A, \ A^* = A. \tag{10.3}$$

Here, A^* is the formal adjoint of A. Notice in particular that this includes the case of a compatible Dirac operator when S is a Clifford module and $\gamma = cl(du)$ is the Clifford multiplication by du. If $\ker A = \{0\}$, consider the spectral boundary condition

$$\varphi \in \mathcal{C}^\infty(X, S), \ \Pi_-(\varphi|_Y) = 0, \tag{10.4}$$

where Π_- is the projection onto the positive spectrum of A. In the case where $\ker A \neq \{0\}$, a unitary involution $\sigma : \ker A \longrightarrow \ker A$ should be chosen such that $\sigma\gamma = -\gamma\sigma$ (such an involution exists), and the boundary condition is then modified to

$$\varphi \in \mathcal{C}^\infty(X, S), \ (\Pi_- + P_-)(\varphi|_Y) = 0, \tag{10.5}$$

where P_- is the orthogonal projection onto $\ker(\sigma + \operatorname{Id})$. The associated operator D_σ is selfadjoint and has pure point spectrum. For this operator, the eta invariant is

$$\eta_X(D_\sigma) = \frac{1}{\sqrt{\pi}} \int_0^\infty s^{-\frac{1}{2}} \operatorname{Tr}(D_\sigma e^{-sD_\sigma^2}) ds. \tag{10.6}$$

To make a link with the cusp calculus, we need to enlarge X by attaching the half-cylinder $\mathbb{R}^+ \times Y$ to the boundary Y of X. The product metric near the boundary extends to this half-cylinder, which makes the resulting manifold a complete Riemannian manifold. Similarly, the operator D has a natural extension to

M using its product structure near the boundary. Denote its L^2 extension (on M) by $\mathcal{D}$. The operator $\mathcal{D}$ is selfadjoint. The eta invariant of $\mathcal{D}$ is

$$\eta_M(\mathcal{D}) = \frac{1}{\sqrt{\pi}} \int_0^\infty s^{-\frac{1}{2}} \int_M tr(E(z, z, s))dzds, \tag{10.7}$$

where $E(z_1, z_2, s)$ is the kernel of $\mathcal{D}e^{-s\mathcal{D}^2}$. One of the main result of [11] is to establish a correspondence between the eta invariants (10.6) and (10.7).

Theorem 10.1 (Müller). *Let* $D : \mathcal{C}^\infty(X, S) \longrightarrow \mathcal{C}^\infty(X, S)$ *be a compatible Dirac operator which, on a neighborhood* $Y \times [0, 1)$ *of* Y *in* X, *takes the form* (10.2), *let* $C(\lambda) : \ker A \longrightarrow \ker A$ *be the associated scattering matrix (see* [11] *for a definition) in the range* $|\lambda| < \mu_1$, *where* μ_1 *is the smallest positive eigenvalue of* A *and put* $\sigma = C(0)$, *then*

$$\eta_X(D_\sigma) = \eta_M(\mathcal{D}).$$

Now, on M, it is possible to relate D to a cusp operator. Extending the variable u to the negative reals gives a neighborhood $Y \times (-\infty, 1) \subset M$ of ∂X in M. The variable

$$x = -\frac{1}{u} \tag{10.8}$$

takes value in $(0, 1)$ and by extending it to $x = 0$, gives a manifold with boundary $\overline{M}$, with x as a boundary defining function so fixing a cusp structure. Denote by D_c the natural extension of D to $\overline{M}$. Near the boundary of $\overline{M}$,

$$D_c = \gamma\left(x^2 \frac{\partial}{\partial x} + A\right) \tag{10.9}$$

and so is clearly a cusp differential operator. If $S = S^+ \oplus S^-$ is the decomposition of S as a superspace, then

$$\hat{D}_{cs}(t) = D_c + it \in \Psi^1_{cs\star}(\overline{M}; S) \tag{10.10}$$

is a suspended cusp operator, where there are no x and εx terms. When D_c is invertible, $\hat{D}_{cs}$ is invertible as well and $\eta_{cu}(\hat{D}_{cs})$ is well defined.

Proposition 10.2. *Let* $X, Y, M, \overline{M}$ *be as above and let* D *be a compatible Dirac operator for some Clifford module* S *on* X, *which, on a neighborhood* $Y \times [0, 1)$ *of* Y *takes the form* (10.2), *suppose that* D *is invertible, and let* D_c *be its extension to* $\overline{M}$, *then*

$$\eta_X(D_\sigma) = \eta_{cu}(\hat{D}_{cs}),$$

where $\hat{D}_{cs} = D_c + it \in \Psi^1_{cs\star}(\overline{M}; S)$ *and* σ *is trivial since* A *is invertible.*

Proof. By the theorem of Müller, it suffices to show that $\eta_{cu}(\hat{D}_{cs}) = \eta_M(\mathcal{D})$. In order to do this, we closely follow the proof of Proposition 5 in [7], which is the same statement but in the case of a manifold without boundary.

Let $E(z_1, z_2, s)$ denote the kernel of $\mathcal{D}e^{-s\mathcal{D}^2}$, where $\mathcal{D}$ is the L^2 extension of $\eth$ on M. In [11], it is shown that $\operatorname{tr}(E(z, z, s))$ is absolutely integrable on M, so set

$$h(s) = \int_M \operatorname{tr}(E(z, z, s))dz \, , s \in [0, \infty). \tag{10.11}$$

Then, (see [11]) for $n = \dim(X)$ even, $h(s) \in \mathcal{C}^\infty([0, \infty))$, while for $n = \dim(X)$ odd, $h(s) \in s^{\frac{1}{2}}\mathcal{C}^\infty([0, \infty))$. Moreover, since $\ker \mathcal{D} = \{0\}$, h is exponentially decreasing as $s \to +\infty$. As in [7], consider

$$g(v, t) = \int_v^\infty e^{-st^2} h(s)ds \, , v \geq 0. \tag{10.12}$$

This is a smooth function of $v^{\frac{1}{2}}$ in $v \geq 0$ and $t \in \mathbb{R}$, and as $|t| \to \infty$, it is rapidly decreasing if $v > 0$. From the fact that $h(s) \in C^\infty([0, \infty))$ for n even, $h(s) \in s^{\frac{1}{2}}C^\infty([0, \infty))$ for n odd, and the exponential decrease, we get

$$\left| \left(t\frac{\partial}{\partial t} \right)^p g(v, t) \right| \leq \frac{C_p}{1 + t^2} \, , v \geq 0, t \in \mathbb{R}. \tag{10.13}$$

So g is uniformly a symbol of order -2 in t as v approaches 0. In fact, when n is odd, it is uniformly a symbol of order -3. Now, using the identity

$$1 = \frac{1}{\sqrt{\pi}} \int_{-\infty}^{+\infty} s^{\frac{1}{2}} \exp(-st^2)dt \, , s > 0, \tag{10.14}$$

$\eta_M(\mathcal{D})$ may be written as a double integral

$$\eta_M(\mathcal{D}) = \frac{1}{\sqrt{\pi}} \int_0^\infty s^{-\frac{1}{2}} h(s)ds = \frac{1}{\pi} \int_0^\infty \left(\int_{-\infty}^{+\infty} e^{-st^2} h(s)dt \right) ds. \tag{10.15}$$

The uniform estimate (10.13) allows the limit and integral to be exchanged so

$$\eta_M(\mathcal{D}) = \frac{1}{\pi} \lim_{v \to 0} \int_{-\infty}^{+\infty} g(v, t)dt. \tag{10.16}$$

For $p \in \mathbb{N}_0$

$$g_p(v, t) := \int_{-t}^t \int_0^{t_p} \cdots \int_0^{t_1} \frac{\partial^p}{\partial r^p} g(v, r)dr dt_1 \ldots dt_p, \tag{10.17}$$

has a uniform asymptotic expansion as $t \to \infty$ and $\eta_M(\mathcal{D})$ is just the limit as $v \to 0$ of the coefficient of t^0 in this expansion.

The kernel $E(z_1, z_2, s)$ can also be thought as the kernel of $D_c e^{-sD_c}$, a cusp operator of order $-\infty$ on $\overline{M}$. This can be checked directly from the explicit construction of $E(z_1, z_2, s)$ given in [11]. Therefore,

$$h(s) = \overline{\operatorname{Tr}}(D_c e^{-sD_c^2}), \tag{10.18}$$

where $\overline{\operatorname{Tr}}$ is the regularized trace defined in [9]. Note however that in this case, it is just the usual trace, that is, the integral of the kernel along the diagonal, since the residue trace vanishes. Consider now the (cusp product-suspended) operator

$$\hat{A}(t) = \int_0^\infty e^{-st^2} D_c e^{-sD_c^2} ds = \frac{D_c}{t^2 + D_c^2} . \tag{10.19}$$

Then, $\overline{\overline{\mathrm{Tr}}}(\hat{A})$ is the coefficient of t^0 in the asymptotic expansion as $t \to \infty$ of

$$\int_{-t}^t \int_0^{t_p} \cdots \int_0^{t_1} \overline{\overline{\mathrm{Tr}}}(\frac{d^p}{dr^p}\hat{A}(r)) dr dt_1 \ldots dt_p \tag{10.20}$$

for $p > n = \dim(\overline{M})$ and

$$\overline{\overline{\mathrm{Tr}}}\left(\frac{d^p}{dt^p}\hat{A}(t)\right) = \overline{\overline{\mathrm{Tr}}}\left(\frac{d^p}{dt^p}\int_0^\infty e^{-st^2} D_c e^{-sD_c^2} ds\right)$$
$$= \frac{d^p}{dt^p}\int_0^\infty e^{-st^2}\overline{\overline{\mathrm{Tr}}}(D_c e^{-sD_c^2}) ds = \frac{d^p}{dt^p} g(0, t), \tag{10.21}$$

so $\eta_M(\mathcal{D}) = \frac{1}{\pi}\overline{\overline{\mathrm{Tr}}}(\hat{A})$. Instead of $\hat{A}$, consider

$$\hat{B}(t) = \int_0^\infty e^{-st^2}(D_c - it)e^{-sD_c^2} ds = \frac{1}{it + D_c} = (\hat{D}_{\mathrm{cs}})^{-1} . \tag{10.22}$$

Since $\hat{B}(t) - \hat{A}(t)$ is odd in t, $\overline{\overline{\mathrm{Tr}}}(\hat{B} - \hat{A}) = 0$, so finally

$$\eta_M(\mathcal{D}) = \frac{1}{\pi}\overline{\overline{\mathrm{Tr}}}((\hat{D}_{\mathrm{cs}})^{-1}) = \frac{1}{2\pi i}\overline{\overline{\mathrm{Tr}}}\left(\frac{\partial}{\partial t}(\hat{D}_{\mathrm{cs}})(\hat{D}_{\mathrm{cs}})^{-1} + (\hat{D}_{\mathrm{cs}})^{-1}\frac{\partial}{\partial t}(\hat{D}_{\mathrm{cs}})\right) \tag{10.23}$$
$$= \eta_{\mathrm{cu}}(\hat{D}_{\mathrm{cs}}). \qquad \square$$

Let $\eth$ be some compatible Dirac operator as in Proposition 10.2. Then near the boundary of $\overline{M}$, its cusp version $\eth_c$ takes the form

$$\eth_c = \gamma(x^2\frac{\partial}{\partial x} + A). \tag{10.24}$$

Here, it is tacitly assumed that near the boundary, S is identified with the pull-back of $S|_{\partial\overline{M}}$ via the projection $\partial\overline{M} \times [0, 1) \longrightarrow \partial\overline{M}$. Since the map

$$T^*(\partial\overline{M}) \ni \xi \longmapsto cl(du)cl(\xi) \in \mathrm{Cl}(\overline{M}), \quad \gamma = cl(du), \quad x = -\frac{1}{u}, \tag{10.25}$$

extends to an isomorphism of algebras

$$\mathrm{Cl}(\partial\overline{M}) \longrightarrow \mathrm{Cl}^+(\overline{M})|_{\partial\overline{M}}, \tag{10.26}$$

where $\mathrm{Cl}(\partial\overline{M})$ and $\mathrm{Cl}(\overline{M})$ are the Clifford algebras of $\partial\overline{M}$ and $\overline{M}$, this gives an action of $\mathrm{Cl}(\partial\overline{M})$ on $S^0 = S^+|_{\partial\overline{M}}$. If $\nu^+ : S^+_{\partial\overline{M}} \longrightarrow S^0$ denotes this identification, $S^-_{\partial\overline{M}}$ can be identified with S^0 via the map

$$\nu^- = \nu^+ \circ cl(du) : S^-|_{\partial\overline{M}} \longrightarrow S^0. \tag{10.27}$$

The combined identification $\nu : S|_{\partial\overline{M}} \longrightarrow S^0 \oplus S^0$ allows us to write $\eth_c$ and γ as

$$\eth_c = \begin{pmatrix} 0 & \eth_0 + x^2\frac{\partial}{\partial x} \\ \eth_0 - x^2\frac{\partial}{\partial x} & 0 \end{pmatrix} \quad \gamma = \begin{pmatrix} 0 & 1 \\ -1 & 0 \end{pmatrix} \tag{10.28}$$

acting on $S^0 \oplus S^0$, where $\eth_0$ is the Dirac operator associated to the $\mathrm{Cl}(\partial \overline{M})$-module S^0. If instead we decompose the bundle $S^0 \oplus S^0$ in terms of the $\pm i$ eigenspaces $S^\pm$ of γ, then $\eth_c$ and γ take the form

$$\eth_c = \begin{pmatrix} ix^2 \frac{\partial}{\partial x} & \eth_0^- \\ \eth_0^+ & -ix^2 \frac{\partial}{\partial x} \end{pmatrix}, \quad \gamma = \begin{pmatrix} i & 0 \\ 0 & -i \end{pmatrix} \qquad (10.29)$$

with $\eth_0^\pm = \pm i \eth_0$, so that $\eth_0^+$ and $\eth_0^-$ are the adjoint of each other. In this notation, the suspended operator $\hat{\eth}_{cs}(t)$ can be written as

$$\hat{\eth}_{cs}(t) = \begin{pmatrix} it + ix^2 \frac{\partial}{\partial x} & \eth_0^- \\ \eth_0^+ & it - ix^2 \frac{\partial}{\partial x} \end{pmatrix}. \qquad (10.30)$$

Thus, its indicial operator $\hat{\eth}_{s(2)}(t,\tau)$ is

$$\hat{\eth}_{s(2)}(t,\tau) = e^{\frac{i\tau}{x}} \hat{D}_c(t) e^{-\frac{i\tau}{x}} \Big|_{x=0} = \begin{pmatrix} it - \tau & \eth_0^- \\ \eth_0^+ & it + \tau \end{pmatrix}. \qquad (10.31)$$

Note that

$$\hat{\eth}_{s(2)}^* \hat{\eth}_{s(2)} = \begin{pmatrix} t^2 + \tau^2 + \eth_0^- \eth_0^+ & 0 \\ 0 & t^2 + \tau^2 + \eth_0^+ \eth_0^- \end{pmatrix} \qquad (10.32)$$

which is invertible everywhere except possibly at $t = \tau = 0$ so $\hat{\eth}_{s(2)}$ is invertible for every $t, \tau \in \mathbb{R}$ if and only if $\eth_0^+$ (and consequently $\eth_0^-$) is invertible.

Now, we wish to relate the determinant bundle associated to the family $\hat{\eth}_{s(2)}$ with the determinant bundle of the boundary Dirac family $\eth_0^+$ using the periodicity of the determinant line bundle discussed in [10]. For Dirac operators on a closed manifold, this can be formulated as follows.

Proposition 10.3. *If $\eth_0^+ \in \mathrm{Diff}^1(N/B; S)$ is a family of Dirac type operators on a fibration of closed manifold $N \longrightarrow B$ with vanishing numerical index then the determinant bundle of $\eth_0^+$ is naturally isomorphic to the determinant bundle of the associated family of twice suspended operators*

$$\eth_{s(2)}(t,\tau) = \begin{pmatrix} it - \tau & \eth_0^- \\ \eth_0^+ & it + \tau \end{pmatrix} \in \Psi_{s(2)}^1(N/B; S \oplus S)$$

where $\eth_0^- = (\eth_0^+)^*$.

In [10], this periodicity is formulated in terms of product-suspended operators instead of suspended operators, since in general, given $P \in \Psi^1(Y; S)$, the family of operators

$$P_{s(2)}(t,\tau) = \begin{pmatrix} it - \tau & P^* \\ P & it + \tau \end{pmatrix}$$

is a twice product-suspended operator but not a suspended operator unless P is a differential operator. In this latter case, which includes Dirac operators, the periodicity of the determinant line bundle can be formulated using only suspended operators.

11. Generalization to product-suspended operators

As written, Theorem 9.1 applies to elliptic families of once-suspended cusp pseudodifferential operators. As we discussed in §10, this includes the result of Dai and Freed [5] for a family of self-adjoint Dirac operators D on a manifold with boundary by passing to the associated family of elliptic cusp operators and then to the elliptic family of once-suspended cusp operators $D + it$, the suspension parameter being t.

More generally, one can consider the case of an arbitrary elliptic family of first-order self-adjoint cusp pseudodifferential operators P. Then $P + it$ is not in general a once-suspended family of cusp operators. Instead we pass to the larger algebra, and related modules $\Psi_{\mathrm{cps}}^{k,l}(M/B; E, F)$, of product-suspended cusp pseudodifferential operators since then

$$P + it \in \Psi_{\mathrm{cps}}^{1,1}(M/B; E). \tag{11.1}$$

The ellipticity of P again implies the corresponding full ellipticity of $P + it$.

Enough of the properties of suspended (cusp) operators extend to the product-suspended case to allow the various definitions of regularized traces and the eta invariant to carry over to the more general case. In this context, the proof of Proposition 8.4 still applies, so the eta invariant is also multiplicative under composition of invertible fully elliptic cusp product-suspended operators.

So, given an elliptic family of cusp product-suspended operators (see Definition .2 in the Appendix), consider the bundle $\mathcal{P}$ of invertible perturbations by elements in

$$\Psi_{\mathrm{cps}}^{-\infty,-\infty}(M/B; E, F) = \Psi_{\mathrm{cs}}^{-\infty}(M/B; E, F).$$

Again the fibres are non-empty since the full ellipticity of P_b implies that it is invertible for large values of the suspension parameter t. Then the same argument as in the proof of Theorem 7.1 of [9] applies, using the contractibility of $G_{\mathrm{cu}}^{-\infty}(M_b; E)$, to show the existence of an invertible perturbation $P_b + Q_b$. At the same time, this shows the existence of an invertible perturbation of the indicial family $I(P_b) \in \Psi_{\mathrm{ps}(2)}^{k,l}(M_b; E_b, F_b)$, and so the associated index bundle and determinant bundle of the indicial family are also well defined, in the latter case using the $*$-product as before.

Consequently, we can formulate the following generalization of Theorem 9.1 with the proof essentially unchanged.

Theorem 11.1. *If $P \in \Psi_{\mathrm{cps}}^{k,l}(M/B; E, F)$ is a fully elliptic family of cusp product-suspended pseudodifferential operators and $\mathcal{P}$ is the bundle of invertible perturbations by elements of $\Psi_{\mathrm{cs\star}}^{-\infty}(M/B : E, F)$, then the τ invariant*

$$\tau = \exp(i\pi\eta_{\mathrm{cu}}) : \mathcal{P} \longrightarrow \mathbb{C}^*$$

descends to a non-vanishing linear function on the determinant line bundle of the indicial family $\tau : \mathrm{Det}(I(P)) \longrightarrow \mathbb{C}$.

As a special case, Theorem 11.1 includes elliptic families of self-adjoint first-order cusp pseudodifferential operators $P \in \Psi^1_{\mathrm{cu}}(M/B; E, F)$ by considering the cusp product suspended family

$$P + it \in \Psi^{1,1}_{\mathrm{cps}}(M/B; E, F).$$

Appendix. Product-suspended operators

In this appendix, we will briefly review the main properties of product-suspended pseudodifferential operators and then discuss the steps needed to extend this notion to the case of the cusp algebra of pseudodifferential operators on a compact manifold with boundary as used in §11. For a more detailed discussion on product-suspended operators see [10].

For the case of a compact manifold without boundary, product-suspended operators are, formally, generalizations of the suspended operators by relaxing the conditions on the the full symbols. This is achieved by replacing the radial compactification $\overline{\mathbb{R}^p} \times T^*X$ by the following blown-up version of it

$$^X\overline{\mathbb{R}^p \times T^*X} = [\overline{\mathbb{R}^p \times T^*X}; \partial(\overline{\mathbb{R}^p} \times X)] \tag{A.2}$$

where X is understood as the zero section of T^*X. If ρ_r and ρ_s denote boundary defining functions for the 'old' boundary and the 'new' boundary (arising from the blow-up) then set

$$S^{z,z'}(^X\overline{\mathbb{R}^p \times T^*X}; \hom(E, F)) = \rho_r^{-z} \rho_s^{-z'} \mathcal{C}^\infty(^X\overline{\mathbb{R}^p \times T^*X}; \hom(E, F)). \tag{A.3}$$

This is the space of 'full symbols' of product-suspended pseudodifferential operators (with possibly complex multiorders). After choosing appropriate metrics and connections, Weyl quantization gives families of operators on X which we interpret as elements of $\Psi^{z,z'}_{\mathrm{ps}(p)}(X; E, F)$, the space of product p-suspended operators of order (z, z'). However this map is not surjective modulo rapidly decaying smoothing operators as is the case for ordinary (suspended) pseudodifferential operators. Rather we need to allow as a subspace

$$\Psi^{-\infty,z'}_{\mathrm{ps}(p)}(X; E, F) = \rho_s^{-z'} \mathcal{C}^\infty(\overline{\mathbb{R}^p} \times X^2; \mathrm{Hom}(E, F), \Omega_R)) \subset \Psi^{-\infty,z'}_{\mathrm{ps}(p)}(X; E, F) \tag{A.4}$$

considered as smoothing operators on X with parameters in $\mathbb{R}^p$. The image of the symbol space (A.3) under Weyl quantization is, modulo such terms, independent of the choices made in its definition. The product-suspended operators form a bigraded algebra.

Property 1. For all k, k', l and l' and bundles E, F and G,

$$\Psi^{k,k'}_{\mathrm{ps}(p)}(X; F, G) \circ \Psi^{l,l'}_{\mathrm{ps}(p)}(X; E, F) \subset \Psi^{k+l,k'+l'}_{\mathrm{ps}(p)}(X; E, F).$$

There are two symbol maps. The usual symbol coming from the leading part of the full symbol in (A.3) at the 'old' boundary (B_σ) and the 'base family' which involves both the leading term of this symbol at the 'new' boundary and the leading term of the smoothing part in (A.4). These symbols are related by a compatibility condition just corresponding to the leading part of the full symbol at the corner.

Property 2. The two symbols give short exact sequences

$$\Psi_{\mathrm{ps}(p)}^{k-1,k'}(X;E,F) \longrightarrow \Psi_{\mathrm{ps}(p)}^{k,k'}(X;E,F) \xrightarrow{\sigma} \rho^{-k'}\mathcal{C}^{\infty}(B_{\sigma};\hom(E,F)), \qquad (A.5)$$

and

$$\Psi_{\mathrm{ps}(p)}^{k,k'-1}(X;E,F) \longrightarrow \Psi_{\mathrm{ps}(p)}^{k,k'}(X;E,F) \xrightarrow{\beta} \Psi^{k}((X \times \mathbb{S}^{p-1})/\mathbb{S}^{p-1};E,F) \qquad (A.6)$$

and the joint range is limited only by the condition

$$\sigma(\beta) = \sigma|_{\partial B_{\sigma}}. \qquad (A.7)$$

Ellipticity of A is the condition of invertibility of $\sigma(A)$ and full ellipticity is in addition the invertibility of $\beta(A)$.

Property 3. If a fully elliptic product-suspended operator $Q \in \Psi_{\mathrm{ps}(p)}^{k,k'}(X;E,F)$ is invertible (i.e., is bijective from $\mathcal{C}^{\infty}(X;E)$ to $\mathcal{C}^{\infty}(X;F)$) then its inverse is an element of $\Psi_{\mathrm{ps}(p)}^{-k,-k'}(X;F,E)$.

In §11, we also make use of the following important properties.

Property 4. For all $k \in \mathbb{Z}$,

$$\Psi_{\mathrm{s}(p)}^{k}(X;E,F) \subset \Psi_{\mathrm{ps}(p)}^{k,k}(X;E,F),$$
$$\Psi_{\mathrm{s}(p)}^{-\infty}(X;E,F) = \Psi_{\mathrm{ps}(p)}^{-\infty,-\infty}(X;E,F).$$

Property 5. If $P \in \Psi^{1}(X;E,F)$ is any first-order pseudodifferential operator, then

$$P + it \in \Psi_{\mathrm{ps}(1)}^{1,1}(X;E,F)$$

where t is the suspension parameter.

Property 6. Given $Q \in \Psi_{\mathrm{ps}(p)}^{k,l}(X;E,F)$,

$$\left(\frac{\partial Q}{\partial t_i}\right) \in \Psi_{\mathrm{ps}(p)}^{k-1,l-1}(X;E,F),$$

where $t = (t_1, \ldots, t_p)$ is the suspension parameter and $i \in \{1, \ldots, p\}$.

Next we extend this to a construction of cusp product-suspended pseudodifferential operators on a compact manifold with boundary Z. Again, for suspended cusp operators, there is a Weyl quantization map from the appropriate space of classical symbols $\rho^{-z}\mathcal{C}^{\infty}(\overline{\mathbb{R}^p \times {}^{\mathrm{cu}}T^*Z};\hom(E,F))$ which is surjective modulo cusp operators of order $-\infty$ decaying rapidly in the parameters. To capture the product-suspended case consider the spaces of symbols analogous to (A.3)

$$\rho_r^{-z}\rho_s^{-z'}\mathcal{C}^{\infty}([\overline{\mathbb{R}^p \times {}^{\mathrm{cu}}T^*Z}; \partial(\overline{\mathbb{R}^p \times Z})];\hom(E,F))$$

with the corresponding 'old' and 'new' boundaries. Then an element of the space

$$A \in \Psi_{\mathrm{cps}(p)}^{z,z'}(Z;E,F)$$

is the sum of the Weyl quantization (for the cusp algebra) of an element of (11) plus an element of the residual space

$$\Psi^{-\infty,z'}_{\mathrm{cps}(p)}(Z;E,F) = \rho^{-z'}\mathcal{C}^{\infty}(\overline{\mathbb{R}^p};\Psi^{-\infty}_{\mathrm{cu}}(Z;E,F)). \tag{A.8}$$

Now, with this definition the properties above carry over to the boundary setting. Property 1 is essentially unchanged. The same two homomorphisms are defined, the symbol and base family, with the latter taking values in families of cusp operators. In addition the indicial family for cusp operators leads to a third homomorphism giving a short exact sequence

$$x\Psi^{m,m'}_{\mathrm{cps}(p)}(Z;E,F) \longrightarrow \Psi^{m,m'}_{\mathrm{cps}(p)}(Z;E,F) \xrightarrow{I_{\mathrm{cu}}} \Psi^{k,l}_{\mathrm{ps}(p),\mathrm{sus}}(\partial Z;E,F)$$

where the image space has the same p product-suspended variables but taking values in the suspend operators on ∂Z. Since the suspended algebra may be realized in terms of ordinary pseudodifferential operators on $\mathbb{R} \times \partial Z$ there is no difficulty in considering these 'mixed-suspended' operators.

Definition .2. A cusp product-suspended operator $A \in \Psi^{m,m'}_{\mathrm{cps}(p)}(Z;E,F)$ is said to be *elliptic* if both its symbol $\sigma(A)$ and its base family $\beta(A)$ are invertible; it is *fully elliptic* if its symbol $\sigma(A)$, its base family $\beta(A)$ and its indicial family are all invertible.

References

[1] P. Albin and R.B. Melrose, *Fredholm realizations of elliptic symbols on manifolds with boundary*, Arxiv: math.DG/0607154.

[2] M.F. Atiyah, V.K. Patodi, and I.M. Singer, *Spectral asymmetry and Riemannian geometry III*, Math. Proc. Cambridge Philos. Soc. **79** (1976), 71–99.

[3] M.F. Atiyah and I.M. Singer, *The index of an elliptic operator IV*, Ann. of Math. **93** (1971), 119–138.

[4] J.-M. Bismut and D. Freed, *The analysis of elliptic families, II*, Comm. Math. Phys. **107** (1986), 103.

[5] X. Dai and D.S. Freed, *η-invariants and determinant lines*, J. Math. Phys. **35** (1994), no. 10, 5155–5194, Topology and physics.

[6] R. Mazzeo and R.B. Melrose, *Pseudodifferential operators on manifolds with fibred boundaries*, Asian J. Math. **2** (1999), no. 4, 833–866.

[7] R.B. Melrose, *The eta invariant and families of pseudodifferential operators*, Math. Res. Lett. **2** (1995), no. 5, 541–561. MR **96h:**58169

[8] R.B. Melrose and V. Nistor, *Homology of pseudodifferential operators I. manifold with boundary*.

[9] R.B. Melrose and F. Rochon, *Families index for pseudodifferential operators on manifolds with boundary*, IMRN (2004), no. 22, 1115–1141.

[10] ―――, *Periodicity of the determinant bundle*, Comm. Math. Phys. **274** (2007), no. 1, 141–186.

[11] W. Müller, *Eta invariants and manifolds with boundary*, J. Diff. Geom. **40** (1994), 311–377.

[12] D. Quillen, *Determinants of Cauchy-Riemann operators over a Riemann surface*, Funct. Anal. Appl. **19** (1985), 31–34.

Richard Melrose
Department of Mathematics
MIT
77 Massachusetts Avenue
Cambridge, MA 02139-4307
USA
e-mail: `rbm@math.mit.edu`

Frédéric Rochon
Department of Mathematics
University of Toronto
40 St. George Street
Toronto, Ontario, M5S 2E4
Canada
e-mail: `rochon@math.utoronto.ca`

C^*-algebras and Elliptic Theory II

Trends in Mathematics, 183–206

© 2008 Birkhäuser Verlag Basel/Switzerland

Elliptic Theory on Manifolds with Corners: I. Dual Manifolds and Pseudodifferential Operators

Vladimir Nazaikinskii, Anton Savin and Boris Sternin

Abstract. In this first part of the paper, we define a natural dual object for manifolds with corners and show how pseudodifferential calculus on such manifolds can be constructed in terms of the localization principle in C^*-algebras. In the second part, these results will be applied to the solution of Gelfand's problem on the homotopy classification of elliptic operators for the case of manifolds with corners.

Mathematics Subject Classification (2000). Primary 58J05; Secondary 47L15, 35S35.

Keywords. Manifold with corners, elliptic operator, localization principle.

Introduction

This paper deals with elliptic theory on manifolds with corners. Such manifolds arise for instance if one supplements the class of closed manifolds by manifolds with boundary and considers products of manifolds. A natural class of operators on such manifolds was introduced by Melrose [7, 8]. Operators on manifolds with corners have been actively studied; see, e.g., [1, 3, 4, 5, 6, 10, 11, 12, 13, 15].

The present paper consists of two parts. In the first part, we define a natural dual object for manifolds with corners and show how pseudodifferential calculus on such manifolds can be constructed in terms of the localization principle in C^*-algebras. In the second part, these results will be applied to the solution of Gelfand's problem on the homotopy classification of elliptic operators for the case of manifolds with corners.

Supported in part by RFBR grants 05-01-00982 and 06-01-00098, by President of the Russian Federation grant MK-1713.2005.1, and by the DFG project 436 RUS 113/849/0-1® "K-theory and Noncommutative Geometry of Stratified Manifolds."

In more detail, the outline of the first part is as follows. In Section 1, we deal with the geometry of manifolds with corners. Specifically,

- In Section 1.1 we recall some facts and definitions concerning manifolds with corners. Most of the material in this section is not new, except possibly in form.
- In Section 1.2 we introduce a new geometric object, the dual manifold $M^{\#}$ of a manifold M with corners, and study some structures on it. The importance of this space lies in the fact that, on the one hand, pseudodifferential operators on manifolds with corners can be naturally defined as operators local with respect to the action of the algebra of continuous functions on the dual manifold. On the other hand, as will be shown in the second part of this paper, under an additional assumption the K-homology of the dual manifold $M^{\#}$ classifies the elliptic theory on M.

In Section 2 we define zero-order pseudodifferential operators (ψDO) in L^2 spaces on manifolds with corners. The definition is based on the localization principle in C^*-algebras (e.g., see [16, Proposition 3.1]), goes by induction over the depth of the manifold, i.e., the maximum codimension of the strata (one starts from smooth manifolds, which have depth zero), and naturally involves parameter-dependent ψDO (which serve as symbols for ψDO at subsequent inductive steps). Hence we need some preliminaries:

- In Section 2.1 we introduce L^2 spaces on manifolds with corners.
- In Section 2.2 we discuss translation-invariant operators in vector bundles over manifolds with corners and their relationship with parameter-dependent operators.
- In Section 2.3 we present the adaptation [14] of the localization principle to operator families. The proofs are either contained in [14] or can be obtained from those in [14] by obvious modifications; hence we omit them altogether.

After that, in Section 2.4 we give the definition of ψDO and prove their properties.

Nomenclature

We shall use the following notation.

$L^2(X, \mu, H)$ is the space of square integrable H-valued functions on a metric space X with respect to a measure μ (where H is a Hilbert space). We omit the argument H if $H = \mathbb{C}$ and also omit μ if it is clear from the context.

$\mathcal{B}H$ and $\mathcal{K}H$ are the algebra of bounded operators and the ideal of compact operators in a Hilbert space H.

$C(X, \mathcal{A})$ is the C^*-algebra of continuous bounded functions on X ranging in a C^*-algebra $\mathcal{A}$, and $C_0(X, \mathcal{A})$ is the subalgebra of functions decaying at infinity. We omit the argument $\mathcal{A}$ if $\mathcal{A} = \mathbb{C}$.

k=1

k=2

k=3

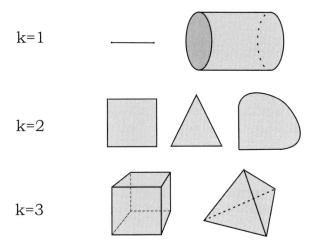

Figure 1. Manifolds with corners of depth k.

1. Geometry

1.1. Manifolds with corners and their faces

Definition 1.1. A *manifold of dimension n with corners* is a Hausdorff topological space M in which each point x has a coordinate neighborhood of the form $\overline{\mathbb{R}}_{+}^{d} \times \mathbb{R}^{n-d}$, $d = d(x) \in \{0, \ldots, n\}$, where x is represented by the origin. Moreover, the transition maps are smooth functions. Unless specified otherwise, we assume that M is connected and compact. The maximum number d is called the *depth* of the manifold and will be denoted by $k = k(M)$.

Some examples of manifolds with corners are shown in Fig. 1.

Open faces. The set

$$M_l = \{x \in M : d(x) = l\}$$

is a smooth manifold of codimension l in M. Its connected components are called *open faces of codimension l*. Let $\Gamma_j^{\circ}(M)$, $j = 0, \ldots, N$, be all possible open faces of M, and let d_j be their codimensions. We assume that $d_0 = 0$ (so that $\Gamma_0^{\circ}(M) = M^{\circ}$ is the interior of M) and $d_j > 0$ for $j > 0$. Thus M is represented as the disjoint union

$$M = \bigsqcup_{j=0}^{N} \Gamma_j^{\circ}(M) \equiv M^{\circ} \sqcup \partial M, \quad \text{where } \partial M = \bigsqcup_{j=1}^{N} \Gamma_j^{\circ}(M) \text{ is the } \textit{boundary} \text{ of } M.$$

Faces of codimension one are called *hyperfaces*.

Local defining functions. By definition, each point $x \in F$ of a face of codimension d has a neighborhood $U \subset M$ with local coordinates $\rho_1, \dots, \rho_n$ such that the manifold is determined in these coordinates by the system of inequalities

$$\rho_1 \geq 0, \dots, \rho_d \geq 0. \tag{1}$$

The coordinates $(\rho_1, \dots, \rho_d)$ are called *local defining functions* of the face F. We point out the difference between the definition of manifold with corners used in this paper and the one in [7], where it is additionally assumed that the defining functions exist globally.

Closed faces.

Proposition 1.2. *There exist canonically defined manifolds $\Gamma_j(M)$ with corners such that $\Gamma_j^\circ(M)$ is the interior of $\Gamma_j(M)$ and the diagram*

$$
\begin{array}{ccc}
\Gamma_j^\circ(M) & \longrightarrow & \Gamma_j(M) \\
& \searrow{\scriptstyle i_j^\circ} & \Big\downarrow{\scriptstyle i_j} \\
& & M,
\end{array}
$$

where the horizontal arrow and i_j° are natural embeddings and i_j is an immersion of manifolds with corners, commutes. The manifold $\Gamma_j(M)$ is called a closed face of M.

Proof is given in the Appendix. $\qquad\square$

Since $\Gamma_j(M)$ is a compact manifold with corners, we have

$$\partial\Gamma_j(M) = \bigsqcup_{l=1}^{L} \Gamma_l^\circ(\Gamma_j(M)).$$

The image under i_j of each open face $\Gamma_l^\circ(\Gamma_j(M))$, $l > 0$, of the manifold $\Gamma_j(M)$ coincides with some open face $\Gamma_r^\circ(M)$, $r = r(l)$, of M with $d_r > d_j$. In this case, we say that the faces Γ_j (or Γ_j°) and Γ_r (or Γ_r°) are *adjacent to each other* and write $\Gamma_j \succ \Gamma_r$. The restriction

$$i_{jl} := i_j|_{\Gamma_l^\circ(\Gamma_j(M))} \colon \Gamma_l^\circ(\Gamma_j(M)) \longrightarrow \Gamma_{r(l)}^\circ(M)$$

is a finite covering whose structure group is a quotient of the homotopy group $\pi_1(\Gamma_{r(l)}^\circ(M))$.

The compressed cotangent bundle. The compressed cotangent bundle T^*M of a manifold M with corners is defined in the usual way (see [7]). We take the subspace $\mathrm{Vect}_b(M)$ of the space $\mathrm{Vect}(M)$ of vector fields on M formed by vector fields tangent to all open faces. The subspace $\mathrm{Vect}_b(M)$ is a locally free $C^\infty(M)$-module.

Indeed, in local coordinates

$$(\rho_1, \dots, \rho_d, y_{d+1}, \dots, y_n) \in \overline{\mathbb{R}}_+^d \times \mathbb{R}^{n-d}$$

a local basis in $\mathrm{Vect}_b(M)$ is formed by the vector fields

$$\rho_1 \frac{\partial}{\partial \rho_1}, \ldots, \rho_d \frac{\partial}{\partial \rho_d}, \frac{\partial}{\partial y_{d+1}}, \ldots, \frac{\partial}{\partial y_n}.$$

Consequently, $\mathrm{Vect}_b(M)$ is the section space of some vector bundle on M, which will be denoted by TM (the *extended cotangent bundle* of M), and the *compressed cotangent bundle* T^*M is now defined as the bundle ($\mathbb{R}$-)dual to TM. In the local coordinates $(\rho_1, \ldots, \rho_d, y_{d+1}, \ldots, y_n)$, a basis in the module of sections of T^*M is given by the forms

$$\rho_1^{-1} d\rho_1, \ldots, \rho_d^{-1} d\rho_d, dy_{d+1}, \ldots, dy_n.$$

Conormal bundles of faces. Let $F = \Gamma_j^\circ(M)$ be an open face of codimension $d = d_j$ in M. We define the *conormal bundle* of F as the subset $N^*F \subset T^*M|_F$ formed by functionals ξ vanishing on any vector $v \in TM|_F$ that can be continued to a vector field second-order tangent to all faces in ∂M. One readily sees that N^*F is indeed a vector bundle; a basis in its fiber consists of the 1-forms

$$\rho_1^{-1} d\rho_1, \ldots, \rho_d^{-1} d\rho_d.$$

This bundle can be canonically extended to a bundle over the closed face $\overline{F}$; the latter bundle is called the *conormal bundle* of $\overline{F}$ and is denoted by $N^*\overline{F}$.

Proposition 1.3. *One has the canonical direct sum decomposition*

$$T^*M|_{\overline{F}} = T^*\overline{F} \oplus N^*\overline{F}$$

(where the bundle on the left-hand side is obtained as the pullback under the immersion of $\overline{F}$ in M).

Proof. The assertion is local, so that we can assume that $\overline{F}$ is embedded in M. Then the embedding $T^*\overline{F} \subset T^*M$ is obtained as the map dual to the restriction

$$\mathrm{Vect}_b(M) \longrightarrow \mathrm{Vect}_b(\overline{F})$$

of vector fields in $\mathrm{Vect}_b(M)$ to $\overline{F}$. Now the desired properties can be verified in coordinates. $\qquad\square$

Normal bundles of faces. Let $F = \Gamma_j^\circ(M)$ be again an open face of codimension $d = d_j$ in M, and let (ρ, y) and $(\widetilde{\rho}, \widetilde{y})$ be two coordinate systems on M in a neighborhood of some point in F. Since $\rho = \widetilde{\rho} = 0$ on F, we see that the change of variables $(\rho, y) \longmapsto (\widetilde{\rho}, \widetilde{y})$ has the form

$$\widetilde{y} = f(y) + O(\rho), \quad \widetilde{\rho} = A(y)\rho + O(\rho^2), \tag{2}$$

where $A(y)$ is a smooth $d \times d$ matrix function. The mapping (2) should take the positive quadrant with respect to the variable ρ to itself, and hence, letting ρ tend to zero, we verify that

$$A(y) = \Pi(y)\Lambda(y),$$

where $\Pi(y)$ is a permutation matrix (and hence is locally constant in y) and

$$\Lambda(y) = \mathrm{diag}\{\lambda_1(y), \ldots, \lambda_d(y)\}$$

is a diagonal matrix with positive entries. The cocycle condition for the matrices $A(y)$ implies that the matrices $\Pi(y)$ themselves satisfy the same cocycle condition, so that we can define the d-dimensional real vector bundle NF over F for which the matrices $\Pi(y)$ are the transition mappings. The change of variables

$$t_j = -\ln \rho_j, \quad j = 1, \ldots, d,$$

clarifies the meaning of this bundle. The second component in (2) becomes

$$\widetilde{t} = \Pi(y)t + \ln \Lambda(y) + O(e^{-2t}) = \Pi(y)t + O(1),$$

$$t_j \to +\infty, \quad j = 1, \ldots, d.$$

Thus NF is the "bundle of logarithms of determining functions" of the submanifold F. We call it the *logarithmic normal bundle* of F. The matrices Π simultaneously specify a bundle of positive quadrants $\overline{\mathbb{R}}_+^d$ over F, which we denote by N_+F and call the *normal bundle* of F. We have the exponential mapping

$$\exp : NF \longrightarrow N_+F,$$

$$(y, t) \longmapsto (y, \exp(-t)) = (y, e^{-t_1}, \ldots, e^{-t_d}),$$

which diffeomorphically maps the first bundle onto the interior of the second.

One readily sees that both bundles naturally extend to bundles $N\overline{F}$ and $N_+\overline{F}$ over the closed face $\overline{F}$.

By construction, the structure group of these bundles is a subgroup $\mathfrak{S}_{\overline{F}}$ of the permutation group $\mathfrak{S}_d$. (Thus the numbering of the coordinates ρ in all charts is chosen in such a way that the transition matrices range in the subgroup $\mathfrak{S}_{\overline{F}}$.)

Remark. We shall assume that the bundles $N\overline{F}$ and $N^*\overline{F}$ are reduced to the minimal possible permutation structure group $\mathfrak{S}_{\overline{F}}$.

Proposition 1.4. *The logarithmic normal bundle $N\overline{F}$ and the conormal bundle $N^*\overline{F}$ are canonically dual.*

Proof. It suffices to write out a natural invariant pairing; this can be done in the coordinates (ρ, y): for a form

$$\omega = \sum a_j \rho_j^{-1} d\rho_j \in N^*\overline{F}$$

and a vector

$$\xi = (b_1, \ldots, b_d) \in N\overline{F},$$

we set

$$\langle \omega, \xi \rangle = \sum a_j b_j.$$

Under changes of coordinates, the components of ξ and ω are subjected to the same permutation, and the defining functions ρ_j are multiplied by nonzero numbers (the diagonal entries of the matrix $\Lambda(y)$), which does not affect the logarithmic derivatives, so that the numbers a_j remain the same. Thus the pairing is independent of the choice of coordinates. $\qquad \square$

Remark. The bundles $N\overline{F}$ and $N^*\overline{F}$ viewed as bundles with the structure group $\mathfrak{S}_{\overline{F}}$ are canonically isomorphic, since permutation matrices are unitary.

Compatible exponential mappings. For each closed face $\Gamma_j(M)$ of a manifold M with corners, we have defined the normal bundle $N_+\Gamma_j(M)$. Just as with submanifolds of smooth manifolds, one can define *exponential mappings* of these bundles into the manifold M itself, which are local diffeomorphisms in a neighborhood of the zero section. Moreover, for adjacent faces these diffeomorphisms will be compatible in some sense. More precisely, the following theorem holds.

Theorem 1.5. *Let $\varepsilon > 0$ be sufficiently small. Then there exist smooth mappings*

$$f_j \colon N_+\Gamma_j(M) \longrightarrow M, \quad j = 1, \ldots, N,$$

defined for $|\rho| \le \varepsilon$, where ρ is the variable in the fiber of the bundle $N_+\Gamma_j(M)$, such that the following conditions hold:

1. *On the zero section, $f_j = i_j$.*
2. *f_j is a local diffeomorphism.*
3. *The restriction $f_j|_U$ of the mapping f_j to some neighborhood of the open face $\Gamma_j^\circ(M)$ in $N_+\Gamma_j(M)$ is a diffeomorphism.*
4. *If $\Gamma_j(M) \succ \Gamma_l(M)$, then the mappings f_j and f_l are locally compatible in the following sense. In a neighborhood of any point $x \in \Gamma_l(M)$, the diagram*

$$
\begin{array}{ccc}
N_+\Gamma_j(M) & \xrightarrow{\;f_l^{-1}\circ f_j\;} & N_+\Gamma_l(M) \\
{\scriptstyle \pi_1}\Big\downarrow & & \Big\downarrow{\scriptstyle \pi_2} \\
\Gamma_j(M) & \xrightarrow[\;f_l^{-1}\circ f_j\;]{} & \varphi(\Gamma_j(M))
\end{array}
\tag{3}
$$

commutes, where π_1 is the natural projection and π_2 is the projection in the fibers of $N_+\Gamma_l(M)$ onto the coordinate subbundle into which $\Gamma_j(M)$ is mapped under the local diffeomorphism $\varphi = f_l^{-1}\circ f_j$, along the complementary coordinate subbundle.

Proof will be given in the Appendix. □

Remark. (a) Let $\Gamma_j(M) \succ \Gamma_r(M)$. Since

$$N_+\Gamma_l^\circ(\Gamma_j(M)) \subset i_{jl}^* N_+\Gamma_r(M), \quad r = r(l),$$

we see that by specifying a compatible tuple of exponential mappings f_j for the faces of M we automatically specify such tuples for the faces of any closed face of M.

(b) The composition

$$\widetilde{f}_j = f_j \circ \{t \mapsto e^{-t}\} \colon N\Gamma_j(M) \longrightarrow M$$

will also be referred to as the exponential map.

Corollary 1.6. *The manifold M can be covered by finitely many coordinate neighborhoods U with coordinates $\rho_U = (\rho_1, \ldots, \rho_n)$ such that M is given in these coordinates by the system of inequalities* (1) *and the following compatibility condition holds. Suppose that two charts U and U' have a nonempty intersection.*

1. *If the number of defining functions in U and U' is the same, then they coincide in $U \cap U'$ up to a permutation.*
2. *Otherwise, the smaller set of defining functions is a subset of the larger set in $U \cap U'$.*

Remark 1.7. This assertion plays the same role in the theory of manifolds with corners as the collar neighborhood theorem does in the theory of manifolds with boundary.

To be definite, we assume in the following that the defining functions specify coordinates in the domains where they are less than $3/2$.

1.2. The dual manifold $M^{\#}$ and the algebra $C(M^{\#})$

Definitions. On the space $\mathbb{R}^k_t \times \mathbb{R}^m_x$, we define the algebra $C(k, m)$ of bounded continuous functions $f(t, x)$ such that

$$f(\omega|t|, x) \longrightarrow F(\omega) \quad \text{as } |t| \to \infty$$

uniformly with respect to x and $\omega = t/|t|$, where $F(\omega)$ is some (continuous) function.

In the algebra of continuous functions on the interior M° of the manifold M, we single out a subalgebra $C(M^{\#})$ as follows. We say that $f \in C(M^{\#})$ if for each coordinate neighborhood $U \simeq \mathbb{R}^k_+ \times \mathbb{R}^{n-k}$ on M the function

$$F(t, x) = f|_U(e^{-t_1}, \ldots, e^{-t_k}, x_{k+1}, \ldots, x_n)$$

can be extended to a function in $C(k, n - k)$.

One can readily see that each function $f \in C(M^{\#})$ is constant on each hyperface of M.

One can readily describe the space $M^{\#}$ of maximal ideals of the algebra $C(M^{\#})$. As a set, it is the disjoint union of the interior M° of the manifold M and the following sets $F^{\#}$ corresponding to faces F of positive codimension.

- To each hyperface F, there corresponds a singleton $F^{\#}$.
- To each face F of codimension $k = \operatorname{codim} F > 1$, there corresponds a set $F^{\#}$ that is the quotient of the open $k - 1$-simplex

$$\overset{\circ}{\triangle}_{k-1} = \left\{ x \in \mathbb{R}^k : x_i > 0,\ i = 1, \ldots, k, \quad \sum x_i = 1 \right\}$$

by the action of the structure group $\mathfrak{S}_F$ of the bundle NF.

We omit the straightforward but cumbersome description of the topology on $M^{\#}$, which can be derived from the preceding. We only note that, for two faces F_1 and F_2, the set $F_1^{\#}$ is adjacent to $F_2^{\#}$ (the adjacency being induced by an embedding of a simplex of smaller dimension in the boundary of a simplex of larger dimension) if and only if F_2 is adjacent to F_1.

Example 1.8. (1) If M is a manifold with boundary, then $C(M^\#)$ is the algebra of functions constant on connected components of the boundary. Hence $M^\#$ is obtained by shrinking each boundary component into a point.

(2) Any polyhedron in $\mathbb{R}^3$ in which exactly three edges (and hence three facets) meet at each vertex is a manifold with corners. For these polyhedra, our notion of duality coincides with the standard definition of the dual polyhedron. In particular, the dual of a cube is an octahedron, and the dual of a dodecahedron is an icosahedron. The tetrahedron proves to be self-dual.

Fibered structure on the dual space. Here we assume that M is a manifold with corners such that the normal bundle of each face is trivial and show that a neighborhood of each simplex $F^\#$ of the dual manifold $M^\#$ is fibered over $F^\#$ with fiber being a cone. This result will be used only in the proof of the classification theorem in the second part of this paper.

Let $F \subset M$ be an open face of codimension j. We shall construct a neighborhood $U^\#$ of the simplex $F^\#$ in the dual manifold $M^\#$.

First, we construct a neighborhood $U \subset N_+F$. It is convenient to use the logarithmic coordinates

$$\begin{aligned}\ln : N_+F \setminus F &\xrightarrow{\;\simeq\;} NF, \\ (x, \rho_F) &\longmapsto (x, y = -\ln \rho_F).\end{aligned} \tag{4}$$

Here $\rho_F = (\rho_1, \ldots, \rho_j)$ is the set of defining functions of F and the logarithm is taken componentwise. By virtue of the triviality assumption, it is globally defined.

The image of the set in which $\rho_l < 1$ for all $1 \leq l \leq j$ will be denoted by $N'_+F \subset NF$. In the coordinates y, it is given by the condition $y > 0$.

We use similar coordinates in neighborhoods of all faces of the face $\overline{F}$. Then in the space N'_+F we obtain the following coordinates: the coordinates $y \in \mathbb{R}^j_+$ in the fibers; the coordinates in the neighborhood $\mathbb{R}^l_+ \times \mathbb{R}^{n-j-l} \subset F$ of codimension l in F, which will be denoted by

$$(x, \omega), \text{ where } (x_1, \ldots, x_l) = (-\ln \rho_1, \ldots, -\ln \rho_l).$$

(The coordinates x are uniquely determined up to permutation; the number of these coordinates is determined by the codimension in F of the face near which the point sits.)

To construct the neighborhood U, on F we define the function $|x| := \sum_s x_s$. This is invariant under permutations of defining functions and hence well defined.

Now we globally define a set $U \subset N_+F$ by the condition

$$U = \Big\{ (y, x, \omega) \in N'_+F \mid \min y > |x| + 1 \Big\}$$

in local coordinates, where $\min y$ is the minimum of the coordinates $y_1, \ldots, y_j$. By way of example, Fig. 2 shows the case in which the manifold with corners is a 1-gon; the set U corresponding to the one-dimensional edge is shown in the lower part of the figure as a dashed infinite domain.

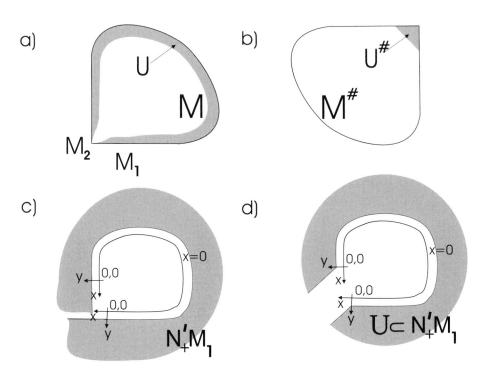

Figure 2. a) the manifold M; b) the dual space $M^\#$; c) the positive quadrant in the normal bundle N'_+M_1; d) the neighborhood $U \subset N'_+M_1$.

Consider the space

$$M^\#_{\geq j} = M^\# \setminus \bigcup_{j'=1}^{j-1} M^\#_{j'},$$

obtained from $M^\#$ by deleting all simplices of dimension $\leq j - 2$.

Lemma 1.9. 1. *The restriction of the projection $p : N_+F \to M$ to U is one-to-one (i.e., U can also be treated as an open set in M; see top left in Fig. 2).*
  2. *The dual space $U^\# \subset M^\#$ is an open neighborhood of the open simplex $F^\#$ in $M^\#_{\geq j}$ (see top right in Fig. 2).*

Proof. Let us prove that p is one-to-one. This can be violated only where distinct parts of F meet each other. We should prove that the projections of components of U corresponding to two adjacent faces are disjoint. Indeed, let U be defined in the first part by the condition

$$\min y > |x| + 1.$$

Then in the second part some of the coordinates x_I are interchanged with some of the coordinates y_I for some nonempty index set I. Then the set U in the second part is described by the inequality

$$\min(x_I, y_{\overline{I}}) > |y_I| + |x_{\overline{I}}| + 1$$

(in the original coordinates). Writing out these two systems componentwise, we see that they are inconsistent, so that the projections of the parts of U into M are disjoint.

The second assertion holds by construction.   □

Now we can prove that the neighborhood $U^\#$ of the stratum $F^\#$ is homeomorphic to the product of $F^\#$ by the cone

$$K_\Omega = [0, 1) \times \Omega / \{0\} \times \Omega.$$

Here the base Ω of the cone is the dual space $\overline{F}^\#$ of the closed face $\overline{F}$. The dual manifold is well defined, since the closed face is a manifold with corners. As a result, we find that $M^\#$ is a stratified manifold with singularities.

Proposition 1.10. *The projection*

$$\widetilde{p} : U^\# \to F^\#, \quad \widetilde{p}(y) := y/|y|,$$

is well defined. Its fiber is the cone $K_{\overline{F}^\#}$, and there is a homeomorphism

$$U^\# \simeq F^\# \times K_{\overline{F}^\#}.$$

Proof. A straightforward computation shows that the projection $\widetilde{p}$ is well defined. We define the map

$$
\begin{array}{ccc}
U & \longrightarrow & F^\# \times (0, 1) \times F, \\
(y, x, \omega) & \longmapsto & \left[\dfrac{y}{|y|}, \dfrac{|x| + 1}{\min y}, (x, \omega) \right],
\end{array}
$$

The inverse map has the form

$$
\begin{array}{ccc}
F^\# \times (0, 1) \times F & \longrightarrow & U, \\
(\theta, r, x, \omega) & \longmapsto & \left[\dfrac{\theta}{\min \theta} \dfrac{|x| + 1}{r}, x, \omega \right],
\end{array}
$$

A routine verification of the fact that these mappings extend to homeomorphisms $U^\# \simeq F^\# \times K_{\overline{F}^\#}$ is left to the reader.   □

2. Pseudodifferential operators

2.1. The space $L^2(M)$

Our ψDO will act in the space $L^2(M)$, which is defined as follows.

Let M be a compact manifold with corners, and let $d\mathrm{vol}$ be a smooth measure on M (obtained, say, via an embedding of M in a compact Riemannian manifold).

Now for each point $x \in M$ we define a measure μ_x in $M^\circ \cap U_x$, where $U_x \simeq V \subset \overline{\mathbb{R}}_+^k \times \mathbb{R}^{n-k}$, $k = d(x)$, is a coordinate neighborhood of x, by setting

$$\mu_x = (\rho_1 \rho_2 \cdots \rho_k)^{-1} dvol, \tag{5}$$

where $\rho_1, \ldots, \rho_k$ are the coordinates in the $\mathbb{R}_+$-factors. Next, we take a finite cover $M = \bigcup_{j=1}^{N'} U_{x_j}$ and a subordinate partition of unity $\{e_j\}$ and set

$$\mu = \sum_{j=1}^{N'} e_j \mu_{x_j}. \tag{6}$$

This measure is up to equivalence independent of the ambiguity in the construction, and hence the space $L^2(M) \overset{\text{def}}{\equiv} L^2(M^\circ, \mu)$ is well defined up to norm equivalence. For the following, we choose and fix such a measure and hence a Hilbert space structure in $L^2(M)$. Note that the interiors of M and $M^{\#}$ are the same, and so $L^2(M)$ can also be viewed as $L^2(M^{\#})$ (with respect to the same measure). Hence it bears the natural structure of a $C(M^{\#})$-module.

2.2. Translation-invariant operators

By $\mathfrak{S}_s$ we denote the permutation group on s elements.

Let M be a connected compact manifold with corners, and let $E \longrightarrow M$ be a vector bundle on M with fiber $\mathbb{R}^s$ and structure group $\mathfrak{S}_s$ acting by permutations of the standard basis vectors. We reduce E to a minimal structure group $\mathcal{G} \subset \mathfrak{S}_s$ (which is uniquely determined up to conjugacy) and consider the principal $\mathcal{G}$-bundle $\pi \colon \widetilde{M} \longrightarrow M$ associated with E. The following assertion is routine.

Proposition 2.1. *The space $\widetilde{M}$ is a connected manifold with corners equipped with the natural action of $\mathcal{G}$ given in any chart $U \times \mathcal{G}$ on $\widetilde{M}$, where U is a chart on M, by the formula $\sigma(z, g) = (z, g\sigma^{-1})$, $\sigma \in \mathcal{G}$. The lift $\pi^* E$ is a trivial bundle, $\pi^* E \simeq \widetilde{M} \times \mathbb{R}^s$, where the trivialization is uniquely determined up to an automorphism of $\mathbb{R}^s$. The natural projection $\hat{\pi} \colon \pi^* E \longrightarrow E$ is given in coordinates by the formula*

$$U \times \mathcal{G} \times \mathbb{R}^s \ni (z, g, y) \longmapsto (z, gy) \in U \times \mathbb{R}^s.$$

The space $L^2(E)$ (where the measure on E is locally chosen as the direct product of the measure on M constructed in the preceding subsection by the standard Lebesgue measure in the fibers) can be identified via the mapping

$$\hat{\pi}^* \colon L^2(E) \longrightarrow L^2(\pi^* E)$$

with the subspace $L^2_{\mathcal{G}}(\pi^* E) \subset L^2(\pi^* E)$ formed by $\mathcal{G}$-invariant functions $u(x, y)$, i.e., functions satisfying the condition

$$u(\sigma x, \sigma y) = u(x, y), \qquad x \in \widetilde{M}, \quad y \in \mathbb{R}^s, \quad \sigma \in \mathcal{G}.$$

Definition 2.2. A bounded operator

$$A \colon L^2(E) \longrightarrow L^2(E)$$

is said to be *translation invariant* if it is the restriction to $L^2_{\mathcal{G}}(\pi^* E) \simeq L^2(E)$ of a bounded operator

$$\widetilde{A} \colon L^2(\pi^* E) \longrightarrow L^2(\pi^* E) \tag{7}$$

such that $\widetilde{A} L^2_{\mathcal{G}}(\pi^* E) \subset L^2_{\mathcal{G}}(\pi^* E)$ and the condition

$$[\widetilde{A} u](x, y + t) = \widetilde{A}[u(x, y + t)], \qquad t \in \mathbb{R}^s,$$

of invariance under translations in $\mathbb{R}^s$ is satisfied.

Proposition 2.3. *If* $A \colon L^2(E) \to L^2(E)$ *is a translation-invariant operator, then the corresponding operator* (7) *is unique and commutes with the action of* $\mathcal{G}$. *Moreover,* $\|A\| = \|\widetilde{A}\|$.

Remark 2.4. Since the trivialization of $\pi^* E$ is uniquely determined up to an automorphism of $\mathbb{R}^s$ (independent of the point of the base $\widetilde{M}$), the notion of a translation-invariant operator in $\widetilde{E}$ is well defined. (Indeed, automorphisms of $\mathbb{R}^s$ take translations to translations.)

Proof of Proposition 2.3. The translation-invariant operator (7) can be represented in the form

$$\widetilde{A} = B\left(-i \frac{\partial}{\partial y}\right), \tag{8}$$

where

$$B(p) \colon L^2(\widetilde{M}) \longrightarrow L^2(\widetilde{M}), \qquad p \in \mathbb{R}^s, \tag{9}$$

is a bounded operator-valued function strongly measurable with respect to p. (This is well known routine fact; one possible proof is given in [14, Proposition 16].)

   In terms of the function $B(p)$, the assertion of Proposition 2.3 acquires the form of the following lemma. (We give a slightly more general statement omitting the assumption that the group $\mathfrak{G}$ acts by permutations of basis vectors.)

Lemma 2.5. *Let a finite group* $\mathfrak{G}$ *act on the space* $L^2(\mathbb{R}^s; H)$, *where* H *is a Hilbert space, by the formula*

$$[T_g f](p) = S_g f(\sigma_g^{-1} p), \quad p \in \mathbb{R}^s,$$

where S *is a representation of* $\mathfrak{G}$ *on* H *and* σ *is a* **faithful** *representation of the same group on* $\mathbb{R}^s_p$ *by linear transformations. Let*

$$B(p) \colon H \longrightarrow H, \qquad p \in \mathbb{R}^s,$$

be a bounded strongly measurable operator-valued function such that the operator

$$B \colon L^2(\mathbb{R}^s; H) \longrightarrow L^2(\mathbb{R}^s; H)$$

induced by the pointwise application of $B(p)$ *preserves the subspace*

$$L^2_{\mathfrak{G}}(\mathbb{R}^s; H) = \{f \in L^2(\mathbb{R}^s; H) \colon T_g f = f \quad \forall g \in \mathfrak{G}\} \tag{10}$$

of $\mathfrak{G}$-*invariant functions. Then*

(i) *The norm of B is equal to the norm of the restriction B° of B to the subspace (10). In particular, if $B(p)f(p) = 0$ for almost all p for any element $f \in L^2_\mathfrak{G}(\mathbb{R}^s; H)$, then $B(p) = 0$ for almost all p.*

(ii) *The operator B is $\mathfrak{G}$-invariant, or, equivalently, the operator function $B(p)$ satisfies $B(p) = S_g^{-1}B(\sigma_g(p))S_g$, $g \in \mathfrak{G}$, for almost all $p \in \mathbb{R}^s$.*

Proof. (i) The norm of B is given by the formula

$$\|B\| = \operatorname*{ess\,sup}_{p \in \mathbb{R}^s} \|B(p)\| = \operatorname*{ess\,sup}_{p \in \Omega} \|B(p)\|,$$

where $\Omega \subset \mathbb{R}^s$ is an arbitrary subset of full measure. For this set we take

$$\Omega = \mathbb{R}^s \setminus \Big(\bigcup_{\substack{g \in \mathfrak{G} \\ g \neq e}} \operatorname{fix}\sigma_g \Big),$$

where $\operatorname{fix}\sigma_g$ is the set of fixed points of σ_g (which is at most a hyperplane, since σ is faithful).

Let $\|B\| = K$ and $\|B^\circ\| = K^\circ$. Then $K^\circ \leq K$. To prove the opposite inequality, note that for each $\varepsilon > 0$ there exists a point $p_0 \in \Omega$ such that

$$\operatorname*{ess\,sup}_{p \in U} \|B(p)\| > K - \varepsilon,$$

where U is an arbitrary neighborhood of p_0. It follows that there exists a function $f \in L^2(\mathbb{R}^s; H)$ supported in U such that

$$\|Bf\| \geq (K - \varepsilon)\|f\|.$$

Let U be so small that

$$\sigma_g(U) \cap \sigma_h(U) = \varnothing \quad \text{for} \quad g \neq h, \quad g, h \in \mathfrak{G}.$$

Then the function

$$\varphi(p) = \begin{cases} S_g f(p) & \text{if } p \in \sigma_g(U) \quad \text{for some } g \in \mathfrak{G}, \\ 0 & \text{otherwise} \end{cases} \tag{11}$$

is well defined and $\mathfrak{G}$-invariant, and so $B\varphi$ is also $\mathfrak{G}$-invariant by assumption. Since B acts by pointwise application of $B(p)$, we have

$$\|Bf\| = \frac{1}{|\mathfrak{G}|}\|B\varphi\| < \frac{1}{|\mathfrak{G}|}K^\circ\|\varphi\| = K^\circ\|f\|,$$

so that $K^\circ > K - \varepsilon$. Since ε is arbitrary, we have $K^\circ = K$.

(ii) The second assertion of the lemma is proved by the same method. Let $U \subset \Omega$ be an arbitrary sufficiently small open set, and let $f \in L^2(\mathbb{R}^s; H)$ be a function supported in U and equal to an arbitrary vector $v \in H$ there. Then the function (11) is well defined and $\mathfrak{G}$-invariant, and by applying to it the operator B we obtain the relation

$$S_h B(\sigma_{h^{-1}}(p))S_h^{-1}v = B(p)v$$

for almost all $p \in U$. Now the desired assertion readily follows.　□

This also completes the proof of our proposition. Note that the requirement for the representation σ to be faithful is important. Without this condition, the lemma fails. (One can only prove the assertion of the lemma on the subspace of elements invariant under S_g for $g \in \ker \sigma$.) $\square$

Definition 2.6. The function (9) is called the *symbol* of the translation-invariant operator B and is denoted by $\sigma(B) \equiv \sigma(B)(p)$.

In a special case, translation-invariant operators were called "suspended operators" [9].

2.3. General local operators and localization principle

Local operators with parameters. Let X be a Hausdorff compact metric space equipped with a nonatomic Borel measure μ such that $\mu(U) > 0$ for any nonempty open set $U \subset X$. We deal with *local operators with a parameter $q \in \mathbb{R}^s$* in the $C(X)$-module $H = L^2(X, d\mu)$. They are defined as operator families $A \in C(\mathbb{R}^s, \mathcal{B}H)$ such that for each $\varphi \in C_0(X)$ the commutator $[A(q), \varphi]$ belongs to the ideal $\mathcal{J} = C_0(\mathbb{R}^s, \mathcal{K}H)$ of compact-valued families decaying in norm as $q \to \infty$. Such families A obviously form a C^*-subalgebra in $C(\mathbb{R}^s, \mathcal{B}H)$, which will be denoted by $\mathcal{A} = \mathcal{A}(\mathbb{R}^s, \mathcal{B}H)$.

Localization principle. For $x \in X$, let $\mathcal{J}_x \subset \mathcal{A}$ be the ideal in $\mathcal{A}$ generated by the maximal ideal $\mathcal{I}_x \subset C(X)$ of functions vanishing at x, and let $p_x : \mathcal{A} \longrightarrow \mathcal{A}_x$ be the natural projection into the *local algebra* $\mathcal{A}_x = \mathcal{A}/\mathcal{J}_x$.

Theorem 2.7 (localization principle; cf. [14, Theorem 3]). *One has $\mathcal{J} = \bigcap_{x \in X} \mathcal{J}_x$, and hence an operator $A \in \mathcal{A}$ is*

1. *Compact with parameter q ($A \in \mathcal{J}$) if and only if all its local representatives $p_x(A) \in \mathcal{A}_x$ are zero.*
2. *Fredholm with parameter q (invertible modulo $\mathcal{J}$) if and only if all its local representatives $p_x(A) \in \mathcal{A}_x$ are invertible.*

The ideals $\mathcal{J}_x$ can be described as follows. For $U \subset X$ and $A \in \mathcal{A}$, set

$$\|A\|_U = \sup_{q \in \mathbb{R}^s} \|A(q)|_{H_U} : H_U \longrightarrow H\|, \quad \text{where } H_U = \{v \in H : \operatorname{supp} v \subset \overline{U}\}. \quad (12)$$

Proposition 2.8 (cf. [14, Proposition 4]). *The ideal $\mathcal{J}_x$ is the set of elements $A \in \mathcal{A}$ such that*

$$\lim_{U \downarrow x} \|A\|_U = 0. \quad (13)$$

(*Here the limit is taken over the filter of neighborhoods of x, i.e., over a sequence of open sets U shrinking to x.*)

Remark 2.9. Condition (13) is stated in [14] in the different form $\lim \|A\varphi\| = 0$, where $|\varphi| \le 1$ and the support of φ shrinks to x; the two forms are easily seen to be equivalent.

Local representatives. Let us describe the range of the family $\{p_x\}_{x\in X}$ of "localizing homomorphisms." Consider a family $\{a_x \in \mathcal{A}_x\}_{x\in X}$. For each x, we arbitrarily pick up some representative $A_x \in a_x$. Proposition 2.8 has an immediate corollary:

Corollary 2.10. *The family $\{a_x\}$ has the form $a_x = p_x(A)$ for some $A \in \mathcal{A}$ if and only if for any $\varepsilon > 0$ each point $x \in X$ has a neighborhood $U(\varepsilon, x)$ such that*

$$\|A_x - A\|_{U(\varepsilon,x)} \le \varepsilon. \tag{14}$$

This is not especially useful, because one has to know A in advance. Fortunately, one can give a criterion that does not resort to A.

Definition 2.11. The family $\{a_x\}$ is said to be *continuous* if for all $\varepsilon > 0$ and $x \in X$ there exist neighborhoods $U(\varepsilon, x)$ such that

$$\|A_y - A_{y'}\|_{U(\varepsilon,y)\cap U(\varepsilon,y')} \le \varepsilon \quad \text{for any } y, y' \in X. \tag{15}$$

One can readily see that the definition of continuity is independent of the choice of $A_x \in a_x$ (but the neighborhoods $U(\varepsilon, x)$ depend on this choice).

Proposition 2.12 (cf. [14, Proposition 7]). (i) *The family $\{a_x\}$ is continuous if and only if it has the form $a_x = p_x(A)$ for some $A \in \mathcal{A}$.*

(ii) *Under the assumptions of (i), if $a_x \in \mathcal{B}/\mathcal{J}$ for all $x \in X$, where $\mathcal{B} \subset \mathcal{A}$ is a C^*-subalgebra containing $\mathcal{J}$, then $A \in \mathcal{B}$.*

Remark. For the general localization principle, the topology on the disjoint union $\bigsqcup_x \mathcal{A}_x$ in which the families $\{p_x(A)\}_{x\in X}$, $A \in \mathcal{A}$, are exactly continuous sections of the projection $\bigsqcup_x \mathcal{A}_x \longrightarrow X$ is described, e.g., in [2, 17]. In our special case, these sections admit the simpler description given above.

Infinitesimal operators. The study of local representatives of an operator $A \in \mathcal{A}$ is also local in the following sense. The class $p_{x_0}(A) \in \mathcal{A}_{x_0}$ remains unchanged if we multiply A (on the left or on the right) by any cutoff function $f \in C_0(X)$ such that $f(x_0) = 1$. (This can readily be derived from the fact that if $K \in \mathcal{J}$, then $\|K\|_U \to 0$ as $U \downarrow x$.) It follows that only what happens in an arbitrarily small neighborhood of x_0 is actually important. Consequently, we can take another space $\widetilde{X}$ and a homeomorphism $f \colon U \longrightarrow \widetilde{U}$ identifying some neighborhood $U \subset X$ of x_0 with a neighborhood $\widetilde{U} \subset \widetilde{X}$ of the point $\widetilde{x}_0 = f(x_0)$; then we can interpret local representatives of A as operators in a Hilbert function space on $\widetilde{X}$.

We systematically use this construction in what follows; the space $\widetilde{X}$ will only reflect local properties of X near x_0 and is usually noncompact. Such local representatives, uniquely determined by certain additional conditions, will also be called *infinitesimal operators* to emphasize the fact that $X \ne \widetilde{X}$.

Example 2.13. If A is a pseudodifferential operator on a smooth manifold X, then one can identify a small neighborhood of x_0 with a small neighborhood of zero in $\widetilde{X} = T_{x_0}X$ via the geodesic exponential mapping and take the operator $\sigma(A)(x_0, -i\partial/\partial y)$ with constant coefficients on $T_{x_0}X$ for a local representative (infinitesimal operator) of A at x_0. (Here $\sigma(A)$ is the principal symbol of A and

$y \in T_{x_0}X$.) This infinitesimal operator is uniquely determined by the condition of invariance with respect to the dilations $y \longmapsto \lambda y$ in $T_{x_0}X$.

2.4. Definition and properties of ψDO

Now we are in a position to define pseudodifferential operators with a parameter $q \in \mathbb{R}^s$ on a manifold M with corners. They will be local operators with a parameter in the sense of Section 2.3 possessing a number of additional properties.

We treat $L^2(M)$ as a module over $C(M^\#)$ (by interpreting elements $u \in L^2(M)$ as functions on $M^\circ = M^{\#\circ}$) and consider the algebra $\mathcal{A}(\mathbb{R}^s, \mathcal{B}L^2(M))$ of operators with parameter $q \in \mathbb{R}^s$ local with respect to the action of $C(M^\#)$.

Interior symbol. Let $A \in \mathcal{A}(\mathbb{R}^s, \mathcal{B}L^2(M))$.

Definition 2.14. Let $x \in M^\circ$ be an interior point of M. We say that the operator A is *Agranovich–Vishik at x* if, under the identification of a neighborhood of x in M° with a neighborhood of the origin in $T_x M$ via a coordinate system near x_0, A has a local representative of the form

$$A_{x_0} = B\left(q, -i\frac{\partial}{\partial y}\right), \quad y \in T_{x_0}M,$$

where $B(q, \xi)$ is a function continuous for $|q| + |\xi| \neq 0$ and zero-order homogeneous:

$$B(\lambda q, \lambda \xi) = B(q, \xi), \quad \lambda \in \mathbb{R}_+.$$

The function $B(q, \xi)$ is called the *interior symbol* of A and is denoted by

$$\sigma_0(A)(x, \xi, q) := B(q, \xi).$$

Basically, the definition says that at the point x the operator A is a parameter-dependent pseudodifferential operator with continuous symbol.

Proposition 2.15. *If A is Agranovich–Vishik at x, then $\sigma_0(A)$ is a well-defined function on $T_x^* M \times \mathbb{R}^s$ outside zero (i.e., its existence and form are independent of the choice of the coordinate system).*

Sketch of proof. The operator

$$\widehat{B} = B\left(q, -i\frac{\partial}{\partial y}\right)$$

behaves as desired under linear changes of the variable y. Thus, essentially, one should prove that if $f : \mathbb{R}^n \to \mathbb{R}^n$ is a diffeomorphism with identity differential at the origin, then $\widehat{B}$ and $(f^*)^{-1}\widehat{B}f^*$ define the same element in the local algebra $\mathcal{A}_0$. To prove this, we approximate B by smooth classical symbols and use the theorem on the change of variables in a classical pseudodifferential operator. $\qquad\square$

Face symbols. From now on, we choose and fix a compatible system of exponential maps from the normal bundles of the faces into their neighborhoods in M. Our definition of face symbols and of ψDO tacitly depends on the choice of this system.

Let again $A \in \mathcal{A}(\mathbb{R}^s, \mathcal{B}L^2(M))$, and let $z \in F^\#$ be a point of the open face $F^\#$ dual to a face F of positive codimension $d \geq 1$ in M. Some neighborhood U of F can be identified via the exponential map with a neighborhood of the zero section in N_+F or (additionally applying the logarithmic map) with a neighborhood of the point at infinity in the positive quadrant in NF. Hence we have the embedding

$$L^2(M)|_U \subset L^2(N_+F) \simeq L^2(NF),$$

which implies that we can treat a local representative of the operator A at a point $z \in F^\#$ as an operator in the space $L^2(NF)$.

Definition 2.16. We say that the operator $A(q)$ has a *translation-invariant infinitesimal operator* at a point $z \in F^\#$ if it has a representative at z which is a translation-invariant operator $A_\infty(q)$ in $L^2(NF)$ (see Definition 2.2). The symbol (Definition 2.6) of $A_\infty(q)$ will be called the *symbol of $A(q)$ at z* and will be denoted by $\sigma_z(A)$.

Theorem 2.17. *If $A(q)$ has a translation-invariant infinitesimal operator, then it is unique. Thus the symbol $\sigma_z(A)$ is well defined. It is a $\mathfrak{S}_F$-invariant operator-valued function on $\mathbb{R}^d \times \mathbb{R}^s$ with values in $\mathcal{B}L^2(\widetilde{F})$, where $\widetilde{F}$ is the principal $\mathfrak{S}_F$-covering over F trivializing NF.*

Proof. It suffices to prove uniqueness for the case in which $A(q) \equiv 0$. Consider a sequence $\varphi_n \in C_0(N\overline{F})$ strongly convergent to the identity operator. Then $\varphi_n A_\infty \varphi_n$ strongly converges to A_∞. Passing to the cover $N\widetilde{F}$, we see that the product $\widetilde{\varphi}_n \widetilde{A}_\infty \widetilde{\varphi}_n$ of the corresponding lifted operators strongly converges to $\widetilde{A}_\infty$.

On the other hand, let us show that $\widetilde{\varphi}_n \widetilde{A}_\infty \widetilde{\varphi}_n = 0$ for all n. Indeed, for some fixed n, let $a_j \in \mathbb{R}^d$ be a sequence of vectors such that the supports of the functions $t^*_{a_j} \widetilde{\varphi}_n$, where t_{a_j} is the shift by the vector $a_j \in \mathbb{R}^d$, lie as $j \to \infty$ in an arbitrarily small neighborhood of some of the preimages z_* of the point z (i.e., go to infinity in the positive quadrant along the corresponding ray). These functions are no longer $\mathfrak{S}_F$-invariant. However, one can show that there exist functions ψ_j bounded by 1 with supports shrinking to z such that their invariant lifts satisfy the condition

$$\widetilde{\psi}_j t^*_{a_j} \widetilde{\varphi}_n = t^*_{a_j} \widetilde{\varphi}_n.$$

Then, according to the properties of the local algebra $\mathcal{A}_z$, we have

$$(t^*_{a_j} \widetilde{\varphi}_n) \widetilde{A}_\infty (t^*_{a_j} \widetilde{\varphi}_n) = (t^*_{a_j} \widetilde{\varphi}_n) \widetilde{\psi}_j \widetilde{A}_\infty \widetilde{\psi}_j (t^*_{a_j} \widetilde{\varphi}_n) = (t^*_{a_j} \widetilde{\varphi}_n) \widetilde{\psi_j A_\infty \psi_j} (t^*_{a_j} \widetilde{\varphi}_n) \to 0$$

as $j \to \infty$ (convergence in norm). Indeed, the extreme factors are uniformly bounded, and the middle factor converges to zero, since A_∞ represents the zero class. Thus

$$(t^*_{a_j} \widetilde{\varphi}_n) \widetilde{A}_\infty (t^*_{a_j} \widetilde{\varphi}_n) = t^*_{a_j} \circ \widetilde{\varphi}_n \widetilde{A}_\infty \widetilde{\varphi}_n \circ t^{*-1}_{a_j} \to 0.$$

(We have used the translation invariance of A.) We see that $\widetilde{\varphi}_n \widetilde{A}_\infty \widetilde{\varphi}_n = 0$ and, passing to the limit as $n \to \infty$, find that $\widetilde{A}_\infty$ and hence A_∞ are zero. □

Pseudodifferential operators. Let M be a manifold with corners.

Definition 2.18. The space $\Psi(M) \equiv \Psi(M, \mathbb{R}^s)$ of pseudodifferential operators consists of operator families $A(q)$ satisfying the following conditions:

1. $A(q) \in \mathcal{A}(\mathbb{R}^s, \mathcal{B}L^2(M))$.
2. For each interior point $x \in M^\circ$, the family $A(q)$ is Agranovich–Vishik at x.
3. For each face F of codimension $d = d(F) > 0$ in M, the family $A(q)$ has a $\mathfrak{S}_F$-invariant symbol $\sigma_z(A)$ in the sense of Definition 2.16 at each point $z \in F^\#$, and $\sigma_z(A)$ is independent of z. Moreover, $\sigma_z(A) \in \Psi(\widetilde{F}, \mathbb{R}^{d+s})$; i.e., the symbol $\sigma_z(A)$ is a $\mathfrak{S}_F$-invariant ψDO with parameters $(q, p) \in \mathbb{R}^s \times \mathbb{R}^d$ on the manifold $\widetilde{F}$ with corners, the covering of $\overline{F}$ trivializing the bundle $N\overline{F}$.

Since the symbol $\sigma_z(A)$ is independent of $z \in F^\#$, it will be denoted by $\sigma_F(A)$ in what follows. The interior symbol will be denoted by $\sigma_0(A)$; it is defined on the interior of $T^*M \times \mathbb{R}^s$ minus the zero section.

Main theorem of the calculus. The localization principle (Theorem 2.7) readily implies the following assertion.

Theorem 2.19 (main theorem of the calculus). *A pseudodifferential operator A on a compact manifold M with corners is uniquely determined modulo the ideal $\mathcal{J}$ of compact operators with parameters by the symbol tuple $(\sigma_0(A), \{\sigma_F(A)\})$, where F runs over all faces of positive codimension. The map*

$$\sigma \colon A \longmapsto (\sigma_0(A), \{\sigma_F(A)\})$$

that takes each ψDO $A \in \Psi(M, \mathbb{R}^s)$ to it symbol tuple is a C^*-algebra homomorphism.

The symbol algebra. Now let us describe the symbol algebra, i.e., the range of the symbol map σ. In other words, we should indicate conditions on the interior symbol and the face symbols on faces of positive codimension necessary and sufficient for the existence of a ψDO with these symbols. To avoid awkward formulas, we first do so for the case in which the normal bundles of all faces are trivial and then indicate the modifications needed in the general case.

Thus let M be a manifold with corners such that the normal bundle NF is trivial for all faces F of M.

Let the following data be given:

- For each interior point $x \in M^\circ$, a continuous zero-order homogeneous function σ_x on $(T_x^*M \times \mathbb{R}^s) \setminus 0$.
- For each face F of codimension $d > 0$, a pseudodifferential operator $\sigma_F \in \Psi(F, \mathbb{R}^{d+s})$.

Theorem 2.20 (description of the symbol algebra). *For the existence of a ψDO $A \in \Psi(M, \mathbb{R}^s)$ such that*

$$\sigma_0(A) = \sigma_x \quad \text{on } (T_x^* M \times \mathbb{R}^s) \setminus 0 \text{ for each } x \in M^\circ, \tag{16}$$

$$\sigma_F(A) = \sigma_F \quad \text{for each face } F \text{ of positive codimension}, \tag{17}$$

the following conditions are necessary and sufficient:

1. *The functions σ_x form a continuous zero-order homogeneous function on the interior of $(T^* M^\circ \times \mathbb{R}^s) \setminus 0$ and extend by continuity to a continuous function (which we denote by σ_0) on the whole space $(T^* M \times \mathbb{R}^s) \setminus 0$.*
2. *The restriction of σ_0 to the boundary satisfies the compatibility conditions*

   $$\sigma_0 \big|_F = \sigma_0(\sigma_F) \quad \text{for each face } F \text{ of positive codimension}, \tag{18}$$

   where the left-hand side is the restriction of σ_0 to $T^* M|_F \oplus \mathbb{R}^s$, naturally identified with $T^* F \oplus N^* F \oplus \mathbb{R}^s = T^* F \oplus \mathbb{R}^{d+s}$.
3. *If $F_1 \succ F_2$ are two adjacent faces of M and Γ is a face of F_1 mapped into F_2 under the immersion of F_1 in M, then*

   $$\sigma_\Gamma(\sigma_{F_1}) = \sigma_{F_2}. \tag{19}$$

Proof. First, note that routine computations based on composition formulas for pseudodifferential operators and standard norm estimates show that, being quantized, the symbols $\sigma_{x_0}(q, p)$ and $\sigma_F(q, \xi)$ give rise to the local representatives $\widehat{\sigma}_x = \sigma_{x_0}(q, -i\partial/\partial x)$ and $\widehat{\sigma}_F = \sigma_F(q, -i\partial/\partial t)$ that belong to $\mathcal{A}(\mathbb{R}^s, \mathcal{B}L^2(M))$.

By Proposition 2.12, to prove the theorem it remains to establish that conditions (1)–(3) are exactly equivalent to the continuity of this family of local representatives in the sense of Definition 2.11.

(a) Let us show that the function σ_x continuously depends on x in the interior of M. Localizing our considerations, we can assume that $M = \mathbb{R}^n$. The family σ_x is continuous if and only if for each $\varepsilon > 0$ each point has a neighborhood $U(\varepsilon, x)$ such that $\|\widehat{\sigma}_x - \widehat{\sigma}_y\|_{U(\varepsilon, x) \cap U(\varepsilon, y)} \leq \varepsilon$ for any x and y. The intersection $U = U(\varepsilon, x) \cap U(\varepsilon, y)$ is necessarily nonempty if $y \in U(\varepsilon, x)$. Since the operator $\widehat{\sigma}_x - \widehat{\sigma}_y$ is dilatation invariant, it follows that

$$\|\widehat{\sigma}_x - \widehat{\sigma}_y\|_U = \|\widehat{\sigma}_x - \widehat{\sigma}_y\| = \max_p |\sigma_x - \sigma_y|$$

(provided that U is nonempty). Combining this with the homogeneity of σ_x in (p, q), we see that the continuity of the family of local representatives in the sense of Definition 2.11 is equivalent to the continuity of the interior symbol. This is of course well known from the theory of ψDO on smooth manifolds.

(b) Let us show that the interior symbol is continuous up to the boundary and satisfies the compatibility conditions (18) there. Fix a point $z_0 \in F$. Multiplying by a cutoff function $f \in C(M)$, we can study the problem assuming that $F = \mathbb{R}^{n-d}$ and $M^\circ = \mathbb{R}^{n-d} \times \mathbb{R}^d$. (Here we use the logarithmic coordinates $y \in \mathbb{R}^d$, see (4), on the fibers of the normal bundle of F.) Let $x_0 \in F$. Applying Corollary 2.10 and using the Fourier transform with respect to $\mathbb{R}^d$, we see that for each $\varepsilon > 0$ there is a neighborhood U_ε of x_0 in F such that

$$\|\widehat{\sigma}_F - \widehat{\sigma}_{x_0}(\sigma_F)\|_{U_\varepsilon \times \mathbb{R}^d} < \varepsilon. \tag{20}$$

(Here $\sigma_{x_0}(\sigma_F)$ is the symbol $\sigma_0(\sigma_F)$ restricted to the fiber over x_0.)

On the other hand, the continuity of the family of local representatives on M near $F^\#$ is equivalent to the existence of a neighborhood W_ε of the point at infinity on the diagonal of the positive quadrant in $\mathbb{R}^d$ such that

$$\|\widehat{\sigma}_F - \widehat{\sigma}_x\|_{(F \times W_\varepsilon) \cap U(\varepsilon, x)} < \varepsilon. \tag{21}$$

For $x \in U_\varepsilon \times W_e$, using the triangle inequality, from (20) and (21) we conclude that

$$\|\widehat{\sigma}_x - \widehat{\sigma}_{x_0}(\sigma_F)\|_U < 2\varepsilon \tag{22}$$

on the nonempty set $U = (U_\varepsilon \times W_e) \cap U(\varepsilon, x)$. Arguing as above, we see that $|\sigma_x - \sigma_{x_0}(\sigma_F)| \le 2\varepsilon$ for these x.

(c) In a similar way, one shows that condition (3) also follows from the continuity of local representatives and finally concludes that conditions (1)–(3) together are equivalent to the continuity. We leave the details to the reader.   □

Remark 2.21. In particular, it follows from the compatibility condition that the symbol on a face of positive codimension determines the symbols on all adjacent faces of larger codimension.

Let us now discuss how the compatibility conditions should be modified if the normal bundles of the faces are not trivial.

Let again $\overline{F}_1 \succ \overline{F}_2$ be two adjacent faces of M, and let Γ be a face of $\overline{F}_1$ covering $\overline{F}_2$ (there can be several such faces). The symbols $\sigma_{F_1}(A)$ and $\sigma_{F_2}(A)$ of A are operators with parameters on the minimum coverings $\widetilde{F}_1$ and $\widetilde{F}_2$ trivializing the bundles NF_1 and NF_2, respectively. Let $\widetilde{\Gamma}$ be the lift of Γ to $\widetilde{F}_1$. The symbol $\sigma_{\widetilde{\Gamma}}(\sigma_{F_1}(A))$ is defined on the covering $\widetilde{\Gamma}$ trivializing the bundle $N\widetilde{\Gamma}$. The composite covering $\widetilde{\widetilde{\Gamma}} \longrightarrow F_2$ trivializes NF_2 (since it trivializes both direct summands $N\widetilde{\Gamma}$ and $NF_1|_\Gamma$). Since the trivializing covering $\widetilde{F}_2 \longrightarrow F_2$ is minimal and hence universal, there exists a unique (modulo permutation of the sheets) covering $\widetilde{\widetilde{\Gamma}} - \rightarrow \widetilde{F}_2$ making the triangle

$$\tag{23}$$

commute.

Let $L^2_{inv}(\pi)$ be the subspace of $L^2(\widetilde{\widetilde{\Gamma}})$ consisting of functions invariant with respect to permutations of the sheets of π. The compatibility condition (19) in this situation is generalized to

$$\sigma_{\widetilde{\Gamma}}(\sigma_{F_1}(A))|_{L^2_{inv}(\pi)} = \sigma_{F_2}(A). \tag{24}$$

The counterpart of the compatibility condition (18) reads

$$\sigma_0(\sigma_F(A)) = \pi_F^* \left[\sigma_0(A)|_{T^*F} \right], \tag{25}$$

where $\pi_F : T^*\widetilde{F} \longrightarrow T^*F$ is the covering associated with the covering $\widetilde{F} \longrightarrow F$.

Appendix A. Proofs of some assertions

Proof of Proposition 1.2. For brevity, we write

$$F = \Gamma^\circ_j(M) \quad d = d_j.$$

Let $U \simeq \overline{\mathbb{R}}^s_+ \times \mathbb{R}^{n-s}$ be a coordinate neighborhood on M. If the intersection $U \cap F$ is nonempty (which can happen only for $s \geq d$), then it consists of finitely many ($\leq \binom{s}{d}$) connected components of the form $V \simeq \mathbb{R}^{s-d}_+ \times \mathbb{R}^{n-s}$, where the open coordinate quadrant $\mathbb{R}^{s-d}_+$ of dimension $s - d$ is singled out in $\overline{\mathbb{R}}^s_+$ by the relations

$$x_{j_1} = \cdots = x_{j_d} = 0, \quad x_{j_{d+1}}, \ldots, x_{j_s} > 0$$

for some (depending on V) permutation $j_1, \ldots, j_s$ of the indices $1, \ldots, s$. If we accordingly permute the standard coordinates $x_1, \ldots x_n$ in U, setting

$$\rho_1 = x_{j_1}, \ldots, \rho_d = x_{j_d}, y_{d+1} = x_{j_{d+1}}, \ldots, y_s = x_{j_s}, y_{s+1} = x_{s+1}, \ldots, y_n = x_n,$$

then the variables $y = (y_{d+1}, \ldots, y_n)$ are coordinates in V and the variables $\rho = (\rho_1, \ldots, \rho_d)$ are defining functions of V for the embedding $V \subset U$ (and local defining functions of $\overline{F}$); i.e., locally the face is given by the conditions $\rho = 0$.

We take a finite cover of M by coordinate neighborhoods U and various connected components $V \subset U \cap F$ and obtain a finite atlas

$$\left\{ \left(V, y : V \longrightarrow \mathbb{R}^{s-d}_+ \times \mathbb{R}^{n-s} \right) \right\}$$

on $\overline{F}$ such that associated with each coordinate neighborhood V of this atlas is a pair (U, V) and coordinates (ρ, y) in U. Let $\widetilde{V}$ be another coordinate neighborhood $(\widetilde{U}, \widetilde{V})$ with coordinates $(\widetilde{\rho}, \widetilde{y})$ in $\widetilde{U}$, and suppose that the intersection $V \cap \widetilde{V}$ is nonempty. The change of variables

$$\widetilde{y} \circ y^{-1} : y(V \cap \widetilde{V}) \longrightarrow \widetilde{y}(V \cap \widetilde{V})$$

is obtained by restriction to $y(V \cap \widetilde{V})$ from the change of coordinates $(\rho, y) \longmapsto (\widetilde{\rho}, \widetilde{y})$ on the intersection of the coordinate neighborhoods U and $\widetilde{U}$ on M and hence has a smooth continuation to the closure of the set $y(V \cap \widetilde{V})$ in $\overline{\mathbb{R}}^{s-d}_+ \times \mathbb{R}^{n-s}$. (The continuation is obtained by restriction of the same change of coordinates to the closure.) These continuations determine the transition functions of some compact manifold $\overline{F}$ with corners whose local models are $\overline{\mathbb{R}}^{s-d}_+ \times \mathbb{R}^{n-s}$ and into which F is naturally embedded as a dense open submanifold. This manifold $\overline{F} =: \Gamma_j(M)$ is the *closed face* of M corresponding to the open face $\Gamma^\circ_j(M)$. The embedding

$\Gamma_j^\circ(M) \subset M$ extends by continuity to $\Gamma_j(M)$; the resulting mapping is in general an immersion with self-intersections.

Proof of Theorem 1.5. We need the following simple lemma.

Lemma A.1. *If smooth mappings*

$$g_j : \mathbb{R}_+^k \longrightarrow \mathbb{R}_+^k, \quad g_j(0) = 0, \ j = 1, \dots, l,$$

are diffeomorphisms in a neighborhood of the origin, if all matrices $g_j'(0)\big(g_i'(0)\big)^{-1}$ are diagonal, and if $\lambda_1, \dots, \lambda_l$ are nonnegative numbers at least one of which is nonzero, then the mapping

$$g \equiv \sum_{j=1}^{l} \lambda_j g_j : \mathbb{R}_+^k \longrightarrow \mathbb{R}_+^k$$

is also a diffeomorphism in a neighborhood of the origin.

Indeed, the only nontrivial assertion is that g is epimorphic, but this can be verified as follows. Since the matrices $g_j'(0)\big(g_i'(0)\big)^{-1}$ are diagonal, it follows that all g_j take any given coordinate quadrant of arbitrary dimension to one and the same coordinate quadrant.

The lemma suggests that one can construct the desired mapping $f = f_j$ specifying it locally by the formula

$$\rho = r, \tag{26}$$

where ρ is a local tuple of defining functions of the face $\Gamma_j(M)$ and r are the corresponding coordinates in the fiber of $N_+\Gamma_j(M)$ and then gluing the local mappings with the use of a partition of unity.

We implement this idea and construct the mapping f, successively extending the set on which it is defined. Suppose that f has already been defined over some set $O \subset \Gamma_j(M)$, and let V be a local chart on $\Gamma_j(M)$ with the corresponding pair $(U = V \times \overline{\mathbb{R}}_+^{d_j}, V)$, so that over V the mapping can be given by formula (26). Let (φ_O, φ_V) be a nonnegative partition of unity on $O \cup V$ subject to the cover by O and V. We construct the map over $O \cup V$ by setting

$$f_{O \cup V} = \begin{cases} f_O & \text{over } O \setminus \operatorname{supp} \varphi_V, \\ \varphi_O f_O + \varphi_V f_V & \text{over } V, \end{cases}$$

where the addition in the second line is carried out in the fibers of $U \longrightarrow V$ (and is well defined in a sufficiently small neighborhood of V). Since $\varphi_O = 1$ on $V \setminus \operatorname{supp} \varphi_V$, it follows that both definitions are compatible on the set where they apply simultaneously, and the lemma now implies that we have defined a mapping with the desired properties over $O \cup V$.

To complete the proof of Theorem 1.5, it suffices to start from an empty set O and successively add to it all charts from a finite atlas on $\Gamma_j(M)$. To obtain compatible (in the sense that diagram (3) commutes) exponential mappings for all faces, one should start from faces of maximal codimension.

References

[1] U. Bunke, *Index theory, eta forms, and Deligne cohomology*, Preprint arXiv: math.DG/0201112.

[2] J. Dauns and K.H. Hofmann, *Representation of rings by sections*, Memoirs of the American Mathematical Society, No. 83, Amer. Math. Soc., Providence, R.I., 1968.

[3] T. Krainer, *Elliptic boundary problems on manifolds with polycylindrical ends*, J. Funct. Anal., **244** (2007), no. 2, 351–386.

[4] R. Lauter and S. Moroianu, *The index of cusp operators on manifolds with corners*, Ann. Global Anal. Geom. **21** (2002), no. 1, 31–49.

[5] P.-Y. Le Gall and B. Monthubert, *K-theory of the indicial algebra of a manifold with corners*, K-Theory **23** (2001), no. 2, 105–113.

[6] P. Loya, *The index of b-pseudodifferential operators on manifolds with corners*, Ann. Global Anal. Geom. **27** (2005), no. 2, 101–133.

[7] R. Melrose, *Analysis on manifolds with corners*, Lecture Notes, MIT, Cambrige, MA, 1988, Preprint.

[8] ———, *Pseudodifferential operators, corners, and singular limits*, Proceedings of the International Congress of Mathematicians, Kyoto (Berlin–Heidelberg–New York), Springer-Verlag, 1990, pp. 217–234.

[9] ———, *The eta invariant and families of pseudodifferential operators*, Math. Research Letters **2** (1995), no. 5, 541–561.

[10] R. Melrose and V. Nistor, *K-theory of C^*-algebras of b-pseudodifferential operators*, Geom. Funct. Anal. **8** (1998), no. 1, 88–122.

[11] R. Melrose and P. Piazza, *Analytic K-theory on manifolds with corners*, Adv. in Math. **92** (1992), no. 1, 1–26.

[12] B. Monthubert, *Groupoids and pseudodifferential calculus on manifolds with corners*, J. Funct. Anal. **199** (2003), no. 1, 243–286.

[13] B. Monthubert and V. Nistor, *A topological index theorem for manifolds with corners*, arXiv: math.KT/0507601, 2005.

[14] V. Nazaikinskii, A. Savin, and B. Sternin, *Pseudodifferential operators on stratified manifolds I and II*, Differ. Equations, (2007) **43**, no. 4, 536–549; **43**, no. 5, 704–716.

[15] V. Nistor, *An index theorem for gauge-invariant families: The case of solvable groups*, Acta Math. Hungarica **99** (2003), no. 2, 155–183.

[16] B.A. Plamenevsky and V.N. Senichkin, *Representations of C^*-algebras of pseudodifferential operators on piecewise-smooth manifolds*, St. Petersbg. Math. J. **13** (2001), no. 6, 993–1032.

[17] N. Vasilevski, *Local principles in operator theory*, Lineinye operatory v funktsionalnykh prostranstvakh. Tez. dokl. Severo-Kavkaz. reg. konf., Groznyi, 1989, pp. 32–33.

Vladimir Nazaikinskii
Institute for Problems in Mechanics, Russian Academy of Sciences
pr. Vernadskogo 101-1, 119526 Moscow, Russia
e-mail: `nazaikinskii@yandex.ru`

Anton Savin and Boris Sternin
Independent University of Moscow, Bol'shoi Vlas'evskii per. 11, 119002 Moscow, Russia
e-mail: `antonsavin@mail.ru`
e-mail: `sternin@mail.ru`

C^*-algebras and Elliptic Theory II
Trends in Mathematics, 207–226

Elliptic Theory on Manifolds with Corners: II. Homotopy Classification and K-Homology

Vladimir Nazaikinskii, Anton Savin and Boris Sternin

Abstract. We establish the stable homotopy classification of elliptic pseudo-differential operators on manifolds with corners and show that the set of elliptic operators modulo stable homotopy is isomorphic to the K-homology group of some stratified manifold. By way of application, generalizations of some recent results due to Monthubert and Nistor are given.

Mathematics Subject Classification (2000). Primary 58J05; Secondary 19K33, 35S35.

Keywords. Manifold with corners, elliptic operator, stable homotopy, K-homology, stratified manifold.

Introduction

Recently there has been considerable progress in understanding the notion of ellipticity on noncompact manifolds and manifolds with singularities. For a wide class of manifolds, ellipticity conditions for operators were established and the corresponding finiteness theorems[1] were proved; the corresponding operator C^*-algebras were constructed. Hence the study of *topological* aspects of the theory of elliptic operators becomes topical. Here one mainly speaks of the classification problem and the index problem. Note that Gelfand's homotopy classification problem for elliptic operators can naturally be restated in modern language as the problem of computing the K-groups of symbol algebras, which prove noncommutative in most cases. Thus Gelfand's problem naturally fits in the framework of topical problems of Connes's noncommutative geometry [4].

Supported in part by RFBR grants 05-01-00982 and 06-01-00098, by President of the Russian Federation grant MK-1713.2005.1, and by the DFG project 436 RUS 113/849/0-1® "K-theory and Noncommutative Geometry of Stratified Manifolds."
[1]Stating that an elliptic operator is Fredholm in certain function spaces.

Aim of the paper. This paper deals with elliptic theory on manifolds with corners. Operators on manifolds with corners have been actively studied, and a number of important interesting results emerged recently. For example, the C^*-closure of symbol algebras was studied in [12], and a spectral sequence converging to the K-theory of the C^*-algebra of symbols was constructed. Monthubert [14] obtained a description of the operator algebra in the spirit of noncommutative geometry in terms of a special groupoid that can be associated with a manifold with corners (see also [9]). Bunke [3] constructed index theory of Atiyah–Patodi–Singer type for Dirac operators and studied cohomological obstructions to elliptic problems (see also [10, 8]); Monthubert and Nistor [15] produced a formula for the boundary map in the K-theory of symbol algebras in topological terms. Krainer [7] studied boundary value problems in this setting.

Although these results permitted finding the group classifying the homotopy classes of elliptic operators in a number of special cases (e.g., see [13] or [16]), the homotopy classification problem remained open.

We solve Gelfand's problem for *manifolds with corners*. Our goal is to obtain a simple explicit formula for the classifying group in terms of Atiyah's K-homology functor [1].

Elliptic operators and K-homology. Note that the idea of classifying elliptic operators by the K-homology functor has long been known. For the reader's convenience, we recall it using operators on a smooth compact manifold M as an example.

The commutator of an elliptic zero-order[2] operator D on M with the operator of multiplication by a continuous function $f \in C(M)$ is compact,

$$[D, f] \in \mathcal{K}. \tag{1}$$

By one definition, the *contravariant K-theory* $K^0(C(M))$ of the algebra $C(M)$ just consists of Fredholm operators for which the commutators (1) are compact. Thus D determines an element of the group $K^0(C(M))$, which is isomorphic to the K-homology group of M:

$$K^0(C(M)) \simeq K_0(M)$$

by the Atiyah–Brown–Douglas–Fillmore–Kasparov theorem. Thus, assigning the corresponding class in the K-homology to each elliptic operator, we obtain a mapping

$$\mathrm{Ell}(M) \longrightarrow K_0(M),$$

where $\mathrm{Ell}(M)$ is the group of elliptic operators in sections of bundles on M modulo stable homotopy and $K_0(M)$ is the even K-homology group of M.

Kasparov [6] showed this mapping to be an isomorphism. In other words, the K-homology group of a smooth manifold classifies elliptic operators on this manifold modulo stable homotopy.

---

[2]Working solely with zero-order operators does not result in loss of generality, since order reduction (say, multiplication by an appropriate power of the Laplace operator) is always available.

This approach to classification also proved fruitful in the case of compact stratified manifolds with singularities. Namely, it was shown in [18] that in this case the even K-homology group of the underlying compact topological space classifies elliptic operators on this manifold.

However, no classification results were known for manifolds with corners of codimension ≥ 2. The classification in the form of the K-homology of the manifold with corners, which suggests itself, is too meagre to be true: one can always smooth the corners, and we see that the K-homology of the manifold with corners is too coarse an invariant, for it does not take into account the structure of a manifold with corners.

Moreover, even the space whose K-homology would classify elliptic operators was unknown.

Main result. We establish the isomorphism

$$\mathrm{Ell}(M) \simeq K_0(M^{\#}), \tag{2}$$

where M is a manifold with corners and $M^{\#}$ is the dual manifold (see Part I), which is a stratified manifold with singularities. More precisely, the isomorphism (2) will be established under the following assumption concerning the combinatorial structure of the faces of our manifold:

The normal bundles of all faces of M are trivial.

If this assumption fails, then, generally speaking, the isomorphism (2) does not hold. In this case, one should abandon the search for a classifying *space* and seek some *algebra* whose K-cohomology classifies elliptic operators. This algebra proves to be noncommutative, and one needs to use ideas of noncommutative geometry. These results will be considered elsewhere.

Note an interesting special case: if a manifold with corners is a polyhedron with a given triangulation of the boundary, then the dual stratified space is also a polyhedron, namely, the one used in the classical proof of Poincaré duality in cohomology! For example, if M is a cube, then $M^{\#}$ is an octahedron. Thus the construction of the dual manifold in the first part of the present paper generalizes the *Poincaré dual polyhedron* to the case of noncontractible faces.

Manifolds with corners and manifolds with multicylindrical ends. Note that there is a different perspective on the theory of operators on manifolds with corners. An application of a logarithmic change of variables in a neighborhood of the boundary taking the boundary to infinity (see Fig. 1, where this is shown for manifolds with boundary) results in the class of so-called *manifolds with multicylindrical ends*. These two pictures give the same operator algebras. Thus the results of the present paper also provide classification on manifolds with multicylindrical ends.

Outline of the paper. This is the second of the two parts of the paper. In the first part, the dual manifold of a manifold with corners was constructed and the calculus of pseudodifferential operators ($\psi\mathrm{DO}$) on manifolds with corners was developed in the C^*-algebraic context.

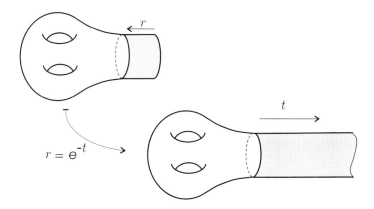

Figure 1. Transition from a neighborhood of the boundary to an infinite cylinder.

The present part has the following structure. In the first section, we recall some information from Part I of the paper. In Section 2 we state the classification theorem. The proof occupies the next three sections. Note that the general scheme of the proof is the same as in [18], and we proceed by induction, passing from a smooth manifold to increasingly complex manifolds with singularities. In Section 6 we discuss the relationship with some results due to Monthubert and Nistor. As a consequence of the classification theorem, we obtain a K-homology criterion for the vanishing of the index and a formula for the K-group of the C^*-algebra of ψDO with zero interior symbol (this algebra corresponds to the C^*-algebra of the groupoid constructed by Monthubert). In the appendix, we prove a higher analog of the relative index theorem, which naturally arises when we obtain the classification of operators.

1. Manifolds with corners and dual stratified manifolds

Manifolds with corners and faces. Here we recall some information given in the first part.

Consider a manifold M with corners of depth k. It has a natural stratification

$$M = \bigcup_{j=0}^{k} M_j,$$

where the stratum $M_0 = M^\circ$ is just the interior of M and each stratum M_j is the union of connected components, open faces $M_{j\alpha}$ of codimension j in M.

Each face $F = M_{ja}$ in the stratum M_j is isomorphic to the interior of a manifold $\overline{F} = \overline{M}_{ja}$, which will be called a closed face of M. Faces of codimension one are called *hyperfaces*.

Main assumption. The main results of the paper will be obtained under the following assumption.

Assumption 1.1. The normal bundle N_+F of each face F is trivial.

In this case, the local defining functions $\rho_1, \ldots, \rho_j$ of F are globally defined as functions on the normal bundle N_+F.

Remark. Assumption 1.1 holds if all hyperfaces are *embedded*, i.e., if there exists a global defining function for each hyperface $F \subset M$. However, it also holds for some manifolds with nonembedded hyperfaces, say, for the raindrop. The simplest example of a manifold with corners that does not satisfy Assumption 1.1 is the mapping cylinder of the raindrop (see Figure 1.2 in Part I) by the involution of the raindrop induced by the reflection in the bisector of the first quadrant.

The dual space. The dual space $M^\#$ of a manifold M with corners was introduced in Part I. If the original manifold is represented as the union

$$M = \bigcup_{j \geq 0} M_j, \qquad M_j = \bigcup_\alpha M_{j\alpha},$$

then $M^\#$ is the union of *dual faces*,

$$M^\# = \bigcup_{j \geq 0} M_j^\#, \qquad M_j^\# = \bigcup_\alpha M_{j\alpha}^\#,$$

each of which is isomorphic to the interior of a simplex,

$$M_{j\alpha}^\# \simeq \Delta_{j-1}^\circ.$$

Here, by definition, $M_0^\# := M_0$ is the interior of M. Thus to each face F of codimension j in M there corresponds a simplex $F^\#$ of dimension $j-1$ in the dual space.

The fibration structure on $M^\#$. It was proved in Part I that a neighborhood $U^\#$ of the stratum $F^\#$ is homeomorphic to the product of $F^\#$ by the cone

$$K_\Omega = [0,1) \times \Omega/\{0\} \times \Omega$$

whose base Ω is the dual space $\overline{F}^\#$ of the closed face $\overline{F}$ (which is well defined, since $\overline{F}$ itself is a manifold with corners). As a result, we find that $M^\#$ is a stratified manifold with singularities.

2. Classification theorem

Let M be a manifold with corners satisfying Assumption 1.1, and let $\Psi(M)$ be the C^*-algebra of zero-order ψDO in the space $L^2(M)$ (see Part I). The notion of a ψDO acting on sections of finite-dimensional vector bundles on M is introduced in the usual way. There is a natural equivalence relation, *stable homotopy*, on the set of elliptic operators. Recall the definition.

Definition 2.1. Two elliptic operators

$$D : L^2(M, E) \to L^2(M, F) \quad \text{and} \quad D' : L^2(M, E') \to L^2(M, F')$$

on M are said to be *stably homotopic* if there exists a continuous homotopy

$$D \oplus 1_{E_0} \sim f^*\big(D' \oplus 1_{F_0}\big)e^*$$

of elliptic operators, where $E_0, F_0 \in \mathrm{Vect}(M)$ are vector bundles and

$$e : E \oplus E_0 \longrightarrow E' \oplus F_0, \qquad f : F' \oplus F_0 \longrightarrow F \oplus E_0$$

are bundle isomorphisms.

Here ellipticity is understood as the invertibility of all components of the symbol of the operator, and only homotopies of ψDO preserving ellipticity are considered.

Even groups $\mathrm{Ell}_0(M)$. Stable homotopy is an equivalence relation on the set of elliptic ψDO acting in sections of vector bundles. By $\mathrm{Ell}_0(M)$ we denote the corresponding quotient set. It is a group with respect to the direct sum of operators, and the inverse in this group is given by the coset of the almost inverse operator (i.e., an inverse modulo compact operators).

Odd groups $\mathrm{Ell}_1(M)$. Odd elliptic theory $\mathrm{Ell}_1(M)$ is defined in a similar way as the group of stable homotopy classes of elliptic self-adjoint operators. Stabilization is defined in terms of the operators $\pm Id$.

Remark 2.2. An equivalent definition of the odd Ell-group can be given in terms of smooth operator families on M with parameter space $\mathbb{S}^1$ modulo constant families.

We shall compute the groups $\mathrm{Ell}_*(M)$ for $* = 0$ and $* = 1$, i.e., find the classification of elliptic operators modulo stable homotopy. Our approach is based on the following fact (see the definition of ψDO in Part I):

ψDO on M can be viewed as local operators in the sense of Atiyah on the dual manifold $M^\#$.

Thus an elliptic ψDO defines a Fredholm module on the space $L^2(M)$ viewed as a $C(M^\#)$-module. (For Fredholm modules and K-theory, see [5] or [2].)

Classification of elliptic operators. The following theorem is the main result of this paper.

Theorem 2.3. *The mapping that takes each elliptic ψDO to the corresponding Fredholm module defines the group isomorphism*

$$\mathrm{Ell}_*(M) \xrightarrow{\simeq} K_*(M^\#). \tag{3}$$

We shall obtain this theorem as a special case of the following more general theorem.

Classification of partially elliptic operators (cf. [18]). Let $\mathrm{Ell}_*\,(M_{\geq j})$ be the group generated by operators whose symbols are invertible on the main stratum and all faces of codimension $\geq j$. Thus we consider operators satisfying the ellipticity condition on part of the faces.

The corresponding dual space

$$M_{\geq j}^{\#} := M^{\#} \setminus \bigcup_{j'=1}^{j-1} M_{j'}^{\#}$$

is obtained from $M^{\#}$ by deleting all simplices of dimension $\leq j - 2$.

Lemma 2.4. *An operator D such that $[D] \in \mathrm{Ell}_*\,(M_{\geq j})$ defines a Fredholm module over the algebra $C_0(M_{\geq j}^{\#})$ of functions on the dual space.*

Proof. We should verify the following properties of a Fredholm module: the expression

$$f(DD^* - 1)$$

is compact for all $f \in C_0\left(M_{\geq j}^{\#}\right)$ (here we assume that D is normalized by the condition $\sigma_s^*(D) = \sigma_s(D)^{-1}$ for $s \geq j$). The compactness follows from the fact that, by construction, on each face $F \subset M$ either the corresponding symbol of our operator is invertible or the functions in the algebra $C_0\left(M_{\geq j}^{\#}\right)$ are zero. $\square$

By Lemma 2.4, the mapping

$$\mathrm{Ell}_*\,(M_{\geq j}) \xrightarrow{\varphi_j} K_*\left(M_{\geq j}^{\#}\right) \tag{4}$$

that takes partially elliptic operators to the corresponding Fredholm modules is well defined (cf. [18]).

Theorem 2.5. *For each $1 \leq j \leq k+1$, the mapping (4) is an isomorphism.*

Theorem 2.3 is the special case of Theorem 2.5 for $j = 1$.

3. Beginning of proof of the classification theorem

We prove Theorem 2.5 by induction on j decreasing from $k + 1$ (where k is the depth of M) to 1.

Inductive assumption. For $j = k+1$, the group $\mathrm{Ell}_*(M_{\geq j})$ classifies elliptic interior symbols and hence is isomorphic to $K_c^*(T^*M)$. Moreover, the mapping taking the symbol to the corresponding operator determines an isomorphism $K_c^*(T^*M) \simeq K_*(M_0)$ (e.g., see [6]). On the other hand, the right-hand side of (4) in this case just contains the group $K_*(M_0)$. Thus the theorem holds for $j = k + 1$.

Inductive step. To justify the inductive step, we need to study exact sequences in K-homology and K-theory permitting one to relate the maps φ_j in (4) for two values of the subscript, j and $j + 1$.

3.1. Exact sequence in K-homology (see [18])

Consider the embedding

$$M_{\geq j}^{\#} \supset M_j^{\#}. \tag{5}$$

The complement $M_{\geq j}^{\#} \setminus M_j^{\#}$ is obviously equal to $M_{\geq j+1}^{\#}$, and we have the exact sequence of the pair (5) in K-homology,

$$\cdots \to K_*(M_j^{\#}) \to K_*(M_{\geq j}^{\#}) \to K_*(M_{\geq j+1}^{\#}) \xrightarrow{\partial} K_{*+1}(M_j^{\#}) \to \cdots. \tag{6}$$

All maps but the boundary map ∂ in this sequence correspond to a change of module structure on the corresponding Fredholm modules. The boundary map ∂ can be reduced to a form convenient for computations by the following standard method.

Let $U^{\#} \subset M_{\geq j}^{\#}$ be the open neighborhood of the stratum $M_j^{\#}$ constructed[3] in Section 1.2 in Part I. We have the homeomorphism

$$U^{\#} \simeq M_j^{\#} \times K_{\overline{M}_j \#}, \tag{7}$$

where the cone K_{Ω_j} is the disjoint union of the cones corresponding to the connected components of the base Ω_j.

Then we have the mappings

$$M_j^{\#} \times (0,1) \xleftarrow{\pi} M_j^{\#} \times (0,1) \times \Omega_j \xleftarrow{\simeq} U^{\#} \setminus M_j^{\#} \xrightarrow{l} M_{\geq j+1}^{\#}$$

(by l we denote the embedding of an open manifold, and π is the projection onto the first two factors), which permit us to represent the boundary map ∂ in (6) as the composition

$$K_*(M_{\geq j+1}^{\#}) \xrightarrow{l^*} K_*(U^{\#} \setminus M_j^{\#}) \xrightarrow{\pi_*} K_*((0,1) \times M_j^{\#}) \xrightarrow{\beta} K_{*+1}(M_j^{\#}) \tag{8}$$

of the restriction l^* of operators to an open set, the push-forward π_*, and the periodicity isomorphism β. This representation follows from the fact that ∂ is natural.

3.2. Exact sequence related to elliptic operators (see [12])

Let M be a manifold with corners of depth $k > 0$, and let j, $1 \leq j \leq k$, be some number. We denote the algebra formed by the symbols $(\sigma_0, \sigma_j, \sigma_{j+1}, \ldots, \sigma_k)$ of all ψDO on M by

$$\Sigma_j = \mathrm{Im}(\sigma_0, \sigma_j, \sigma_{j+1}, \ldots, \sigma_k).$$

Then we have the short exact sequence of C^*-algebras

$$0 \to J \to \Sigma_j \to \Sigma_{j+1} \to 0. \tag{9}$$

Here the ideal J consists of the symbols $(\sigma_0, \sigma_j, \sigma_{j+1}, \ldots, \sigma_k)$ in which all components but σ_j are zero. From the compactness criterion for ψDO and compatibility conditions for symbols (see Part I), we see that under these conditions the symbol

---

[3]The neighborhood U is defined as the union of neighborhoods of all simplices $F^{\#} \subset M_j^{\#}$. By construction, these neighborhoods are disjoint.

σ_j is a tuple of compact-valued families decaying at infinity, so that one has the isomorphism

$$J \simeq \bigoplus_{F \subset M_j} C_0(\mathbb{R}^j, \mathcal{K}L^2(F)),$$

where the sum is taken over faces F of codimension j in M.

By virtue of this isomorphism, we can write out the exact sequence in K-theory corresponding to the short sequence (9) in the form

$$\cdots \to K_*(J) \to K_*(\Sigma_j) \to K_*(\Sigma_{j+1}) \xrightarrow{\delta} K_{*+1}(J) \to \cdots . \qquad (10)$$

Clearly,

$$K_*(J) \simeq K_*(C_0(\mathbb{R}^j)) \oplus K_*(C_0(\mathbb{R}^j)) \oplus \cdots = \mathbb{Z}^l,$$

where l is the number of connected components in M_j. In terms of this isomorphism, the boundary map δ can be represented (for $* = 1$) in the following simple form. An arbitrary class

$$[\sigma] \in K_1(\Sigma_{j+1})$$

is realized by an invertible symbol

$$\sigma = (\sigma_0, \sigma_{j+1}, \sigma_{j+2}, \ldots, \sigma_k).$$

(From now on, for brevity we carry out the computations only for K-theory elements representable by scalar operators; the consideration of the matrix case differs only in the awkwardness of formulas.) Take an arbitrary symbol σ_j compatible with σ. The symbol σ_j defines an elliptic family with parameters in $\mathbb{R}^j$, and the index of that family is a well-defined element of the K-group with compact supports of the parameter space. One has

$$\delta[\sigma] = \operatorname{ind} \sigma_j \in \bigoplus_{F \subset M_j} K_0(C_0(\mathbb{R}^j)).$$

There is a similar expression for the boundary map for the case $* = 0$. (To obtain it, one can pass to the suspension.)

3.3. Comparison of exact sequences

Let us show that the sequences (6) and (10) can be combined into the commutative diagram

$$
\begin{array}{ccccccc}
\cdots \to K_*(J) & \to K_*(\Sigma_j) \to & K_*(\Sigma_{j+1}) & \xrightarrow{\delta} & K_{*+1}(J) & \to \cdots \\
\downarrow \varphi_0 & \downarrow \varphi_j & \downarrow \varphi_{j+1} & & \downarrow \varphi_0 \\
\cdots \to K_{*+1}(M_j^{\#}) & \to K_{*+1}(M_{\geq j}^{\#}) \to & K_{*+1}(M_{\geq j+1}^{\#}) & \xrightarrow{\partial} & K_*(M_j^{\#}) & \to \cdots
\end{array}
$$

$$(11)$$

(The construction of this diagram and the verification of its commutativity will be finished in Section 4.3.)

First, we define the vertical maps in the diagram. Without loss of generality, we can assume that M has no connected components with empty boundary, the classification on such components being well known. Then for all j we have the isomorphism [17][4]

$$K_*(\Sigma_j) \simeq \mathrm{Ell}_{*+1}(M_{\geq j}) \oplus K_*(C(M)). \tag{12}$$

Hence we define the maps φ_j, $j \geq 1$, in diagram (11) as the composition

$$K_*(\Sigma_j) \longrightarrow \mathrm{Ell}_{*+1}(M_{\geq j}) \longrightarrow K_{*+1}(M_{\geq j}^{\#})$$

of the projection onto the Ell-group and the quantization (4). Thus these maps are induced by quantization, which takes symbols to operators.

It remains to define the map φ_0. Just as the other vertical arrows in the diagram, it is defined by quantization, namely, by quantization of symbols $\sigma = \sigma_j$ in the ideal J. The quantization of elements of the ideal differs from the quantization of general elements of the algebra Σ_j only in that the operator is considered in the L^2 space in a small neighborhood U of the stratum M_j in M constructed in Lemma 1.9 of Part I. We denote the operator by $\widehat{\sigma}_j$.

Let us define a module structure on $L^2(U)$. To this end, recall that U can be considered also as a subset of the positive quadrant $N'_+ M_j$ of the logarithmic normal bundle. Thus, this space is naturally a $C_0(M_j^{\#})$-module. (Elements of $C_0(M_j^{\#})$ act on $N'_+ M_j$ as operators of multiplication by radially constant functions $f(y)$, in logarithmic coordinates $y = -\ln \rho$.) The verification of locality of operators with respect to this module structure (i.e., proving that the operator $\widehat{\sigma}_j$ commutes with operators of multiplication by functions modulo compact operators) is immediate, and hence for the element $[\sigma_j] \in K_{*+1}(J)$ we define the element

$$\varphi_0[\sigma_j] := [\widehat{\sigma}_j] \in K_*(M_j^{\#}). \tag{13}$$

Diagram (11) commutes. The commutativity of the middle square of the diagram follows directly from definitions.

Lemma 3.1. *The left square of diagram* (11) *commutes.*

Proof. It is easy to see that the compositions of mappings passing through the right upper corner and lower left corner of the square

$$
\begin{array}{ccc}
K_*(J) & \to & K_*(\Sigma_j) \\
\downarrow \varphi_0 & & \downarrow \varphi_j \\
K_{*+1}(M_j^{\#}) & \to & K_{*+1}(M_{\geq j}^{\#})
\end{array}
$$

both take an elliptic symbol σ_j to the operator $\widehat{\sigma}_j$ in the space $L^2(U)$, on a neighborhood U of the stratum $M_j^{\#}$ acting as the identity operator (modulo compact operators) outside $M_j^{\#}$, but the module structures are different. More precisely,

---

[4]This isomorphism generalizes the well-known expansion $K^1(S^*M) \simeq \mathrm{Ell}(M) \oplus K^1(M)$ on a smooth closed manifold M on which there exists a nonzero vector field. Elimination of closed components permits us to claim that there exists a nonzero vector field in our situation.

let us take as U a neighborhood as in Proposition 1.10 of Part I. Then in the first case we have a natural $C_0(M^{\#}_{\geq j})$-module structure, while in the second case the module structure is induced by the composition

$$C_0(M^{\#}_{\geq j}) \xrightarrow{i^*} C_0(M^{\#}_j) \xrightarrow{\pi^*} C(U^{\#}) \to \mathcal{B}(L^2(U)),$$

where $\pi : U^{\#} \to M^{\#}_j$ is a projection and $i : M^{\#}_j \subset M^{\#}_{\geq j}$ is the embedding.

The two module structures are homotopic. Indeed, $\pi^* i^* = (i\pi)^*$ and the composition $i\pi$ is homotopic to the identity homeomorphism (in the notation of Proposition 1.10 of Part I the homotopy is $(\theta, r, x, \omega) \mapsto (\theta, \varepsilon r, x, \omega)$, where ε is the parameter of the homotopy).

Since the module structures are homotopic, we obtain, by virtue of homotopy invariance of K-homology, the equality of the two elements in $K_{*+1}(M^{\#}_{\geq j})$.

The commutativity of the square is thereby established. $\square$

Verification of the commutativity of the square containing the boundary maps is rather cumbersome, and so we make it in a separate section.

4. Boundary and coboundary maps

In this section, we establish the commutativity of the square

$$
\begin{array}{ccc}
K_*(\Sigma_{j+1}) & \xrightarrow{\delta} & K_{*+1}(J) \\
\downarrow \varphi_{j+1} & & \downarrow \varphi_0 \\
K_{*+1}(M^{\#}_{\geq j+1}) & \xrightarrow{\partial} & K_*(M^{\#}_j)
\end{array}
\tag{14}
$$

containing the boundary maps in diagram (11). The scheme of proof is as follows. We

1. Compute the composition $\varphi_0 \circ \delta$.
2. Compute the composition $\partial \circ \varphi_{j+1}$.
3. Compare the resulting expressions.

4.1. Composition $\varphi_0 \circ \delta$

Let $[\sigma] \in K_*(\Sigma_{j+1})$ be the element defined by some symbol $\sigma = (\sigma_0, \sigma_{j+1}, \ldots, \sigma_k)$. Take a symbol σ_j on M_j compatible with σ and denote by

$$\widehat{\sigma}_j : L^2(NM_j) \to L^2(NM_j) \tag{15}$$

the corresponding translation-invariant infinitesimal operator. (It is conjugate to σ_j by the Fourier transform.)

Representing the space $N'_+ M_j$ as the product $N'_+ M_j \simeq M_j \times \mathbb{R}^j_+$, we see that $L^2(NM_j)$ is a $C_0(\bigsqcup \mathbb{R}^j_+)$-module. Here $\bigsqcup \mathbb{R}^j_+$ is the disjoint union of as many open quadrants as there are faces of codimension j in M. The operator $\widehat{\sigma}_j$ is local

with respect to this module structure. We denote the corresponding element of the
K-homology group by

$$[\widehat{\sigma}_j] \in K_{*+1}\left(\bigsqcup \mathbb{R}^j_+\right). \tag{16}$$

Lemma 4.1. *The element* $[\sigma] \in K_*(\Sigma_{j+1})$ *satisfies the chain of relations*

$$\varphi_0 \delta[\sigma] = \varphi_0(\operatorname{ind} \sigma_j) = \beta[\widehat{\sigma}_j] \in K_*(M_j^\#), \tag{17}$$

where $\beta : K_{*+1}(\bigsqcup \mathbb{R}^j_+) = K_{*+1}(M_j^\# \times \mathbb{R}_+) \longrightarrow K_*(M_j^\#)$ *is the Bott periodicity
isomorphism, and the index is understood as the index*

$$\operatorname{ind} \sigma_j \in K^{*+1}\left(\bigsqcup \mathbb{R}^j\right) \simeq K_{*+1}(J)$$

of the elliptic operator-valued symbol σ_j.

Proof. For brevity, we assume that M_j consists of exactly one stratum. In this
case, we have $M_j^\# \times \mathbb{R}_+ \simeq \mathbb{R}^j_+$.

The first relation in (17) follows from definitions (since the boundary map in
K-theory of algebras is the index map).

1. Let us establish the second relation $\varphi_0(\operatorname{ind} \sigma_j) = \beta[\widehat{\sigma}_j]$. The proof is based
on the diagram

$$
\begin{array}{ccc}
K_{*+1}(\Sigma_{j+1}) & \xrightarrow{\operatorname{ind} \sigma_j} & K_*(J) \\
& \searrow \quad \downarrow q & \quad \searrow \varphi_0 \\
& K_*(M_j^\# \times \mathbb{R}_+) \xrightarrow{\beta} & K_{*+1}(M_j^\#),
\end{array} \tag{18}
$$

where the map $K_{*+1}(\Sigma_{j+1}) \to K_*(M_j^\# \times \mathbb{R}_+)$ is induced by the map that takes
the symbol σ to the operator $\widehat{\sigma}_j$ in (16). Finally, the group $K_*(J) \simeq K^*(\mathbb{R}^j)$
is interpreted as the K-group of the cotangent bundle to $\mathbb{R}^j_+$, and the map q is
induced by standard quantization (to a symbol on the cosphere bundle, one assigns
a pseudodifferential operator).

2. We claim that diagram (18) commutes. Indeed, let us verify the commutativity of the left triangle, i.e., the relation

$$[\widehat{\sigma}_j] = q[\operatorname{ind} \sigma_j]. \tag{19}$$

Note that the operator $\widehat{\sigma}_j$ is given over the product $NM_j = M_j \times \mathbb{R}^j$. Moreover, it
can be viewed as a ψDO on $\mathbb{R}^j$ with operator-valued symbol $\sigma_j = \sigma_j(\xi)$, $\xi \in \mathbb{R}^j$.
This symbol is independent of $x \in \mathbb{R}^j$. Without loss of generality, it can be assumed
to be smooth with respect to the parameter ξ (since $\sigma_j(\xi)$, just as any ψDO with
a parameter, can be arbitrarily closely approximated by a smooth ψDO with a
parameter; see Part I). Hence $\sigma_j(\xi)$ is an operator-valued symbol in the sense
of [11], i.e., has a compact variation with respect to ξ and all of its derivatives
starting from the first decay at infinity. Now relation (19) follows by analogy with
the generalized Luke theorem in [18].

The commutativity of the right triangle follows (see Corollary 5.2 in the appendix) from the higher relative index theorem.

3. The commutativity of diagram (18) implies the second relation (17). (The right-hand side is obtained if from the left top corner of the diagram we go directly to the group $K_*(M_j^\# \times \mathbb{R}_+)$ and then apply the periodicity isomorphism β.) $\quad\square$

4.2. Composition $\partial \circ \varphi_{j+1}$

Space $N'_+ M_j$ as a manifold with corners. (See Section 1.2 of Part I for the notation.) The image of the positive quadrant $N'_+ M_j$ under the inverse of the logarithmic map is the set $M_j \times [0,1)^j \subset M_j \times \overline{\mathbb{R}}_+^j = N_+ M_j$. Hence we treat $N'_+ M_j$ as the interior of a manifold with corners, denoted by $\overline{N'_+ M_j}$. We denote the corresponding dual space by $\overline{N'_+ M_j}^\#$. On the complement $\overline{N'_+ M_j}^\# \setminus M_j^\#$, there is a well-defined projection

$$
\begin{array}{ccc}
\overline{N'_+ M_j}^\# \setminus M_j^\# & \xrightarrow{\ \pi\ } & M_j^\# \times \mathbb{R}_+ \\[4pt]
(y, x, \omega) & \longmapsto & \left(\dfrac{y}{|y|}, \dfrac{|x|+1}{|y|} \right),
\end{array}
\tag{20}
$$

whose fiber is the space $(\overline{M}_j)^\#$ (see Proposition 1.10 of Part I).

Reduction into a neighborhood of the edge. We have the diagram of embeddings

$$
\begin{array}{ccccc}
M_j^\# & \subset & U^\# & \subset & M_{\geq j}^\# \\
 & & \cap & & \\
 & & \overline{N'_+ M}_j^\#. & &
\end{array}
\tag{21}
$$

Let $[\sigma] \in K_{*+1}(\Sigma_{j+1})$ be the element determined by the symbol σ as above. By passing to the corresponding operators, we obtain the element

$$
\varphi_{j+1}[\sigma] \in K_*(M_{\geq j+1}^\#).
$$

On the other hand, the infinitesimal operator $\widehat{\sigma}_j$ compatible with σ (see (15)) defines the element

$$
[\widehat{\sigma}_j]' \in K_*(\overline{N_+ M}_j^\# \setminus M_j^\#).
$$

This element is well defined, since the components of its symbol are elliptic on the corresponding strata. We use primes to distinguish this element from the element (16): although they are determined by one and the same operator, the module structures on the L^2-spaces are different.

The naturality of the boundary map in K-homology results in the following lemma.

Lemma 4.2. *One has*

$$
\partial \varphi_{j+1}[\sigma] = \partial'[\widehat{\sigma}_j]',
$$

where $\partial' : K_*(\overline{N_+ M}_j^\# \setminus M_j^\#) \to K_{*+1}(M_j^\#)$ *is the boundary map for the pair* $M_j^\# \subset \overline{N_+ M}_j^\#$.

Proof. Let D be some operator on M with symbol σ.

1. The infinitesimal operator $\hat{\sigma}_j$ is obtained from D by localization to the set $M_j^{\#}$. Hence the restrictions $D|_U$ and $\hat{\sigma}_j|_U$ of these operators to a small neighborhood U of M_j are connected by a linear homotopy; i.e., one has

$$[D|_U] = [\hat{\sigma}_j|_U] \in K_*(U^{\#}). \tag{22}$$

2. By applying the naturality of the boundary map in K-homology to the embedding diagram (21), we obtain

$$\partial \varphi_{j+1}[\sigma] \equiv \partial[D] = \partial''[D|_U],$$

where ∂'' is the boundary map for the pair $M_j^{\#} \subset U^{\#}$. Now if on the right-hand side of the last relation we replace the element $[D|_U]$ according to (22) and once more use the naturality of the boundary map, then we obtain the desired relation

$$\partial \varphi_{j+1}[\sigma] = \partial''[\hat{\sigma}_j|_U] = \partial'[\hat{\sigma}_j]'. \qquad \square$$

Thus in what follows, when computing the composition $\partial \circ \varphi_{j+1}$, we can (and will) work with the operator $\hat{\sigma}_j$ on $N'_+ M_j$.

Homotopy of the module structure. By (8), the boundary map ∂' in Lemma 4.2 can be represented as the composition

$$K_*(\overline{N'_+ M}_j^{\#} \setminus M_j^{\#}) \xrightarrow{\pi_*} K_*(M_j^{\#} \times \mathbb{R}_+) \xrightarrow{\beta} K_{*+1}(M_j^{\#}) \tag{23}$$

of the push-forward with respect to the projection π and the periodicity isomorphism.

Unfortunately, although the classes $[\hat{\sigma}_j]$ and $\pi_*[\hat{\sigma}_j]'$ are determined by the same operator $\hat{\sigma}_j$, they have different module structures on the space $L^2(NM_j)$: in the first case, the structure is independent of the coordinate x, while in the second case it depends on it (see (23)).

Let us make a homotopy of module structures. To this end, we define a homotopy

$$\pi^{\varepsilon} : \overline{N'_+ M}_j^{\#} \setminus M_j^{\#} \longrightarrow M_j^{\#} \times \mathbb{R}_+$$

of projections by the formula (cf. (20))

$$\pi^{\varepsilon}(y, x, \omega) := \left(\frac{y}{|y|}, \frac{\varepsilon|x| + 1}{|y|} \right).$$

This formula defines a continuous family of maps for $\varepsilon > 0$. However, the family is not continuous as $\varepsilon \to 0$.[5] Nevertheless, continuity takes place for the Fredholm modules, as shown by the following lemma.

Lemma 4.3. *The family $\pi_*^{\varepsilon}(\hat{\sigma}_j)'$ of Fredholm modules obtained by the change of module structure defines a homotopy in the sense of KK-theory, and one has*

$$\lim_{\varepsilon \to 0} \pi_*^{\varepsilon}(\hat{\sigma}_j)' = \pi_*^0(\hat{\sigma}_j)', \tag{24}$$

whence it follows that $\pi_*[\hat{\sigma}_j]' = [\pi_*^0(\hat{\sigma}_j)]' \in K_*(M_j^{\#} \times \mathbb{R}_+)$.

---

[5]And hence the map π_*^0 is not defined on the K-group.

Proof. For brevity, we assume that M_j consists of a single face, i.e., is connected. Then the homotopy in the sense of KK-theory means (e.g., see [2]) that for each function $f \in C_0(\mathbb{R}^j_+)$ the family

$$g^\varepsilon = (\pi^\varepsilon)^*(f) : L^2(N'_+ M_j) \longrightarrow L^2(N'_+ M_j)$$

of operators of multiplication by the functions $(\pi^\varepsilon)^*(f)$ is strongly $*$-continuous and that the operator families

$$[g^\varepsilon, \widehat{\sigma}_j], \quad g^\varepsilon(\widehat{\sigma}_j \widehat{\sigma}_j^{-1} - 1)$$

in $L^2(N M_j)$ are continuous families of compact operators as $\varepsilon \to 0$.

It suffices to prove all these facts for (a dense set of) smooth functions f. If f is smooth, then one can smooth the family g^ε and use the composition formulas, which provide the desired compactness and continuity. $\square$

4.3. Comparison of the compositions $\varphi_0 \circ \delta$ and $\partial \circ \varphi_{j+1}$

Now let us use Lemmas 4.1–4.3. We obtain the chain of relations

$$\partial \varphi_{j+1}[\sigma] \overset{\text{Lemma 4.2}}{=} \partial'[\widehat{\sigma}_j]' \overset{\text{formula (23)}}{=} \beta \pi^1_*[\widehat{\sigma}_j]' \overset{\text{Lemma 4.3}}{=} \beta[\pi^0_* \widehat{\sigma}_j]' =$$

$$= \beta[\widehat{\sigma}_j] \overset{\text{Lemma 4.1}}{=} \varphi_0 \delta[\sigma].$$

The equality at the end of the first row corresponds to the identical coincidence of the corresponding Fredholm modules.

Thus the square (14) commutes, and we have established the commutativity of diagram (11).

5. Higher relative index theorem

To prove Theorem 2.5, we need to show that the mapping φ_0 in (11) is an isomorphism. This is done in this section.

Consider the map (see Eq. (13))

$$\varphi_0 : K_j(C_0(\mathbb{R}^j)) \to K_{j-1}(\Delta^\circ_{j-1}),$$

induced by the map taking a symbol $\sigma(\xi)$, $\xi \in \mathbb{R}^j$, to the corresponding translation-invariant operator

$$\widehat{\sigma} : L^2(\mathbb{R}^j) \longrightarrow L^2(\mathbb{R}^j). \tag{25}$$

Here the space $L^2(\mathbb{R}^j)$ is equipped with the following module structure over the algebra $C_0(\Delta^\circ_{j-1})$ of functions on the interior of the simplex: a function $f \in C_0(\Delta^\circ_{j-1})$ is viewed as a radially constant function equal to zero outside the positive quadrant $\mathbb{R}^j_+$.

We shall prove that φ_0 is an isomorphism. The first step is the following proposition.

Proposition 5.1. *For the index pairing of the element* $\varphi_0[\sigma] \in K_{j-1}(\Delta^\circ_{j-1})$ *with an arbitrary element*

$$[a] \in K_{j-1}(C_0(\Delta^\circ_{j-1})) \simeq \widetilde{K}^{j-1}(\mathbb{S}^{j-1}),$$

where $\widetilde{K}$ *is the reduced K-group, one has the formula*

$$\langle \varphi_0[\sigma], a \rangle = \mathrm{ind}_t \Big([\sigma] \times [a] \Big), \tag{26}$$

where the product $[\sigma] \times [a]$ *is defined as the composition*

$$K^j(\mathbb{R}^j) \times \widetilde{K}^{j-1}(\mathbb{S}^{j-1}) \to K^1(S^*\mathbb{R}^j) \to K^0(T^*\mathbb{R}^j),$$

and $\mathrm{ind}_t : K^0(T^*\mathbb{R}^j) \to \mathbb{Z}$ *is the topological Atiyah–Singer index for* $\mathbb{R}^j$. (*We use topological K-groups with compact supports.*)

Remark. For $j = 1$, this assertion is reduced to the relative index theorem for operators on manifolds with conical points. We mean the formula for the difference of indices of operators with equal interior symbols, or, equivalently, for the index of elliptic operators of the form $1 + G$

$$\mathrm{ind}(1 + G) = w(1 + g),$$

where the interior symbol of G is zero and $w(1 + g)$ is the winding number of the conormal symbol $1 + g$. Hence the index formula (26) in the general case can be referred to as the *higher relative index formula*.

Proof. To be definite, we consider the case of even j. (The odd case can be considered in a similar way.)

1. The element

$$[\sigma] \in K_0(C_0(\mathbb{R}^j))$$

is determined by some projection-valued function $p(x)$ on $\mathbb{R}^j$ equal to the diagonal projection $\mathrm{diag}(1,0)$ at infinity. Conversely, the element

$$[a] \in K_{j-1}(C_0(\Delta^\circ_{j-1})) \simeq \widetilde{K}^{j-1}(\mathbb{S}^{j-1})$$

is determined by some invertible function $a(x)$ on the sphere $\mathbb{S}^{j-1}$. To simplify the notation, we assume that this is a scalar function. The matrix case can be considered in a similar way.

2. In this notation, the index pairing $\langle \varphi_0[\sigma], a \rangle$ is by definition equal to the index of the Toeplitz operator (e.g., see [5])

$$\widehat{p}a : \mathrm{Im}\,\widehat{p} \longrightarrow \mathrm{Im}\,\widehat{p}, \tag{27}$$

where $\widehat{p} : L^2(\mathbb{R}^j) \longrightarrow L^2(\mathbb{R}^j)$ – is the projection determined by the symbol p, as in (25).

3. To compute the index of the operator (27), we make the Fourier transform. Then the operator $\widehat{p}$ becomes the projection $p(x)$, and the space $\mathrm{Im}\,\widehat{p}$ becomes the space of sections of the bundle given by the range of $p(x)$. Conversely, the operator

of multiplication by a passes into a translation-invariant ψDO in $\mathbb{R}^j$ with principal symbol $a = a(\xi)$. Hence we obtain

$$\langle \varphi_0[\sigma], a \rangle = \operatorname{ind}(p\widehat{a} : \operatorname{Im} p \longrightarrow \operatorname{Im} p). \tag{28}$$

4. The last operator coincides at infinity with a direct sum of the invertible operator $\widehat{a}$ acting on functions. By the index locality property, the difference of their indices is given by the Atiyah–Singer formula, which gives the desired expression (26). $\qquad\square$

Let us rewrite this index formula in the equivalent form.

Corollary 5.2. *The following triangle commutes:*

$$K_*(C_0(\mathbb{R}^j)) \tag{29}$$

$$
\begin{array}{ccc}
 & \varphi_0 & \\
q \downarrow & \searrow & \\
K_*(\Delta^{\circ}_{j-1} \times (0,\infty)) & \xrightarrow{\ \beta\ } & K_{*+1}(\Delta^{\circ}_{j-1}),
\end{array}
$$

where β is the Bott periodicity isomorphism and

$$q : K_*(C_0(\mathbb{R}^j)) = K^*(T^*\mathbb{R}^j_+) \longrightarrow K_*(\Delta^{\circ}_{j-1} \times (0,\infty))$$

is the standard isomorphism induced by the pseudodifferential quantization in $\mathbb{R}^j_+ = \Delta^{\circ}_{j-1} \times (0,\infty)$. Hence, φ_0 is an isomorphism.

Proof. All groups in the triangle (29) are isomorphic to $\mathbb{Z}$ and have natural Bott generators. It is known that the maps q and β take Bott elements to Bott elements. Hence to verify the commutativity of the diagram it suffices to verify this property for φ_0. But this readily follows from the index formula (26). $\qquad\square$

6. End of proof of the classification theorem

By virtue of the isomorphism (12), we can single out and cancel the summand $K_*(C(M))$ in diagram (11) in the terms $K_*(\Sigma_j)$ and $K_*(\Sigma_{j+1})$. We obtain the diagram

$$
\begin{array}{ccccccccc}
\cdots \to & K_*(J) & \to & \mathrm{Ell}_{*+1}(M_{\geq j}) & \to & \mathrm{Ell}_{*+1}(M_{\geq j+1}) & \xrightarrow{\ \delta\ } & K_{*+1}(J) & \to \cdots \\
 & \downarrow \varphi_0 & & \downarrow \varphi_j & & \downarrow \varphi_{j+1} & & \downarrow \varphi_0 & \\
\cdots \to & K_{*+1}(M^{\#}_j) & \to & K_{*+1}(M^{\#}_{\geq j}) & \to & K_{*+1}(M^{\#}_{\geq j+1}) & \xrightarrow{\ \partial\ } & K_*(M^{\#}_j) & \to \cdots
\end{array}
$$
$$\tag{30}$$

The map φ_{j+1} is an isomorphism by the inductive assumption. The map φ_0 is also an isomorphism (see Corollary 5.2). Since the diagram commutes, we can apply the 5-lemma and obtain the desired justification of the induction step in Theorem 2.5: if the map φ_{j+1} is isomorphic on the Ell-group, then so is the map φ_j.

The proof of Theorem 2.5 is complete.

7. Application to the Monthubert–Nistor index

Let us discuss the relationship with the problems considered by Monthubert and Nistor [15]. In the notation of the present paper, for the case of manifolds with embedded corners they considered the short exact sequence

$$0 \to J \longrightarrow \Psi(M) \xrightarrow{\sigma_0} C(S^*M) \to 0, \tag{31}$$

where σ_0 is the interior symbol map and the ideal J consists of operators with zero interior symbol. They studied the boundary map corresponding to this sequence:

$$\delta : K_*(C(S^*M)) \longrightarrow K_{*+1}(J).$$

For a closed manifold, J is the ideal of compact operators (hence $K_*(J) \simeq \mathbb{Z}$) and the boundary map coincides with the analytic index. Moreover, Monthubert and Nistor showed that in the general case this map has an important topological meaning: it gives the obstruction to the existence of an invertible operator with a given interior symbol. For these reasons, Monthubert and Nistor call this map the *analytic index of manifolds with corners*.

We claim that the classification theorem readily implies a K-homology criterion for the vanishing of the analytic index. Indeed, consider the diagram

$$
\begin{array}{ccccc}
K_{*+1}(M^{\#}) & \longrightarrow & K_{*+1}(M_0) & \xrightarrow{\partial} & K_*(M^{\#} \setminus M_0) \\
\varphi_1 \uparrow & & \uparrow \varphi_{k+1} & & \\
K_*(\Psi(M)) & \longrightarrow & K_*(C(S^*M)) & \xrightarrow{\delta} & K_{*+1}(J),
\end{array} \tag{32}
$$

where the lower row is the sequence induced by the short exact sequence (31) and the upper row is the exact sequence of the pair $M^{\#} \setminus M_0 \subset M^{\#}$ in K-homology. The maps φ_1 and φ_{k+1} are induced by quantization of elliptic symbols on $M^{\#}$ and M_0, respectively (cf. (11)). The diagram is obviously commutative.

From the exactness of the sequences and the obvious commutativity of the diagram, we obtain the following assertion. Let us assume for simplicity that M has no connected components with empty boundary.

Proposition 7.1. *The analytic index $\delta(x) \in K_{*+1}(J)$ of $x \in K_*(C(S^*M))$ vanishes if and only if $\partial\varphi_{k+1}(x) = 0$.*

Proof. 1. There are splittings (cf. (12))

$$K_*(\Psi(M)) \simeq \widetilde{\mathrm{Ell}}_{*+1}(M) \oplus K_*(C(M)), \quad K_*(C(S^*M)) \simeq \mathrm{Ell}_{*+1}(M_0) \oplus K_*(C(M)),$$

where $\widetilde{\mathrm{Ell}}$ is the *reduced* Ell-group generated by operators of index zero. Moreover, the direct summands $K_*(C(M))$ can be cancelled in (32). This does not affect the boundary map. Hence, we obtain the commutative diagram

$$
\begin{array}{ccccc}
\widetilde{K}_{*+1}(M^{\#}) & \longrightarrow & K_{*+1}(M_0) & \xrightarrow{\partial} & \widetilde{K}_*(M^{\#} \setminus M_0) \\
\varphi_1 \uparrow & & \uparrow \varphi_{k+1} & & \\
K_*(\Psi(M))/K_*(C(M)) & \longrightarrow & K_*(C(S^*M))/K_*(C(M)) & \xrightarrow{\delta} & K_{*+1}(J),
\end{array}
$$
$$\tag{33}$$

where $\widetilde{K}_*$ is the reduced K-homology group generated by operators of index zero.

3. By the classification theorem, the quantization maps φ in (33) induce isomorphisms. Hence, the commutativity of the diagram readily shows that vanishing of δ is equivalent to the vanishing of the boundary map ∂ in K-homology. $\square$

The reader can readily rewrite this formula in a more explicit form as a condition on the interior symbol σ_0. There is also a cohomological form of this condition. Needless to say, the cohomological formula is only valid modulo torsion.

Remark. One actually has the group isomorphism

$$K_*(J) \xrightarrow{\simeq} \widetilde{K}_*(M^\# \setminus M_0)$$

determined by quantization of operators with zero interior symbol. (One can readily obtain this isomorphism by reproducing the proof of our classification theorem. In the proof, only the inductive assumption is changed: now for $j = k+1$ we claim that $0 = 0$.)

References

[1] M.F. Atiyah, *Global theory of elliptic operators*, Proc. of the Int. Symposium on Functional Analysis (Tokyo), University of Tokyo Press, 1969, pp. 21–30.

[2] B. Blackadar, *K-theory for operator algebras*, Mathematical Sciences Research Institute Publications, no. 5, Cambridge University Press, 1998, Second edition.

[3] U. Bunke, *Index theory, eta forms, and Deligne cohomology*. Preprint arXiv: math.DG/0201112.

[4] A. Connes, *Noncommutative geometry*, Academic Press Inc., San Diego, CA, 1994.

[5] N. Higson and J. Roe, *Analytic K-homology*, Oxford University Press, Oxford, 2000.

[6] G. Kasparov, *Equivariant KK-theory and the Novikov conjecture*, Inv. Math. **91** (1988), no. 1, 147–201.

[7] T. Krainer, *Elliptic boundary problems on manifolds with polycylindrical ends*, J. Funct. Anal., **244** (2007), no. 2, 351–386.

[8] R. Lauter and S. Moroianu, *The index of cusp operators on manifolds with corners*, Ann. Global Anal. Geom. **21** (2002), no. 1, 31–49.

[9] P.-Y. Le Gall and B. Monthubert, *K-theory of the indicial algebra of a manifold with corners*, K-Theory **23** (2001), no. 2, 105–113.

[10] P. Loya, *The index of b-pseudodifferential operators on manifolds with corners*, Ann. Global Anal. Geom. **27** (2005), no. 2, 101–133.

[11] G. Luke, *Pseudodifferential operators on Hilbert bundles*, J. Diff. Equations **12** (1972), 566–589.

[12] R. Melrose and V. Nistor, *K-theory of C^*-algebras of b-pseudodifferential operators*, Geom. Funct. Anal. **8** (1998), no. 1, 88–122.

[13] R. Melrose and P. Piazza, *Analytic K-theory on manifolds with corners*, Adv. in Math. **92** (1992), no. 1, 1–26.

[14] B. Monthubert, *Groupoids and pseudodifferential calculus on manifolds with corners*, J. Funct. Anal. **199** (2003), no. 1, 243–286.

[15] B. Monthubert and V. Nistor, *A topological index theorem for manifolds with corners*, arXiv: math.KT/0507601, 2005.

[16] V. Nistor, *An index theorem for gauge-invariant families: The case of solvable groups*, Acta Math. Hungarica **99** (2003), no. 2, 155–183.

[17] A. Savin, *Elliptic operators on singular manifolds and K-homology*, K-theory **34** (2005), no. 1, 71–98.

[18] V.E. Nazaikinskii, A.Yu. Savin, and B.Yu. Sternin, *On the homotopy classification of elliptic operators on stratified manifolds*, Izvestiya: Mathematics, **71** (2007), no. 6, 91–118.

Vladimir Nazaikinskii
Institute for Problems in Mechanics
Russian Academy of Sciences
pr. Vernadskogo 101-1
119526 Moscow, Russia
e-mail: `nazaikinskii@yandex.ru`

Anton Savin and Boris Sternin
Independent University of Moscow,
Bol'shoi Vlas'evskii per. 11
119002 Moscow, Russia
e-mail: `antonsavin@mail.ru`
e-mail: `sternin@mail.ru`

C^*-algebras and Elliptic Theory II

Trends in Mathematics, 227–237

© 2008 Birkhäuser Verlag Basel/Switzerland

Dixmier Traceability for General Pseudo-differential Operators

Fabio Nicola and Luigi Rodino

Abstract. For Hörmander's classes $OPS(m, g)$ of pseudo-differential operators associated with a weight m and a metric g we prove (under an additional technical condition) that, if m is in the space L^1-weak, all operators in that class have finite Dixmier trace.

Mathematics Subject Classification (2000). Primary 35S05; Secondary 58J42.

Keywords. Pseudo-differential operators, Dixmier trace, Lorentz-Marcinkiewicz space.

1. Introduction

A compact operator A in a separable Hilbert space H is in the Schatten-von Neumann class $S_p(H)$, with $1 \leq p < \infty$, if the sequence of its singular values $\mu_j(A) \searrow 0$ is in l^p, that is

$$\sum_{j=1}^{\infty} \mu_j(A)^p < \infty.$$

Recall that the singular values of A are defined to be the eigenvalues of the positive self-adjoint compact operator $|A| = (A^*A)^{1/2}$.

The elements of $S_2(H)$ are Hilbert-Schmidt operators, while $S_1(H)$ is the algebra of trace class operators. Let us also write $S_\infty(H)$ for the class of bounded operators in H. Finally, A is in the Dixmier class $\mathcal{L}^{(1,\infty)}(H)$ if

$$\sigma_N(A) = \sum_{j=1}^{N} \mu_j(A) = O(\log N) \quad \text{as } N \to +\infty.$$

Equivalently, we may require that $\mu_j(A) = O(\frac{1}{j})$ as $j \to \infty$, or also assume

$$\lambda \sum_{\mu_j(A) \geq \lambda} 1 = \#\{\mu_j(A) \geq \lambda\}\lambda = O(1) \quad \text{as } \lambda \to 0^+.$$

We obviously have

$$S_1(H) \subset \mathcal{L}^{(1,\infty)}(H) \subset \bigcap_{p>1} S_p(H). \tag{1.1}$$

As well as for $A \in S_1(H)$, also for $A \in \mathcal{L}^{(1,\infty)}(H)$ a trace can be defined, called *Dixmier trace* (see Dixmier [9], Connes [5] and Schrohe [21] for the precise definition); operators in $\mathcal{L}^{(1,\infty)}(H)$ are also called *Dixmier traceable*.

In this paper we discuss the case when $H = L^2(\mathbb{R}^n)$ and A is a pseudo-differential operator with symbol $a(x,\xi)$ defined in $\mathbb{R}^{2n}$. The natural problem is to read on the symbol $a(x,\xi)$ whether $A \in S_p$, $1 \le p < \infty$, or $A \in \mathcal{L}^{(1,\infty)}$. In the following, we shall survey some recent results for the case of the Schatten-von Neumann classes S_p, and give a new result for $\mathcal{L}^{(1,\infty)}$.

Let us begin by observing that the pseudo-differential operator corresponding to $a(x,\xi)$ can be defined in different ways. We recall the quantization of Kohn and Nirenberg

$$Au(x) = (2\pi)^{-n} \int e^{ix\xi} a(x,\xi) \hat{u}(\xi) \, d\xi.$$

In the following, we shall refer to Weyl quantization

$$Au(x) = (a^w u)(x) = (\text{Op}^w(a)u)(x) = (2\pi)^{-n} \int e^{ix\xi} a\left(\frac{x+y}{2}, \xi\right) u(y) \, dy \, d\xi,$$

see the next section for details. We also mention the Wick, anti-Wick quantizations, see for example Boggiatto, Buzano and Rodino [1].

We now describe the class of symbols we will consider. The standard setting is given by the Weyl-Hörmander classes, where one assumes $a \in S(m, g)$, g being a Hörmander metric and m a g-admissible weight function, see [16], Chapter XVIII and the next Section 2 for precise definitions. The non-expert reader may think of to the special case of a metric g of the form

$$g_{x,\xi}(y,\eta) = \frac{|dy|^2}{\phi(x,\xi)^2} + \frac{|d\eta|^2}{\psi(x,\xi)^2}, \tag{1.2}$$

with $\phi > 0$, $\psi > 0$. Then, $a \in S(m, g)$ means that

$$|\partial_x^\alpha \partial_\xi^\beta a(x,\xi)| \le C m(x,\xi) \phi(x,\xi)^{-|\alpha|} \psi(x,\xi)^{-|\beta|}.$$

In order to have good properties for composition, adjoints, etc., one assumes, among other things, the so-called *uncertainty principle*, i.e., that the Planck function $h(x,\xi)$ associated with g satisfies $h(x,\xi) \le 1$. For the metric (1.2) one has $h = (\phi\psi)^{-1}$. To understand the role of h one should recall, for example, that the symbol of the product $a^w b^w$, with $a \in S(m_1, g)$, $b \in S(m_2, g)$, has an asymptotic expansion $\sum_{j\ge0} c_j$, with $c_j \in S(m_1 m_2 h^j, g)$, $c_0 = ab$. The uncertainty principle therefore guarantees that each term c_j is in $S(m_1 m_2, g)$ too, and so is the symbol of $a^w b^w$ (see Theorem 2.1 below). Moreover, the uncertainty principle is also essential for symbols $a \in S(1, g)$ to give rise to bounded operators in $L^2(\mathbb{R}^n)$ (Calderòn-Vaillancourt Theorem). In some of the results below one actually assumes the *strong uncertainty principle*, that is

$$h(x,\xi) \leq C(1 + |x| + |\xi|)^{-\delta},$$

for some $C > 0$, $\delta > 0$. Under this hypothesis one has a full symbolic calculus with asymptotic expansions and remainders with kernels in the Schwartz space $\mathcal{S}(\mathbb{R}^n \times \mathbb{R}^n)$.

As main examples, we observe that Hörmander's classes $S_{\rho,\delta}^\mu$ correspond to $m(x,\xi) = (1 + |\xi|)^\mu$ and to the metric (1.2) with[1] $\phi(x,\xi) = \langle\xi\rangle^{-\delta}$, $\psi(x,\xi) = \langle\xi\rangle^\rho$; hence $h(x,\xi) = \langle\xi\rangle^{\delta-\rho}$ and the uncertainty principle is satisfied if and only if $\delta \leq \rho$.
Shubin's classes G^μ (see [22]) correspond to $m(x,\xi) = (1 + |x| + |\xi|)^\mu$ and to the metric (1.2) with $\phi(x,\xi) = \psi(x,\xi) = (1 + |x|^2 + |\xi|^2)^{1/2}$; hence $h(x,\xi) = 1 + |x|^2 + |\xi|^2$ and the strong uncertainty principle is satisfied.
Finally the so-called SG or scattering classes S^{k_1,k_2} (see Schulze [19]) correspond to $m(x,\xi) = (1 + |x|)^{k_2}(1 + |\xi|)^{k_1}$, and to the metric (1.2) with $\phi(x,\xi) = \langle x\rangle$ and $\psi(x,\xi) = \langle\xi\rangle$; therefore $h(x,\xi) = \langle x\rangle^{-1}\langle\xi\rangle^{-1}$ and the strong uncertainty principle is still verified.

Let us now come to the problem of characterizing the decay of the eigenvalues of a pseudo-differential operator in terms of its symbol. All the results are inspired by the rough conjecture that $A \in S_p(L^2(\mathbb{R}^n))$ should be equivalent to $a \in L^p(\mathbb{R}^{2n})$.
As basic results in this direction we recall that a function $a \in L^1$ yields a compact operator a^w in $L^2(\mathbb{R}^n)$. Moreover $a \in L^2$ is *equivalent* to $a^w \in S_2$ (see, e.g., Folland [10]). By interpolation one then obtains that $a \in L^p$ implies $a^w \in S_{p'}$, $1 \leq p \leq 2$, $1/p + 1/p' = 1$. When $p > 2$ it is no longer true that a^w is a bounded operator (Simon [20]), so that it is clear that some additional condition has to be imposed on a symbol $a \in L^p$ for a^w to be in S_p. Observe that, up to now, the function a is not assumed to be smooth. Robert [18] instead showed that if $\partial_x^\alpha \partial_\xi^\beta a \in L^\infty \cap L^1$ ($|\alpha| + |\beta| \leq N$, N large enough) then $a^w \in S_1$. Another theorem due to Hörmander [15] states that if the weight m is in L^1 then $OPS(m,g) \subset S_1$, namely all the operators with symbols in $S(m,g)$ are trace class. This latter result was also generalized to the case $p > 1$ by Buzano and Nicola [3], where it was proved that $m \in L^p$ is equivalent to $OPS(m,g) \subset S_p$ if one assumes the strong uncertainty principle. Subsequently Toft [24] proved the same result without assuming the strong uncertainty principle.
There is also a variety of related results in the framework of modulations spaces, proved by techniques from Time-Frequency Analysis (see Gröchenig [13]). We recall in particular from Gröchenig [12] that any function $a \in L^1$ whose Fourier transform is in L^1 gives rise to an operator $a^w \in S_1$. Also, it is proved by Gröchenig and Heil [14] that $a \in L_s^2 \cap H^s$, $s \geq 0$, implies $a^w \in S_p$ with $p > 2n/(n+s)$ (H^s is the Sobolev space of order s while L_s^2 is the L^2 space with respect to the density $(1 + |x|^2 + |\xi|^2)^s \, dx d\xi$). Important results when a is assumed in the modulation

---

[1] As usual, $\langle\xi\rangle = (1 + |\xi|^2)^{1/2}$.

spaces $M^{p,q}$ were also obtained by Cordero and Gröchenig [7]. Generalizations within the framework of ultradistributions have recently appeared in Cordero, Pilipović, Rodino and Teofanov [8].

We now present our result for the Dixmier ideal $\mathcal{L}^{(1,\infty)}$. Inspired by its role of threshold with respect to the summability of the singular values (see (1.1)) and taking into account the above-mentioned results, it is quite natural to look at the Lorentz-Marcinkiewicz space L_w^1 of L^1-weak functions as a candidate for the following rough conjecture: $a \in L_w^1(\mathbb{R}^{2n}) \Longrightarrow a^w \in \mathcal{L}^{(1,\infty)}(L^2(\mathbb{R}^n))$.

We recall that the space $L_w^1(\mathbb{R}^N)$, also denoted also $L^{1,\infty}(\mathbb{R}^N)$, is defined by

$$L_w^1(\mathbb{R}^N) = \{f : \mathbb{R}^N \to \mathbb{C} \text{ measurable such that } \sup_{s>0} s \cdot \text{meas}(\{|f| > s\}) < +\infty\},$$

(see Stein and Weiss [23]). For example, the functions $|x|^{-N}$ and $(1 + |x|^2)^{-N/2}$ are in $L_w^1(\mathbb{R}^N)$.

Here is our main result.

Theorem 1.1. *Let g be a Hörmander metric and m a g-admissible weight function. Assume that, for some constants $C > 0$, $\delta > 0$,*

$$h(x,\xi) \leq C(1 + |x| + |\xi|)^{-\delta}. \tag{1.3}$$

Then, if $m \in L_w^1(\mathbb{R}^{2n})$, it turns out

$$OPS(m,g) \subset \mathcal{L}^{(1,\infty)}(L^2(\mathbb{R}^n)). \tag{1.4}$$

We consider now two simple examples which Theorem 1.1 applies to (see above for the definitions of the Shubin and SG classes).
We have that any symbol a in the Shubin class G^{-2n} gives rise to an operator $a^w \in \mathcal{L}^{(1,\infty)}$. Indeed, the corresponding weight $(1+|x|+|\xi|)^{-2n}$ is just in $L_w^1(\mathbb{R}^{2n})$. Similarly, we see that any symbol in the SG classes $S^{-n,-\mu}$ or in $S^{-\mu,-n}$, with $\mu > n$, gives rise to an operator in $L^{(1,\infty)}$ too, since the weights $\langle x \rangle^{-n} \langle \xi \rangle^{-\mu}$ and $\langle x \rangle^{-\mu} \langle \xi \rangle^{-n}$ are in $L_w^1(\mathbb{R}^{2n})$. We observe that, for these two classes of pseudo-differential operators, connections between the Dixmier trace and suitably defined Wodzicki's residues were studied in Boggiatto and Nicola [2] and Nicola [17] respectively.

Finally we observe that the characterization of the pseudo-differential operators which are in $L^{(1,\infty)}$ is a subject of interest in noncommutative geometry (see Connes [4, 5], and Gayral, Gracia-Bondía, Iochum, Schücker, Vàrilly [11]). Indeed, in the standard setting [5] one mostly deals with pseudo-differential operators on compact manifolds. Due to the requirements of invariance, the class of admissible operators is therefore essentially limited to the classical one or its anisotropic version, when the manifold is foliated (see Connes [6]). Instead, the case of operators in $\mathbb{R}^n$ is much more rich and one can consider a wide variety of classes as above. Notice that, similarly, pseudo-differential operators in $\mathbb{R}^n$ play an important role in noncompact noncommutative geometry, as it is shown in [11].

2. Preliminaries

In order to fix notation, we recall some basic definitions concerning the Weyl-Hörmander calculus; see Chapter XVIII of [16] for details.

A metric is a measurable function $g : (x, \xi) \mapsto g_{x,\xi}$ of $\mathbb{R}^{2n}$ into the set of positive definite quadratic forms on $\mathbb{R}^{2n}$. With any metric $g_{x,\xi}$ it is associated the so-called Planck function $h(x, \xi)$, defined by

$$h(x, \xi) := \left(\sup_{(t, \tau)} \frac{g_{x,\xi}(t, \tau)}{g^{\sigma}_{x,\xi}(t, \tau)} \right)^{1/2},$$

where g^{σ} is the *dual quadratic form*:

$$g^{\sigma}_{x,\xi}(t, \tau) := \sup_{g_{x,\xi}(y,\eta)=1} \sigma\big((t, \tau); (y, \eta)\big)^2,$$

with respect to the *standard symplectic 2-form* $\sigma = \sum_{i=1}^{n} d\xi_i \wedge dx_i$ in $\mathbb{R}^{2n}$.

A *Hörmander metric* is a metric which is

- *slowly varying*, i.e., there exists $C > 0$ such that, with $X, Y \in \mathbb{R}^{2n}$,

$$g_X(Y - X) \leq C^{-1} \Longrightarrow C^{-1} g_Y(Z) \leq g_X(Z) \leq C g_Y(Z) \quad \forall Z \in \mathbb{R}^{2n};$$

- σ-*temperate*, i.e., there exist constants $C > 0$, $N \in \mathbb{Z}_+$ such that

$$g_X(Z) \leq C g_Y(Z)(1 + g^{\sigma}_X(Y - X))^N \quad \forall X, Y, Z \in \mathbb{R}^{2n};$$

- satisfying the *uncertainty principle*, namely

$$h(x, \xi) \leq 1 \quad \forall (x, \xi) \in \mathbb{R}^{2n}.$$

A *g-admissible weight* is a positive measurable function $m : \mathbb{R}^{2n} \to \mathbb{R}_+$, which is

- *g-continuous*, i.e., there exists $C > 0$ such that

$$g_X(Y - X) \leq C^{-1} \Longrightarrow C^{-1} m(Y) \leq m(X) \leq C m(Y);$$

- (σ, g)-temperate, i.e., there exist constant $C > 0$, $N \in \mathbb{Z}_+$ such that

$$m(X) \leq C m(Y)(1 + g^{\sigma}_X(Y - X))^N \quad \forall X, Y \in \mathbb{R}^{2n}.$$

We denote by $S(m, g)$ the set of the smooth functions $a : \mathbb{R}^{2n} \to \mathbb{C}$ satisfying

$$\sup_{(x,\xi)} \frac{|a|^g_k(x, \xi)}{m(x, \xi)} < \infty, \qquad \text{for all } k \in \mathbb{Z}_+,$$

where $|a|^g_0(x, \xi) := |a(x, \xi)|$ and

$$|a|^g_k(x, \xi) := \sup_{T_j \in \mathbb{R}^{2n}} \frac{|a^{(k)}\big((x, \xi); T_1, \ldots, T_k\big)|}{g_{x,\xi}(T_1)^{1/2} \cdots g_{x,\xi}(T_k)^{1/2}}, \qquad \text{for } k \geq 1,$$

where $a^{(k)}(X, \cdot)$ denotes the k-multi-linear form given by the differential of order k of a at $X \in \mathbb{R}^{2n}$.

The space $S(m, g)$ is equipped with the Fréchet topology given by the semi-norms

$$\|a\|_{k;S(m,g)} := \sup_{(x,\xi)} \frac{\|a\|_k^g(x,\xi)}{m(x,\xi)}, \qquad k \in \mathbb{Z}_+,$$

where

$$\|a\|_k^g(x,\xi) := \sup_{j \leq k} |a|_j^g(x,\xi).$$

Given a symbol $a \in S(m, g)$ we define its Weyl quantization as the linear operator

$$(\mathrm{Op}^w(a)u)(x) = (2\pi)^{-n} \iint e^{i(x-y)\xi} a\left(\frac{x+y}{2}, \xi\right) u(y) \, dy \, d\xi,$$

first for u in the Schwartz space $\mathcal{S}(\mathbb{R}^n)$ and then on the space $\mathcal{S}'(\mathbb{R}^n)$ of temperate distributions. Sometimes it is also denoted by a^w. Moreover $OPS(m, g)$ stands for the space of such operators, whose symbol is in $S(m, g)$.

Finally we recall the composition formula of two pseudo-differential operators (Theorem 18.5.4 of [16]).

Theorem 2.1. *Given two symbols $a \in S(m_1, g)$ and $b \in S(m_2, g)$, we have that $a^w b^w$ is a pseudo-differential operator with symbol*

$$a \# b \in S(m_1 m_2, g)$$

such that

$$R_N(a, b) := a \# b - \sum_{j=1}^{N} \frac{\{a, b\}_j}{(2i)^j j!} \in S(m_1 m_2 h^{N+1}, g),$$

for all $N \in \mathbb{Z}_+$, where $\{a, b\}_0 = ab$, and

$$\{a, b\}_j = \left[\left(\sum_{r=1}^{n} \left(\frac{\partial}{\partial \xi_r} \frac{\partial}{\partial y_r} - \frac{\partial}{\partial x_r} \frac{\partial}{\partial \eta_r} \right) \right)^j a(x, \xi) b(y, \eta) \right]_{y=x, \eta=\xi},$$

for $j > 0$. More precisely, for each $N, k \in \mathbb{Z}_+$ there exist an integer l and a constant $C > 0$ such that

$$\|R_N(a, b)\|_{k;S(m_1, m_2 h^{N+1}, g)} \leq C \|a\|_{l;S(m_1,g)} \|b\|_{l;S(m_2,g)}$$

for all $a \in S(m_1, g)$ and all $b \in S(m_2, g)$.

3. Proof of Theorem 1.1

We assume the hypotheses of Theorem 1.1. The proof consists of several steps and we start with the following result.

Lemma 3.1. *We have $m(x, \xi) \to 0$ as $(x, \xi) \to \infty$.*

Proof. Let us suppose that m does not tend to 0 at ∞. Then there exist $\alpha > 0$ and $X_j \in \mathbb{R}^{2n}$, $j \in \mathbb{N}$, such that $m(X_j) \geq \alpha$ for every j. Since m is g-continuous, it follows that, with some constant $C > 0$ it turns out $m(Y) > \alpha C^{-1}$ if $Y \in B_j :=$ $\{Y : g_{X_j}(Y - X_j) < C^{-1}\}$. In order to compute the measure of B_j we observe that, by Lemma 18.6.4. of [16] for any given j we can perform a linear symplectic change of coordinates so that

$$g_{X_j}(Y) = \sum_{k=1}^{n} \lambda_k (y_k^2 + \eta_k^2),$$

with $Y = (y, \eta)$ and $\max_{k=1,\dots,n} \lambda_k = h(X_j)$. It follows that there exists $c > 0$ such that

$$\text{meas}(B_j) = \left(\prod_{k=1}^{n} \lambda_k^{-1}\right) \text{meas} \left\{\sum_{k=1}^{n} y_k^2 + \eta_k^2 < C^{-1}\right\} \geq ch(X_j)^{-n} \quad \forall j \in \mathbb{Z}_+.$$

We therefore deduce, by (1.3),

$$\text{meas}(B_j) \to +\infty \quad \text{as } j \to +\infty.$$

Since

$$\cup_{j=1}^{+\infty} B_j \subset \{m > \alpha C^{-1}\},$$

this contradicts the fact that $m \in L_w^1$. $\qquad\square$

We observe that the fact that $m \to 0$ at ∞ is *necessary* for (1.4) to hold, since it is equivalent to saying that all operators in $OPS(m, g)$ are compact in $L^2(\mathbb{R}^n)$ (see [16], Theorem 18.6.6).

Lemma 3.2. *There exists an operator $A = \text{Op}^w(a)$, with a real symbol $a \in S(m^{-1}, g)$ such that*

(i) *$a \geq Cm^{-1}$ for some constant $C > 0$;*
(ii) *A is self-adjoint as an operator in $L^2(\mathbb{R}^n)$ with domain $\mathcal{D} = \{u \in L^2(\mathbb{R}^n) : Au \in L^2(\mathbb{R}^n)\}$;*
(iii) *A has a spectrum made of a sequence of eigenvalues $1 \leq \lambda_j \nearrow +\infty$.*
(iv) *A^{-1} is a compact pseudo-differential operator in $OPS(m, g)$.*

Proof. Consider the operator $\text{Op}^w(m^{-1})$ (we are assuming, without loss of generality, that m is smooth; see [15], page 143). Since m tends to 0 at ∞ and (1.3) holds, it follows from Theorem 3.4 of [15] that it is a self-adjoint operator in $L^2(\mathbb{R}^n)$, with a spectrum made of a sequence of eigenvalues bounded from below. It follows that $A = \text{Op}^w(m^{-1} + c)$ is bounded from below, say, by 1 if c is large enough. Consider then a parametrix $B \in OPS(m, g)$: $AB = I + R$, where R has kernel in $\mathcal{S}(\mathbb{R}^{2n})$ (cf. Lemma 3.1 of [15] and the subsequent remark). We have $B = A^{-1} + A^{-1}R$ on $\mathcal{S}(\mathbb{R}^n)$ (notice that $\mathcal{S}(\mathbb{R}^n) \subset \mathcal{D}$, and $A^{-1} : \mathcal{D} \to L^2(\mathbb{R}^n)$). Since A is globally hypoelliptic, $A^{-1} : \mathcal{S}(\mathbb{R}^n) \to \mathcal{S}(\mathbb{R}^n)$, continuously by the closed graph theorem. It follows that $A^{-1}R$ is an operator with kernel in $\mathcal{S}(\mathbb{R}^{2n})$. Hence, $A^{-1} \in OPS(m, g)$. Finally, A^{-1} is compact in view of Theorem 18.6.6 of [16] so that, in particular, the sequence of eigenvalues of A tends to $+\infty$. $\qquad\square$

We observe that Theorem 1.1 is proved if we verify that $A^{-1} \in \mathcal{L}^{(1,\infty)}$. Indeed, given any operator $P \in \mathrm{OPS}(m,g)$ we can write $P = PAA^{-1}$. Since $PA \in \mathrm{OPS}(1,g)$ is bounded in $L^2(\mathbb{R}^n)$ and $\mathcal{L}^{(1,\infty)}$ is an ideal in the space of bounded operators we deduce that $P \in \mathcal{L}^{(1,\infty)}$. Thus we are reduced to prove that

$$\lambda_j^{-1} = O(1/j) \quad \text{as } j \to +\infty. \tag{3.1}$$

Theorem 3.3. *The operator* e^{-tA}, $t \geq 0$ *can be written as*

$$e^{-tA} = b_t^w + S(t),$$

where b_t *is bounded family of symbols in* $S(1,g)$ *for* $t \geq 0$, *satisfying*

$$|b_t(x,\xi)| \leq Ce^{-ta(x,\xi)/2}, \quad \forall t \geq 0, \forall(x,\xi) \in \mathbb{R}^{2n},$$

and $S(t)$ *is a trace class operator with*

$$\|S(t)\|_{\mathrm{Tr}} \leq Ct, \quad \forall t \geq 0. \tag{3.2}$$

Proof. We search $b_t \in S(1,g)$, satisfying

$$\begin{cases} (\partial_t + a^w)b_t^w = K(t) \\ b_0^w = I, \end{cases} \tag{3.3}$$

for some trace class operator K. Precisely, we look for b_t in the form $b_t(x,\xi) = \sum_{j=0}^N u_j(t,x,\xi)$ with $\partial_t^l u_j \in S(m^{-l}h^j, g)$, for $l \geq 0$. From (3.3) and Theorem 2.1 we obtain the following transport equations in $S(m^{-1}h^j, g)$:

$$\begin{cases} \partial_t u_j + \sum_{k+l=j} \frac{\{a, u_k\}_l}{(2i)^l l!} = 0, \\ u_0(0,x,\xi) = 1, \\ u_j(0,x,\xi) = 0, \quad\quad \text{if } j > 0. \end{cases} \tag{3.4}$$

For $j = 0$ we obtain $u_0 = e^{-ta}$. For $j > 0$, we have to solve the equations

$$\begin{cases} \partial_t u_j + a u_j + \sum_{\substack{k+l=j \\ k<j}} \frac{\{a, u_k\}_l}{(2i)^l l!} = 0 \\ u_j(0,x,\xi) = 0, \end{cases} \tag{3.5}$$

from which one easily verifies by induction that u_j can be written as $u_j = e^{-ta}p_j$ where $p_j(t,x,\xi)$ satisfies the estimates

$$|\partial_t^l p_j|_q^g \leq C \sum_{s=1}^{q+2j} t^s a^{s+l} h^j. \tag{3.6}$$

From these estimates it follows that

$$|\partial_t^l u_j|_k^g \leq Ce^{-ta/2} a^l h^j.$$

We now choose N so large that $m^{-1}h^{N+1} \in L^1(\mathbb{R}^{2n})$ (this is possible, for m is temperate and h satisfies (1.3)), and we set $b_t(x,\xi) = \sum_{j=0}^N u_j(t,x,\xi)$. Then b_t^w satisfies (3.3) with $K(t) \in \mathrm{OPS}(m^{-1}h^{N+1}, g)$ *uniformly with respect to* $t \geq 0$. It follows therefore from Theorem 3.9 of [15] that $K(t)$ is trace class and $\|K(t)\|_{\mathrm{Tr}} \leq C$ for every $t \geq 0$.

In order to verify (3.2) we observe that, since the operator e^{-tA} solves (3.3) with $K = 0$ we have

$$b_t^w - e^{-tA} = \int_0^t e^{-(t-s)A} K(s)\, ds. \tag{3.7}$$

Then

$$\|b_t^w - e^{-tA}\|_{\mathrm{Tr}} \leq \int_0^t \|e^{-(t-s)A} K(s)\|_{\mathrm{Tr}} ds \leq \int_0^t \|e^{-(t-s)A}\|_{\mathcal{L}} \|K(s)\|_{\mathrm{Tr}} ds \leq Ct.$$

This concludes the proof. $\qquad \square$

Proposition 3.4. *We have*

$$\sum_{j=1}^{+\infty} e^{-t\lambda_j} = O(t^{-1}), \quad as\ t \searrow 0.$$

Proof. We have

$$\sum_{j=1}^{+\infty} e^{-t\lambda_j} = \mathrm{Tr}\ e^{-tA} = \|e^{-tA}\|_{\mathrm{Tr}} \leq \|b_t^w\|_{\mathrm{Tr}} + \|S(t)\|_{\mathrm{Tr}}. \tag{3.8}$$

It follows from Theorem 3.9 of [15] that, for N and k large enough,

$$\|b_t^w\|_{\mathrm{Tr}} \leq C(\|b_t\|_{L^1} + \|h^N\|_{L^1} \|b_t\|_{k;S(1,g)}). \tag{3.9}$$

The second term in the right-hand side (3.9) is $O(1)$ as $t \searrow 0$. As regards the first one, we have

$$\|b_t\|_{L^1} \leq C \int e^{-ta(x,\xi)/2} dx\, d\xi = C \int_0^{+\infty} e^{-ts} d\lambda(s), \tag{3.10}$$

where we set

$$\lambda(s) = \mathrm{meas}(\{a/2 \leq s\}).$$

Now we have $\lambda(s) = 0$ for s in a right neighborhood of 0 and $\lambda(s) \leq Cs$, since $a^{-1} \in L_w^1(\mathbb{R}^{2n})$; this follows from (i) in Lemma 3.2, for $m \in L_w^1(\mathbb{R}^{2n})$. Thus, integrating by parts in (3.10) yields

$$\int_0^{+\infty} e^{-ts} d\lambda(s) = t \int_0^{+\infty} e^{-ts} \lambda(s) ds \leq Ct \int_0^{+\infty} e^{-ts} s\, ds = Ct^{-1}.$$

This shows that $\|b_t^w\|_{\mathrm{Tr}} = O(t^{-1})$ as $t \searrow 0$, which together with (3.8) and (3.2) concludes the proof. $\qquad \square$

Let now $N(\lambda) = \#\{\lambda_j : \lambda_j \leq \lambda\}$ be the so-called counting function of A (recall, the eigenvalues are repeated according to their multiplicity).

Proposition 3.5. *We have* $N(x) = O(x)$ *as* $x \to +\infty$.

Proof. Upon setting $\phi_t(\lambda) = N(t^{-1}\lambda)$ it turns out

$$\int_0^{+\infty} e^{-t\lambda} dN(\lambda) = \int_0^{+\infty} e^{-\lambda} d\phi_t(\lambda)$$

and

$$N(t^{-1}) = \int_0^{t^{-1}} dN(\lambda) = \int_0^{+\infty} \chi_{[0,1]}(\lambda) d\phi_t(\lambda).$$

Hence we have

$$N(t^{-1}) \le e \int_0^{+\infty} e^{-t\lambda} dN(\lambda). \qquad (3.11)$$

On the other hand, the right-hand side of (3.11) is exactly $e \sum_{j=1}^{+\infty} e^{-t\lambda_j}$, which is $O(t^{-1})$ as $t \searrow 0$ in view of Proposition 3.4. $\qquad \square$

We finally prove (3.1).

Proposition 3.6. *We have $\lambda_j^{-1} = O(1/j)$ as $j \to +\infty$.*

Proof. We know from Proposition 3.5 that $N(x) \le Cx$ for $x \ge 0$. Now, given any j, take $j_1 \ge j$ such that $\lambda_j = \lambda_{j_1} < \lambda_{j_1+1}$. Then $N(\lambda_{j_1}) = j_1$ so that

$$j \le j_1 \le C\lambda_{j_1} = C\lambda_j.$$

This concludes the proof. $\qquad \square$

Theorem 1.1 is therefore proved.

References

[1] P. Boggiatto, E. Buzano, L. Rodino, *Global Hypoellipticity and Spectral Theory.* Akademie Verlag, Berlin, 1996.

[2] P. Boggiatto, F. Nicola, *Non-commutative residues for anisotropic pseudo-differential operators in $\mathbb{R}^n$.* J. Funct. Anal. **203** (2003), 305–320.

[3] E. Buzano, F. Nicola, *Pseudodifferential operators and Schatten-von Neumann classes.* In: P. Boggiatto, R. Ashino, M.W. Wong (Eds.), *Advances in Pseudodifferential Operators* (Proceedings ISAAC, Toronto 2003), Operator Theory Adv. Appl., vol. 155, Birkhäuser, Basel, 2004, 117–130.

[4] A. Connes, *The action functional in non-commutative geometry.* Comm. Math. Physics **117** (1988), 673–683.

[5] A. Connes, *Noncommutative Geometry.* Academic Press, New York, London, Tokyo, 1994.

[6] A. Connes, H. Moscovici, *The local index formula in noncommutative geometry.* Geom. Funct. Anal. **5** (1995), 174–243.

[7] E. Cordero, K.H. Gröchenig, *Time-frequency analysis of localization operators.* J. Funct. Anal. **205** (2003), 107–131.

[8] E. Cordero, S. Pilipović, L. Rodino, N. Teofanov, *Localization operators and exponential weights for modulation spaces.* Mediterr. J. Math. **2** (2005), 381–394.

[9] J. Dixmier, *Existence de traces non normales*. C.R. Acad. Sc. Paris, Série A, **262** (1966), 1107–1108.

[10] G.B. Folland, *Harmonic Analysis in Phase Space*. Princeton University Press, 1989.

[11] V. Gayral, J.M. Gracia-Bondía, B. Iochum, T. Schücker, J.C. Vàrilly, *Moyal planes are spectral triples*. Comm. Math. Phys. **246** (2004), 569–623.

[12] K.H. Gröchenig, *An uncertainty principle related to the Poisson summation formula*. Studia Math. **121** (1996), 87–104.

[13] K.H. Gröchenig, *Foundation of Time-Frequency Analysis*. Birkhäuser, Boston, 2001.

[14] K.H. Gröchenig, C. Heil, *Modulation spaces and pseudo-differential operators*. Integral Equations Operator Theory **34** (1999), 439–457.

[15] L. Hörmander, *On the asymptotic distribution of the eigenvalues of pseudo-differential operators in $\mathbb{R}^n$*. Arkiv för Mat. **17** (1979), 297–313.

[16] L. Hörmander, *The analysis of linear partial differential operators* III. Springer-Verlag, Berlin, 1985.

[17] F. Nicola, *Trace functionals for a class of pseudo-differential operators in $\mathbb{R}^n$*. Math. Phys. Anal. Geom. **6** (2003), 89–105.

[18] D. Robert, *Autour de l'approximation semi-classique*. Birkhäuser, Boston, 1987.

[19] B.-W. Schulze, *Boundary value problems and singular pseudo-differential operators*. Pure and Applied Mathematics, John Wiley & Sons, Chichester, England, 1998.

[20] B. Simon, *The Weyl transfrom and L^p functions on phase space*. Proc. Amer. Math. Soc. **116** (1992), 1045–1047.

[21] E. Schrohe, *Wodzicki's noncommutative residue and traces for operator algebras on manifolds with conical singularities*. In L. Rodino, editor, *Microlocal Analysis and Spectral Theory*, 1997 Kluwer Academic Publishers, Printed in the Netherlands, 1997, 227–250.

[22] M.A. Shubin, *Pseudo-differential operators and spectral theory*. Springer-Verlag, Berlin, 1987.

[23] E.M. Stein, G. Weiss, *Introduction to Fourier Analysis on Euclidean Spaces*. Princeton University Press, 1971.

[24] J. Toft, Schatten-von Neumann properties in the Weyl calculus, and calculus of metrics on symplectic vector spaces. Preprint 2004.

Fabio Nicola
Dipartimento di Matematica, Politecnico di Torino,
Corso Duca degli Abruzzi, 24
10129 Torino, Italy
e-mail: `fabio.nicola@polito.it`

Luigi Rodino
Dipartimento di Matematica, Università di Torino,
Via Carlo Alberto, 10
10123 Torino, Italy
e-mail: `luigi.rodino@unito.it`

C^*-algebras and Elliptic Theory II

Trends in Mathematics, 239–250

© 2008 Birkhäuser Verlag Basel/Switzerland

Topological Invariants of Bifurcation

Jacobo Pejsachowicz

Abstract. I will shortly discuss an approach to bifurcation theory based on elliptic topology. The main goal is a construction of an index of bifurcation points for C^1-families of Fredholm maps derived from the index bundle of the family of linearizations along the trivial branch. As illustration, I will present an application to bifurcation of homoclinic solutions of non-autonomous differential equations from a branch of stationary solutions.

Mathematics Subject Classification (2000). Primary 58E07, 58J55; Secondary 47A35, 34C23.

Keywords. Bifurcation, Fredholm maps, Index bundle, J-homomorphism.

1. Introduction

The classical topological approach to bifurcation of zeroes of parametrized families of maps is essentially of local nature [13, 14, 2, 3]. Sufficient condition for bifurcation are obtained by analyzing the behavior of the linearized family in a small enough neighborhood of an isolated point of the trivial branch at which the linearization fails to be invertible. Here, instead, I would like to discuss an alternative, non-local approach based on elliptic topology. I will consider families of C^1-Fredholm maps of index 0 parametrized by topologically nontrivial spaces and will use the non-vanishing of a global invariant associated to the family of linearizations in order to find on the trivial branch at least one bifurcation point. This kind of argument, applied to families of linear Fredholm operators, was successfully used in many places (see for example [12], [23] among others). My point here is that, after adding one more tool, essentially the same method works for nonlinear Fredholm maps as well. Although some of the results stated here were already proved for families of Fredholm maps of special type in [20, 19], the general case is an ongoing work and full details will appear in [21].

In what follows I will describe more precisely what I mean by bifurcation from the trivial branch and the topological invariants under consideration.

Let X, Y be Banach spaces and let P be an n-dimensional compact connected smooth manifold. Let $f : P \times \mathcal{O} \to Y$ be a continuously differentiable map defined on the product of P with an open neighborhood $\mathcal{O}$ of the origin in X. Assume that $f(p, 0) = 0$ for all p in P. Solutions of the equation $f(p, x) = 0$ of the form $(p, 0)$ are called *trivial* and the set $P \times \{0\}$ is called the *trivial branch*. In what follows I will identify the parameter space P with the set of trivial solutions and will write the parameter variable as a subscript. Accordingly I will denote by $f_p : U \to Y$ the map defined by $f_p(x) = f(p, x)$.

A *bifurcation point* for solutions of the equation $f(p, x) = 0$ is a point p_* in P such that every neighborhood of $(p_*, 0)$ contains nontrivial solutions of this equation.

Let $L_p = Df_p(0)$ be the Fréchet derivative of the map f_p at 0. The map L sending $p \in P$ to L_p is called the *family of linearizations along the trivial branch*. By the Implicit Function Theorem, bifurcation cannot occur at points where the operator L_p is an isomorphism. However, in general, the set $Bif(f)$ of bifurcation points of f is only a proper closed subset of the set $\Sigma(L) = \{p \in P \mid L_p \text{ is singular}\}$. Assuming that L_p is a Fredholm operator of index 0 for all $p \in P$, sufficient conditions for the existence of bifurcation points can be obtained from homotopy invariants of the family of linearizations along the trivial branch.

Since bifurcation arises only at points of $\Sigma(L)$, the first invariant that comes in mind is the obstruction to the existence of a homotopy deforming the family L into a family of isomorphisms. This obstruction is given by an element Ind L of the reduced Grothendieck group of virtual vector bundles $\widetilde{KO}(P)$, called *family index* or *index bundle* [4, 16]. However, in dealing with nonlinear perturbations of the family L one has to consider a stronger invariant and, quite naturally, our bifurcation invariant is not Ind L but rather its image $J(\text{Ind } L)$ under the generalized J-homomorphism which associates to each vector bundle the stable fiberwise homotopy class of its unit sphere bundle.

Our main result asserts that if $J(\text{Ind } L)$ does not vanish and $\Sigma(L)$ is a proper subset of P then there exists at least one bifurcation point from the trivial branch for solutions of the equation $f(p, x) = 0$. The corresponding theorem together with some consequences and generalizations are stated in Section 2. Section 3 is devoted to a localized version of the basic invariant. To each admissible open subset U of the parameter space is assigned a bifurcation index $\sigma(f, U)$ belonging to the finite group $J(P)$ defined in [5] which gives a measure of the number of bifurcation points of f in U. It is related to the global invariant $J(\text{Ind } L)$ much in the same way as the local fixed-point index is related to the Lefschetz number and provides an interpolation between the global invariant and the Alexander-Ize invariant at isolated bifurcation points [2]. The precise relation with the Alexander-Ize invariant is discussed in Section 4. In Section 5 the previous theory is used in order to show how the topology of the parameter space leads to the appearance of homoclinic trajectories of non-autonomous differential equations emanating from a stationary solution.

Few comments to related work: one-parameter families of C^k-Fredholm mappings of index 0 were studied in [8, 18] among others. The well-known Global Bifurcation Theorem of P.Rabinowitz was extended to one parameter families of C^1 Fredholm maps in [22] using an appropriate degree theory (see also [11]). For a special class of bifurcation problems involving Fredholm maps a different method was developed by Zvyagin in [24] using a device due to Ize. Krasnosel'skij-Rabinowitz theory was carried to the setting of several-parameter families of compact perturbations of identity mainly by the work of Alexander and Ize [2, 3, 13, 14, 10]. The review paper [15] contains a complete reference list for this topic. In [6, 7] a different approach to local bifurcation index in the semilinear case was developed by Bartsch.

2. The main result

Let us recall the definition of the index bundle. I will use here a construction slightly different from the one given by Atiyah in [4] but, of course, both approaches give the same element in K-theory.

If P is a compact space, the Grothendieck group $KO(P)$ is the group completion of the abelian semigroup $\mathrm{Vect}(P)$ of all isomorphisms classes of vector bundles over P. As a group $KO(P) = \mathrm{Vect}(P) \times \mathrm{Vect}(P)/\Delta$ where Δ is the diagonal subsemigroup. The elements of $KO(P)$ are called virtual bundles. Each virtual bundle can be written as a difference $[E] - [F]$, where $[E]$ is the equivalence class of $(E, 0)$.

Let X, Y be real Banach spaces, let $\Phi(X,Y)$ be the space of all Fredholm operators. With $\Phi_k(X,Y)$ I will denote the space of operators of index k. Given a continuous family $L\colon P \to \Phi(X,Y)$ of Fredholm operators parametrized by a compact topological space P, using compactness of P one can find a finite-dimensional subspace V of Y transverse to the family L, i.e., such that

$$\mathrm{Im}\ L_p + V = Y \quad \text{for any}\ \ p \in P. \tag{2.1}$$

It follows from 2.1 that the finite-dimensional spaces $E_p = L_p^{-1}(V)$ are fibers of a vector bundle E over P. By definition the index bundle

$$\mathrm{Ind}\ L = [E] - [\Theta(V)] \in KO(P),$$

where $\Theta(V) = P \times V$ denotes the trivial vector bundle over P with fiber V. That the above virtual bundle is independent from the choice of V follows easily from the identity $[E] - [F] = [E \oplus H] - [F \oplus H]$, which holds in $KO(P)$.

It is easy to see that $\mathrm{Ind}\ L$ depends only on the homotopy class of L. It vanishes whenever L can be deformed by a homotopy to a family of invertible operators. The index bundle is functorial under the change of parameter space and moreover it has the logarithmic property of the ordinary index. Namely, $\mathrm{Ind}\ (LM) = \mathrm{Ind}\ L + \mathrm{Ind}\ M$.

I will need also the generalized J-homomorphism. Given a vector bundle E, let $S[E]$ be the sphere bundle with respect to some chosen scalar product on E.

Two vector bundles E, F are said to be *stably fiberwise homotopy equivalent* if for some n the sphere bundle $S[E \oplus \Theta(\mathbf{R}^n)]$ is fiberwise homotopy equivalent to the sphere bundle $S[F \oplus \Theta(\mathbf{R}^n)]$. Let $\widetilde{KO}(P)$ be the kernel of the rank homomorphism $rk \colon KO(P) \to \mathbf{Z}$ and let $T(P)$ be the subgroup of $\widetilde{KO}(P)$ generated by elements $[E] - [F]$ such that $S[E]$ and $S[F]$ are stably fiberwise homotopy equivalent. Put $J(P) = \widetilde{KO}(P)/T(P)$. The projection to the quotient $J \colon \widetilde{KO}(P) \to J(P)$ is the *generalized J-homomorphism.*

The groups $J(P)$ were introduced by Atiyah in [5] who also proved that $J(S^n)$ coincides with the image in π^s of the stable j-homomorphism. It follows from this that $J(P)$ is a finite group for any compact CW-complex P. Since Stiefel-Whitney characteristic classes can be obtained from the Thom class using Steenrod squares they depend only on the stable fiber homotopy type of the associated sphere bundle and hence are well defined on elements of $J(P)$.

Theorem 2.1. [21] *Let P be a compact connected orientable n-dimensional manifold and let $f \colon P \times \mathcal{O} \to Y$ be a C^1-family of Fredholm maps of index 0 parametrized by P such that $f(p, 0) = 0$. Assume that the linearization $L_p = Df_p(0)$ at the points of the trivial branch is nonsingular at some point $p_0 \in P$ and that $J(\operatorname{Ind} L) \neq 0$ in $J(P)$, then the family f possesses at least one bifurcation point from the trivial branch.*

The next theorem uses Stiefel-Whitney classes in order to estimate the covering dimension of the set of bifurcation points.

Theorem 2.2. [21, 6] *Let $f \colon P \times X \to Y$ be as in Theorem 2.1 and let $m = \min\{k/\ \omega_k(\operatorname{Ind} L) \neq 0\}$, then the Lebesgue covering dimension of the set $Bif(f)$ of all bifurcation points of f is at least $n - m$.*

The proof uses the previous theorem and Poincaré duality.

Remark 2.3. For $P = S^1$ this reduces to the bifurcation theorems proved in [11] and [22] by other means. However, in [11, 22] was proved that the bifurcating branch is global.

In the remaining part of the paper, except when differently stated, Fredholm means Fredholm of index 0.

Theorem 2.1 is a particular case of a slightly more general result. In order to formulate it I will need the degree theory constructed in [22]. The construction of the degree in [22] is based on a homotopy invariant of paths of Fredholm operators called parity. Given a path $L \colon [a, b] \to \Phi_0(X, Y)$ with invertible end points and transverse to the one-codimensional analytic variety Σ of all non-invertible Fredholm operators, its *parity* $\sigma(L) \in \mathbf{Z}_2$ is defined by $\sigma(L) = \#(L \cap \Sigma)$-mod 2. This definition can be extended to general paths with invertible end points using approximation by transversal paths (see [11] for this and for a different construction using parametrices).

Let $f\colon X \to Y$ be a C^1-Fredholm map of index 0 that is proper on closed bounded subsets. In order to assign to each regular point of the map f an orientation $\epsilon(x) = \pm 1$, with properties analogous to the sign of the Jacobian determinant in finite dimensions, we choose a fixed regular point b called *base point* and define "ad arbitrium" $\epsilon(b) = \pm 1$. With this said, the multiplicity $\epsilon(x)$ at any regular point x is uniquely defined by the requirement $\epsilon(x) = (-1)^{\sigma(Df \circ \gamma)} \epsilon(b)$, where γ is any path in X joining b to x. The independence from the choice of the path follows from the homotopy invariance of the parity. If Ω is an open bounded subset of X such that $0 \notin f(\partial\Omega)$ and is a regular value of the restriction of f to Ω, then the *base point degree of f* is defined by $\deg_p(f, \Omega, 0) = \sum_{x \in f^{-1}(0)} \epsilon(x)$.

It was proved in [22] that when 0 is not a regular value of f the degree can still be defined using approximation (although not by regular values since the Sard-Smale theorem does not extend to C^1-Fredholm maps of index 1). Moreover, the above assignment defines an integral-valued degree theory for C^1-Fredholm maps which are proper on closed bounded sets. The base point degree is invariant under homotopies only up to sign and, as a matter of fact, no degree theory for general Fredholm maps can be homotopy invariant. However, the change in sign can be determined as follows: let $h\colon I \times X \to Y$ be a homotopy and let Ω be an open bounded subset of X such that $0 \notin h([0,1] \times \partial\Omega)$. Assume (for simplicity) that b is a regular point both of h_0 and h_1, then

$$\deg_b(h_0, \Omega, 0) = (-1)^{\sigma(H)} \deg_b(h_1, \Omega, 0), \qquad (2.2)$$

where H is the path $t \to Dh_t(b)$.

If f is a C^1-Fredholm map and x_0 is an isolated but necessarily regular zero its *multiplicity $m(f, x_0)$* is defined by $m(f, x_0) = |\deg_b(f, B(x_0, \delta), 0)|$, where $B(x_0, \delta)$ open ball centered at x_0 and small enough radius δ and b is any regular point of f. Notice that properness need not be assumed since all Fredholm maps are locally proper. Finally, let us denote by $Z[\frac{1}{m}]$ the ring of all rational numbers whose denominator is a power of m.

Theorem 2.4. *Let P be as in 2.1, let $\mathcal{O} = B(0, \delta)$ be an open ball in X and let $f\colon P \times \mathcal{O} \to Y$ be a C^1-family of Fredholm maps parametrized by P. Assume that the only solutions of $f(p, x) = 0$ are those of the form $(p, 0)$. Suppose moreover that for some (and hence all) $q \in P$ we have that $m = m(f_q, 0) \neq 0$, then*

i) *the index bundle $\mathrm{Ind}\, L$ is orientable.*
ii) *$J(\mathrm{Ind}\, L) = 0$ in $J(P) \otimes Z[\frac{1}{m}]$.*

Theorem 2.1 follows the above theorem with $m = 1$.

Sketch of proof of Theorem 2.4. For the first claim one must show that $\omega_1(\mathrm{Ind}\, L)$ vanishes. For this it is enough to check that for any closed path with $\gamma(0) = q = \gamma(1)$, $\langle \omega_1(\mathrm{Ind}\, L); \gamma_*[S^1] \rangle = \langle \omega_1(\mathrm{Ind}\, L \circ \gamma); [S^1] \rangle = 0$.

By Proposition 2.7 of [9], $\langle \omega_1(\mathrm{Ind}\, L \circ \gamma), [S^1] \rangle = \sigma(L \circ \gamma)$. Consider the homotopy $h(t, x) = f(\gamma(t), x)$ and choose a regular base point $b \in \mathcal{O}$ for f_q (there must be at least one since $m \neq 0$). The parity of the path $t \to Dh_t(b)$ equals

$\sigma(L \circ \gamma)$. On the other hand, since there are no zeroes of h on $I \times \partial \mathcal{O}$ we can apply the homotopy property (2.2) of the base point degree from which we obtain that $\sigma(L \circ \gamma) = 0$ being $m \neq 0$. This proves the first claim.

The proof of the second claim is roughly speaking as follows: using a modified version of the Caccioppoli reduction one shows that the zero-set of the map f coincides with the zero-set of a map $\bar{f}$ defined on a finite-dimensional fiber bundle M over P with values in $\mathbb{R}^s$ and such that $\bar{f}$ has degree $\pm m$ on each fiber. By construction, the bundle E of tangents to the fibers of M at the points of the trivial branch represents the index bundle. Composing $\bar{f}$ with the fiberwise exponential map produces a map g from the sphere bundle $S(E)$ to S^{s-1} of degree $\pm m$ on each fiber. With this, the second assertion follows from the first and the mod-k Dold's theorem of Adams [1].

Corollary 2.5. *If $J(\mathrm{Ind}\ L) \neq 0$ and for some $q \in P$, the multiplicity $m = m(f_q, 0)$ is defined and is prime to the order of $J(P)$, then $Bif(f) \neq \emptyset$.*

Proof. Assume that there are no bifurcation points. By $ii)$ of theorem 2.4 for some k, $m^k J(\mathrm{Ind}\ L) = 0$. Hence the order of $J(\mathrm{Ind}\ L)$, divides both m^k and the order of $J(P)$. $\qquad\square$

3. The local bifurcation index

In this section I will assume that the range of the family, Y is a Kuiper space, i.e., that $GL(Y)$ is contractible. Let U be an open subset of a compact connected manifold P and let $f \colon U \times X \to Y$ be a family of C^1-Fredholm maps parametrized by U such that $f(p, 0) = 0$ for any $p \in U$. The pair (f, U) will be called *admissible* if the singular set $\Sigma(L)$ of the family L of linearizations along the trivial branch is a compact, proper subset of U.

Theorem 3.1. *There exists a function assigning to each admissible pair (f, U) an element $\sigma(f, U) \in J(P)$, called* bifurcation index, *verifying the following properties:*

$\mathcal{P}1)$ Existence – *If $\sigma(f, U) \neq 0$ then the family f has a bifurcation point.*

$\mathcal{P}2)$ Normalization – *If $U = P$ then $\sigma(f, U) = J(\mathrm{Ind}\ L)$.*

$\mathcal{P}3)$ Homotopy invariance – *Let $h \colon [0, 1] \times U \times X \to Y$ be a C^1-Fredholm map of index 1 such that the set $\{(t, p)/Dh_{(t,p)}(0)$ is singular $\}$ is compact, then $\sigma(h_0, U) = \sigma(h_1, U)$.*

$\mathcal{P}4)$ Additivity – *Let $U \subset \bigcup U_i$. Put $f_i = f|_{U_i}$ and $\Sigma_i = \Sigma(f) \cap U_i$. If (f_i, U_i) are admissible and $\Sigma_i \cap \Sigma_j = \emptyset$, then $\sigma(f, U) = \sum_i \sigma(f_i, U_i)$.*

The construction of the local bifurcation index follows the approach of the previous section. If Y is a Kuiper space, then $GL(X, Y)$, when nonempty, is an open contractible subset of a Banach space. By a theorem of Borsuk any continuous map with values in $GL(X, Y)$ defined on a closed subset of a metric space can be extended. I will use this fact in order to define for any family $L : U \to \Phi_0(X, Y)$ such that $\Sigma(L)$ is a compact subset of U a localized form of the index bundle $\mathrm{Ind}(L, U)$ belonging to $\widetilde{KO}(P)$.

For this, let V be any neighborhood of $\Sigma(L)$ in U such that $\Sigma(L) \subset V \subset \bar{V} \subset U$. The restriction $L|_{\partial V}$ of L to the boundary of V can be extended to a family $L' \colon P - V \to GL(X, Y)$. Patching L' with L gives a family $\tilde{L}$ of Fredholm operators parametrized by P which coincides with L in a neighborhood of $\Sigma(L) = \Sigma(\tilde{L})$. It is easy to see that Ind($\tilde{L}$) is independent of the choice of V and the extension. The *index bundle* of the family L on U is defined by $\mathrm{Ind}(L, U) = \mathrm{Ind}(\tilde{L}) \in \widetilde{KO}(P)$. Now if (f, U) is admissible we define its bifurcation index $\sigma(f, U) \in J(P)$ by

$$\sigma(f, U) = J\left(\mathrm{Ind}(L, U)\right). \tag{3.1}$$

The verification of $\mathcal{P}2 - \mathcal{P}4$ is quite standard. The existence property $\mathcal{P}1$ follows from Theorem 2.1 applied to an appropriate extension of the map f to $P \times X$. This is the only point where the assumption that Y is a Kuiper space is essential since $\mathrm{Ind}(L, U)$ can be alternatively constructed via K-theory of locally compact spaces.

Remark 3.2. It follows easily from the results in [8] that if $P = S^1$, viewed as a one point compactification of the real line R and $U = (a, b)$, then the local index of bifurcation points $\sigma(f, U)$ coincides with the parity of the path L.

4. Comparison with the Alexander-Ize invariant

Now let us discuss the relation of the local bifurcation index with the Alexander-Ize invariant. I will consider here only the stable version defined in [2].

Let $g \colon \mathrm{R}^k \times \mathrm{R}^n \to \mathrm{R}^n$ be a C^1-family of maps, parametrized by R^k, such that $g_p(0) = 0$. Assume that p_0 is an isolated point in the set $\Sigma(L)$. The homotopy class of the restriction of L to the boundary of a small closed disk D centered at p_0 defines an element in the homotopy group $\pi_{k-1}(GL(n))$. Stabilizing this element through the inclusion of $GL(n)$ into $GL(m)$, $n \leq m$, one obtains the Alexander-Ize invariant γ_g belonging to the homotopy group $\pi_{k-1}(GL(\infty))$. Let $\pi_{k-1}^s(S^0) = \lim_{m \to \infty} \pi_{m+k-1}(S^m)$ be the $(k-1)$-stable homotopy group of S^0. In [2] it is shown that the point p_0 is a bifurcation point of f provided the image of γ_f by the classical j-homomorphism $j \colon \pi_{k-1}(GL(\infty)) \to \pi_{k-1}^s(S^0)$ does not vanish in $\pi_{k-1}^s(S^0)$.

In the remaining part of this section I will take as parameter space the sphere S^n, viewed as one point compactification of R^n. The definition of γ_g can be easily extended to parametrized families of C^1-Fredholm maps.

For this, let $f \colon \bar{D} \times X \to Y$ be a C^2-Fredholm family such that $f(p, 0) = 0$. Here $D = D(p_0, r) \subset \mathrm{R}^n$ be an open disk centered at p_0. Put $L_p = Df_p(0)$ and assume that $\Sigma(L) = \{p_0\}$. Being D contractible by Theorem 1.6.3 of [8], the family L has a parametrix. In other words, there exists a family of isomorphisms $A \colon \bar{D} \to GL(Y, X)$ such that $A_p L_p = Id + K_p$ for any p in $\bar{D}$, where $K \colon \bar{D} \to K(Y)$ is a family of operators such that the image of K_p is contained in a fixed finite-dimensional subspace V of Y. Let T_p be the restriction of $Id + K_p$ to V. By the preceding discussion the family T defines an element γ_f in $\pi_{n-1}(GL(\infty))$ which can be easily shown to be independent from the choice of the parametrix.

Theorem 4.1. *If $f\colon \bar{D} \subset S^n \times X \to Y$, p_0 and D are as above, then, upon identification of the group $J(S^n)$ with Im j,*

$$\sigma(f, D) = j(\gamma_f). \tag{4.1}$$

From the above theorem and the computation of $j(\gamma_f)$ in terms of the nth Radon-Hurwitz number c_n obtained in [3] it follows:

Corollary 4.2. *Let $f\colon S^n \times X \to Y$ be a C^1 family of Fredholm maps such that $f(p,0) = 0$. Assume that there exist $\epsilon > 0$, $\delta > 0$ such that $|L_p x| \geq \epsilon |p||x|$, for $|p| \leq \delta$. If $D = D(0, \delta)$, then, for $n \equiv 1, 2, 4, 8 \mod 8$, dim ker L_0 is divisible by c_n. Moreover if dim ker $L_0 = k\, c_n$ with k an odd integer then $\sigma(f, D) \neq 0$ in $J(S^n)$.*

5. Bifurcation of homoclinic trajectories

This section is devoted to the application of the previous results to bifurcation of homoclinic solutions of systems of time dependent ordinary differential equations from the stationary solution.

Let $g\colon \Lambda \times \mathbf{R} \times \mathbf{R}^n \to \mathbf{R}^n$ be a smooth family of time dependent vector fields on $\mathbf{R}^n$ parametrized by a compact connected manifold Λ of dimension m. I will assume that $g(\lambda, t, 0) = 0$, (thus $u(t) \equiv 0$ is a stationary solution of $u'(t) - g(\lambda, t, u(t)) = 0$) and I will look for conditions on the linearization of g_λ at $u \equiv 0$ which entails the appearance of nonvanishing (but close to zero) solutions to the problem:

$$\begin{cases} u'(t) - g(\lambda, t, u(t)) = 0, \\ \lim_{t \to \infty} u(t) = \lim_{t \to -\infty} u(t) = 0. \end{cases} \tag{5.1}$$

Nontrivial solutions of (5.1) are precisely the trajectories homoclinic to 0. The linearization of (5.1) at 0 is

$$\begin{cases} u'(t) - A(\lambda, t)u(t) = 0, \\ \lim_{t \to \infty} u(t) = 0 = \lim_{t \to -\infty} u(t) \end{cases} \tag{5.2}$$

where $A(\lambda, t) = D_u g(\lambda, t, 0)$.

I will assume that g and $D_u g$ are bounded and that the following asymptotic condition holds true:

(A1) As $t \to \pm\infty$ the family $A(\lambda, t)$ converges, to a family of matrices $A(\lambda, \pm\infty)$, such that $A(\lambda, \pm\infty)$ has no eigenvalues on the imaginary axis.

As a consequence of $(A1)$, the map $\lambda \to A(\lambda, \pm\infty)$ is continuous and by perturbation theory [17] the projectors onto the real part of the spectral subspaces of $A(\lambda, \pm\infty)$ corresponding to the eigenvalues with negative (respectively positive) real part are continuous as well. It follows from this that the generalized eigenspaces $E^s(\lambda, \pm\infty)$ and $E^u(\lambda, \pm\infty)$ corresponding to the part of the spectrum of $A(\lambda, \pm\infty)$ on the left and right half-plane respectively, are fibers of a pair of vector bundles $E^s(\pm\infty)$ and $E^u(, \pm\infty)$ over Λ which decompose the trivial bundle $\Theta(\mathrm{R}^n)$ with fiber R^n into a direct sum:

$$E^s(\pm\infty) \oplus E^u(\pm\infty) = \Theta(\mathrm{R}^n). \tag{5.3}$$

The bundles E^s, E^u are called *stable and unstable bundle* at $\pm\infty$. They can be alternatively described by

$$E^s(\lambda, \pm\infty) = \{v \in \mathrm{R}^n \mid \lim_{t\to\infty} e^{tA(\lambda,\pm\infty)}v \to 0\}, \tag{5.4}$$

$$E^u(\lambda, \pm\infty) = \{v \in \mathrm{R}^n \mid \lim_{t\to-\infty} e^{tA(\lambda,\pm\infty)}v \to 0\}. \tag{5.5}$$

My final assumption is

(A2) For some $\lambda_0 \in \Lambda$ both (5.2) and the adjoint problem

$$\begin{cases} u'(t) + A^*(\lambda_0, t)u(t) = 0, \\ \lim_{t\to\infty} u(t) = 0 = \lim_{t\to-\infty} u(t) \end{cases} \tag{5.6}$$

admit only the trivial solution $u \equiv 0$.

Let $\omega(E) = \omega_1(E) + \cdots + \omega_n(E)$ be the total Stiefel-Whitney class of E.

Theorem 5.1. *If the system* (5.1) *verifies* (A1), (A2) *and if*

$$\omega(E^s(+\infty)) \neq \omega(E^s(-\infty)), \tag{5.7}$$

then, at some $\lambda_* \in \Lambda$, *bifurcation of homoclinic trajectories from the stationary solution occurs. More precisely there is a sequence* (λ_n, u_n) *where* $u_n \neq 0$ *is solution of* (5.1) *with* $\lambda_n \to \lambda_*$ *and* $u_n \to 0$ *in the space* $C_0^1(\mathrm{R}; \mathrm{R}^n)$ *of all* C^1 *functions vanishing at infinity together with its derivative.*

Moreover if $k = \min\{i \mid \omega_i(E^s(+\infty)) \neq \omega_i(E^s(-\infty))\}$ *then the set of all bifurcation points has dimension not less than* $m - k$.

Proof. The space $H^1(\mathrm{R}; \mathrm{R}^n)$ of all absolutely continuous functions $u \in L^2(\mathrm{R}; \mathrm{R}^n)$ with square integrable derivative is a natural function space for our problem since any function $u \in H^1(\mathrm{R}; \mathrm{R}^n)$ has the property that $\lim_{t\to\pm\infty} u(t) = 0$. Let $X = H^1(\mathrm{R}; \mathrm{R}^n), Y = L^2(\mathrm{R}; \mathrm{R}^n)$ and let us consider the family of maps $f: P \times X \to Y$ defined by

$$[f(\lambda, u)](t) = u'(t) - g(\lambda, t, u(t)). \tag{5.8}$$

Because of the continuous embedding of $H^1(\mathrm{R}; \mathrm{R}^n)$ into $C(\mathrm{R}; \mathrm{R}^n)$ it follows that upon assumption (A1) the map f is C^1 and such that $f(\lambda, 0) = 0$. Moreover the Fréchet derivative $D_u f(\lambda, 0)$ is the operator $L_\lambda: X \to Y$ defined by

$$[L_\lambda u](t) = u'(t) - A(\lambda, t)u(t). \tag{5.9}$$

The next proposition shows that f is a C^1-Fredholm map of index 0 and computes the index bundle of the family L defined by (5.9) in terms of the asymptotic bundles.

Proposition 5.2. *The family L defined by* (5.9) *verifies*

i) $L_\lambda \in \Phi_0(X, Y)$ *for all* $\lambda \in \Lambda$

ii) Ind $L = [E^s(+\infty)] - [E^s(-\infty)] \in \widetilde{KO}(\Lambda)$

Proof. Let us split R into $R = R^+ \cup R^-$ with $R^\pm = [0, \pm\infty)$ and denote with $X^\pm, Y^\pm$ the spaces $H^1(R_\pm; \mathbb{R}^n)$ and $L^2(R_\pm; \mathbb{R}^n)$ respectively. Consider the operators $L_\lambda^\pm \colon X^\pm \to Y^\pm$ defined as in (5.9) by the restrictions of A_λ to $R_\pm$. I will show that $L_\lambda^\pm$ are Fredholm and compute their index bundles. Notice that, if $M_\lambda^\pm \colon X^\pm \to Y^\pm$ are defined by

$$[M_\lambda^\pm u](t) = u'(t) - A(\lambda, \pm\infty)u(t), \qquad (5.10)$$

then $K_\lambda^\pm = M_\lambda^\pm - L_\lambda^\pm$ is a compact operator for each $\lambda \in \Lambda$. Indeed, if ϕ_m is a smooth function in R_+ such that $\phi_m \equiv 1$ on $[0, m-1]$ and $\phi_m \equiv 0$ on $[m, +\infty)$, then K_λ^+ is limit of

$$[K_\lambda^m u](t) = \phi_m(t)[A(\lambda, +\infty) - A(\lambda, t)](u(t). \qquad (5.11)$$

Moreover the operator K_λ^m is compact because it can be factorized through the inclusion $H^1([0, m]; \mathbb{R}^n) \subset L^2(\mathbb{R}^+; \mathbb{R}^n)$ which is compact. On the other hand it is well known that M_λ is surjective with $\ker M_\lambda = E^s(\lambda, +\infty)$. Indeed the second assertion is clear while for the first it is enough to observe that a right inverse for M_λ is is given by

$$S_\lambda(v)(t) = \int_0^t P_\lambda e^{(s-t)A_\lambda(s)} v(s) ds + \int_t^\infty (\mathrm{id} - P_\lambda) e^{(t-s)A_\lambda(s)} v(s) ds$$

where P_λ is the projector onto $E^s(\lambda, +\infty)$.

Thus M_λ^+ and hence also L_λ^+ are Fredholm operators whose numerical index equals dim $E^s(\lambda, +\infty)$. Moreover by homotopy invariance of the index bundle

$$\text{Ind } L^+ = \text{Ind } M^+ = [E^s(+\infty)]. \qquad (5.12)$$

Similarly we have that L_λ^- is Fredholm of index dim $E^u(\lambda, -\infty)$ by $(A3)$ and Ind $L^- = [E^u(-\infty)]$.

In order to compute the index of L let us observe that the restriction map $I \colon Y \to Y^- \oplus Y^+$ defined by $Iv = (v_{|R^-}, v_{|R^+})$ is an isomorphism, while the analogous map $J \colon H \to X^- \oplus X^+$ is injective with

$$\text{Im } J = \{(u^-, u^+)/u^-(0) = u^+(0)\}.$$

Thus Im $J = \ker \psi$ where $\psi(u^-, u^+) = u^-(0) - u^+(0)$ and hence J is a Fredholm operator of index $-n$. Moreover there is a commutative diagram

$$
\begin{array}{ccc}
 & L_\lambda^- \oplus L_\lambda^+ & \\
X^- \oplus X^+ & \longrightarrow & Y^- \oplus Y^+ \\
J \uparrow & & I \uparrow \\
X & \longrightarrow & Y \\
 & L_\lambda &
\end{array}
\qquad (5.13)
$$

It follows from (5.13) that L_λ is Fredholm. Moreover, by assumption (A2), the index of L_λ must be zero. This proves i). Now ii) follows from (5.13) by the logarithmic property of the index bundle. Indeed, considering I and J as constant families, Ind $I = 0$, Ind $J = -\Theta(\mathbf{R}^n)$. Hence, by (5.3),

$$\text{Ind } L = [E^u(-\infty)] + [E^s(+\infty)] - [\Theta(\mathbf{R}^n)] = [E^s(+\infty)] - [E^s(-\infty)],$$

as claimed. $\qquad\square$

Theorem 5.1 follows from Theorem 2.2 and the above proposition. Indeed, since L takes values in $\Phi_0(X;Y)$ it follows that, for δ small enough, the restriction of f to $\Lambda \times B(0,\delta)$ is a family of C^1-Fredholm maps such that $f(\lambda, 0) = 0$. By hypothesis, the total Stiefel-Whitney class $\omega(\text{Ind } L) \neq 0$ and hence by Theorem 2.2 the set of bifurcation points of H^1-solutions of (5.1) must be of dimension at least $m \geq 0$. Being $H^1(\mathbf{R};\mathbf{R}^n) \subset C(\mathbf{R};\mathbf{R}^n)$ the regularity and convergence in $C_0^1(\mathbf{R};\mathbf{R}^n)$ are easily obtained by bootstrap. $\qquad\square$

References

[1] J.F. Adams, *On the groups J(X) – I*. Topology **2** (1963), 181–195.

[2] J.C. Alexander, *Bifurcation of zeroes of parametrized functions*. J. of Funct. Anal, **29** (1978), 37–53.

[3] J.C. Alexander, James Yorke, *Calculating bifurcation invariants as elements in the homotopy of the general linear group*. J. of Pure and Appl. Algebra, **13** (1978), 1–9.

[4] M.F. Atiyah, *K-Theory*. Benjamin, 1967.

[5] M.F. Atiyah, *Thom complexes*. Proc. Lond. Math. Soc. **11** (1961), 291–310.

[6] T. Bartsch, *The global structure of the zero set of a family of semilinear Fredholm maps*. Nonlinear Analysis **17** (1991), 313–331.

[7] T. Bartsch, *A global index for bifurcation of fixed points*. J. reine angew. Math. **391** (1988), 181–197.

[8] P.M. Fitzpatrick, J. Pejsachowicz, *Fundamental group of the space of Fredholm operators and global analysis of non linear equations*. Contemporary Math. **72** (1988), 47–87.

[9] P.M. Fitzpatrick, J. Pejsachowicz, *Nonorientability of the index bundle and several-parameter bifurcation*. J. of Functional Anal. **98** (1991), 42–58.

[10] P.M. Fitzpatrick, I. Massabò, J. Pejsachowicz, *Global several parameter bifurcation and continuation theorems*. Math. Ann. **263** (1983), 61–73.

[11] P.M. Fitzpatrick, J. Pejsachowicz, P.J. Rabier, *The degree for proper C^2-Fredholm mappings* I . J. reine angew. Math. **424** (1992), 1–33.

[12] N. Hitchin,*Harmonic spinors*, Adv. in Math. **14** (1974), 1–55.

[13] J. Ize, *Bifurcation theory for Fredholm operators*. Mem. Am. Math.Soc. **174** (1976).

[14] J. Ize, *Necessary and sufficient conditions for multiparameter bifurcation*. Rocky Mountain J. of Math **18** (1988), 305–337.

[15] J. Ize, *Topological bifurcation*. Topological Nonlinear Analysis, Birkhäuser, Progress in nonlinear differential equations, **15** (1995), 341–463.

[16] K. Jänich, *Vektorraumbündel und der Raum der Fredholm-Operatoren.* Mathematische Annalen, **161** (1965),129–142.

[17] T. Kato, *Perturbation Theory for Linear Operators.* Springer-Verlag, 1976.

[18] H. Kielhöfer, *Multiple eigenvalue bifurcation for Fredholm mappings.* J. reine angew. Math. **358** (1985), 104–124.

[19] J. Pejsachowicz, *K-theoretic methods in bifurcation theory.* Contemporary Math., **72** (1988), 193–205.

[20] J. Pejsachowicz , *The Leray-Schauder Reduction and Bifurcation for Parametrized Families of Nonlinear Elliptic Boundary Value Problems.* TMNA **18** (2001), 243–268.

[21] J. Pejsachowicz, *The index bundle and bifurcation theory of Fredholm maps.* In preparation.

[22] J. Pejsachowicz, P.J. Rabier , *Degree theory for C^1-Fredholm mappings of index 0.* Journal d'Analyse Mathématique **76** (1998), 289–319.

[23] C. Vafa, E. Witten, *Eigenvalue inequalities for fermions in gauge theories.* Comm. Math. Phys. **95** (1984), 257–276.

[24] V.G. Zvyagin, *On oriented degree of a certain class of perturbations of Fredholm mappings and on bifurcations of solutions of a nonlinear boundary value problem with noncompact perturbations.* Mat. USSR Sbornik **74** (1993), 487–512.

Jacobo Pejsachowicz
Dipartimento di Matematica
Politecnico di Torino
Torino, To, Italy
e-mail: jacobo.pejsachowicz@polito.it

C^*-algebras and Elliptic Theory II
Trends in Mathematics, 251–265
© 2008 Birkhäuser Verlag Basel/Switzerland

Modified Hochschild and Periodic Cyclic Homology

Nicolae Teleman

Abstract. The Hochschild and (periodic) cyclic homology of Banach algebras are either trivial or not interesting, see Connes [2], [4], [6]. To correct this deficiency, Connes [3] had produced the *entire cyclic cohomology* (see also [4], [6], [5]). The entire cyclic cochains are elements of the infinite product (b, B) cohomology bi-complex which satisfy a certain bidegree asymptotic growth condition. The entire cyclic cohomology is a natural target for the asymptotic Chern character of θ-summable Fredholm modules.

More recently, Puschnigg [15] introduced the *local cyclic cohomology* based on precompact subsets of the algebra in an inductive limits system setting.

The main purpose of this paper is to create an analogue of the Hochschild and periodic cyclic homology which gives the right result (i.e., the ordinary $\mathbb{Z}_2$-graded Alexander-Spanier co-homology of the manifold) when applied, at least, onto the algebra of continuous functions on topological manifolds and CW-complexes. This is realized by replacing the Connes periodic bi-complex (b, B), see Connes [2], [4] and Loday [12], by the bi-complex $(\tilde{b}, d)$, where the operator $\tilde{b}$ is obtained by blending the Hochschild boundary b with the Alexander-Spanier boundary d; the operator $\tilde{b}$ anti-commutes with the operator d. The homologies of these complexes will be called *modified Hochschild*, resp. *modified periodic cyclic homology*.

Our construction uses in addition to the algebraic structure solely the *locality* relationship extracted from the topological structure of the algebra.

The modified periodic cyclic homology is invariant under *continuous* homotopies, while the others are invariant at *smooth* homotopies (diffeotopies) only.

The modified Hochschild and periodic cyclic homology are directly connected to the Alexander-Spanier cohomology.

Mathematics Subject Classification (2000). 13D03, 16E40.

Keywords. Hochschild homology, cyclic homology, Alexander-Spanier co-homology.

1. Introduction

The Hochschild complex has the critical limitation that in the case of the C^*-algebra of continuous functions on a compact topological space, its Hochschild homology vanishes in positive degrees. The purpose of the present paper is to correct this deficiency.

The main idea of our procedure consists of replacing the Hochschild boundary operators b by operators of the form $\tilde{b} = bUb$. Such operators will be called *modified Hochschild operators* and the corresponding homology will be called *modified Hochschild homology*. The pair $(\tilde{b}, d)$ is an example of S^1 chain complex, see Burghelea [1], later renamed mixed complex by Kassel [10]. The effect of our construction is that in the modified Hochschild complex there are more $\tilde{b}$-cycles and less $\tilde{b}$-boundaries than those in the Hochschild complex and therefore the modified Hochschild homology is larger than the Hochschild homology.

In addition, whilst the Hochschild boundary does not behave well with respect to the Alexander-Spanier boundary operator, the new operator $\tilde{b}$ anti-commutes with it. This crucial commutativity relation should facilitate the investigation of connections between non-commutative differential geometry and the classical differential geometry. As a next step in this direction, we replace the (b, B) Connes bi-complex by the *modified periodic cyclic bi-complex* $(\tilde{b}, d)$; its total homology will be called *modified periodic cyclic homology*.

The boundary operator $\tilde{b}$ will be realized by blending the Hochschild boundary b with the Alexander-Spanier boundary. In order to reach this objective we will have to restrict the Alexander-Spanier complex to *germs* of functions. In the sequel we will be careful to perform only operations which are compatible with the *locality* phenomena, to encompass at least the case of scalar functions, sections in vector bundles and quasi-local operators. We stress on the observation that the Hochschild homology is local in nature, see Connes [2] for compact manifolds and Teleman [17] for paracompact manifolds.

The construction of the operator $\tilde{b}$ is based on the Karoubi operator [9] σ defined by the formula

$$db + bd = 1 - \sigma, \tag{1.1}$$

discussed in Sections 4–5.

Although the next considerations are general, to the end, in order to get interesting results, we will have to involve *locality*. As long as we perform algebraic operations only with the Alexander-Spanier and Hochschild operators, this objective is not obstructed.

In this paper we show that the modified periodic cyclic homology of the algebra of continuous functions on a smooth manifold is isomorphic to the $\mathbb{Z}_2$-graded Alexander-Spanier cohomology of the manifold. As a collateral result we obtain that the modified periodic Hochschild homology of the algebra of continuous functions is not trivial.

One of the main ideas of the non-commutative geometry consists of the delocalization of the classical differential geometry objects. The delocalized objects, realized in the non-commutative geometry as operators, allow one to bypass one of the barriers of the classical analysis presented by the problem of multiplying distributions. It is known, however, that the Hochschild homology of commutative algebras is based on derivations, which involve, in change, too much locality. This is the reason why the Hochschild homology of the algebra of continuous functions is trivial. We expect that the modified Hochschild complex should provide the correct definition of non-commutative differential forms for algebras of functions which possess few or no non-trivial derivations. The homotopy invariance of the Alexander-Spanier homology implies that the modified periodic cyclic homology is a continuous homotopy invariant.

We stress that all our considerations do not make any kind of commutativity assumption on the associative algebra or the ground ring of the algebra.

Although some of the technicalities presented in this paper are closely related to different constructions/presentations of the non-commutative de Rham homology appearing in the literature, the author chose to make this paper self-contained, stressing that the main objective of the present paper is to provide a non-commutative topologically natural homological environment which enables one to extract the right homology at least from the algebra of continuous functions.

In a subsequent note we intend to show that the modified periodic cyclic homology allows one to extract the Chern character from continuous direct connections (for direct connections see Teleman [18], [19], Kubarski-Teleman [11]) in continuous vector bundles. For applications of linear direct connections, used as a tool, see Connes-Moscovici [6] and Mishchenko-Teleman [14].

The author thanks J.-M. Lescure for some corrections. The author thanks the referee for useful remarks which led to the improvement of this paper.

Acknowledgement. The present research was funded by the Italian Ministry for University and Research grant Nr. 2005010942/2005.

2. Alexander-Spanier complex

Let $\mathcal{A}$ be an arbitrary associative algebra with unit 1 over an arbitrary ring K. Any commutativity assumption is made neither on the algebra $\mathcal{A}$ nor on the ground ring K. We assume of course that $\mathcal{A}$ is a K-bimodule.

In what follows we require only that the unit 1 commutes with all elements of the ring K and we assume that the tensor products are *circular* over K, i.e.,

$$f_0 \otimes_K f_1 \otimes_K \cdots \otimes_K f_k.\alpha = \alpha.f_0 \otimes_K f_1 \otimes_K \cdots \otimes_K f_k \qquad (2.2)$$

for any $\alpha \in K$. If K is a field, any tensor product over K is automatically circular.

For any non negative integer r define

$$C_k(\mathcal{A}) := \otimes_K^{k+1} \mathcal{A}; \qquad (2.3)$$

its elements are called non commutative chains of degree k over $\mathcal{A}$.

In the sequel the tensor product $\otimes$ is understood to mean $\otimes_K$. The formula

$$\alpha(a_0 \otimes a_1 \otimes \cdots \otimes a_k)\beta := (\alpha a_0) \otimes a_1 \otimes \cdots \otimes (a_k \beta) \tag{2.4}$$

defines an $\mathcal{A}$ bi-module structure on $C_k(\mathcal{A})$.

We define the Alexander-Spanier co-boundary face map

$$d_i : C_k(\mathcal{A}) \to C_{k+1}(\mathcal{A})$$

by the formulas

$$d_i(a_0 \otimes a_1 \otimes \cdots \otimes a_k) := a_0 \otimes \cdots \otimes 1 \otimes a_i \otimes \cdots \otimes a_k, \quad \text{for} \quad 0 \le i \le k, \tag{2.5}$$

and

$$d_{k+1}(a_0 \otimes a_1 \otimes \cdots \otimes a_k) := a_0 \otimes a_1 \otimes \cdots \otimes a_k \otimes 1, \quad \text{for} \quad i = k+1. \tag{2.6}$$

The Alexander-Spanier boundary is defined by

$$d := \sum_{i=0}^{i=k+1} (-1)^i d_i; \tag{2.7}$$

it agrees with the classical Alexander-Spanier co-boundary operator, see, e.g., Spanier [16]. It satisfies $d^2 = 0$.

In particular, for any $a \in \mathcal{A}$ one has

$$da = 1 \otimes a - a \otimes 1 \quad \text{and} \quad d1 = 0. \tag{2.8}$$

If $\alpha \in K$ and $a \in \mathcal{A}$, and as $\alpha.1 = 1.\alpha$ then

$$d(\alpha.a) = 1 \otimes (\alpha.a) - (\alpha.a) \otimes 1 = 1.\alpha \otimes a - (\alpha a) \otimes 1 = \alpha.da \tag{2.9}$$

The product

$$\times : C_r(\mathcal{A}) \otimes C_s(\mathcal{A}) \to C_{r+s}(\mathcal{A}) \tag{2.10}$$

of chains over A is defined precisely as in the case of the classical Alexander-Spanier co-chains

$$(a_0 \otimes a_1 \otimes \cdots \otimes a_r) \times (b_0 \otimes b_1 \otimes \cdots \otimes b_s) := a_0 \otimes a_1 \otimes \cdots \otimes (a_r b_0) \otimes b_1 \otimes \cdots \otimes b_s. \tag{2.11}$$

The complex $\mathcal{C}_*(\mathcal{A}) := \{\sum_{r=0}^{\infty} C_r(\mathcal{A}), d\}$ is a graded differential complex: for any $\omega \in C_r(\mathcal{A})$ and $\sigma \in C_s(\mathcal{A})$ one has

$$d(\omega \times \sigma) = (d\omega) \times \sigma + (-1)^r \omega \times (d\sigma). \tag{2.12}$$

If $\rho : \mathcal{A} \to K$ is a K-homomorphism and $\rho(1_\mathcal{A}) = 1_K$, then $h : C_r(\mathcal{A}) \to C_{r-1}(\mathcal{A})$, defined by the formula

$$h(a_0 \otimes a_1 \otimes \cdots \otimes a_r) = \rho(a_0)a_1 \otimes \cdots \otimes a_r \tag{2.13}$$

satisfies the identity

$$dh + hd = 1 \tag{2.14}$$

and hence the complex $\{\mathcal{C}_*(\mathcal{A}), d\}$ is acyclic. In the case of the classical Alexander-Spanier complex the homomorphism ρ is given by the valuation of functions at one point.

If the algebra $\mathcal{A}$ has a locally convex topology, it is natural and customary (see Connes [2]) to replace the algebraic tensor product $\mathcal{C}_i(\mathcal{A})$ by a topological tensor product completion $\hat{C}_i(\mathcal{A})$. The elements of $\hat{C}_r(\mathcal{A})$ are called continuous Alexander-Spanier co-chains.

In the particular case of the algebra $\mathcal{A} = C^\infty(M)$, endowed with the Fréchet topology, where M is a smooth manifold, for the projective tensor product completion, the continuous Alexander-Spanier co-chains consist of all smooth functions on various powers of M. The homology of this complex is acyclic, as explained above. If, however, the complex of continuous Alexander-Spanier chains is replaced by the complex of germs of such functions about the diagonals, the classical Alexander-Spanier theorem, see Spanier [16], states that its homology is canonically isomorphic to the de Rham cohomology.

It is very important to recall that the Alexander-Spanier theorem holds if smooth functions are replaced by arbitrary functions, or by special classes of functions (like measurable, Lipschitz, etc.); such generalizations hold if M is merely a CW-complex.

The main objective of this paper is to create an analogue of the Hochschild and periodic cyclic homology which does not give trivial results on algebras of functions as the algebra of continuous functions. This will be realized by blending the Hochschild boundary b with the Alexander-Spanier boundary d.

Although the following considerations are general, in order to get interesting results, we will have to involve *locality*.

3. Recall of Hochschild and periodic cyclic homology

In this section we recall some basic notions and results due to A. Connes [2], [4] which lay to the foundations of non-commutative geometry.

We keep the hypotheses and notations from the previous section.

Let $b_r : C_k(\mathcal{A}) \to C_{k-1}(\mathcal{A})$, be the Hochschild boundary face operator defined on generators by

$$b_r(f_0 \otimes f_1 \otimes \cdots \otimes f_{k-1} \otimes f_k) = f_0 \otimes f_1 \otimes \cdots \otimes (f_r.f_{r+1}) \otimes \cdots \otimes f_k,$$
$$for \quad 0 \leq r \leq k-1 \tag{3.15}$$

and

$$b_k(f_0 \otimes f_1 \otimes \cdots \otimes f_{k-1} \otimes f_k) = (f_k f_0) \otimes f_1 \otimes \cdots \otimes f_{k-1}, \quad for \quad r = k. \tag{3.16}$$

Two boundary operators, b' and $b : C_k(\mathcal{A}) \to C_{k-1}(\mathcal{A})$ are introduced by the formulas

$$b' = \sum_{r=0}^{r=k-1} (-1)^r b_r \tag{3.17}$$

and

$$b = b' + (-1)^k b_k. \tag{3.18}$$

It is true that $(b')^2 = b^2 = 0$.

The complex $\{C_*(\mathcal{A}), b'\}$ is the bar complex; if the algebra $\mathcal{A}$ is unitary, as assumed, the bar complex is acyclic; it provides the so called bar resolution of the algebra $\mathcal{A}$, see [13]. The acyclicity of the bar resolution is provided by the homotopy

$$\chi(f_0 \otimes f_1 \otimes \cdots \otimes f_{k-1} \otimes f_k) = 1 \otimes (f_0 \otimes f_1 \otimes \cdots \otimes f_{k-1} \otimes f_k). \quad (3.19)$$

The complex $\{C_*(\mathcal{A}), b\}$ is the Hochschild complex of the algebra $\mathcal{A}$ with coefficients in the bi-module $\mathcal{A}$; its homology, denoted $HH_*(\mathcal{A})$, is the Hochschild homology of the algebra $\mathcal{A}$ with coefficients in itself.

If $\mathcal{A}$ is a topological real or complex algebra, the homology of the complex $\{\hat{C}_*(\mathcal{A}), b\}$ is the continuous Hochschild homology of the algebra $\mathcal{A}$, see A. Connes [2], [4].

The following theorem, proven by A. Connes [2] on compact manifolds was extended by N. Teleman [17] to paracompact manifolds.

Theorem 3.1. *For any smooth paracompact manifold*

$$HH_k(C^\infty(M)) \approx \Omega_k(M), \quad (3.20)$$

where $\Omega_k(M)$ denotes the space of k forms.

4. The operator σ

Definition 4.1. The operator σ given by the formula

$$db + bd := 1 - \sigma. \quad (4.21)$$

is due to Karoubi [9]; see also [8] and [7].

A general remark shows that σ commutes both with d and b, that is σ is a chain homomorphism both in the Alexander-Spanier and in the Hochschild complex. Consequently, the range of the operator σ, and its (fixed) powers, are subcomplexes both in the Alexander-Spanier and Hochschild complexes. Additionally, as σ is homotopic to the identity, the inclusions of these subcomplexes into the Alexander-Spanier, resp. Hochschild, complexes induce isomorphisms between their respective homologies.

Lemma 4.2.
(i) b' *anticommutes with* d

$$b'd + db' = 0 \quad (4.22)$$

(ii) σ *verifies the formula*

$$\sigma(f_0 \otimes f_1 \otimes \cdots \otimes f_k) = (-1)^{k+1}((df_k)f_0) \otimes f_1 \otimes \cdots \otimes f_{k-1} \quad (4.23)$$

(iii) *the kth power of σ has the explicit expression*

$$\sigma^k(f_0 \otimes f_1 \otimes \cdots \otimes f_k) = df_1.df_2 \cdots df_k.f_0. \quad (4.24)$$

Proof of Lemma 4.2. (i) The Alexander-Spanier and Hochschild boundary face operators satisfy the following relations on $C_k(\mathcal{A})$

$$d_i b_j = b_{j+1} d_i \quad mboxfor \quad 0 \leq i \leq j \leq k-1 \tag{4.25}$$

$$d_i b_j = b_j d_{i+1} \quad \text{for} \quad 0 \leq j < i \leq k \tag{4.26}$$

$$d_i b_i = b_{i-1} d_i = Id \quad for \quad 0 \leq i \leq k \tag{4.27}$$

$$b_0 d_0 = b_k d_{k+1} = Id \tag{4.28}$$

By virtue of the relations (4.25), (4.26) one has

$$db' = \sum_{0 \leq i \leq j \leq k-1} (-1)^{i+j} d_i b_j + \sum_{0 \leq j < i \leq k} (-1)^{i+j} d_i b_j$$

$$= \sum_{0 \leq i \leq j \leq k-1} (-1)^{i+j} b_{j+1} d_i + \sum_{0 \leq j < i \leq k} (-1)^{i+j} b_j d_{i+1}. \tag{4.29}$$

On the other hand

$$b'd = \sum_{0 \leq i < j+1 \leq k} (-1)^{i+(j+1)} b_{j+1} d_i + \sum_{0 \leq j < i \leq k} (-1)^{(i+1)+j} b_j d_{i+1} \tag{4.30}$$

$$+ \sum_{0 \leq j = i \leq k} (-1)^{(j+i)} b_j d_j + \sum_{0 \leq j \leq k, i = j+1} (-1)^{(j+i)} b_j d_{j+1} \tag{4.31}$$

(relations (4.27), (4.28))

$$= \sum_{0 \leq i < j+1 \leq k} (-1)^{i+(j+1)} b_{j+1} d_i + \sum_{0 \leq j < i \leq k} (-1)^{(i+1)+j} b_j d_{i+1} \tag{4.32}$$

$$+ \sum_{0 \leq j = i \leq k} (-1)^{(j+i)} Id + \sum_{0 \leq j \leq k, i = j+1} (-1)^{j+(j+1)} Id. \tag{4.33}$$

$$= \sum_{0 \leq i < j+1 \leq k} (-1)^{i+(j+1)} b_{j+1} d_i + \sum_{0 \leq j < i \leq k} (-1)^{(i+1)+j} b_j d_{i+1} = -db' \tag{4.34}$$

by virtue of relations (4.30).

(ii) Where confusion might occur, the underscript (k) will specify that the corresponding operator is acting on chains of degree k. We have

$$(bd + db)_{(k)} = (b' + (-1)^{k+1} b_{k+1}) d + d(b' + (-1)^k b_k) \tag{4.35}$$

$$= (-1)^{k+1} (b_{k+1} d - db_k). \tag{4.36}$$

A direct calculation shows that

$$(-1)^{k+1} (b_{k+1} d - db_k)(f_0 \otimes f_1 \otimes \cdots \otimes f_k) \tag{4.37}$$

$$= f_0 \otimes f_1 \otimes \cdots \otimes f_k + (-1)^k ((df_k) f_0) \otimes f_1 \otimes \cdots \otimes f_{k-1}. \tag{4.38}$$

Therefore

$$(1 - \sigma)(f_0 \otimes f_1 \otimes \cdots \otimes f_k) \tag{4.39}$$

$$= f_0 \otimes f_1 \otimes \cdots \otimes f_k + (-1)^k ((df_k) f_0) \otimes f_1 \otimes \cdots \otimes f_{k-1}. \tag{4.40}$$

We have proved that

$$\sigma(f_0 \otimes f_1 \otimes \cdots \otimes f_k) = (-1)^{k+1}((df_k).f_0) \otimes f_1 \otimes \cdots \otimes f_{k-1}, \qquad (4.41)$$

which completes the proof of (ii).

By iterating σ, one gets

$$\sigma^k(f_0 \otimes f_1 \otimes \cdots \otimes f_k) = df_1.df_2 \cdots df_k.f_0, \qquad (4.42)$$

which completes the proof of the lemma. $\qquad\qquad\qquad\qquad\qquad\qquad$ □

Definition 4.3. We introduce the operator

$$\Pi_{(k)} := (1 - bd)\sigma_{(k)}^k. \qquad (4.43)$$

Proposition 4.4.

(i) *The operator $\Pi_{(k)}$ has the explicit formula*

$$\Pi_{(k)}(f_0 \otimes f_1 \otimes \cdots \otimes f_k) = f_0 . df_1 \cdots df_k \qquad (4.44)$$

(ii) *$\Pi_{(k)}$ is a projector on $C_k(\mathcal{A})$*

$$(\Pi_{(k)})^2 = \Pi_{(k)} \qquad (4.45)$$

(iii) *the operators d commute with the projectors $\Pi_{(k)}$ and hence they keep the ranges of $\Pi_{(k)}$ invariant*

$$d \, \Pi_{(k)} = \Pi_{(k+1)} \, d \qquad (4.46)$$

(iv) *the operators b commute with the projectors $\Pi_{(k)}$ and hence they keep the ranges of $\Pi_{(k)}$ invariant*

$$b \, \Pi_{(k)} = \Pi_{(k-1)} \, b. \qquad (4.47)$$

Proof of Proposition 4.4. (i) From the Definition 4.3 and formula (4.42) we get

$$\Pi_{(k)}(f_0 \otimes f_1 \otimes \cdots \otimes f_k) = (1 - bd) \, df_1 \cdots df_k \, f_0 \qquad (4.48)$$

$$= df_1 \cdots df_k \, f_0 - (-1)^k b \, (df_1 \cdots df_k \, df_0) \qquad (4.49)$$

$$= df_1 \cdots df_k \, f_0 - (-1)^k b \, [df_1 \cdots df_k \, (1 \otimes f_0 - f_0 \otimes 1)] \qquad (4.50)$$

$$= f_0 \, df_1 \cdots df_k \qquad (4.51)$$

because the Hochschild boundary faces b_i contract the factors $1 \otimes f_{i-1} - f_{i-1} \otimes 1$ into zero, for $0 \le i \le k - 1$.

(ii) We recall that in the literature a chain of the form $f_0 \otimes f_1 \otimes \cdots \otimes f_k$ in which at least one of the factors $f_i = 1$, $1 \le i \le k$, is called degenerate. Then, it is easy to see that

$$\Delta(f_0 \otimes f_1 \otimes \cdots \otimes f_k) := (1 - \Pi_{(k)})(f_0 \otimes f_1 \otimes \cdots \otimes f_k) \qquad (4.52)$$

$$= f_0 \otimes f_1 \otimes \cdots \otimes f_k - f_0 \, df_1 \cdots df_k \qquad (4.53)$$

is a finite sum of degenerate chains.

As $d \, 1 = 0$, $\Pi_{(k)}$ carries any degenerate chain into zero. Therefore,

$$0 = \Pi_{(k)}\Delta = \Pi_{(k)}(1 - \Pi_{(k)}), \qquad (4.54)$$

which proves the assertion.

It follows also that the range of the complementary projector $1 - \Pi_{(k)}$ consists precisely of degenerate chains.

(iii) In view of the identity (i) already proven, one has

$$d \, \Pi_{(k)}(f_0 \otimes f_1 \otimes \cdots \otimes f_k) = d(f_0 \, . \, df_1 \cdots df_k) \qquad (4.55)$$

$$= df_0 \, . \, df_1 \cdots df_k = 1.df_0 \, . \, df_1 \cdots df_k \qquad (4.56)$$

$$= \Pi_{(k+1)} \, d_0 \, (f_0 \otimes f_1 \otimes \cdots \otimes f_k) = \Pi_{(k+1)} \, d \, (f_0 \otimes f_1 \otimes \cdots \otimes f_k) \qquad (4.57)$$

because $\Pi_{(k+1)} \, d_i \, (f_0 \otimes f_1 \otimes \cdots \otimes f_k) = 0$ for $1 \le i \le k+1$.

(iv) $\qquad b \, \Pi_{(k)} = b \, (1 - bd)\sigma_{(k)}^k = b \, (1 - bd)(1 - bd - db)\sigma_{(k)}^{k-1} \qquad (4.58)$

$$= b \, (1 - bd - db)\sigma_{(k)}^{k-1} = b \, (1 - db)\sigma_{(k)}^{k-1} \qquad (4.59)$$

$$= (1 - bd) \, b \, \sigma_{(k)}^{k-1} = (1 - bd) \, \sigma_{(k-1)}^{k-1} \, b = \Pi_{(k-1)}b. \qquad (4.60)$$

$$\square$$

Definition 4.5. Define the complex

$$\widetilde{C}_k(\mathcal{A}) \ := \ \Pi_{(k)}(C_k(\mathcal{A})) \qquad (4.61)$$

$$= \left\{ \ \sum_{\text{finite/series}} f_0 \, df_1 \, df_2 \cdots df_k, \ f_i \in \mathcal{A} \ \right\}. \qquad (4.62)$$

Proposition 4.6.

(i) *One has the inclusion of complexes*

$$\{\widetilde{C}_*(\mathcal{A}), d\} \quad \text{is a sub-complex of} \quad \{C_*(\mathcal{A}), d\} \qquad (4.63)$$

(ii) *one has the inclusion of complexes*

$$\{\widetilde{C}_*(\mathcal{A}), b\} \quad \text{is a sub-complex of} \quad \{C_*(\mathcal{A}), b\} \qquad (4.64)$$

(iii) *on the sub-complex $\widetilde{C}_k(\mathcal{A})$ one has the identity*

$$(1 - bd)\sigma_{(k)}^k = 1 \qquad (4.65)$$

(iv) *the above sub-complex inclusions induce isomorphisms in the Alexander-Spanier, resp. Hochschild, homology.*

In the literature the Hochschild complex modulo degenerate chains is known as the normalized Hochschild complex, see, e.g., Loday's book [12]. It is also well known [12] that the Hochschild and the normalized Hochschild complexes have isomorphic homologies. Our considerations show also that the normalized Hochschild complex is precisely the sub-complex $\widetilde{C}_k(\mathcal{A})$. The formula Proposition 4.6 (iii) is the formula (2.6.8.1–2) from Loday's book [12], page 85 formulated onto the normalized complex. Although the normalized complex and the sub-complex $\widetilde{C}_k(\mathcal{A})$ coincide, we stress that from the very beginning we avoided working into the normalized complex for the purpose of getting exact formulas in the non-normalized Hochschild and Alexander-Spanier complexes.

Proof of Proposition 4.6. The defining formula for σ shows that it (and any of its powers) is chain homotopic (both, with respect to d or b) to the identity. If $\omega \in C_k(\mathcal{A})$ and $d\omega = 0$, then $\sigma_{(k)}^k(\omega)$ is co-homologous to ω.

Moreover, $(\sigma_{(k)} + db)\sigma_{(k)}^k(\omega) = (1 - bd)\sigma_{(k)}^k(\omega)$ is co-homologous to $\sigma_{(k)}^k(\omega)$; hence any homology class in the Alexander-Spanier complex is co-homologous to its projection into the subcomplex $\{\widetilde{C}_*(\mathcal{A}), d\}$. Therefore, the homology of the sub-complex is at least as large as the Alexander-Spanier homology.

The same argument works for the Hochschild complex, using the original definition of the projection $\Pi_{(k)}$.

On the other hand, suppose ω is a d-cycle in $\widetilde{C}_*(\mathcal{A})$ and that $\omega = d\phi$, where $\phi \in C_{k-1}(\mathcal{A})$. Then, by virtue of Proposition 4.4 (iii)

$$\omega = \Pi_{(k)}(\omega) = \Pi_{(k)}(d\phi) = d\Pi_{(k-1)}(\phi), \tag{4.66}$$

which shows that ω is a boundary in the subcomplex $\widetilde{C}_*(\mathcal{A})$.

The same argument works for the Hochschild complex. $\qquad\square$

Corollary 4.7.

  (i) *The Alexander-Spanier, resp. Hochschild, complex decomposes in a direct sum of complexes*

$$\{\widetilde{C}_*(\mathcal{A}), d\} \oplus \{degenerate\ chains, d\} \tag{4.67}$$

  respectively,

$$\{\widetilde{C}_*(\mathcal{A}), b\} \oplus \{degenerate\ chains, b\} \tag{4.68}$$

  (ii) *the Alexander-Spanier, resp. Hochschild, sub-complex of degenerate chains is acyclic.*

5. Modified Hochschild boundary

From now on we will be working only on the sub-spaces $\{\widetilde{C}_*(\mathcal{A})\}$, which are Alexander-Spanier and Hochschild sub-complexes.

In the previous section we introduced the operator σ defined by the formula

$$db + bd = 1 - \sigma. \tag{5.69}$$

We have shown that σ satisfies on $\{\widetilde{C}_*(\mathcal{A})\}$ the identity

$$\sigma^k = 1 + bd\sigma^k. \tag{5.70}$$

Replacing

$$\sigma = 1 - (db + bd) \tag{5.71}$$

in the above identity one gets

$$1 = (1 - bd)\sigma^n = (1 - bd)[1 - (bd + db)]^n \tag{5.72}$$

$$= (1 - bd)[1 + \sum_{k=1}^{n}(-1)^k C_n^k (bd + db)^k] \tag{5.73}$$

$$= (1 - bd)\{1 + \sum_{k=1}^{n}(-1)^k C_n^k[(bd)^k + (db)^k]\} \tag{5.74}$$

$$= 1 + \sum_{k=1}^{n}(-1)^k C_n^k[(bd)^k + (db)^k] - bd - \sum_{k=1}^{n}(-1)^k C_n^k(bd)^{k+1}, \tag{5.75}$$

and hence

$$0 = bd + \sum_{k=1}^{n}(-1)^{k-1} C_n^k[(bd)^k + (db)^k] + \sum_{k=1}^{n}(-1)^k C_n^k(bd)^{k+1} \tag{5.76}$$

$$= bd + \sum_{k=1}^{n}(-1)^{k-1} C_n^k(bd)^k + \sum_{k=1}^{n}(-1)^k C_n^k(bd)^{k+1} + \sum_{k=1}^{n}(-1)^{k-1} C_n^k(db)^k \tag{5.77}$$

$$= (1+n)bd + \sum_{k=2}^{n}(-1)^{k-1} C_n^k(bd)^k + \sum_{k=1}^{n}(-1)^k C_n^k(bd)^{k+1} \\ + \sum_{k=1}^{n}(-1)^{k-1} C_n^k(db)^k \tag{5.78}$$

$$= (1+n)bd + \sum_{k=2}^{n}(-1)^{k-1} C_n^k(bd)^k + \sum_{k=1}^{n-1}(-1)^k C_n^k(bd)^{k+1} \\ + (-1)^n C_n^n(bd)^{n+1} + \sum_{k=1}^{n}(-1)^{k-1} C_n^k(db)^k \tag{5.79}$$

$$= (1+n)bd + \sum_{k=2}^{n}(-1)^{k-1} C_n^k(bd)^k + \sum_{k=2}^{n}(-1)^{k-1} C_n^{k-1}(bd)^k \\ + (-1)^n C_n^n(bd)^{n+1} + \sum_{k=1}^{n}(-1)^{k-1} C_n^k(db)^k \tag{5.80}$$

$$= (1+n)bd + \sum_{k=2}^{n}(-1)^{k-1}(C_n^k + C_n^{k-1})(bd)^k + (-1)^n C_n^n(bd)^{n+1} \\ + \sum_{k=1}^{n}(-1)^{k-1} C_n^k(db)^k \tag{5.81}$$

$$= (C_n^k + C_n^{k-1} = C_{n+1}^k)$$

$$= (1+n)bd + \sum_{k=2}^{n}(-1)^{k-1} C_{n+1}^k(bd)^k + (-1)^n C_{n+1}^{n+1}(bd)^{n+1} \\ + \sum_{k=1}^{n}(-1)^{k-1} C_n^k(db)^k \tag{5.82}$$

or

$$0 = \sum_{k=1}^{n+1}(-1)^{k-1}C_{n+1}^k(bd)^k + \sum_{k=1}^{n}(-1)^{k-1}C_n^k(db)^k. \tag{5.83}$$

The operator $\tilde{b}_n$, acting on n-forms, is defined by the formula

$$\tilde{b}_n := b\sum_{k=1}^{n}(-1)^{k-1}C_n^k(db)^{k-1}. \tag{5.84}$$

The above relation (6.84) becomes

$$0 = \tilde{b}_{n+1}d_n + d_{n-1}\tilde{b}_n \tag{5.85}$$

and hence, the operators $\tilde{b}$ and d anti-commute.

As the operator $\tilde{b}_n$ is of the form $\tilde{b}_n = bTb$, it follows that $\tilde{b}\tilde{b} = 0$ and hence is a boundary operator. The complex obtained by replacing the operator b by $\tilde{b}$ will be called *modified Hochschild complex*, denoted $C_*(\tilde{b}, d)$. We intend to study its homology.

Given the above mentioned structure of the operator $\tilde{b}_n$, it follows immediately that in the modified Hochschild complex the Hochschild cycles remain cycles and no new boundaries appear. Therefore, the homology of the modified Hochschild complex, called *modified Hochschild homology,* is larger than the Hochschild homology.

6. Modified periodic cyclic homology

In analogy with the periodic cyclic complex due to A. Connes [4], see also J.-L. Loday [12] Sect. 5.1.7., page 159, we introduce the *modified periodic cyclic bi-complex* of the (topological) algebra $\mathcal{A}$, by

$$\widetilde{C}^{\lambda,\mathrm{per}}(\mathcal{A}) = \{\widetilde{C}_{p,q}\}_{(p,q)\in\mathbf{Z}\times\mathbf{Z}}, \tag{6.86}$$

where

$$\widetilde{C}_{p,q} = \widetilde{C}_{q-p}(\mathcal{A}) \ \ for \ \ p \le q, \ \ and \ \ \widetilde{C}_{p,q} = 0 \ \ for \ \ q < p, \tag{6.87}$$

while the bi-complex operators $\tilde{b}$ and d are acting as such

$$\tilde{b} : \widetilde{C}_{p,q} \to \widetilde{C}_{p,q-1}$$

$$d : \widetilde{C}_{p,q} \to \widetilde{C}_{p-1,q};$$

as seen in the previous section, they anti-commute and hence they are legitimate bi-complex operators.

The *modified periodic cyclic homology* of the algebra $\mathcal{A}$, $\widetilde{H}_*^{\lambda,\mathrm{per}}(\mathcal{A})$, is by definition the homology of the total complex associated to the modified cyclic periodic complex $\tilde{C}^{\lambda,\mathrm{per}}(\tilde{b}, d)$.

The author is indebted to the referee for the following

Remark. *The above formulas are valid in the simplicial category Δ^{op} to which the extra degeneracy is added (for its definition see [12], Sect. 6.1) and therefore the construction applies to any functor from this category to the category of modules, not only to the functor $\mathcal{L}(\mathcal{A}) : [n] \to \mathcal{A}^{\otimes n+1}$.*

As applications of the above considerations, in the next subsections 6.1, 6.2 we are going to compute the modified periodic cyclic homology both in the case of the algebra of smooth functions and the algebra of continuous functions, on smooth manifolds. We recall the expression of the boundary operator $\tilde{b}$

$$\tilde{b}_r := b \sum_{k=1}^{r} (-1)^{k-1} C_n^k (db)^{k-1} \tag{6.88}$$

and the expression of the Alexander-Spanier boundary operator d

$$(df_r)(x_0, x_1, \ldots, x_r, x_{r+1}) = \sum_{i=0}^{i=r+1} (-1)^i f_r(x_0, \ldots, \hat{x}_i, \ldots, x_{r+1}). \tag{6.89}$$

In both application we consider d to be the Alexander-Spanier co-boundary acting on germs of smooth, resp. continuous, functions defined on neighborhoods of the diagonal in the different powers of the base space.

6.1. The smooth case

It is clear that the operator d is well defined on germs of functions about the diagonals. It is also important to notice that the Hochschild boundary is also well defined on germs and that the Hochschild homology depends only on the quotient complex consisting of germs, see Teleman [17].

We use the spectral sequence associated to the first filtration of the bicomplex (with respect to the d-degree of chains) to compute the homology of the total complex. We are going to prove that the corresponding terms $E_{p,q}^1$ and $E_{p,q}^2$ of the spectral sequence, in the case of the algebra $A := C^\infty(M)$, are

$$E_{p,q}^1 = H_p(C_{*,q}, d) \cong H_{dR}^{q-p}(M)$$

and

$$E_{p,q}^2 = H_q(E_{p,*}^1, \tilde{b}) \cong H_{dR}^{q-p}(M).$$

For the computation of the term E^2 we recall that any Alexander-Spanier cohomology class may be represented by a totally skew-symmetric function. The Hochschild boundary b of such a representative is clearly zero, and hence, a fortiori, $\tilde{b}\gamma = 0$. From here it follows that $d^1 = 0$ and therefore $E_{p,q}^2 = E_{p,q}^1 = H_{dR}^{q-p}(M)$.

We have proven the

Theorem 6.1. *In the case of the algebra of smooth functions $C^\infty(M)$ on the smooth manifold M, the modified periodic cyclic homology and the periodic cyclic homology defined by A. Connes [4] coincide*

$$\tilde{H}_k^{\lambda, \text{per}}(C^\infty(M)) = H_k^{\lambda, \text{per}}(C^\infty(M)) = \oplus^{r \equiv k \pmod 2} H_{dR}^r(M).$$

6.2. The continuous case

In this subsection we compute the modified periodic cyclic homology of the algebra of continuous functions $C(M)$ on the *smooth* manifold M.

For its computation we proceed along the same lines as in the case of smooth functions. To begin with, we observe that the modified periodic cyclic bicomplex of the algebra of smooth functions is a sub-bicomplex of the modified periodic cyclic bicomplex of the algebra of continuous functions. Given that the inclusion of the Alexander-Spanier complex of smooth functions into the Alexander-Spanier complex of continuous functions induces isomorphism in homology, one infers that the bicomplex inclusion induces isomorphisms between the corresponding terms E^1 and therefore E^2

$$E^1_{p,q} = H_p(C_{*,q}, d) \cong H^{q-p}_{dR}(M) \tag{6.90}$$

$$E^2_{p,q} = H_q(E^1_{p,*}, \tilde{b}) \cong H^{q-p}_{dR}(M). \tag{6.91}$$

This proves the

Theorem 6.2. *The modified periodic cyclic homology of the algebra of real-valued continuous functions $C(M)$ on the smooth manifold M, coincides with the periodic cyclic homology of the algebra of smooth functions*

$$H^{\lambda,\mathrm{per}}_k(C(M)) = \oplus^{r \equiv k \pmod 2} H^r_{dR}(M) \cong \oplus^{r \equiv k \pmod 2} H^r(M, \mathbf{R}). \tag{6.92}$$

Theorem 6.3. *The modified periodic cyclic homology of the algebra of continuous functions $C(M)$ on a smooth manifold M is a continuous homotopy invariant.*

The statement of the theorem follows from the above spectral sequence argument along with the property of the Alexander-Spanier cohomology of being homotopy invariant.

We expect, of course, the same result to hold if M is merely a topological manifold, or a CW-complex.

Corollary 6.4. *The modified Hochschild homology of the algebra $C(M)$ is not trivial.*

This result should be used as the correct definition of non-commutative differential forms in the case of algebras of continuous, or more singular, functions.

References

[1] D. Burghelea, *Cyclic homology and the algebraic K-theory of spaces. I.*, in *Applications of algebraic K-theory to algebraic geometry and number theory*, Part I, II Boulder, Colorado, 1983, 89–115. Contemp. Math., 55, Amer. Math. Soc., Providence, RI, 1986.

[2] A. Connes, *Noncommutative differential geometry*, Publ. Math. IHES 62 (1985), pp. 257–360.

[3] A. Connes, *Entire cyclic cohomology of Banach algebras and characters of θ-summable Fredholm modules*, K-Theory 1 (1988), 519–548.

[4] A. Connes, *Noncommutative Geometry*, Academic Press, (1994).

[5] A. Connes, M. Gromov, H. Moscovici, *Conjectures de Novikov et fibrés presque plats*, C. R. Acad. Sci. Paris Sér. A-B 310 (1990), 273–277.

[6] A. Connes, H. Moscovici, *Cyclic Cohomology, The Novikov Conjecture and Hyperbolic Groups*, Topology Vol. 29, pp. 345–388, 1990.

[7] J. Cuntz, *Cyclic Theory, Bivariant K-Theory and the Bivariant Chern-Connes Character*, in *Operator Algebras and Non-Commutative Geometry II*, Encyclopedia of Mathematical Sciences, Vol. 121, 1–71, Springer Verlag, 2004.

[8] J. Cuntz, D. Quillen, *Operators on noncommutative differential forms and cyclic homology,*, in *Geometry, Topology and Physics*, 77–111, International Press, Cambridge, MA, 1995.

[9] M. Karoubi, *Homologie cyclique et K-Théorie.* Astérisque No. 149 (1987), 147 pp.

[10] C. Kassel, *Cyclic homology, comodules and mixed complexes.* Journal of Algebra, 107 (1987), 195–216.

[11] J. Kubarski, N. Teleman, *Linear Direct Connections*, Proceedings 7th Conference on "Geometry and Topology of Manifolds – The Mathematical Legacy of Charles Ehresmann", Będlewo, May 2005 (to appear).

[12] J.-L. Loday, *Cyclic Homology*, Grundlehren in mathematischen Wissenschaften 301, Springer Verlag, Berlin Heidelberg, 1992.

[13] S. McLane *Homology*, Third Ed., Grundlehren der mathematischen Wissenschaften in Einzeldarstellung Band 114, Springer Verlag, Heidelberg, 1975.

[14] A.S. Mishchenko, N. Teleman, *Almost flat bundles and almost flat structures.* in Topological Methods in non Linear Analysis, Vol. 26, Nr. 1, pp. 75–88, 2005 (Volume dedicated to Olga Ladyzhenkaya).

[15] M. Puschnigg, *Diffeotopy Functors of Ind-Algebras and Local Cyclic Cohomology*, Documenta Mathematica Vol. 8, 143–245, 2003.

[16] E.H. Spanier, *Algebraic Topology*, McGraw-Hill Series in Higher Mathematics, New York, 1966.

[17] N. Teleman, *Microlocalisation de l'Homologie de Hochschild*, Compt. Rend. Acad. Scie. Paris, Vol. 326, 1261–1264, 1998.

[18] N. Teleman, *Distance Function, Linear quasi connections and Chern Character*, IHES Prepublications M/04/27, June 2004.

[19] N. Teleman N, *Direct Connections and Chern Character*, in Proceedings of the International Conference in Honour of Jean-Paul Brasselet, Luminy – 2005. In "Singularity Theory", Eds. D. Chéniot, N. Dutertre, C. Murolo, D. Trotman, A. Pichon. World Scientific Publishing Company, 2007.

Nicolae Teleman
Università Politecnica delle Marche
Facoltà di Ingegneria
Dipartimento di Scienze Matematiche
Via Brecce Bianche
60131 Ancona, Italy
e-mail: `teleman@dipmat.univpm.it`

C^*-algebras and Elliptic Theory II

Trends in Mathematics, 267–280

© 2008 Birkhäuser Verlag Basel/Switzerland

L^2-invariants and Rank Metric

Andreas Thom

Abstract. We introduce a notion of rank completion for bi-modules over a finite tracial von Neumann algebra. We show that the functor of rank completion is exact and that the category of complete modules is abelian with enough projective objects. This leads to interesting computations in the L^2-homology for tracial algebras. As an application, we also give a new proof of a Theorem of Gaboriau on invariance of L^2-Betti numbers under orbit equivalence.

Mathematics Subject Classification (2000). 46L10, 37A20.

1. Preliminaries

1.1. Introduction

The aim of this article is to unify approaches to several results in the theory of L^2-invariants of groups, see [Lüc02, Gab02a], and tracial algebras, see [CS05]. The new approach allows us to sharpen several results that were obtained in [Tho06b]. We also give a new proof of D. Gaboriau's Theorem on invariance of L^2-Betti numbers under orbit equivalence. In order to do so, we introduce the concept of rank metric and rank completion of bi-modules over a finite tracial von Neumann algebra.

All von Neumann algebras in this article have a separable pre-dual. Recall, a von Neumann algebra is called finite and tracial, if it comes with a fixed positive, faithful and normal trace. Every finite (i.e., Dedekind finite) von Neumann algebra admits such a trace, but we assume that a choice of a trace is fixed.

The rank is a natural measure of the size of the support of an element in a bi-module over a finite tracial von Neumann algebra. The induced metric endows each bi-module with a topology, such that all bi-module maps are contractions. The main utility of completion with respect to the rank metric is revealed by the observation that the functor of rank completion is exact and that the category of complete modules is abelian with enough projective objects.

Employing the process of rank completion, we aim to prove two main results. First of all, we will show that certain L^2-Betti number invariants of von Neumann algebras coincide with those for arbitrary weakly dense sub-C^*-algebras. The particular case of the first L^2-Betti number was treated in [Tho06b]. The general result required a more conceptual approach and is carried out in this article. The importance of this result was pointed out to the author by D. Shlyakhtenko. Indeed, according to A. Connes and D. Shlyakhtenko, K-theoretic methods might be used to relate the L^2-Euler characteristic of a group C^*-algebra to the ordinary Euler characteristic of the group. This could finally lead to a computation of the L^2-Betti numbers for certain von Neumann algebras and would resolve some longstanding conjectures, as for example the non-isomorphism conjecture for free group factors, see [Voi05]. However, a concrete implementation of this idea is not in reach and a lot of preliminary work has still to be carried out.

Secondly, inspired by ideas of R. Sauer from [Sau03], we will give a new and self-contained proof of invariance of L^2-Betti numbers of groups under orbit equivalence. The idea here is very simple. We show that L^2-Betti number invariants cannot see the difference between an $L^\infty(X)$-algebra and its rank completion. Then, we observe the following: If free measure preserving actions of Γ_1 and Γ_2 on a probability space X induce the same equivalence relation, then $L^\infty(X) \rtimes_{\mathrm{alg}} \Gamma_1$ and $L^\infty(X) \rtimes_{\mathrm{alg}} \Gamma_2$ have isomorphic rank completions as bi-modules (with respect to the diagonal left action) over $L^\infty(X)$. It remains to carry out several routine calculations in homological algebra.

I thank the unknown referee for helpful comments that improved the exposition.

1.2. Dimension theory

In his pioneering work, W. Lück was able to describe a lot of the analytic properties of the category of Hilbert-modules over a finite tracial von Neumann algebras (M, τ) in purely algebraic terms. This allowed to employ the machinery of homological algebra in the study L^2-invariants and lead to substantial results and a conceptional understanding from an algebraic point of view. One important ingredient in his work is a dimension function which is defined for all M-modules, see [Lüc02]. Due to several ring-theoretic properties of M, the natural dimension function for projective modules has an extension to all modules and shares several convenient properties. In particular, it was shown in [Lüc02], that the sub-category of zero-dimensional modules is a Serre sub-category, i.e., is closed under extensions. This implies that there is a 5-Lemma for dimension isomorphisms. The following lemma is immediate from this. (See [Wei94] for the necessary definitions.)

Lemma 1.1. Let (M, τ) be a finite tracial von Neumann algebra. Let $\mathcal{A}$ be an abelian category with enough projective objects and let $F, G \colon \mathcal{A} \to \mathrm{Mod}^M$ be right exact functors into the category Mod^M of M-modules. If there exists a natural transformation $h \colon F \to G$ which consists of dimension isomorphisms, then the induced natural transformations

$$h_i \colon L_i(F) \to L_i(G)$$

of left-derived functors consist of dimension isomorphisms too.

L^2-Betti numbers for certain group-actions on spaces were introduced by M. Atiyah in [Ati76]. The domain of definition was extended by J. Cheeger and M. Gromov in [CG86]. For references and most of the main results, see [Lüc02]. An important result of Lück was the following equality, which we take as a basis for our computations in Section 4:

$$\beta_k^{(2)}(\Gamma) = \dim_{L\Gamma} \mathrm{Tor}_k^{\mathbb{C}\Gamma}(L\Gamma, \mathbb{C}). \tag{1}$$

The following observation concerning a characterization of zero-dimensional modules is due to R. Sauer, see [Sau03], and will be of major importance in the sequel.

Theorem 1.2 (Sauer). *Let (M, τ) be a finite tracial von Neumann algebra and let L be an M-module. The following conditions are equivalent:*

1. *L is zero-dimensional.*
2. *$\forall \xi \in L, \forall \varepsilon > 0, \exists p \in \mathrm{Proj}(M) \colon \quad p\xi = \xi \quad and \quad \tau(p) \leq \varepsilon.$*

The second condition is usually referred to as a local criterion of zero-dimensionality. In the next section, we want to exploit this observation further and study completions of bi-modules with respect to a certain metric that measures the size of the support.

Let $a \in M$ be an arbitrary element in a finite tracial von Neumann algebra (M, τ). We denote by $s(a)$ is support projection and by $r(a)$ its range projection. Note that the equality $\tau(s(a)) = \tau(r(a))$ always holds. We denote by $\mathrm{Proj}(M)$ the set of projections of M. Note that $\mathrm{Proj}(M)$ is a complete, complemented modular lattice. We denote the operations of meet, join and complement by $\wedge, \vee$ and $\perp$.

2. Completion of bi-modules

2.1. Definition

Let us denote the category of M-bi-modules by Bimod^M. We will loosely identify M-bi-modules with $M \otimes M^o$-modules. In the sequel we regard $M \otimes M^o$ as a bi-module over M, acting by multiplication with $M \otimes M^o$ on the left.

Let (M, τ) be a finite tracial von Neumann algebra and let L be an M-bi-module. Associated to an element $\xi \in L$, there is a real-valued quantity that measures the size of the support. Let us set

$$[\xi] = \inf \left\{ \tau(p) + \tau(q) \colon p, q \in \mathrm{Proj}(M), p^\perp \xi q^\perp = 0 \right\} \in [0, 1].$$

Obviously, for the bi-module M and $x \in M$, we get that $[x]$ equals the trace of the support projection $s(x)$ of x. Indeed, if $p^\perp x q^\perp = 0$, then $\tau(p^\perp) + \tau(s(x)) + \tau(q^\perp) \leq 2$ and thus

$$\tau(p) + \tau(q) \geq \tau(s(x)).$$

We conclude that $[x] \geq \tau(s(x))$. The reverse inequality is obvious.

Lemma 2.1. *Let L be an M-bi-module and let $\xi_1, \xi_2 \in L$. The inequality*

$$[\xi_1 + \xi_2] \leq [\xi_1] + [\xi_2]$$

holds.

Proof. Let $\varepsilon > 0$ be arbitrary. We find projections p_1, q_1, p_2, q_2, such that $\tau(p_i) + \tau(q_i) \leq [\xi_i] + \varepsilon$ and $p_i^\perp \xi_i q_i^\perp = 0$, for $i = 0, 1$. Since $(p_1^\perp \wedge p_2^\perp)(\xi_1 + \xi_2)(q_1^\perp \wedge q_2^\perp) = 0$, we get that

$$[\xi_1 + \xi_2] \leq \tau(p_1 \vee p_2) + \tau(q_1 \vee q_2) \leq \tau(p_1) + \tau(p_2) + \tau(q_1) + \tau(q_2) \leq [\xi_1] + [\xi_2] + 2\varepsilon.$$

Since ε was arbitrary, the claim follows. $\square$

Note, there is no reason to assume that $[\xi] = 0 \implies \xi = 0$. Indeed, one can easily construct examples where this fails.

Definition 2.2. *Let L be an M-bi-module. The quantity $d(\xi, \zeta) \overset{\text{def}}{=} [\xi - \zeta] \in \mathbb{R}$ defines a quasi-metric on L, which we call* rank metric.

Lemma 2.3. *Let $\phi \colon L \to L'$ be a homomorphism of M-bi-modules.*
  1. *The map ϕ is a contraction in the rank-metric.*
  2. *Let $\varepsilon > 0$ be arbitrary. If ϕ is surjective and $\xi' \in L'$, then there exists $\xi \in L$, such that $\phi(\xi) = \xi'$ and $[\xi] \leq [\xi'] + \varepsilon$.*

Proof. (1) Let $\xi \in L$, $\varepsilon > 0$ and $p, q \in \operatorname{Proj}(M)$, such that $\tau(p) + \tau(q) \leq [\xi] + \varepsilon$ and $p^\perp \xi q^\perp = 0$. Clearly, $p^\perp \phi(x) q^\perp = 0$, and hence $[\phi(x)] \leq \tau(p) + \tau(q) \leq [\xi] + \varepsilon$. Since ε was arbitrary, the assertion follows.

(2) Let $\xi' \in L'$. There exists $p, q \in \operatorname{Proj}(M)$, such that $\tau(p) + \tau(q) \leq [\xi'] + \varepsilon$ and $p^\perp \xi' q^\perp = 0$. Let $\xi'' \in L$ be any lift of ξ' and set $\xi = \xi'' - p^\perp \xi'' q^\perp$. We easily see that $\phi(\xi) = \xi'$ and that $p^\perp \xi q^\perp = 0$. Hence, ξ is a lift and $[\xi] \leq [\xi'] + \varepsilon$ as required. $\square$

Definition 2.4. *Let L be an M bi-module. The rank metric endows L with a uniform structure.*
  1. *We denote by $CS(L)$ the linear space of Cauchy sequences in L, by $ZS(L) \subset CS(L)$ the sub-space of sequences that converge to $0 \in L$. Finally, we set $c(L) = CS(L)/ZS(L)$ and call it the* completion *of L.*
  2. *There is a natural map $L \to c(L)$ which sends an element to the constant sequence. The bi-module L is called* complete, *if it is an isomorphism. We denote by Bimod_c^M the full sub-category of complete $M \otimes M^o$-modules.*

Lemma 2.5. *Let M be a finite tracial von Neumann algebra.*
  1. *The completion $M \hat{\otimes} M^o$ of $M \otimes M^o$ as an M-bimodule is a unital ring containing $M \otimes M^o$.*
  2. *Let L be a M-bi-module. The completion is naturally an $M \hat{\otimes} M^o$-module and in particular an M-bi-module.*
  3. *The assignment $L \mapsto c(L)$ extends to a functor from the category of $M \otimes M^o$-modules to the category of Bimod_c^M of complete $M \otimes M^o$-modules.*

Proof. Let L be an M-bi-module. Let us first show that the $M \otimes M^o$-module structure extends to $c(L)$. Let $\xi \in M \otimes M^o$. We consider the map $\lambda_\xi : L \to L$ which is defined to be left-multiplication by ξ. λ_ξ is not a module-homomorphism but still to some extend compatible with the rank metric. Let $\eta \in L, \varepsilon > 0$ and $p, q \in \mathrm{Proj}(M)$ with $(p^\perp \otimes q^{\perp o})\eta = 0$ and $\tau(p) + \tau(q) \leq [\eta] + \varepsilon$.

Specifying to $\xi = a \otimes b^o$, we get:

$$\lambda_\xi(\eta) = (a \otimes b^o)\eta = (a \otimes b^o)(1 \otimes 1 - p^\perp \otimes q^{\perp o})\eta = (ap \otimes b^o)\eta + (ap^\perp \otimes (qb)^o)\eta.$$

We compute: $(r(ap)^\perp \otimes 1)(ap \otimes b^o)\eta = 0$ and hence $[(ap \otimes b^o)\eta] \leq \tau(r(ap)) = \tau(s(ap)) \leq \tau(p)$. Similarly, $[(ap^\perp \otimes (qb)^o)\eta] \leq \tau(q)$ and hence $[(a \otimes b^o)\eta] \leq \tau(p) + \tau(q) \leq [\eta] + \varepsilon$. Again, since $\varepsilon > 0$ was arbitrary, we conclude $[\lambda_{a \otimes b^o}(\eta)] \leq [\eta]$.

If $\xi = a_1 \otimes b_1 + \cdots + a_n \otimes b_n$, we get from Lemma 2.1, that

$$[\lambda_\xi(\eta)] \leq n \cdot [\eta], \quad \forall \eta \in L.$$

We conclude that λ_ξ is Lipschitz for all $\xi \in M \otimes M^o$. Hence, there is an extension $\lambda_\xi : CS(L) \to CS(L)$ which preserves $ZS(L)$. Hence, there exists a bi-linear map $m' : (M \otimes M^o) \times c(L) \to c(L)$ which defines a module structure that is compatible with the module structure on L.

It is clear that $[m'(\xi, \eta)] \leq [\xi]$. Indeed, $(p^\perp \otimes q^{\perp o})\xi = 0$ implies $(p^\perp \otimes q^{\perp o})m(\xi, \eta) = 0$. Hence m' has a natural extension

$$m : (M \hat\otimes M^o) \times c(L) \to c(L).$$

Obviously, if $L = M \otimes M^o$, then m defines a multiplication that extends the multiplication on $M \otimes M^o$, i.e., the natural inclusion $M \otimes M^o \hookrightarrow M \hat\otimes M^o$ is a ring-homomorphism. This shows (1) and (2). Assertion (3) is obvious. □

2.2. Completion is exact

Lemma 2.6. *The functor of completion is exact.*

Proof. Let

$$0 \to J \to L \xrightarrow{\pi} K \to 0$$

be an exact sequence of M-bi-modules. We have to show that the induced sequence

$$0 \to c(J) \to c(L) \to c(K) \to 0$$

is exact.

First, we consider the exactness at $c(K)$. Let $(\xi_n)_{n \in \mathbb{N}}$ be a Cauchy sequence in K. Without loss of generality, we can assume that $[\xi_n - \xi_{n+1}] \leq 2^{-n}$. Lemma 2.3 implies that we can lift $(\xi_n)_{n \in \mathbb{N}}$ to a sequence $(\xi'_n)_{n \in \mathbb{N}} \subset L$ with $[\xi'_n - \xi'_{n+1}] \leq 2^{1-n}$, hence a Cauchy sequence. This shows surjectivity of $c(\pi)$.

We consider now the exactness at $c(L)$. Obviously, $\mathrm{im}(c(J)) \subset \ker(c(\pi))$. Let $(\xi_n)_{n \in \mathbb{N}}$ be a Cauchy sequence in L which maps to zero in $c(K)$. This says that $(\pi(\xi_n))_{n \in \mathbb{N}}$ tends to zero. Again, by Lemma 2.3, we can lift $(\pi(\xi_n))_{n \in \mathbb{N}}$ to a zero-sequence $(\xi'_n)_{n \in \mathbb{N}} \subset L$. Now, $(\xi_n - \xi'_n)_{n \in \mathbb{N}}$ defines a Cauchy sequence in J, that is equivalent to the sequence $(\xi_n)_{n \in \mathbb{N}}$ in the completion of L. Hence $\ker(c(\pi)) \subset \mathrm{im}(c(J))$ and the argument is finished.

The exactness at $c(J)$ is obvious since $J \subset L$ is a contraction in the rank metric by Lemma 2.3. This finishes the proof. $\qquad\square$

Theorem 2.7. *Let M be a finite tracial von Neumann algebra. Consider the category* Bimod^M *of M-bi-modules and the* full *sub-category* Bimod_c^M *of complete modules.*

1. *The completion functor $c\colon \mathrm{Bimod}^M \to \mathrm{Bimod}_c^M$ is left-adjoint to the forgetful functor from Bimod_c^M to Bimod^M, i.e., whenever K is complete:*
$$\hom_{\mathrm{Bimod}^M}(c(L), K) = \hom_{\mathrm{Bimod}^M}(L, K).$$

2. *The category Bimod_c^M is abelian and has enough projective objects.*
3. *The completion functor $c\colon \mathrm{Bimod}^M \to \mathrm{Bimod}_c^M$ preserves projective objects.*
4. *The kernel of the comparison map $L \to c(L)$ is $\{\xi \in L\colon [\xi] = 0\}$.*

Proof. (1) First of all, note that there is no need to distinguish morphisms between complete modules in Bimod^M from morphisms in Bimod_c^M, since Bimod_c^M is defined to be a full sub-category of Bimod^M. If K is complete, the natural map $K \to c(K)$ is an isomorphism, so that applying the functor c defines a natural map
$$\hom_{\mathrm{Bimod}^M}(L, K) \to \hom_{\mathrm{Bimod}^M}(c(L), K).$$
A map in the inverse direction is provided by pre-composition with the map $L \to c(L)$. Assertion (1) follows easily by Lemma 2.3 since $\mathrm{im}(L)$ is dense in $c(L)$ and $\{\xi \in K, [\xi] = 0\} = \{0\}$.

(2) It follows from Lemma 2.6, that Bimod_c^M is abelian. Indeed, by exactness, kernels and co-kernels can be formed in Bimod^M and hence all properties of those remain to be true in Bimod_c^M. Let $L = \oplus_\alpha M \otimes M^o$ be a free $M \otimes M^o$-module. By (1), $c(L)$ is a projective object in Bimod_c^M. If K is complete and $\oplus_\alpha M \otimes M^o \xrightarrow{\pi} K$ is, using Lemma 2.6, $c(\pi)$ is also a surjection onto K. I.e., there are enough projective objects. (3) follows from (1). (4) is obvious. $\qquad\square$

2.3. Completion is dimension-preserving

It is major importance to understand the behaviour of the functor $M \overline{\otimes} M^o \underset{M \otimes M^o}{\otimes} ?$, from the category Bimod^M to $M \overline{\otimes} M^o$-modules. Here, $M \overline{\otimes} M^o$ denote the spatial tensor product of von Neumann algebras. We equip $M \overline{\otimes} M^o$ with the trace $\tau \otimes \tau$. The next Lemma relates rank completion in Bimod^M to dimension isomorphisms in the category of $M \overline{\otimes} M^o$-modules. It will be useful to prove the main result of this section, which is Theorem 2.9.

Lemma 2.8. *Let L be an M bi-module. The natural map*
$$M \overline{\otimes} M^o \otimes_{M \otimes M^o} L \to M \overline{\otimes} M^o \otimes_{M \otimes M^o} c(L)$$
is a dimension isomorphism.

Proof. By Lemma 2.6, it suffices to show that
$$\ker\left(M \overline{\otimes} M^o \otimes_{M \otimes M^o} L \to M \overline{\otimes} M^o \otimes_{M \otimes M^o} c(L)\right)$$

is zero-dimensional for all bi-modules L. Indeed $\operatorname{coker}(L \to c(L))$ has vanishing completion, and knowing the assertion for $\operatorname{coker}(L \to c(L))$ in place of L implies that

$$\operatorname{coker}(M \overline{\otimes} M^o \otimes_{M \otimes M^o} L \to M \overline{\otimes} M^o \otimes_{M \otimes M^o} c(L))$$

has dimension zero.

We want to apply the local criterion of Theorem 1.2. Let $\theta = \sum_{i=1}^n \eta_i \otimes \xi_i \in M \overline{\otimes} M^o \otimes_{M \otimes M^o} L$ be in the kernel. This is to say that there exists some $l \in \mathbb{N}$, $\zeta_i \in M \overline{\otimes} M^o$ and zero-sequences $(\alpha_{i,k})_{k \in \mathbb{N}}$, for $1 \le i \le l$, such that

$$\sum_{i=1}^n \eta_i \otimes \xi_i = \sum_{i=1}^l \zeta_i \otimes \alpha_{i,k}, \quad \forall k \in \mathbb{N}.$$

Indeed, the map factorizes through the split-injection

$$M \overline{\otimes} M^o \otimes_{M \otimes M^o} L \to M \overline{\otimes} M^o \otimes_{M \otimes M^o} CS(L)$$

and hence $\operatorname{im}(M \overline{\otimes} M^o \otimes_{M \otimes M^o} ZS(L)) \subset M \overline{\otimes} M^o \otimes_{M \otimes M^o} CS(L)$ contains the image of

$$\ker(M \overline{\otimes} M^o \otimes_{M \otimes M^o} L \to M \overline{\otimes} M^o \otimes_{M \otimes M^o} c(L)).$$

Since $\alpha_{i,k} \to 0$, for all $1 \le i \le l$, for every $\varepsilon > 0$ there exists k big enough and projections $p_i, q_i \in \operatorname{Proj}(M)$, such that $(p_i^\perp \otimes q_i^{\perp o})\alpha_{i,k} = 0$ and $\tau(p_i) + \tau(q_i) \le \varepsilon/l$.

Let f_i be projections in $M \overline{\otimes} M^o$, such that $f_i \zeta_i (1 \otimes 1 - p_i^\perp \otimes q_i^{\perp o}) = \zeta_i (1 \otimes 1 - p_i^\perp \otimes q_i^{\perp o})$. One can choose f_i to satisfy $\tau(f_i) \le \tau(p_i) + \tau(q_i) \le \varepsilon/l$, for all $1 \le i \le l$. We compute as follows:

$$\left(\bigwedge_{i=1}^l f_i^\perp\right)\theta = \left(\bigwedge_{i=1}^l f_i^\perp\right)\sum_{i=1}^l \zeta_i \otimes \alpha_{i,k} = \left(\bigwedge_{i=1}^l f_i^\perp\right)\sum_{i=1}^l \zeta_i \otimes (1 \otimes 1 - p_i^\perp \otimes q_i^{\perp o})\alpha_{i,k}$$

$$= \left(\bigwedge_{i=1}^l f_i^\perp\right)\sum_{i=1}^l \zeta_i(1 \otimes 1 - p_i^\perp \otimes q_i^{\perp o}) \otimes \alpha_{i,k}$$

$$= \left(\bigwedge_{i=1}^l f_i^\perp\right)\sum_{i=1}^l f_i \zeta_i(1 \otimes 1 - p_i^\perp \otimes q_i^{\perp o}) \otimes \alpha_{i,k}$$

$$= 0.$$

Thus $\tau(\vee_{i=1}^l f_i) \le \varepsilon$ and $(\vee_{i=1}^l f_i)\theta = \theta$. Hence

$$\ker(M \overline{\otimes} M^o \otimes_{M \otimes M^o} L \to M \overline{\otimes} M^o \otimes_{M \otimes M^o} c(L))$$

is zero-dimensional by Theorem 1.2. $\qquad\square$

Theorem 2.9. *Let $\phi\colon L \to L'$ be a morphism of M-bi-modules. If $c(\phi)$ is an isomorphism, then*

$$\operatorname{Tor}_i^{M \otimes M^o}(M \overline{\otimes} M^o, L) \xrightarrow{\phi_*} \operatorname{Tor}_i^{M \otimes M^o}(M \overline{\otimes} M^o, L')$$

is a dimension isomorphism.

Proof. The exactness of $c \colon \mathrm{Bimod}^M \to \mathrm{Bimod}_c^M$ implies the following natural identification among left-derived functors:

$$L_i(M \,\overline{\otimes}\, M^o \otimes_{M \otimes M^o} ?) \circ c = L_i(M \,\overline{\otimes}\, M^o \otimes_{M \otimes M^o} c(?)).$$

Indeed, this follows from the fact that $c \colon \mathrm{Bimod}^M \to \mathrm{Bimod}_c^M$ maps free modules in Bimod^M to projective objects in Bimod_c^M. This implies the existence of a Grothendieck spectral sequence (see [Wei94, pp. 150]) that yields the desired result.

Lemma 2.8 together with Lemma 1.1 implies the existence of a natural map

$$\mathrm{Tor}_i^{M \otimes M^o}(M \,\overline{\otimes}\, M^o, ?) \to L_i(M \,\overline{\otimes}\, M^o \otimes_{M \otimes M^o} c(?))$$

which is a dimension isomorphism. Combining the preceding two observations, we conclude that

$$\dim_{M \overline{\otimes} M^o} \mathrm{Tor}_i^{M \otimes M^o}(M \,\overline{\otimes}\, M^o, L) = 0, \quad \forall i \geq 0,$$

whenever $c(L) = 0$. This implies the claim, since $c(\ker(\phi)) = 0$ and $c(\mathrm{coker}(\phi)) = 0$ by Lemma 2.6. $\qquad\square$

3. L^2-Betti numbers for tracial algebras

3.1. Preliminaries

In [CS05], A. Connes and D. Shlyakhtenko introduced a notion of L^2-homology and L^2-Betti numbers for tracial algebras, compare also earlier work of W.L. Paschke in [Pas97]. The definition works well in a situation where the tracial algebra (A, τ) is contained in a finite von Neumann algebra M, to which the trace τ extends. More precisely, using the dimension function of W. Lück, see [Lüc02], they set:

$$\beta_k^{(2)}(A, \tau) = \dim_{M \overline{\otimes} M^o} \mathrm{Tor}_k^{A \otimes A^o}(M \,\overline{\otimes}\, M^o, A)$$

Several results concerning these L^2-Betti numbers where obtained in [CS05] and [Tho06a, Tho06b]. In particular, it was shown in [CS05] that

$$\beta_k^{(2)}(\mathbb{C}\Gamma, \tau) = \beta_k^{(2)}(\Gamma),$$

where the right side denotes the L^2-Betti number of a group in the sense of Atiyah, see [Ati76] and Cheeger-Gromov, see [CG86].

It is conjectured in [CS05] that $\beta_k^{(2)}(M, \tau)$ is an interesting invariant for the von Neumann algebra. Several related quantities where studied in [CS05] as well. In particular,

$$\Delta_k(A, \tau) = \dim_{M \overline{\otimes} M^o} \mathrm{Tor}_k^{M \otimes M^o}(M \,\overline{\otimes}\, M^o, M \otimes_A M)$$

was studied for $k = 1$.

3.2. Pedersen's Theorem

Lemma 3.1. *Let (M, τ) be a finite tracial von Neumann algebra and let $A_1, A_2 \subset M$ be $*$-sub-algebras of M. If A_1 and A_2 have the same closure with respect to the rank metric, then*

$$\Delta_k(A_1, \tau) = \Delta_k(A_2, \tau), \quad \forall k \geq 0.$$

Proof. Without loss of generality, $A_1 \subset A_2$. We show that $\pi \colon M \otimes_{A_1} M \to M \otimes_{A_2} M$ induces an isomorphism after completion with respect to the rank metric. The claim follows then from Theorem 2.9.

By Lemma 2.6, it suffices to show that the kernel of π has vanishing completion. An element ξ in the kernel can be written as

$$\xi = \sum_{i=1}^{n} c_i a_i \otimes d_i - c_i \otimes a_i d_i,$$

for some $c_i, d_i \in M$ and $a_i \in A_2$. Since A_1 is dense in A_2 there exists $a_i' \in A_1$ with $[a_i - a_i'] \leq \varepsilon/n$, for all $1 \leq i \leq n$.

The following equality holds in $M \otimes_{A_1} M$:

$$\xi = \sum_{i=1}^{n} c_i a_i \otimes d_i - c_i \otimes a_i d_i = \sum_{i=1}^{n} c_i(a_i - a_i') \otimes d_i - c_i \otimes (a_i - a_i')d_i,$$

Arguing as before, we see that there exists projections p, q of trace less than ε, such that $p^\perp \xi q^\perp = 0$. This implies that $[\xi] \leq \varepsilon$. Since ε was arbitrary, we get that $[\xi] = 0$ for all $\xi \in \ker(\pi \colon M \otimes_{A_1} M \to M \otimes_{A_2} M)$ and thus $c(\ker(\pi)) = 0$. $\quad\square$

The following result by G. Pedersen, see [Ped79, Thm. 2.7.3], is a deep result in the theory of operator algebras, which required a detailed analysis of the precise position of a weakly dense C^*-algebra inside a von Neumann algebra. It is a generalization of a more classical theorem of Lusin in the commutative case.

Theorem 3.2 (Pedersen). *Let (M, τ) be a finite tracial von Neumann algebra and let $A \subset M$ be a weakly dense sub-C^*-algebra. The algebra A is dense in M with respect to the rank metric.*

Corollary 3.3. *Let (M, τ) be a finite tracial von Neumann algebra and let $A \subset M$ be a weakly dense sub-C^*-algebra.*

$$\Delta_k(A, \tau) = \Delta_k(M, \tau), \quad \forall k \geq 0.$$

Remark 3.4. In [Tho06b], it was shown that $\beta_1^{(2)}(A, \tau) = \beta_1^{(2)}(M, \tau)$, whenever A is a weakly dense sub-C^*-algebra. In view of the factorization

$$\mathrm{Tor}_1^{A \otimes A^o}(M \overline{\otimes} M^o, A) \to \mathrm{Tor}_1^{M \otimes M^o}(M \overline{\otimes} M^o, M \otimes_A M) \to \mathrm{Tor}_1^{M \otimes M^o}(M \overline{\otimes} M^o, M),$$

the proof also shows that $\Delta_1(A, \tau) = \Delta_1(M, \tau)$ holds. Hence we can view Corollary 3.3 as a generalization of this result from [Tho06b].

4. Equivalence relations and Gaboriau's Theorem

4.1. Equivalence relations and completion

Most of the proofs in the section are parallel to proofs in Section 2 and 3 and hence we will give less detail and point to the relevant parts of Section 2 and 3. Let X be a standard Borel space and let μ be a probability measure on X. Given a discrete measurable equivalence relation (see [FM77a, FM77b] for the necessary definitions)

$$R \subset X \times X,$$

we can form a *relation ring* $R(X)$ as follows:

$$R(X) = \left\{ \sum_{i=1}^{n} f_i \phi_i : f_i \in L^\infty(X), \phi_i \text{ local isomorphism from } R \right\} \subset L^\infty(R).$$

Here, $L^\infty(R)$ denotes the generated von Neumann algebra, see [FM77a]. Note that $R(X)$ is a $L^\infty(X)$-bi-module with respect to the diagonal left action. (All $L^\infty(X)$-modules are bi-modules in this way.) The following observation is the key to our results.

Proposition 4.1. *Let Γ be a discrete group and let $\rho\colon \Gamma \times X \to X$ be a measure preserving free action of Γ on X. We denote by R_ρ the induced measurable equivalence relation on X. The natural inclusion $\iota\colon L^\infty(X) \rtimes_{\mathrm{alg}} \Gamma \to R_\rho(X)$ induces an isomorphisms after completion.*

Proof. According to the foundational work in [FM77a, FM77b], each local isomorphism ϕ which is implemented by the equivalence relation R_ρ can be decomposed as a infinite sum of local isomorphism ϕ_i, each of which is a cut-down of an isomorphism which is implemented by the action of a group element. The sizes of the supports of the cut-down local isomorphisms ϕ_i in this decomposition sum to the size of the support of ϕ. It clearly implies, that ϕ can be approached by elements of $L^\infty(X) \rtimes \Gamma$ in rank metric. This finishes the proof. □

Definition 4.2. *Two group Γ_1 and Γ_2 are called* orbit equivalent, *if there exists a probability space X and free, measure preserving actions of Γ_1 and Γ_2 on X that induce the same equivalence relation.*

For an excellent survey on the properties of orbit equivalence and related notions, see [Gab02b].

Lemma 4.3. *Let $R \subset X \times X$. We denote the completion of $R(X)$ by $\widehat{R}(X)$. $\widehat{R}(X)$ is a unital $L^\infty(X)$-algebra that contains $R(X)$ as a $L^\infty(X)$-sub-algebra.*

Proof. First of all, $\widehat{R}(X)$ is a $R(X)$-module. Indeed, left multiplication by $\sum_{i=1}^{n} f_i \phi_i$ is easily seen to be Lipschitz with constant n. In particular, there exists a map $m'\colon R(X) \times \widehat{R}(X) \to \widehat{R}(X)$.

As before, we easily see that m' has an extension to an associative and separately continuous multiplication:

$$m\colon \widehat{R}(X) \times \widehat{R}(X) \to \widehat{R}(X).$$ □

Lemma 4.4. *Let L be a $L^\infty(X) \rtimes_{\mathrm{alg}} \Gamma$-module. The completion of L, with respect to the diagonal $L^\infty(X)$-bi-module structure is naturally a $\widehat{R}(X)$-module and in particular a $R(X)$-module.*

Proof. By Proposition 4.1 $L^\infty(X) \rtimes_{\mathrm{alg}} \Gamma$ is dense in $R(X)$ and hence in $\widehat{R}(X)$. Again, since $x = \sum_{i=1}^n f_i \phi_i$ acts with Lipschitz constant n on L, the action extends to $c(L)$. Let x_n be a Cauchy sequence in $L^\infty(X) \rtimes \Gamma$ and $\xi \in c(L)$. The rank of $(x_n - x_m)\xi$ is less that the rank of $x_n - x_m$ and if $x_n \to 0$, then $x_n \xi \to 0$. This finishes the proof. □

4.2. Proof of Gaboriau's Theorem

The proof of Gaboriau's Theorem which is presented in this section uses the technology of rank completion. It is very much inspired by a proof of Gaboriau's Theorem given by R. Sauer in [Sau03]. In his proof, the local criterion was a crucial ingredient to make the arguments work. We hope that the concept of rank completion will provide a good understanding of why the Theorem is true.

Lemma 4.5. *Let Γ be a discrete group and let $\rho\colon \Gamma \times X \to X$ be a measurable and measure preserving action on a probability space X.*

1. *$\mathbb{C}\Gamma \subset L^\infty(X) \rtimes_{\mathrm{alg}} \Gamma$ is flat.*
2. *$L\Gamma \subset L^\infty(X) \rtimes \Gamma$ is flat and dimension preserving.*

Proof. The first assertion is obvious. Indeed, $L^\infty(X) \rtimes_{\mathrm{alg}} \Gamma$ is a free $\mathbb{C}\Gamma$-module. The second assertion follows from the fact that $L\Gamma$ is semi-hereditary, see [Sau05]. □

Theorem 4.6. *Let Γ_1 and Γ_2 are orbit equivalent groups, then*

$$\beta_k^{(2)}(\Gamma_1) = \beta_k^{(2)}(\Gamma_2), \quad \forall k \geq 0.$$

Proof. It suffices to write $\beta_k^{(2)}(\Gamma_1)$ entirely in terms of the equivalence relation it generates. Using Lemma 4.5, we rewrite:

$$(L^\infty(X) \rtimes \Gamma) \otimes_{L\Gamma} \mathrm{Tor}_k^{\mathbb{C}\Gamma}(L\Gamma, \mathbb{C}) = \mathrm{Tor}_k^{\mathbb{C}\Gamma}(L^\infty(X) \rtimes \Gamma, \mathbb{C})$$
$$= \mathrm{Tor}_k^{L^\infty(X) \rtimes_{\mathrm{alg}} \Gamma}(L^\infty(X) \rtimes \Gamma, L^\infty(X)).$$

Here, the second equality follows since

$$(L^\infty(X) \rtimes_{\mathrm{alg}} \Gamma) \otimes_{\mathbb{C}\Gamma} \mathbb{C} = L^\infty(X)$$

as $L^\infty(X) \rtimes_{\mathrm{alg}} \Gamma$-module. By Lemma 4.5 and Equation 1, we conclude that

$$\beta_k^{(2)}(\Gamma) = \dim_{L^\infty(X) \rtimes \Gamma} \mathrm{Tor}_k^{L^\infty(X) \rtimes_{\mathrm{alg}} \Gamma}(L^\infty(X) \rtimes \Gamma, L^\infty(X)). \tag{2}$$

There exists an exact functor which completes the category of $L^\infty(X) \rtimes_{\mathrm{alg}} \Gamma$-modules with respect to the diagonal left $L^\infty(X)$-bi-module structure. We have shown in Lemma 4.4 that the resulting full subcategory of those $L^\infty(X) \rtimes_{\mathrm{alg}} \Gamma$-modules which are complete, is naturally a category of $R(X)$-modules. Let us denote the functor of completion by

$$c\colon \mathrm{Mod}^{L^\infty(X) \rtimes_{\mathrm{alg}} \Gamma} \to \mathrm{Mod}_c^{R(X)}.$$

Proposition 4.7. *Let L be a $L^\infty(X) \rtimes_{\mathrm{alg}} \Gamma$-module. The completion map induces an dimension isomorphism:*

$$(L^\infty(X) \rtimes \Gamma) \otimes_{L^\infty(X) \rtimes_{\mathrm{alg}} \Gamma} L \to (L^\infty(X) \rtimes \Gamma) \otimes_{R(X)} c(L).$$

Proof. The map can be factorized as

$$(L^\infty(X) \rtimes \Gamma) \otimes_{L^\infty(X) \rtimes_{\mathrm{alg}} \Gamma} L \to (L^\infty(X) \rtimes \Gamma) \otimes_{L^\infty(X) \rtimes_{\mathrm{alg}} \Gamma} c(L)$$
$$\to (L^\infty(X) \rtimes \Gamma) \otimes_{R(X)} c(L).$$

We show that each of the maps is a dimension isomorphism. Let us start with the first one. Again, by exactness of c, it suffices to show that

$$\ker\left(L^\infty(X) \rtimes \Gamma) \otimes_{L^\infty(X) \rtimes_{\mathrm{alg}} \Gamma} L \to L^\infty(X) \rtimes \Gamma) \otimes_{L^\infty(X) \rtimes_{\mathrm{alg}} \Gamma} c(L)\right)$$

is zero-dimensional. As in the proof of Lemma 2.8, an element θ in kernel is of the form:

$$\theta = \sum_{i=1}^{l} \zeta_i \otimes \alpha_{i,k}, \quad \forall k \in \mathbb{N},$$

for some zero-sequences $(\alpha_{i,k})_{k \in \mathbb{N}} \subset L$. The proof proceeds as the proof of Lemma 2.8.

The second map can be seen to be a dimension isomorphism as follows. Clearly, the map is surjective and it remains to show that the kernel is zero-dimensional. An element of the kernel is of the form:

$$\theta = \sum_{i=1}^{n} \xi_i \eta_i \otimes \zeta_i - \xi_i \otimes \eta_i \zeta_i,$$

for some $\xi_i \in L^\infty(X) \rtimes \Gamma, \eta_i \in R(X)$ and $\zeta_i \in L$. Approximating η_i by elements in $L^\infty(X) \rtimes_{\mathrm{alg}} \Gamma$ we can assume (as in the proof of Lemma 3.1) that $[\eta_i] \le \varepsilon/(2n)$. The first summands are smaller than $\varepsilon/(2n)$, since support and range projection have the same trace. The second summand are also smaller than $\varepsilon/(2n)$ by the same argument and since projections in $L^\infty(X)$ can be moved through the tensor product. Hence $[\theta] \le \varepsilon$. Since θ and ε were arbitrary, we conclude by Theorem 1.2 that

$$\ker\left((L^\infty(X) \rtimes \Gamma) \otimes_{L^\infty(X) \rtimes_{\mathrm{alg}} \Gamma} c(L) \to (L^\infty(X) \rtimes \Gamma) \otimes_{R(X)} c(L)\right)$$

is zero-dimensional. This finishes the proof. $\square$

To conclude the proof of Theorem 4.6, we note that by Lemma 1.1 we get an induced map

$$\mathrm{Tor}_i^{L^\infty(X) \rtimes_{\mathrm{alg}} \Gamma}(L^\infty(X) \rtimes \Gamma, ?) \to L_i((L^\infty(X) \rtimes \Gamma) \otimes_{R(X)} c(?))$$

which is a dimension isomorphism. The right-hand side applied to $L^\infty(X)$ depends only on the generated equivalence relation. Indeed, as in the proof of Theorem 2.9, a Grothendieck spectral sequence shows

$$L_i((L^\infty(X) \rtimes \Gamma) \otimes_{R(X)} c(?)) = L_i((L^\infty(X) \rtimes \Gamma) \otimes_{R(X)} ?) \circ c.$$

Here, we use implicitly that the category of complete $R(X)$-modules is abelian with enough projective objects. The proof of this fact can be taken verbatim from the proof of Theorem 2.7 and the adjointness relations:

$$\hom_{L^\infty(X)\rtimes\Gamma}(L, K) = \hom_{L^\infty(X)\rtimes\Gamma}(c(L), K) = \hom_{R(X)}(c(L), K).$$

The projective objects are completions of free $L^\infty(X) \rtimes \Gamma$-modules. $\square$

Remark 4.8. There is a second major result of Gaboriau's on proportionality of L^2-Betti numbers for weakly orbit equivalent groups, see [Gab02b]. Sauer has shown in [Sau03], how homological methods and properties of the dimension function allow to deduce this result. The same arguments apply to our setting.

References

[Ati76]   M.F. Atiyah. Elliptic operators, discrete groups and von Neumann algebras. In *Colloque "Analyse et Topologie" en l'Honneur de Henri Cartan (Orsay, 1974)*, pages 43–72. Astérisque, No. 32–33. Soc. Math. France, Paris, 1976.

[CG86]   Jeff Cheeger and Mikhael Gromov. L_2-cohomology and group cohomology. *Topology*, 25(2):189–215, 1986.

[CS05]   Alain Connes and Dimitri Shlyakhtenko. L^2-homology for von Neumann algebras. *J. Reine Angew. Math.*, 586:125–168, 2005.

[FM77a] Jacob Feldman and Calvin C. Moore. Ergodic equivalence relations, cohomology, and von Neumann algebras. I. *Trans. Amer. Math. Soc.*, 234(2):289–324, 1977.

[FM77b] Jacob Feldman and Calvin C. Moore. Ergodic equivalence relations, cohomology, and von Neumann algebras. II. *Trans. Amer. Math. Soc.*, 234(2):325–359, 1977.

[Gab02a] Damien Gaboriau. Invariants l^2 de relations d'équivalence et de groupes. *Publ. Math. Inst. Hautes Études Sci.*, (95):93–150, 2002.

[Gab02b] Damien Gaboriau. On orbit equivalence of measure preserving actions. In *Rigidity in dynamics and geometry (Cambridge, 2000)*, pages 167–186. Springer, Berlin, 2002.

[Lüc02] Wolfgang Lück. *L^2-invariants: theory and applications to geometry and K-theory*, volume 44 of *Ergebnisse der Mathematik und ihrer Grenzgebiete. 3. Folge. A Series of Modern Surveys in Mathematics [Results in Mathematics and Related Areas. 3rd Series. A Series of Modern Surveys in Mathematics]*. Springer-Verlag, Berlin, 2002.

[Pas97] William L. Paschke. L^2-homology over traced *-algebras. *Trans. Amer. Math. Soc.*, 349(6):2229–2251, 1997.

[Ped79] Gert K. Pedersen. *C^*-algebras and their automorphism groups*, volume 14 of *London Mathematical Society Monographs*. Academic Press Inc. [Harcourt Brace Jovanovich Publishers], London, 1979.

[Sau03] Roman Sauer. Power series over the group ring of a free group and applications to Novikov-Shubin invariants. In *High-dimensional manifold topology*, pages 449–468. World Sci. Publishing, River Edge, NJ, 2003.

[Sau05]   Roman Sauer. L^2-Betti numbers of discrete measured groupoids. *Internat. J. Algebra Comput.*, 15(5-6):1169–1188, 2005.

[Tho06a]   Andreas Thom. L^2-Betti numbers for sub-factors. *to appear in the Journal of Operator algebras*, arXiv:math.OA/0601408, 2006.

[Tho06b]   Andreas Thom. L^2-cohomology for von Neumann algebras. *to appear in GAFA*, arXiv:math.OA/0601447, 2006.

[Voi05]   Dan Voiculescu. Free probability and the von Neumann algebras of free groups. *Reports on Mathematical Physics*, 55(1):127–133, 2005.

[Wei94]   Charles A. Weibel. *An introduction to homological algebra*, volume 38 of *Cambridge Studies in Advanced Mathematics*. Cambridge University Press, Cambridge, 1994.

Andreas Thom
Mathematisches Institut der Universität Göttingen
Bunsenstr. 3-5
D-37073 Göttingen, Germany
e-mail: thom@uni-math.gwdg.de
URL: http://www.uni-math.gwdg.de/thom

C^*-algebras and Elliptic Theory II

Trends in Mathematics, 281–295

© 2008 Birkhäuser Verlag Basel/Switzerland

Group Bundle Duality, Invariants for Certain C^*-algebras, and Twisted Equivariant K-theory

Ezio Vasselli

Abstract. A duality is discussed for Lie group bundles vs. certain tensor C^*-categories with non-simple identity, in the setting of Nistor-Troitsky gauge-equivariant K-theory. As an application, we study C^*-algebra bundles with fibre a fixed-point algebra of the Cuntz algebra: a classification is given, and a cohomological invariant is assigned, representing the obstruction to perform an embedding into a continuous bundle of Cuntz algebras. Finally, we introduce the notion of twisted equivariant K-theory.

Mathematics Subject Classification (2000). Primary 19L47; Secondary 46L05; 46M15.

1. Introduction

An important branch of abstract harmonic analysis is group duality, whose task is to establish (possibly one-to-one) correspondences between groups and suitable tensor categories. For the case of a compact group G, the basic idea is to consider the tensor category of finite-dimensional, unitary representations of G; on the converse, given a tensor category $\mathcal{T}$ (carrying some additional structure), one has to construct an embedding functor $F : \mathcal{T} \hookrightarrow \mathbf{hilb}$ into the category of finite-dimensional Hilbert spaces, and define G as the group of natural transformations of F.

The existence of F is a crucial step. In fact, in general $\mathcal{T}$ may be not presented as a subcategory of the one of Hilbert spaces. For example, this happens in the case of algebraic quantum field theory, where the involved category is the one of superselection sectors of a C^*-algebra of localized quantum observables. The duality for compact groups proved by Doplicher and Roberts in [7] was motivated by such physical applications, and is the starting point for the work presented in the present paper.

The main feature of a tensor category $\mathcal{T}$ is given by the fact that objects ρ, σ of $\mathcal{T}$ can be multiplied by means of the tensor product $\rho, \sigma \mapsto \rho\sigma$: the existence of a unit object ι is postulated, in such a way that $\rho\iota = \iota\rho = \rho$ for every object ρ. One of the properties assumed for the duality proved in [7] is that ι is simple, i.e., that the algebra of arrows (ι, ι) reduces to the complex numbers; in general, the tensor categories considered in the above-cited reference are such that (ι, ι) is a commutative C^*-algebra. This is an important point w.r.t. duality, because in such a case we have to look for an embedding functor into the category $\mathbf{mod}(\iota, \iota)$ of (ι, ι)-bimodules, instead of the one of Hilbert spaces.

In the present work, we study a class of tensor categories with non-simple unit object: we give a classification, and discuss the existence of an embedding functor into $\mathbf{mod}(\iota, \iota)$. In particular, we show that such embedding functors may not exist, in contrast to the case $(\iota, \iota) \simeq \mathbb{C}$.

The Serre-Swan equivalence implies that we may equivalently consider the category of vector bundles over the spectrum of (ι, ι), instead of $\mathbf{mod}(\iota, \iota)$. Since structural properties of the categories that we consider are encoded by geometrical invariants, we adopt this last point of view.

Our motivation arises from possible physical applications in low-dimensional quantum field theory. Anyway, at a purely mathematical level we note that actions in the setting of vector bundles have been object of interesting research ([12]).

The present paper is intended to be more self-contained as possible, and most of the material exposed here is a friendly exposition of [17]. The only original part is Section 7, where we introduce a twisted equivariant topological K-functor.

Our work is organized as follows. In Section 2, we associate a K-theory group to a given C^*-category; our construction provides usual C^*-algebra K-theory in the case in which the given C^*-category is the one of finitely generated Hilbert modules over a fixed C^*-algebra. In Section 3, we expose a simple cohomological construction for compact Lie groups, which will be applied in the following sections. In Section 4, we expose a version of the Serre-Swan equivalence for bundles of C^*-algebras. In Section 5, we give a procedure to construct tensor categories with non-simple units, in the way as follows. Let $d \in \mathbb{N}$, $\mathbb{U}(d)$ denote the unitary group, and $G \subseteq \mathbb{U}(d)$ a compact group. We denote by $NG \subseteq \mathbb{U}(d)$ the normalizer of G in $\mathbb{U}(d)$, and by $QG := NG/G$ the quotient group. If X is a compact Hausdorff space, we associate to every principal QG-bundle $\mathcal{Q}$ a tensor category $T\mathcal{Q}$ with unit ι, such that $(\iota, \iota) \simeq C(X)$; roughly speaking, $T\mathcal{Q}$ is a bundles of tensor categories over X, with fibre the category of tensor powers of the defining representation $G \hookrightarrow \mathbb{U}(d)$. Let G_{ab} denote the (Abelian) quotient of G by the adjoint action of NG. We define a cohomology class $\delta(\mathcal{Q}) \in H^2(X, G_{ab})$, encoding the obstruction to construct an embedding functor $F : T\mathcal{Q} \hookrightarrow \mathbf{mod}(\iota, \iota)$. If F does exist, then there is a G-bundle $\mathcal{G} \to X$ such that $T\mathcal{Q}$ is the category of tensor powers of a $\mathcal{G}$-equivariant vector bundle in the sense of [12]. Existence of F is equivalent to the condition that there is a principal NG-bundle $\mathcal{N}$ such that $\mathcal{Q} = \mathcal{N} \mod G$. In Section 6, we provide a translation of the results of Section 5 in terms of C^*-algebra bundles. We consider the Cuntz algebra $\mathcal{O}_d$, and the fixed point algebra $\mathcal{O}_G$ w.r.t.

a natural G-action; the invariant $\delta(\mathcal{Q})$ is now interpreted as the cohomological obstruction to embed a given $\mathcal{O}_G$-bundle into some $\mathcal{O}_d$-bundle. In Section 7, we associate to $\mathcal{Q}$ a K-theory group $K^0_\mathcal{Q}(X)$. If $\mathcal{Q} = \mathcal{N} \mod G$ for some principal NG-bundle $\mathcal{N}$, then there exists a G-bundle $\mathcal{G} \to X$ such that $K^0_\mathcal{Q}(X)$ can be embedded into the gauge equivariant K-theory $K^0_\mathcal{G}(X)$ defined in the sense of [12].

Notation and background references. Let X be a compact Hausdorff space. We denote by $C(X)$ the C^*-algebra of bounded, continuous, $\mathbb{C}$-valued functions. If $x \in X$, we denote by $C_x(X)$ the closed ideal of functions vanishing on x. If $\{X_i\}$ is an open cover of X, we define $X_{ij} := X_i \cap X_j$, $X_{ijk} := X_{ij} \cap X_k$. In the present work, we adopt the convention that the set $\mathbb{N}$ of natural numbers includes 0. If A is any set, then id_A denotes the identity map on A. For topological K-theory, we refer to [10], while for (principal) group bundles our reference is [9]. About C^*-algebras and Hilbert modules, we refer to [3].

2. C^*-categories and K-theory

A C^*-category $\mathcal{C}$ is given by a collection of objects, denoted by **obj** $\mathcal{C}$, such that for every pair $\rho, \sigma \in$ **obj** $\mathcal{C}$ there is an associated Banach space (ρ, σ), called the space of arrows. A bilinear composition is defined, $(\sigma, \tau) \times (\rho, \sigma) \to (\rho, \tau)$, $t', t \mapsto t' \circ t$, in such a way that $\|t' \circ t\| \leq \|t'\| \|t\|$, and an involution $* : (\rho, \sigma) \to (\sigma, \rho)$ is assigned, in such a way that the C^*-identity $\|t^* \circ t\| = \|t\|^2$ is satisfied. For every $\rho \in$ **obj** $\mathcal{C}$, we assume the existence of an identity $1_\rho \in (\rho, \rho)$ such that $1_\rho \circ t = t$, $t' \circ 1_\rho = t'$, $t \in (\sigma, \rho)$, $t' \in (\rho, \tau)$. It is customary to postulate the existence of a *zero object* $\mathbf{o} \in$ **obj** $\mathcal{C}$, such that $(\mathbf{o}, \rho) = \{0\}$, $\rho \in$ **obj** $\mathcal{C}$. For basic properties about C^*-categories, we refer the reader to [7, §1].

As an immediate consequence of the above definition, we find that every (ρ, ρ) is a unital C^*-algebra w.r.t. the Banach space structure, composition of arrows and involution; moreover, every (ρ, σ) is a Hilbert (σ, σ)-(ρ, ρ)-bimodule w.r.t. composition of arrows and the (ρ, ρ)-valued scalar product $t, t' \mapsto t^* \circ t'$, $t, t' \in (\rho, \sigma)$. If $\rho, \sigma \in$ **obj** $\mathcal{C}$, we say that ρ is *unitarily equivalent* to σ if there exists $u \in (\rho, \sigma)$ with $u \circ u^* = 1_\sigma$, $u^* \circ u = 1_\rho$. We denote by $\{\rho\} \subseteq$ **obj** $\mathcal{C}$ the set of objects which are unitarily equivalent to ρ.

Let $\mathcal{C}$, $\mathcal{C}'$ be C^*-categories. A *functor* $\phi : \mathcal{C} \to \mathcal{C}'$ is given by a map $F :$ **obj** $\mathcal{C} \to$ **obj** $\mathcal{C}'$ and a family $\{F_{\rho,\sigma} : (\rho, \sigma) \to (F(\rho), F(\sigma))\}$ of bounded linear maps preserving composition and involution. This implies that every $F_{\rho,\rho}$, $\rho \in$ **obj** $\mathcal{C}$, is a C^*-algebra morphism.

We give some examples. A C^*-category with a single object is clearly a unital C^*-algebra. If $\mathcal{A}$ is a unital C^*-algebra, then the category with objects right Hilbert $\mathcal{A}$-modules and arrows adjointable, right $\mathcal{A}$-module operators is a C^*-category, denoted by $\mathbf{mod}(\mathcal{A})$. If X is a compact Hausdorff space, the category $\mathbf{vect}(X)$ with objects vector bundles over X and arrows morphisms of vector bundles is a C^*-category (actually, by the Serre-Swan theorem [10, Thm. I.6.18] we may identify $\mathbf{vect}(X)$ with $\mathbf{mod}(C(X))$).

We now define some further structure. We say that $\mathcal{C}$ has *subobjects* if for every $\rho \in \mathbf{obj}\,\mathcal{C}$ and projection $E = E^2 = E^* \in (\rho, \rho)$, there is $\sigma \in \mathbf{obj}\,\mathcal{C}$ and $S \in (\rho, \sigma)$ such that $1_\sigma = S \circ S^*$, $E = S^* \circ S$. For every C^*-category $\mathcal{C}$, we may construct a 'larger' C^*-category $\mathcal{C}_s$ with subobjects, by defining $\mathbf{obj}\,\mathcal{C}_s := \{E = E^2 = E^* \in (\rho, \rho), \rho \in \mathbf{obj}\,\mathcal{C}\}$, $(E, F) := \{t \in (\rho, \sigma) : t = t \circ E = F \circ t\}$. $\mathcal{C}_s$ is called the *closure for subobjects of $\mathcal{C}$*.

$\mathcal{C}$ has *direct sums* if for every $\rho, \sigma \in \mathbf{obj}\,\mathcal{C}$ there exists $\tau \in \mathbf{obj}\,\mathcal{C}$ with $\psi_\rho \in (\rho, \tau)$, $\psi_\sigma \in (\sigma, \tau)$ such that $\psi_\rho \circ \psi_\rho^* + \psi_\sigma \circ \psi_\sigma^* = 1_\tau$, $\psi_\rho^* \circ \psi_\rho = 1_\rho$, $\psi_\sigma^* \circ \psi_\sigma = 1_\sigma$. The object τ is called *a direct sum* of ρ and σ, and is unique up to unitary equivalence. If a C^*-category $\mathcal{C}$ does not have direct sums, we can construct the *additive completion* $\mathcal{C}_+$, having objects n-ples $\underline{\rho} := (\rho_1, \ldots, \rho_n)$, $n \in \mathbb{N}$, and arrows spaces of matrices $(\underline{\rho}, \underline{\sigma}) := \{(t_{ij}) : t_{ij} \in (\rho_j, \sigma_i)\}$. In the case in which $\mathcal{C}$ has direct sums, we define $\mathcal{C}_+ := \mathcal{C}$. Note that the operations $\mathcal{C} \mapsto \mathcal{C}_s$ and $\mathcal{C} \mapsto \mathcal{C}_+$ do not commute, so that $\mathcal{C}_{+,s}$ is not isomorphic to $\mathcal{C}_{s,+}$: in general, there is an immersion

$$\mathcal{C}_{s,+} \hookrightarrow \mathcal{C}_{+,s} \ ,$$

in fact every formal direct sum $E := (E_1, \ldots, E_n)$, $E_k \in \mathbf{obj}\,\mathcal{C}_s$, $k = 1, \ldots, n$, is a projection in the additive completion $\mathcal{C}_+$.

Let $\mathcal{C}$ be a C^*-category with direct sums. Then, a semigroup $S(\mathcal{C})$ is associated with $\mathcal{C}$, in the following way: for every $\rho, \sigma \in \mathcal{C}$, we consider a direct sum τ and the classes $\{\rho\}$, $\{\sigma\}$; then, we define $\{\rho\} + \{\sigma\} := \{\tau\}$. Since $\{\tau\}$ does not depend on the order of ρ, σ, we find that $(S(\mathcal{C}), +)$ is an Abelian semigroup with identity $\{\mathbf{o}\}$.

Definition 2.1. Let $\mathcal{C}$ be a C^*-category. The K-theory group of $\mathcal{C}$ is defined as the Grothendieck group $K_0(\mathcal{C})$ associated with $S(\mathcal{C}_+)$.

We give a class of examples: if $\mathcal{A}$ is a unital C^*-algebra, then $K_0(\mathbf{mod}(\mathcal{A}))$ coincides with the usual K-theory group $K_0(\mathcal{A})$. In particular, if $\mathcal{A} = C(X)$, then $K_0(\mathbf{mod}(C(X)))$ coincides with the topological K-theory $K^0(X)$.

3. A cohomological construction for certain Lie groups

Let L be a compact group. A L-cocycle over X is given by a pair $\mathcal{L} := (\{X_i\}, \{g_{ij}\})$, where $\{X_i\}$ is a finite open cover and $\{g_{ij}\}$ is a family of continuous maps $g_{ij} : X_{ij} \to L$ such that $g_{ik}(x) = g_{ij}(x)g_{jk}(x)$, $x \in X_{ijk}$. L-cocycles $(\{X_i\}, \{g_{ij}\})$, $(\{X_l'\}, \{g_{lm}'\})$ are said *equivalent* if there are continuous maps $u_{il} : X_i \cap X_l' \to L$ such that $g_{ij}(x) = u_{il}(x)g_{lm}'(x)u_{jm}(x)^{-1}$, $x \in X_{ij} \cap X_{lm}'$. The set of equivalence classes of L-cocycles is called *cohomology set*, and is denoted by $H^1(X, L)$. It is well known that $H^1(X, L)$ classifies principal L-bundles over X ([9, Chp. 4]); if L is Abelian, then $H^1(X, L)$ has a well-defined group structure, and coincides with the first cohomology group with coefficients in the sheaf of germs of continuous L-valued maps. $H^1(X, \cdot)$ satisfies natural functoriality properties: if $\phi : L \to L'$ is a group morphism, then the pair $(\{X_i\}, \{\phi \circ g_{ij}\})$ define a L'-cocycle; thus, a map

$$\phi_* : H^1(X, L) \to H^1(X, L') \tag{3.1}$$

is defined. If L, L' are Abelian, then ϕ_* is a group morphism. As an example, let us consider the quotient L_{ab} of L w.r.t. the adjoint action $L \to \mathbf{aut}L$. L_{ab} is an Abelian group, and there is a natural epimorphism $\pi_L : L \to L_{ab}$, inducing a map $\pi_{L,*} : H^1(X, L) \to H^1(X, L_{ab})$.

The following construction appeared in [17, §4]. Let $d \in \mathbb{N}$, $\mathbb{U}(d)$ denote the unitary group, and $G \subseteq \mathbb{U}(d)$ a compact group. We define NG as the normalizer of G in $\mathbb{U}(d)$, and $QG := NG/G$. Both NG and QG are compact Lie group; there is an epimorphism $p : NG \to QG$ with kernel G, and a monomorphism $i_{NG} : G \hookrightarrow NG$. In general, the induced map

$$p_* : H^1(X, NG) \to H^1(X, QG) \tag{3.2}$$

is not surjective. We now define a cohomological class measuring the obstruction for surjectivity of p_*. Elementary computations (see [17, §4] for details) show that there is a commutative diagram

$$
\begin{array}{ccccccccc}
1 & \longrightarrow & G & \xrightarrow{i} & NG & \xrightarrow{p} & QG & \longrightarrow & 1 \\
 & & \downarrow{\pi_G} & & \downarrow{\pi_{NG}} & & \downarrow{\pi_{QG}} & & \\
1 & \longrightarrow & G_{ab} & \xrightarrow{i_{ab}} & NG_{ab} & \xrightarrow{p_{ab}} & QG_{ab} & \longrightarrow & 1
\end{array}
\tag{3.3}
$$

where 1 denotes the trivial group. By functoriality of $H^1(X, \cdot)$, and by applying the long exact sequence in sheaf cohomology, we obtain the commutative diagram

$$
\begin{array}{ccc}
H^1(X, NG) & \xrightarrow{p_*} & H^1(X, QG) \\
\downarrow{\pi_{NG,*}} & & \downarrow{\pi_{QG,*}} \\
H^1(X, NG_{ab}) \xrightarrow{p_{ab,*}} & H^1(X, QG_{ab}) \xrightarrow{\delta_{ab}} & H^2(X, G_{ab})
\end{array}
\tag{3.4}
$$

Definition 3.1. Let $d \in \mathbb{N}$, and $G \subseteq \mathbb{U}(d)$ be a compact (Lie) group. For every principal QG-bundle $\mathcal{Q} \in H^1(X, QG)$, the Dixmier-Douady class of $\mathcal{Q}$ is defined by

$$\delta(\mathcal{Q}) := \delta_{ab} \circ \pi_{QG,*} \ \in \ H^2(X, G_{ab}) \ .$$

If $\mathcal{Q}$ belongs to the image of the map $p_* : H^1(X, NG) \to H^1(X, QG)$, then $\delta(\mathcal{Q}) = 0$.

We conclude the present section with some notation. NG acts on G via the adjoint action, and is naturally embedded in $\mathbb{U}(d)$. Thus, we have maps

$$\mathbf{ad} : NG \to \mathbf{aut}G \ , \quad i_{\mathbb{U}(d)} : NG \hookrightarrow \mathbb{U}(d) \tag{3.5}$$

which induce maps

$$
\begin{cases}
\mathbf{ad}_* : H^1(X, NG) \to H^1(X, \mathbf{aut}G) \\
i_{\mathbb{U}(d),*} : H^1(X, NG) \to H^1(X, \mathbb{U}(d)) \ .
\end{cases}
\tag{3.6}
$$

4. $C(X)$-algebras and C^*-bundles

In the present section, we expose some basic properties of C^*-algebra bundles. Such properties will be applied in Section 6.

Let X be a compact Hausdorff space. A unital C^*-algebra $\mathcal{A}$ is said $C(X)$-*algebra* if there is a unital morphism $C(X) \to \mathcal{A}' \cap \mathcal{A}$; in the sequel, elements of $C(X)$ will be identified with their image in $\mathcal{A}$, so that $C(X)$ may be regarded as a unital C^*-subalgebra of $\mathcal{A}' \cap \mathcal{A}$. For every $x \in X$, we consider the ideal $I_x := \{fa : f \in C_x(X), a \in \mathcal{A}\}$, and define the *fibre* $\mathcal{A}_x := \mathcal{A}/I_x$ with the epimorphism $\pi_x : \mathcal{A} \to \mathcal{A}_x$; this allows to regard at every $a \in \mathcal{A}$ as a vector field $\widehat{a} := (\pi_x(a))_{x \in X} \in \prod_x \mathcal{A}_x$, $a \in \mathcal{A}$. In general, the *norm function* $n_a(x) := \|\pi_x(a)\|$, $x \in X$, is upper-semicontinuous. In the case in which every n_a, $a \in \mathcal{A}$, is continuous, then $\mathcal{A}$ is called *continuous bundle* over X. A $C(X)$-*algebra morphism* $\phi : \mathcal{A} \to \mathcal{B}$ is a C^*-algebra morphism such that $\phi(fa) = f\phi(a)$, $f \in C(X)$, $a \in \mathcal{A}$. If $\mathcal{F}$ is a fixed C^*-algebra, we denote by $\mathbf{bun}(X, \mathcal{F})$ the set of $C(X)$-isomorphism classes of $C(X)$-algebras having fibres isomorphic to $\mathcal{F}$. Good references about $C(X)$-algebras are [4, 11].

There is categorical equivalence between $C(X)$-algebras and a class of topological objects, called C^*-bundles (see [8, Thm. 5.13], [8, §10.18]). A C^*–*bundle* is given by a Hausdorff space Σ endowed with a surjective, open, continuous map $Q : \Sigma \to X$ such that every fibre $\Sigma_x := Q^{-1}(x)$, $x \in X$, is homeomorphic to a unital C^*-algebra with identity 1_x. Σ is required to be full, i.e., for every $\sigma \in \Sigma$ there exists a continuous section $a : X \hookrightarrow \Sigma$, $a \circ Q = id_X$, such that $a \circ Q(\sigma) = \sigma$. Let $Q' : \Sigma' \to X$ be a C^*-bundle. A C^*-*bundle morphism* from Σ into Σ' is given by a continuous map $\phi : \Sigma \to \Sigma'$ such that

1. $Q' \circ \phi = Q$; this implies that $\phi(\Sigma_x) \subseteq \Sigma'_x$, $x \in X$;
2. $\phi_x := \phi|_{\Sigma_x} : \Sigma_x \to \Sigma'_x$ is a C^*-algebra morphism for every $x \in X$.

ϕ is said **isomorphism** if every ϕ_x, $x \in X$, is a C^*-algebra isomorphism.

The set $S_X(\Sigma)$ of continuous sections of Σ is endowed with a natural structure of $C(X)$-algebra: for every $a, a' \in \mathcal{A}$, we define $a + a'$, aa', a^* as the maps $(a + a')(x) := a(x) + a'(x)$, $(aa')(x) := a(x)a'(x)$, $a^*(x) := a(x)^*$, $x \in X$, and introduce the norm $\|a\| := \sup_x \|a(x)\|$. Moreover, for every $f \in C(X)$, it turns out that the map $f1(x) := f(x)1_x$, $x \in X$, belongs to $S_X(\Sigma)$.

On the converse, if $\mathcal{A}$ is a unital $C(X)$-algebra, then the set $\widehat{\mathcal{A}} := \bigsqcup_{x \in X} \mathcal{A}_x$ is endowed with a natural surjective map $Q : \widehat{\mathcal{A}} \to X$, $Q \circ \pi_x(a) := x$. For every open $U \subseteq X$, $a \in \mathcal{A}$, $\varepsilon > 0$, we consider the following subset of $\widehat{\mathcal{A}}$:

$$T_{U,a,\varepsilon} := \left\{ \sigma \in \widehat{\mathcal{A}} \ : \ Q(\sigma) \in U \text{ and } \left\| \sigma - \pi_{Q(\sigma)}(a) \right\| < \varepsilon \right\} . \qquad (4.1)$$

The family $\{T_{U,a,\varepsilon}\}$ provides a basis for a topology on $\widehat{\mathcal{A}}$. It can be proved that $Q : \widehat{\mathcal{A}} \to X$ is a C^*-bundle, and that there is an isomorphism $\mathcal{A} \to S_X(\widehat{\mathcal{A}})$, $a \mapsto \widehat{a} := \{X \ni x \mapsto \pi_x(a)\}$.

Theorem 4.1. *Let X be a compact Hausdorff space. The map $\mathcal{A} \mapsto \widehat{\mathcal{A}}$ provides an equivalence between the category of $C(X)$-algebras (with arrows $C(X)$-algebra morphisms) and the category of C^*-bundles over X (with arrows C^*-bundle morphisms).*

Some examples follow. If $\mathcal{F}$ is a unital C^*-algebra, then $\mathcal{A} := C(X) \otimes \mathcal{F}$ is a $C(X)$-algebra with C^*-bundle $\widehat{\mathcal{A}} = X \times \mathcal{F}$, endowed with the projection $Q(x, b) := x$, $x \in X$, $b \in \mathcal{F}$; $\widehat{\mathcal{A}}$ is called the *trivial* C^*-bundle. Let $C(X) \hookrightarrow C(Y) =: \mathcal{A}_Y$ be an inclusion of unital, Abelian C^*-algebras; then, a surjective map $q : Y \to X$ is defined, and the C^*-bundle $\widehat{\mathcal{A}}_Y \to X$ has fibres $\mathcal{A}_x = C(q^{-1}(x))$, $x \in X$. Let $d \in \mathbb{N}$, and $\mathcal{E} \to X$ denote a rank d vector bundle; then, the C^*-algebra $L(\mathcal{E})$ of endomorphisms of $\mathcal{E}$ is a continuous bundle of C^*-algebras over X, with fibres isomorphic to the matrix algebra $\mathbb{M}_d$.

We introduce some notation and terminology. Let L denote a compact group acting by automorphisms on $\mathcal{F}$. Moreover, let $\mathcal{A}$, $\mathcal{A}'$ be $C(X)$-algebras with fibre $\mathcal{F}$ (i.e., $\mathcal{A}_x \simeq \mathcal{A}'_x \simeq \mathcal{F}$, $x \in X$), and $\beta : \mathcal{A} \to \mathcal{A}'$ a $C(X)$-algebra isomorphism. Then, for every $x \in X$ an automorphism $\beta_x \in \mathbf{aut}\mathcal{F}$ is defined, in such a way that $\beta_x \circ \pi_x(a) = \pi'_x \circ \beta(a)$, $a \in \mathcal{A}$. If β_x belongs to the image of the L-action for every $x \in X$, then we say that β is *L-covariant*, and use the notation

$$\beta : \mathcal{A} \to_L \mathcal{A}' . \tag{4.2}$$

Let now $I : \mathcal{F}_0 \hookrightarrow \mathcal{F}$ be an inclusion of unital C^*-algebras. Suppose that there are $C(X)$-algebras $\mathcal{A}_0 \in \mathbf{bun}(X, \mathcal{F}_0)$, $\mathcal{A} \in \mathbf{bun}(X, \mathcal{F})$, and a a $C(X)$- monomorphism $\phi : \mathcal{A}_0 \to \mathcal{A}$ such that $\phi_x = I$ for every $x \in X$. Then, we say that ϕ is *I-covariant*, and use the notation

$$\phi : \mathcal{A}_0 \hookrightarrow_I \mathcal{A} . \tag{4.3}$$

We conclude the present section by presenting a construction for continuous bundles having as fibre a fixed C^*-dynamical system. Let L be a compact group, and $\mathcal{F}$ a unital C^*-algebra with an automorphic action $\alpha : L \to \mathbf{aut}\mathcal{F}$. For every L-cocycle $\mathcal{L} := (\{X_i\}, \{g_{ij}\})$, we define a C^*-bundle $Q : \widehat{\mathcal{A}}_{\mathcal{L}} \to X$ as the *clutching* of the family of trivial bundles $\{X_i \times \mathcal{F}\}$ w.r.t. the maps $\{\alpha \circ g_{ij} : X_{ij} \to \mathbf{aut}\mathcal{F}\}$ (in the same way as in [10, I.3.2]). We denote by $\mathcal{A}_{\mathcal{L}}$ the $C(X)$-algebra of continuous sections associated with $\widehat{\mathcal{A}}_{\mathcal{L}}$. If $\mathcal{H}$ is a L-cocycle equivalent to $\mathcal{L}$, then it is easily verified that there is an isomorphism $\widehat{\mathcal{A}}_{\mathcal{L}} \simeq \widehat{\mathcal{A}}_{\mathcal{H}}$. In such a way, we defined a map

$$\alpha_* : H^1(X, L) \to \mathbf{bun}(X, \mathcal{F}) , \quad \mathcal{L} \mapsto \mathcal{A}_{\mathcal{L}} . \tag{4.4}$$

In general α_* is not injective, unless $L \simeq \mathbf{aut}\mathcal{F}$; in fact, cocycle equivalence gives rise to an L-covariant isomorphism.

5. Tensor C^*-categories

A *tensor C^*-category* is a C^*-category $\mathcal{T}$ endowed with a bifunctor $\otimes : \mathcal{T} \times \mathcal{T} \to \mathcal{T}$, called the *tensor product*. In explicit terms, for every pair $\rho, \sigma \in \mathbf{obj}\,\mathcal{T}$ there is an object $\rho\sigma \in \mathbf{obj}\,\mathcal{T}$; for every $\rho', \sigma' \in \mathbf{obj}\,\mathcal{T}$, there are bilinear maps $(\rho, \sigma) \times (\rho', \sigma')$

$\to (\rho\rho', \sigma\sigma')$, $t, t' \mapsto t \otimes t'$. The existence of an *identity object* $\iota \in \mathbf{obj}\ \mathcal{T}$ is postulated, in such a way that $\iota\rho = \rho\iota = \rho$, $\rho \in \mathbf{obj}\ \mathcal{T}$, $t = t \otimes 1_\iota = 1_\iota \otimes t$, $t \in (\rho, \sigma)$. The data of a tensor C^*-category with identity object ι will be denoted by the triple $(\mathcal{T}, \otimes, \iota)$. For basic notions on tensor C^*-categories, we refer to [7, §1].

It is a consequence of the above definition that the C^*-algebra (ι, ι) is Abelian; we will denote by X^ι, the (compact, Hausdorff) spectrum of (ι, ι), so that there is an identification $(\iota, \iota) \simeq C(X^\iota)$.

Well-known examples of tensor C^*-categories are the one of Hilbert spaces endowed with the usual tensor product, denoted by **hilb**, and the one of vector bundles over a compact Hausdorff space X, that we denote by **vect**(X). In the first case X^ι reduces to a single point ($\iota = \mathbb{C}$, so that $(\iota, \iota) \simeq \mathbb{C}$); in the second case $X^\iota = X$ ($\iota = X \times \mathbb{C}$).

The eventual commutativity up-to-unitary-equivalence of the tensor product is described by the property of *symmetry*. A tensor C^*-category $(\mathcal{T}, \otimes, \iota)$ is said *symmetric* if for each $\rho, \sigma \in \mathbf{obj}\ \mathcal{T}$ there is a unitary 'flip' $\varepsilon_{\rho\sigma} \in (\rho\sigma, \sigma\rho)$ such that $\varepsilon_{\sigma\sigma'} \circ (t \otimes t') = (t' \otimes t) \circ \varepsilon_{\rho\rho'}$, $t \in (\rho, \sigma)$, $t' \in (\rho', \sigma')$. A symmetric tensor C^*-category is denoted by $(\mathcal{T}, \otimes, \iota, \varepsilon)$.

The tensor C^*-categories **hilb**, **vect**(X) are symmetric. Another basic example is given by the *dual* of a compact group G, i.e., the category with objects unitary, finite-dimensional representations of G.

5.1. Duals of compact Lie groups

Let G be a compact group endowed with a faithful representation over a rank d Hilbert space H_d, $d \in \mathbb{N}$. We regard at G as a compact Lie subgroup of the unitary group $\mathbb{U}(d)$. We define $\widehat{G}$ as the tensor C^*-category with objects the r-fold tensor powers H_d^r, $r \in \mathbb{N}$ (for $r = 0$ we define $\iota := H_d^0 := \mathbb{C}$), and arrows the spaces $(H_d^r, H_d^s)_G$ of linear operators $t : H_d^r \to H_d^s$ such that $g_s \circ t \circ g_r^* = t\ \forall g \in G$, where

$$g_r := \otimes^r g \in H_d^r . \tag{5.1}$$

It is well known that by performing the closure for subobjects and the additive completion of $\widehat{G}$, we obtain all the finite-dimensional representations of G. $\widehat{G}$ is symmetric, in fact it is endowed with the flip operators $\theta_{r,s} \in (H_d^{r+s}, H_d^{r+s})_G$, $\theta_{r,s}(\psi \otimes \psi') := \psi' \otimes \psi$, $\psi \in H_d^r$, $\psi \in H_d^s$. Thus, we have a symmetric tensor C^*-category $(\widehat{G}, \otimes, \iota, \theta)$.

A *symmetric autofunctor* of $\widehat{G}$ is given by a family F of Banach space isomorphisms

$$F^{r,s} : (H_d^r, H_d^s)_G \to (H_d^r, H_d^s)_G\ ,\quad r, s \in \mathbb{N}$$

such that $F^{r,s}(t \circ t') = F^{l,s}(t) \circ F^{r,l}(t')$, $F^{r,s}(t^*) = F^{r,l}(t)^*$, $F^{r+r',s+s'}(t \otimes t'') = F^{r,s}(t) \otimes F^{r',s'}(t'')$, $F^{r+s,r+s}(\theta_{r,s}) = \theta_{r,s}$, $t \in (H_d^r, H_d^s)_G$, $t' \in (H_d^{r'}, H_d^{s'})_G$, $t'' \in (H_d^l, H_d^s)_G$. The set $\mathbf{aut}_\theta \widehat{G}$ of symmetric autofunctors of $\widehat{G}$ is endowed with a group structure w.r.t. the composition $F, G \mapsto G \circ F := \{G^{r,s} \circ F^{r,s}\}$ and inverse $F^{-1} := \{(F^{r,s})^{-1}\}$.

Now, every $u \in NG$ defines maps

$$\widehat{u}^{r,s} : (H_d^r, H_d^s)_G \to (H_d^r, H_d^s)_G \ , \quad \widehat{u}^{r,s}(t) := u_s \circ t \circ u_r^*$$

(the term u_s is defined according to (5.1)). A direct check show that that the family $\widehat{u} := \{\widehat{u}^{r,s}\}_{r,s}$ defines an element of $\mathbf{aut}_\theta \widehat{G}$. In particular, we find

$$\widehat{g}^{r+s,r+s}(\theta_{r,s}) = \theta_{r,s} \tag{5.2}$$

(see [6, §2]). Since by definition

$$\widehat{g}^{r,s}(t) = t \ , \quad g \in G \ , \ t \in (H_d^r, H_d^s)_G \ , \tag{5.3}$$

we conclude that

$$\widehat{u} = \widehat{ug} \ , \ u \in NG \ , \ g \in G \ , \tag{5.4}$$

so that $\{u \mapsto \widehat{u}\}$ factorizes through a map

$$QG \to \mathbf{aut}_\theta \widehat{G} \ , \quad y \mapsto \widehat{y} \ . \tag{5.5}$$

5.2. Special categories and group bundles

Let X be a compact Hausdorff space, $d \in \mathbb{N}$, $G \subseteq \mathbb{U}(d)$ a compact group. A *special category* is given by a symmetric tensor C^*-category $(\mathcal{T}, \otimes, \iota, \varepsilon)$ with objects the positive integers $r \in \mathbb{N}$, and arrows the Banach $C(X)$-bimodules $\mathcal{M}_{r,s}$ of continuous sections of vector bundles $\mathcal{E}_{r,s} \to X$ with fibre $(H_d^r, H_d^s)_G$, $r, s \in \mathbb{N}$ (the left $C(X)$-action of $\mathcal{M}_{r,s}$ is assumed to coincide with the right one). The tensor product is defined as follows:

$$\begin{cases} r, s \mapsto r + s \ , \quad r, s \in \mathbb{N} \\ t, t' \mapsto t \otimes_X t' \in \mathcal{M}_{r+r',s+s'} \ , \quad t \in \mathcal{M}_{r,s} \ , \ t' \in \mathcal{M}_{r',s'} \ , \end{cases}$$

where $\otimes_X$ denote the tensor product in the category of Banach $C(X)$-bimodules ([3, Chp.VI]). Note that $\iota := 0 \in \mathbb{N}$ is the identity object, with $(\iota, \iota) = C(X)$. We denote by $\mathbf{tens}(X, \widehat{G})$ the set of isomorphism classes of special categories having spaces of arrows with fibres $(H_d^r, H_d^s)_G$, $r, s \in \mathbb{N}$.

As an example, we consider the *trivial special category* $X \times \widehat{G}$ with arrows $\mathcal{M}_{0,r,s} = C(X) \otimes (H_d^r, H_d^s)_G$, $r, s \in \mathbb{N}$, and symmetry $\theta^X := \{\theta_{r,s}^X := 1_X \otimes \theta_{r,s}\}$, where $1_X \in C(X)$ denotes the identity. Note that for every $r, s \in \mathbb{N}$, it turns out that $\mathcal{M}_{0,r,s}$ is the module of sections of $\mathcal{E}_{0,r,s} := X \times (H_d^r, H_d^s)_G$.

We now give a simple procedure to construct special categories. Let $\mathcal{Q}$ be a principal QG-bundle with associated cocycle $(\{X_i\}, \{y_{ij}\})$. For every $r, s \in \mathbb{N}$, we denote by $\mathbf{aut}(H_d^r, H_d^s)_G$ the (topological) group of isometric linear maps of $(H_d^r, H_d^s)_G$. By composing with the isomorphism (5.5), we obtain maps

$$\widehat{y}_{ij}^{r,s} : X_{ij} \to \mathbf{aut}(H_d^r, H_d^s)_G \ , \quad x \mapsto \widehat{y}_{ij}(x)^{r,s} \ ,$$

which define $\mathbf{aut}(H_d^r, H_d^s)_G$-cocycles. We denote by $\mathcal{E}_{\mathcal{Q},r,s} \to X$ the vector bundle with fibre $(H_d^r, H_d^s)_G$ and transition maps $\{\widehat{y}_{ij}^{r,s}\}$. Now, for every $r, s, r', s' \in \mathbb{N}$, it turns out

$$\widehat{y}_{ij}(x)^{r,s} \otimes \widehat{y}_{ij}(x)^{r',s'} = \widehat{y}_{ij}(x)^{r+r',s+s'} \ , \quad x \in X_{ij} \ ;$$

this implies that there are inclusions

$$\mathcal{E}_{\mathcal{Q},r,s} \otimes \mathcal{E}_{\mathcal{Q},r',s'} \subseteq \mathcal{E}_{\mathcal{Q},r+r',s+s'} \quad , \tag{5.6}$$

where $\otimes$ stands for the tensor product in $\mathbf{vect}(X)$. We denote by $\mathcal{M}_{\mathcal{Q},r,s}$ the module of continuous sections of $\mathcal{E}_{\mathcal{Q},r,s}$, and define the following C^*-category:

$$T\mathcal{Q} := \begin{cases} \mathbf{obj}\, T\mathcal{Q} := \mathbb{N} \\ (r,s)_{\mathcal{Q}} := \mathcal{M}_{\mathcal{Q},r,s} \, , \quad r,s \in \mathbb{N} \; . \end{cases}$$

The relations (5.6) imply that $\mathcal{M}_{\mathcal{Q},r,s} \otimes_X \mathcal{M}_{\mathcal{Q},r',s'} \subseteq \mathcal{M}_{\mathcal{Q},r+r',s+s'}$, so that $T\mathcal{Q}$ is a tensor C^*-category. Note that $(0,0)_{\mathcal{Q}} = C(X)$; more in general, for every $r \in \mathbb{N}$ it turns out that $\mathcal{M}_{\mathcal{Q},r,r}$ is a $C(X)$-algebra with fibre $(H_d^r, H_d^r)_G$. About the symmetry, let us consider the constant maps $\varepsilon_{r,s}^i(x) := \theta_{r,s}$, $x \in X_i$; then, (5.2) implies that

$$\widehat{y}_{ij}^{r,s}\, (\varepsilon_{r,s}^i)|_{X_{ij}} = \varepsilon_{r,s}^j|_{X_{ij}} \; .$$

The previous relations imply that we can glue the local sections $\varepsilon_{r,s}^i : X_i \to X_i \times (H_d^{r+s}, H_d^{r+s})_G$ by using the transition maps $\widehat{y}_{ij}^{r,s}$, and obtain elements $\varepsilon_{r,s} \in \mathcal{M}_{\mathcal{Q},r+s,r+s}$. Some routine computations show that the family $\{\varepsilon_{r,s}\}$ defines a symmetry for $T\mathcal{Q}$. Thus, $T\mathcal{Q} \in \mathbf{tens}(X, \widehat{G})$. It is easy to prove that if $\mathcal{Q}' \in H^1(X, QG)$ is cocycle equivalent to $\mathcal{Q}$ then there is an C^*-category isomorphism $T\mathcal{Q} \simeq T\mathcal{Q}'$ preserving symmetry and tensor product. Thus, we defined a map

$$T : H^1(X, QG) \to \mathbf{tens}(X, \widehat{G}) \; .$$

Example (Tensor powers of a vector bundle). Let us consider the trivial group $G := \mathbb{I}_d$, so that $NG = QG = \mathbb{U}(d)$. Every $\mathbb{U}(d)$-cocycle $\mathcal{U} := (\{X_i\}, \{u_{ij}\})$ can be regarded as the set of transition maps of a rank d vector bundle $\mathcal{E} \to X$. For every $r \in \mathbb{N}$, we denote by $\mathcal{E}^r$ the r-fold tensor power of $\mathcal{E}$. The tensor C^*-category $T\mathcal{U}$ has spaces of arrows the bimodules of continuous sections of vector bundles $\mathcal{E}_{\mathcal{U},r,s} \to X$, having fibre (H_d^r, H_d^s) and transition maps $\widehat{u}_{ij}^{r,s}$. It is well known that every $\mathcal{E}_{\mathcal{U},r,s}$ can be identified with the vector bundle of morphisms from $\mathcal{E}^r$ into $\mathcal{E}^s$ (see [10, I.4.8(c)]). Thus, $T\mathcal{U}$ is isomorphic to the full C^*-subcategory of $\mathbf{vect}(X)$ with objects the tensor powers of $\mathcal{E}$. We denote by $\widehat{\mathcal{E}}$ such tensor C^*-category.

Example (The dual of a Lie group bundle). We recall the reader to the notation of Section 3 (in particular, the maps (3.5,3.6)). Let $\mathcal{N} := (\{X_i\}, \{u_{ij}\}) \in H^1(X, NG)$. We consider the vector bundle $\mathcal{E} \to X$ with transition maps $\mathcal{U} := i_{\mathbb{U}(d),*}\mathcal{N} \in H^1(X, \mathbb{U}(d))$, and the QG-cocycle $\mathcal{Q} := p_*(\mathcal{N}) \in H^1(X, QG)$. The symmetric tensor C^*-category $T\mathcal{Q}$ has arrows the bimodules of continuous sections of vector bundles $\mathcal{E}_{\mathcal{Q},r,s} \to X$, which have fibre $(H_d^r, H_d^s)_G$ and transition maps $p \circ \widehat{u}_{ij}^{r,s}$. Let now $\mathcal{E}_{\mathcal{U},r,s} \to X$ be the vector bundles associated with $\mathcal{U}$, according to the previous example. Then, we may regard at each $\mathcal{E}_{\mathcal{Q},r,s}$ as a vector subbundle of $\mathcal{E}_{\mathcal{U},r,s}$; in fact, by (5.4), every set of transition maps $\{\widehat{u}_{ij}^{r,s}\}$ acts in the same way as $\{p \circ \widehat{u}_{ij}^{r,s}\}$ on the trivial bundles $X_i \times (H_d^r, H_d^s)_G \subseteq X_i \times (H_d^r, H_d^s)$. Thus, $T\mathcal{Q}$ is a symmetric tensor C^*-subcategory of $\widehat{\mathcal{E}}$. Let now $\mathcal{G} \to X$ be the group bundle with fibre G and transition maps $\mathbf{ad}_*\mathcal{N}$. Since $\mathbf{ad}(u) = \widehat{u}^{1,1}$, $u \in NG$, there is an inclusion $\mathcal{G} \subset \mathcal{E}_{\mathcal{U},1,1}$. Since $\mathcal{E}_{\mathcal{U},1,1}$ is the vector bundle of endomorphisms of $\mathcal{E}$, we find that $\mathcal{E}$ is $\mathcal{G}$-equivariant in the sense of [12, §1], i.e., there is an action

$$\mathcal{G} \times_X \mathcal{E} \to \mathcal{E} \ , \quad (g, v) \mapsto g(v) \ , \tag{5.7}$$

where $\mathcal{G} \times_X \mathcal{E}$ denotes the fibered cartesian product. Note that in (5.7) the base space of $\mathcal{E}$ coincides with the base space of $\mathcal{G}$; in [12], the base space of $\mathcal{E}$ is a topological bundle carrying a $\mathcal{G}$-action. Now, for every $r, s \in \mathbb{N}$ there is an action

$$\mathcal{G} \times_X \mathcal{E}_{\mathcal{U},r,s} \to \mathcal{E}_{\mathcal{U},r,s} \ , \quad (g, t) \mapsto \widehat{g}^{r,s}(t) \ .$$

In particular, $t \in \mathcal{E}_{\mathcal{Q},r,s}$ if and only if $\widehat{g}^{r,s}(t) = t$ for every $g \in \mathcal{G}$, in the same way as in (5.3); in such a case, we say that t is $\mathcal{G}$-*equivariant*. We conclude that $T\mathcal{Q}$ is the tensor C^*-subcategory of $\widehat{\mathcal{E}}$ with arrows the bimodules of $\mathcal{G}$-equivariant morphisms between tensor powers of $\mathcal{E}$. We denote by $\widehat{\mathcal{E}}_{\mathcal{G}}$ such tensor C^*-category.

Let $T\mathcal{Q}$ be the special category associated with $\mathcal{Q} \in H^1(X, Q G)$. An *embedding functor* is given by a rank d vector bundle $\mathcal{E} \to X$ and a C^*-monofunctor $F : T\mathcal{Q} \hookrightarrow \widehat{\mathcal{E}}$, preserving tensor product and symmetry. In the following results, we characterize tensor C^*-categories $T\mathcal{Q}$ admitting an embedding functor, and give a cohomological obstruction for the existence. This provides a duality between group bundles and special categories admitting an embedding functor.

Theorem 5.1 ([17], Thm.7.1, Thm.7.3). *Let $G \subseteq \mathbb{SU}(d)$. The following are equivalent:*

1. *there exists an embedding functor $F : T\mathcal{Q} \hookrightarrow \widehat{\mathcal{E}}$, and a fibred G-bundle $\mathcal{G} \to X$ such that $\mathcal{E}$ is $\mathcal{G}$-equivariant and $T\mathcal{Q} = \widehat{\mathcal{E}}_{\mathcal{G}}$;*
2. *there exists a principal NG-bundle $\mathcal{N}$ such that $p_*\mathcal{N} = \mathcal{Q}$.*

The interplay between the above objects is the following: if there exists $\mathcal{N} \in H^1(X, NG)$ with $p_*\mathcal{N} = \mathcal{Q}$, then $\mathcal{E}$ is defined by the cocycle $i_{\mathbb{U}(d),*}\mathcal{N} \in H^1(X, \mathbb{U}(d))$, and $\mathcal{G}$ is defined by the cocycle $\mathbf{ad}_*\mathcal{N} \in H^1(X, \mathbf{aut}G)$. The condition $G \subseteq \mathbb{SU}(d)$ is motivated by the fact that the proof of the previous theorem lies on the notion of special object ([7, §3, Lemma 6.7]); it is our opinion that it should suffice to assume $G \subseteq \mathbb{U}(d)$, and this point is object of a work in progress. As a direct consequence of the previous theorem, we obtain

Theorem 5.2. *For every $T\mathcal{Q} \in \mathbf{tens}(X, \widehat{G})$, we define*

$$\delta(T\mathcal{Q}) := \delta(\mathcal{Q}) \in H^2(X, G_{ab}) \quad \text{(the Dixmier-Douady class of } T\mathcal{Q}) \ .$$

If there exists an embedding functor $F : T\mathcal{Q} \hookrightarrow \widehat{\mathcal{E}}$, then $\delta(T\mathcal{Q}) = 0$. In particular, if $T\mathcal{Q}$ is the trivial special category (i.e., $\mathcal{Q} \simeq X \times QG$) then $\delta(T\mathcal{Q}) = 0$.

It is not difficult to construct special categories with $\delta(T\mathcal{Q}) \neq 0$. For example, let us suppose that the epimorphism $\pi_{QG} : QG \to QG_{ab}$ admits a left inverse $S : QG_{ab} \hookrightarrow QG$, $\pi_{QG} \circ S = id_{QG_{ab}}$, and let X be a space such that there is $z \in H^1(X, QG_{ab})$ with $\delta_{ab}(z) \neq 0$. We define $\mathcal{Q} := S_*z \in H^1(X, QG)$; by construction, $\delta(T\mathcal{Q}) = \delta(\mathcal{Q}) = \delta_{ab}(z) \neq 0$. Explicit examples are given in [17, §7.0.8].

6. C^*-bundles with fibre $\mathcal{O}_G$

The following construction appeared in [6], and can be interpreted as a C^*-algebraic version of the Tannaka duality. Let $d \in \mathbb{N}$; we denote by $\mathcal{O}_d$ the Cuntz algebra ([5]) generated by a multiplet $\{\psi_i\}_{i=1}^d$ of isometries satisfying the relations

$$\psi_i^* \psi_j = \delta_{ij} 1 \quad , \quad \sum_i \psi_i \psi_i^* = 1 \quad . \tag{6.1}$$

Let us denote by $H_d \subset \mathcal{O}_d$ the Hilbert space spanned by $\{\psi_i\}$, endowed with the scalar product $\langle \psi, \psi' \rangle 1 := \psi^* \psi'$, $\psi, \psi \in H_d$; then, it is clear that H_d is isomorphic to the canonical rank d Hilbert space. Let now $I := \{i_1, \dots, i_r\}$ be a multiindex with length $|I| := r \in \mathbb{N}$; we introduce the notation

$$\psi_I := \prod_{k=1}^r \psi_{i_k} \in \mathcal{O}_d \quad ,$$

and denote by H_d^r the vector space spanned by $\{\psi_I\}$, which we identify with the r-fold tensor power of H_d. With the above notation, the Banach space

$$(H_d^r, H_d^s) := \text{span } \{\psi_I \psi_J^* \ , \ |I| = s, |J| = r\} \ , \ r, s \in \mathbb{N} \ , \tag{6.2}$$

can be naturally identified with the set of linear operators from H_d^r into H_d^s. By identifying H_d with the canonical rank d Hilbert space, we get a natural $\mathbb{U}(d)$-action $\mathbb{U}(d) \times H_d \to H_d$, $u, \psi \mapsto u\psi$. Universality of the Cuntz algebra implies that there is an automorphic action

$$\mathbb{U}(d) \to \mathbf{aut}\mathcal{O}_d \ , \quad u \mapsto \widehat{u} \ , \tag{6.3}$$

$\widehat{u}(\psi) := u\psi$, $u \in \mathbb{U}(d)$, $\psi \in H_d$. If we restrict (6.4) to elements of a compact group $G \subseteq \mathbb{U}(d)$, then we get an action

$$G \to \mathbf{aut}\mathcal{O}_d \ , \quad g \mapsto \widehat{g} \ . \tag{6.4}$$

We denote by $\mathcal{O}_G$ the fixed-point algebra of $\mathcal{O}_d$ w.r.t. the action (6.4), and by

$$i_G : \mathcal{O}_G \hookrightarrow \mathcal{O}_d$$

the natural inclusion. Since $\widehat{g}(t) = g_s \circ t \circ g_r^*$, $g \in G$, $t \in (H_d^r, H_d^s)$, we may identify $(H_d^r, H_d^s)_G$ with $(H_d^r, H_d^s) \cap \mathcal{O}_G$.

Let us now consider the normalizer $NG \subseteq \mathbb{U}(d)$. Then, every $u \in NG$ defines an automorphism $\widehat{u} \in \mathbf{aut}\mathcal{O}_G$: in fact, if $g \in G$, $t \in \mathcal{O}_G$, then $\widehat{g}(\widehat{u}(t)) = \widehat{u} \circ \widehat{g'}(t) = \widehat{u}(t)$, where $g' \in G$. It is clear that $\widehat{u} = \widehat{ug}$ for every $g \in G$, thus there is an automorphic action

$$QG \to \mathbf{aut}\mathcal{O}_G \ , \quad y \mapsto \widehat{y} \ . \tag{6.5}$$

We now apply the construction (4.4) to the actions (6.3), (6.5), and obtain maps

$$H^1(X, \mathbb{U}(d)) \to \mathbf{bun}(X, \mathcal{O}_d) \ , \quad \mathcal{U} \mapsto \mathcal{O}_\mathcal{U} \ , \tag{6.6}$$

$$H^1(X, QG) \to \mathbf{bun}(X, \mathcal{O}_G) \ , \quad \mathcal{Q} \mapsto \mathcal{O}_\mathcal{Q} \ . \tag{6.7}$$

Let $\mathcal{E} \to X$ be the vector bundle with associated $\mathbb{U}(d)$-cocycle $\mathcal{U}$, and $S\mathcal{E}$ the Hilbert $C(X)$-bimodule of continuous sections of $\mathcal{E}$. It is proved in [15, Prop. 4.2] that $\mathcal{O}_\mathcal{U}$ is the Cuntz-Pimsner algebra associated with $S\mathcal{E}$. The map (6.6) is not injective: if X is a finite-dimensional CW-complex, in order to obtain an isomorphism $\mathcal{O}_\mathcal{U} \simeq \mathcal{O}_\mathcal{V}$ it suffices that $\mathcal{U}, \mathcal{V} \in H^1(X, \mathbb{U}(d))$ are transition maps of vector bundles having the same class in $K^0(X)$ ([16, Prop.10]).

The following result is a translation of Thm. 5.1, Thm. 5.2 in terms of $C(X)$-algebras (note that we use the notation (4.2,4.3)).

Proposition 6.1 ([17, Prop.7.11]). $\mathcal{Q} = \mathcal{Q}' \in H^1(X, QG)$ *if and only if there is a QG-covariant $C(X)$-isomorphism $\alpha : \mathcal{O}_\mathcal{Q} \to_{QG} \mathcal{O}_{\mathcal{Q}'}$. Moreover, the class*

$$\delta(\mathcal{O}_\mathcal{Q}) := \delta(\mathcal{Q}) \in H^2(X, G_{ab})$$

measures the obstruction to find a vector bundle $\mathcal{E} \to X$ with transition maps $\mathcal{U} \in H^1(X, \mathbb{U}(d))$ implementing a $C(X)$-monomorphism $\mathcal{O}_\mathcal{Q} \hookrightarrow_{i_G} \mathcal{O}_\mathcal{U}$: if $\delta(\mathcal{O}_\mathcal{Q}) \neq 0$, then such a vector bundle does not exist.

7. Twisted equivariant K-theory

Let X be a compact Hausdorff space, $d \in \mathbb{N}$, and $G \subseteq \mathbb{U}(d)$ a closed group. For every $\mathcal{Q} \in H^1(X, QG)$, we consider the associated special C^*-category $T\mathcal{Q} \in \mathbf{tens}(X, \widehat{G})$, and define the Abelian group

$$K^0_\mathcal{Q}(X) := K_0(T\mathcal{Q}_{+,s}) \ , \tag{7.1}$$

according to Def.2.1. $K^0_\mathcal{Q}(X)$ is called the *twisted equivariant K-theory of X*. Note that we close for subobjects *after* having performed the additive completion. This implies that there is an immersion $\mathbf{vect}(X) \hookrightarrow T\mathcal{Q}_{+,s}$, in fact every $\mathcal{E} \in \mathbf{vect}(X)$ appears as a subobject of some $\iota_n := (\iota, \dots, \iota) \in T\mathcal{Q}_+$. Thus, there is a morphism $K^0(X) \to K^0_\mathcal{Q}(X)$. We briefly discuss the relationship between $K^0_\mathcal{Q}(X)$ and well-known K-theory groups.

1. If $X := \{x\}$ reduces to a single point, then the unique element of $\mathbf{tens}(\{x\}, \widehat{G})$ is the dual $\widehat{G}$. It is well known that by closing $\widehat{G}$ w.r.t. direct sums and subobjects we get all the finite-dimensional representations of G. Thus, $K_\mathcal{Q}(\{x\})$ coincides with (the additive group) of the representation ring $R(G)$.

2. If $G = \mathbb{I}_d$ is the trivial subgroup of $\mathbb{U}(d)$ (so that $QG = \mathbb{U}(d)$) and $\mathcal{Q} \in H^1(X, \mathbb{U}(d))$, then $T\mathcal{Q} = \widehat{\mathcal{E}}$ for some rank d vector bundle $\mathcal{E} \to X$. Thus, $K^0_\mathcal{Q}(X)$ coincides with $K^0(X)$.

3. Let X be a trivial G-space. We denote by $\mathcal{Q}_0$ the trivial principal QG-bundle over X. The tensor C^*-category $T\mathcal{Q}_0$ is isomorphic to the trivial special category $X \times \widehat{G}$. Now, every finite-dimensional G-Hilbert space M appears as an object of the category $\widehat{G}_{+,s}$ obtained by closing $\widehat{G}$ w.r.t. direct sums and subobjects. Moreover, every G-vector bundle $\mathcal{E} \to X$ in the sense of [14] is a direct summand of some trivial bundle $X \times M$ ([14, Prop. 2.4]). Thus, $\mathcal{E} \in \mathbf{obj}\,(X \times \widehat{G})_{+,s}$, and $K^0_{\mathcal{Q}_0}(X)$ is isomorphic to the equivariant K-theory $K_G(X)$ in the sense of [14]. Note that for a trivial G-space, we find $K^0_{\mathcal{Q}_0}(X) \simeq K_G(X) \simeq R(G) \otimes K^0(X)$ (see [14, Prop. 2.2].

4. Let $\mathcal{Q} \in H^1(X, QG)$ such that $\mathcal{Q} = p_*\mathcal{N}$ for some $\mathcal{N} \in H^1(X, NG)$, $\mathcal{N} :=$ $(\{X_i\}, \{u_{ij}\})$. Then, there exists a vector bundle $\mathcal{E} \to X$ with associated $\mathbb{U}(d)$-cocycle $i_{\mathbb{U}(d),*}\mathcal{N} \in H^1(X, \mathbb{U}(d))$, and a G-fibre bundle $\mathcal{G} \to X$ with associated $\mathbf{aut}G$-cocycle $\mathbf{ad}_*\mathcal{N} \in H^1(X, \mathbf{aut}G)$. By Ex. 5.2, we find that $\mathcal{E}$ is a $\mathcal{G}$-equivariant vector bundle in the sense of [12, §1]. Now, $T\mathcal{Q}$ is isomorphic to the category $\widehat{\mathcal{E}}_{\mathcal{G}}$ introduced in Ex. 5.2, thus there is a morphism

$$K_{\mathcal{Q}}^0(X) \to K_{\mathcal{G}}^0(X) \ , \tag{7.2}$$

where $K_{\mathcal{G}}^0(X)$ denotes the gauge-equivariant K-theory of X in the sense of [12, §3]. At the present moment, it is not clear whether (7.2) is one-to-one. In fact, there is no evidence that by closing $\widehat{\mathcal{E}}_{\mathcal{G}}$ w.r.t. direct sums and subobjects we get all the $\mathcal{G}$-equivariant vector bundles over X. Since it is possible to reconstruct $\mathcal{G}$ starting from the dual category $\widehat{\mathcal{E}}_{\mathcal{G}}$ (see [17, Thm. 7.3]), it is natural to conjecture that (7.2) is an isomorphism. This point is object of a work in progress.

In the case in which $\mathcal{Q}$ does not belong to the image of the map (3.2), the group $K_{\mathcal{Q}}^0(X)$ cannot be interpreted in terms of usual (equivariant) K-theory; up to direct sums, the elements of $K_{\mathcal{Q}}^0(X)$ arise from projections belonging to the $C(X)$-algebras $\mathcal{M}_{\mathcal{Q},r,r}$, $r \in \mathbb{N}$ (see Section 5.2). These $C(X)$-algebras have fibre $(H_d^r, H_d^r)_G$, anyway cannot be interpreted in terms of equivariant endomorphisms of the r-fold tensor power of some rank d vector bundle, as claimed in Thm. 5.2. Something similar happens in the setting of twisted K-theory considered in [1]: given an element $\mathcal{P} \in H^2(X, \mathbb{T}) \simeq H^3(X, \mathbb{Z})$, a K-theory group $K_{\mathcal{P}}^0(X)$ is constructed. The elements of $K_{\mathcal{P}}^0(X)$ are projections of a $C(X)$-algebra $\mathcal{A}_{\mathcal{P}}$ with fibre the C^*-algebra $\mathcal{K}$ of compact operators; if $\mathcal{P} \neq 0$, then $\mathcal{A}_{\mathcal{P}}$ is not isomorphic to $C(X) \otimes \mathcal{K}$, and cannot be interpreted as the algebra of compact endomorphisms of some bundle of Hilbert spaces. $\mathcal{P}$ is called the Dixmier-Douady class of $\mathcal{A}_{\mathcal{P}}$, and this is the reason why we adopted the same terminology for the invariant $\delta(\mathcal{Q})$.

Basic properties and applications of $K_{\mathcal{Q}}^0(X)$ are objects of a work in progress. We just mention the fact the $K_{\mathcal{Q}}^0(X)$ has a natural ring structure arising from the tensor product of $T\mathcal{Q}$; such a ring structure plays an important role in the computation of the K-theory of the C^*-algebra $\mathcal{O}_{\mathcal{Q}}$.

References

[1] M.F. Atiyah, G. Segal: *Twisted K-Theory*, arXiv:math.KT/0407054 (2004).

[2] M.F. Atiyah: *K-Theory*, Benjamin, New York, 1967.

[3] B. Blackadar: *K-Theory of Operator Algebras*, MSRI Publications, 1995.

[4] E. Blanchard: *Déformations de C*-algèbres de Hopf*, Bull. Soc. math. France **124** (1996), 141–215.

[5] J. Cuntz: *Simple C*-algebras Generated by Isometries*, Comm. Math. Phys. **57** (1977), 173–185.

[6] S. Doplicher, J.E. Roberts: *Duals of Compact Lie Groups Realized in the Cuntz Algebras and Their Actions on C*-Algebras*, J. Funct. Anal. **74** (1987) 96–120.

[7] S. Doplicher, J.E. Roberts: *A New Duality Theory for Compact Groups*, Inv. Math. **98** (1989), 157–218.

[8] G. Gierz: *Bundles of topological vector spaces and their duality*, Lecture Notes in Mathematics, **955**, Springer-Verlag, 1982.

[9] D. Husemoller: *Fibre Bundles*, Mc Graw-Hill Series in Mathematics, 1966.

[10] M. Karoubi: *K-Theory*, Springer Verlag, Berlin-Heidelberg-New York, 1978.

[11] M. Nilsen: *C*-Bundles and $C_0(X)$-algebras*, Indiana Univ. Math. J. **45** (1996), 463–477.

[12] V. Nistor, E. Troitsky: *An index for gauge-invariant operators and the Dixmier-Douady invariant*, Trans. AMS. **356** (2004), 185–218

[13] M. Pimsner: *A Class of C*-algebras Generalizing both Cuntz-Krieger algebras and Cross Product by $\mathbb{Z}$*, in: Free Probability Theory, D.-V. Voiculescu Ed., AMS, 1993.

[14] G. Segal, *Equivariant K-theory*, Inst. Hautes Études Sci. Publ. Math. **34** (1968), 129–151.

[15] E. Vasselli: *Continuous Fields of C*-algebras Arising from Extensions of Tensor C*-Categories*, J. Funct. Anal. **199** (2003), 122–152.

[16] E. Vasselli: *The C*-algebra of a vector bundle and fields of Cuntz algebras*, J. Funct. Anal. **222(2)** (2005), 491–502.

[17] E. Vasselli: *Bundles of C*-categories and Duality*, arXiv math.CT/0510594, (2005); *Bundles of C*-categories*, J. Funct. Anal. **247** (2007), 351–377.

Ezio Vasselli
Dipartimento di Matematica
Università La Sapienza di Roma
P.le Aldo Moro, 2
00185 Roma, Italy
e-mail: `vasselli@mat.uniroma2.it`

C^*-algebras and Elliptic Theory II

Trends in Mathematics, 297–309

A New Topology on the Space of Unbounded Selfadjoint Operators, K-theory and Spectral Flow

Charlotte Wahl

Abstract. We define a new topology, weaker than the gap topology, on the space of selfadjoint unbounded operators on a separable Hilbert space. We show that the subspace of selfadjoint Fredholm operators represents the functor K^1 from the category of compact spaces to the category of abelian groups and prove a similar result for K^0. We define the spectral flow of a continuous path of selfadjoint Fredholm operators generalizing the approach of Booss-Bavnek–Lesch–Phillips.

Mathematics Subject Classification (2000). 58J30; 46L80; 47A53.

Keywords. Spectral flow, classifying space, K-theory, unbounded operators.

Introduction

The space of bounded Fredholm operators on a separable Hilbert space endowed with the norm topology is a classifying space for the functor K^0 from the category of compact spaces to the category of abelian groups [J][A]. The index map realizes an isomorphism between the K-theory of a point and $\mathbb{Z}$. Furthermore a particular connected component of the space of selfadjoint bounded Fredholm operators with the norm topology represents the functor K^1 [AS]. An isomorphism $K^1(S^1) \cong \mathbb{Z}$ is given by the spectral flow, which was introduced in [APS].

These results can be applied to unbounded Fredholm operators by using the bounded transform $D \mapsto D(1 + D^*D)^{-\frac{1}{2}}$. However, since many important geometric applications involve unbounded operators, it is more convenient to work directly with the space of unbounded selfadjoint Fredholm operators. The gap topology on the space of unbounded selfadjoint operators is the weakest topology such that the maps $D \mapsto (D \pm i)^{-1}$ are continuous. Gap continuity is weaker than

This research was supported by a grant of AdvanceVT.

continuity of the bounded transform. Booss-Bavnek–Lesch–Phillips defined the
spectral flow for gap continuous paths [BLP] and Joachim proved that the space
of unbounded selfadjoint Fredholm operators endowed with the gap topology is a
classifying space for K^1 and the space of Fredholm operators with the subspace
topology (see §1) is a classifying space for K^0 [Jo].

In the first part of this paper we define and study a new topology on the
space of unbounded selfadjoint operators. In this topology a path $(D_t)_{t\in[0,1]}$ is
continuous if and only if the resolvents $(D_t \pm i)^{-1}$ depend in a strongly continuous
way on t and if there is an even function $\phi \in C_c^\infty(\mathbb{R})$ with $\operatorname{supp}\phi = [-\varepsilon,\varepsilon]$ and
$\phi'|_{(-\varepsilon,0)} > 0$ for some $\varepsilon > 0$ such that $\phi(D_t)$ is continuous in t. This topology is
weaker than the gap topology. Compared with the latter it has some additional
useful properties: The bounded transform of a continuous path is again continuous. If $(D_t)_{t\in[0,1]}$ is a continuous path of Fredholm operators and $(U_t)_{t\in[0,1]}$ is
a strongly continuous path of unitaries, then $(U_t D_t U_t^*)_{t\in[0,1]}$ is again a continuous path of Fredholm operators. We show that the space of selfadjoint Fredholm
operators endowed with this topology represents K^1 and the space of Fredholm
operators with the subspace topology represents K^0. Furthermore we illustrate
with an example that families of Fredholm operators that are continuous with
respect to this topology but not gap continuous arise naturally from differential
operators on noncompact manifolds. Along the way we indicate how these results
generalize to regular Fredholm operators on a Hilbert C^*-module.

In the second part we define and study the spectral flow of a continuous
path of selfadjoint Fredholm operators generalizing the approach of Booss-Bavnek–
Lesch–Phillips and relate it to the winding number. The definition of the spectral
flow given here is for paths with invertible endpoints equivalent to the definition
of the noncommutative spectral flow in [W] applied to a separable Hilbert space.
However, in [W] we used the theory of Hilbert C^*-modules in an essential way.
One aim of this paper is to recover the results of [W] for a Hilbert space using
classical functional analysis. We refer to [W] for applications.

1. A new topology on the space of unbounded selfadjoint operators

Let H be a separable Hilbert space.

Recall that a closed densely defined operator D on H is called Fredholm if
its bounded transform $F_D := D(1 + D^*D)^{-\frac{1}{2}}$ is Fredholm.

We denote the set of selfadjoint unbounded operators on H by $S(H)$ and the
set of selfadjoint unbounded Fredholm operators on H by $SF(H)$.

As usual, $B(H)$ is the space of bounded operators on H endowed with the
norm topology and $K(H)$ is the subspace of compact operators.

Throughout let B be a compact space.

For a Banach space V we denote by $C(B,V)$ the Banach space of continuous
functions from B to V equipped with the supremum norm. We write $C(\mathbb{R})$ for
$C(\mathbb{R},\mathbb{C})$. For $b \in B$ the evaluation map is $\operatorname{ev}_b : C(B,V) \to V$, $f \mapsto f(b)$.

For a map $D : B \to S(H)$ we define

$$\text{Dom } D := \{f \in C(B, H) \mid f(b) \in \text{dom } D(b) \text{ for all } b \in B \text{ and } Df \in C(B, H)\} .$$

Here $Df : B \to H$ is defined as $b \mapsto D(b)f(b)$.

First we note some useful facts about the functional calculus of selfadjoint operators.

Proposition 1.1. *Let $D : B \to S(H)$ be a map.*
The following conditions are equivalent:

1. *At each point $b \in B$ the set $\text{ev}_b(\text{Dom } D) \subset \text{dom } D(b)$ is a core for $D(b)$.*
2. *The resolvents $(D(b) \pm i)^{-1}$ depend in a strongly continuous way on $b \in B$.*
3. *For each $\phi \in C(\mathbb{R})$ the operator $\phi(D(b))$ depends in a strongly continuous way on $b \in B$.*

Proof. Set $R_\lambda(b) = (D(b) + \lambda)^{-1}$.

We show that (1) implies (2): Let $\lambda = \pm i$. Since $R_\lambda(b)$ is uniformly bounded, it is enough to prove that $\text{Dom } R_\lambda$ is dense in $C(B, H)$. Let $f \in C(B, H)$ and let $\varepsilon > 0$. The assumption implies that the set $\text{ev}_b((D + \lambda)(\text{Dom } D))$ is dense in H for any $b \in B$. Hence by compactness there is a finite open covering $\{U_j\}_{j \in I}$ of B and functions $g_j \in \text{Dom } D$, $j \in I$, such that $\|(D(b) + \lambda)g(b)_j - f(b)\| < \varepsilon$ for all $b \in U_j$. Let $\{\chi_j\}_{j \in I}$ be a partition of unity subordinate to the covering $\{U_j\}_{j \in I}$ and set $f_j(b) = (D(b) + \lambda)g_j(b)$. Then $\sum_{j \in I} \chi_j f_j \in \text{Dom } R_\lambda$ and $\|f - \sum_{j \in I} \chi_j f_j\| < \varepsilon$.

(2) $\Rightarrow$ (3): Let $\phi \in C(\mathbb{R})$. Since the algebra generated by the functions $(x + i)^{-1}$ and $(x - i)^{-1}$ is dense in $C_0(\mathbb{R})$, the assertion holds for all $\psi \in C_0(\mathbb{R})$, in particular for $\psi(x) = \phi(x)(x + i)^{-1}$. Hence $\phi(D)f \in C(B, H)$ for $f \in R_i C(B, H)$. By a similar argument as above, (2) implies that $R_i C(B, H)$ is dense in $C(B, H)$. Since $\phi(D(b))$ is uniformly bounded, this implies the assertion.

(3) $\Rightarrow$ (2) $\Rightarrow$ (1) is clear. $\qquad\square$

Lemma 1.2. *Let $D : B \to S(H)$ be a map such that the resolvents $(D(b) \pm i)^{-1}$ depend in a strongly continuous way on $b \in B$. Assume that for each $b \in B$ there is given a symmetric operator $K(b)$ with $\text{dom } D(b) \subset \text{dom } K(b)$ such that $K(b)(D(b) + i)^{-1}$ is compact and depends continuously on b. Then for each $\phi \in C_0(\mathbb{R})$*

$$\phi(D) - \phi(D + K) \in C(B, K(H)) .$$

Proof. It is enough to prove the assertion for the functions $(x \pm i)^{-1}$. Since $\text{Dom } D = \text{Dom}(D + K)$, the previous proposition implies that $(D(b) + K(b) \pm i)^{-1}$ depends in a strongly continuous way on b.

Hence $(D(b) + K(b) \pm i)^{-1} - (D(b) \pm i)^{-1} = -(D(b) + K(b) \pm i)^{-1}K(b)(D(b) \pm i)^{-1}$ is compact and depends continuously on $b \in B$. $\qquad\square$

Lemma 1.3. *Let X be a topological space. Let $D : X \to S(H)$ be a map such that the resolvents $(D(x) \pm i)^{-1} \in B(H)$ depend continuously on $x \in X$. Then $\phi(D(x))$ depends continuously on x for any $\phi \in C_0(\mathbb{R})$.*

Proof. This follows again from the fact that the functions $(x+i)^{-1}$ and $(x-i)^{-1}$ generate a dense subalgebra of $C_0(\mathbb{R})$. $\qquad\square$

In particular, if $D : X \to B(H)$ is a continuous map such that $D(x)$ is selfadjoint for each $x \in X$, then $f(D(x))$ depends continuously on x for all $f \in C(\mathbb{R})$.

Recall that the gap topology on $S(H)$ is the weakest topology such that the maps
$$S(H) \to B(H), \ D \mapsto (D+i)^{-1} \ ,$$
$$S(H) \to B(H), \ D \mapsto (D-i)^{-1}$$
are continuous. We denote by $S(H)_{\text{gap}}$ resp. $SF(H)_{\text{gap}}$ the set $S(H)$ resp. $SF(H)$ equipped with the gap topology. We refer to [BLP] for its properties.

In the following we introduce a new topology on $S(H)$.

Let $\phi \in C_c^{\infty}(\mathbb{R})$ be an even function with $\operatorname{supp} \phi = [-1,1]$ and with $\phi'(x) > 0$ for $x \in (-1,0)$. Define $\phi_n \in C_c^{\infty}(\mathbb{R})$ by $\phi_n(x) := \phi(nx)$ for $n \in \mathbb{N}$.

Let $\mathfrak{S}_n(H)$ be the set $S(H)$ endowed with the weakest topology such that the maps
$$\mathfrak{S}_n(H) \to H, \ D \mapsto (D+i)^{-1}x \ ,$$
$$\mathfrak{S}_n(H) \to H, \ D \mapsto (D-i)^{-1}x \ ,$$
$$\mathfrak{S}_n(H) \to B(H), \ D \mapsto \phi_n(D)$$
are continuous for all $x \in H$.

For any even function $\psi \in C_c(\mathbb{R})$ with $\operatorname{supp} \psi \subset (-\frac{1}{n}, \frac{1}{n})$ there is $g \in C_c(\mathbb{R})$ with $g(0) = 0$ such that $\psi = g \circ \phi_n$. Hence $\mathfrak{S}_n(H) \to B(H), \ D \mapsto \psi(D)$ is continuous. We will often make use of this property. It implies that the inclusion $\mathfrak{S}_m(H) \to \mathfrak{S}_n(H)$ is continuous for $m \leq n$. Let $\mathfrak{S}(H)$ be the set $S(H)$ endowed with the direct limit topology.

Define
$$SF_n(H) := \{D \in SF(H) \mid \phi_n(D) \in K(H)\}$$
and denote by $\mathfrak{S}\mathfrak{F}_n(H)$ the set $SF_n(H)$ endowed with the subspace topology of $\mathfrak{S}_n(H)$. Let $\mathfrak{S}\mathfrak{F}(H)$ be the inductive limit of the spaces $\mathfrak{S}\mathfrak{F}_n(H)$. An operator $D \in S(H)$ is Fredholm if and only if F_D is invertible in $B(H)/K(H)$, and this is equivalent to $\phi_n(D) \in K(H)$ for n big enough. Hence the underlying set of $\mathfrak{S}\mathfrak{F}(H)$ is $SF(H)$.

If $D : B \to \mathfrak{S}(H)$ is continuous, then $f(D) : B \to \mathfrak{S}(H)$ is continuous for any odd non-decreasing continuous function $f : \mathbb{R} \to \mathbb{R}$ with $f^{-1}(0) = \{0\}$. This can be seen as follows: Assume that $D : B \to \mathfrak{S}_n(H)$ is continuous. Since $(f \pm i)^{-1} \in C(\mathbb{R})$, we get from Prop. 1.1 that $(f(D) \pm i)^{-1}x : B \to H$ is continuous for any $x \in H$. Furthermore for m big enough $\operatorname{supp}(\phi_m \circ f) \subset (-\frac{1}{n}, \frac{1}{n})$, hence $\phi_m(f(D)) : B \to B(H)$ is continuous.

In particular the bounded transform $B \to \mathfrak{S}(H), \ b \mapsto F_{D(b)}$ is continuous. The example of Fuglede presented in [BLP] shows that the bounded transform of a gap continuous family is in general not gap continuous.

We need the following technical lemmata.

Lemma 1.4. *Assume that $D : B \to \mathfrak{S}\mathfrak{F}_n(H)$ is continuous. Then for $\psi \in C_c(\mathbb{R})$ with $\operatorname{supp} \psi \subset (-\frac{1}{n}, \frac{1}{n})$ we have that $\psi(D) : B \to K(H)$ is continuous.*

Proof. This follows from an elementary argument in the theory of Hilbert C^*-modules:

Let $B(C(B, H))$ be the algebra of strongly continuous families of bounded operators on H with parameter space B and with adjoint depending in a strongly continuous way on the parameter. Endowed with the supremum norm this is a C^*-algebra and $C(B, K(H))$ defines a closed ideal in $B(C(B, H))$.

Let $\pi : B(C(B, H)) \to B(C(B, H))/C(B, K(H))$ be the projection. Let $g \in C(\mathbb{R})$ with $g(0) = 0$ be such that $\psi^2 = g \circ \phi_n$. We have that $\pi(\phi_n(D)) = 0$, hence $\pi(\psi(D))^2 = g(\pi(\phi_n(D))) = 0$. Since $\pi(\psi(D))$ is selfadjoint in the C^*-algebra $B(C(B, H))/C(B, K(H))$, it follows that $\pi(\psi(D)) = 0$, hence $\psi(D) \in C(B, K(H))$. □

Lemma 1.5. *If $(F_b)_{b \in B}$ is a strongly continuous family of bounded selfadjoint operators such that $(b \mapsto F_b^2 - 1) \in C(B, K(H))$, then for any function $\phi \in C(\mathbb{R})$ with $\phi(1) = \phi(-1) = 1$ we have that $(b \mapsto \phi(F_b) - 1) \in C(B, K(H))$.*

Proof. The argument is similar to the proof of Lemma 1.4. We use its notation.

We have that $\pi((F_b)_{b \in B})^2 = 1$; hence the spectrum of $\pi((F_b)_{b \in B})$ is a subset of $\{-1, 1\}$; thus $\phi(\pi((F_b)_{b \in B})) - 1 = 0$. Since $\phi(\pi((F_b)_{b \in B})) = \pi((\phi(F_b))_{b \in B})$, it follows that $(b \mapsto \phi(F_b) - 1) \in C(B, K(H))$. □

The following two properties of the space $\mathfrak{S}\mathfrak{F}(H)$ are useful:

- Assume that $D : B \to \mathfrak{S}\mathfrak{F}(H)$ is continuous and that $B \ni b \mapsto U(b)$ is a map with values in the group of unitaries of $B(H)$ such that $U(b)$ depends in a strongly continuous way on b. Then $U D U^* : B \to \mathfrak{S}\mathfrak{F}(H)$ is continuous.
- If $D : B \to \mathfrak{S}\mathfrak{F}(H)$ is continuous, then $f(D) : B \to \mathfrak{S}\mathfrak{F}(H)$ is continuous for any non-decreasing continuous function $f : \mathbb{R} \to \mathbb{R}$ with $f^{-1}(0) = \{0\}$.

The first property follows from the fact that the composition of a continuous family of compact operators with of a strongly continuous family of bounded operators is again continuous if the parameter space is compact. Furthermore since U is bounded below, the adjoint depends also in a strongly continuous way on b.

Note that the second property does not assume the function to be odd. Taking Lemma 1.4 into account one proves the property analogously to the corresponding one for $\mathfrak{S}(H)$ from above.

Lemma 1.6. *Let $D : B \to \mathfrak{S}\mathfrak{F}(H)$ be continuous. Then there is an odd non-decreasing function $\chi \in C(\mathbb{R})$ with $\chi^{-1}(0) = \{0\}$ and $\lim_{x \to \infty} \chi(x) = 1$ such that $\chi(D)^2 - 1 : B \to K(H)$ is continuous.*

Proof. There is $n \in \mathbb{N}$ such that $D : B \to \mathfrak{S}\mathfrak{F}_n(H)$ is continuous. Then any non-decreasing $\chi \in C(\mathbb{R})$ with $\chi^{-1}(0) = \{0\}$ and such that $\operatorname{supp}(\chi^2 - 1) \subset (-\frac{1}{n}, \frac{1}{n})$ works. □

Definition 1.7. Let $D : B \to \mathfrak{S}\mathfrak{F}(H)$ be continuous. Then a function χ fulfilling the conditions of the previous lemma is called a *normalizing function* for D.

The terminology is borrowed from [HR]. The definition in [HR] is different since it applies to a different class of operators, but the underlying idea is the same.

The definition of the spaces $\mathfrak{S}(H)$ and $\mathfrak{S}\mathfrak{F}(H)$ generalizes in a straightforward way to the case where H is a Hilbert C^*-module. In this case we assume the unbounded operators to be regular.

The spaces $\mathfrak{S}(H)$ of $S(H)_{\text{gap}}$ share many properties as we will see in the following. We omit some details since the arguments resemble those in [BLP].

First we note that $\mathfrak{S}\mathfrak{F}(H)$ is path-connected since $SF(H)_{\text{gap}}$ is path-connected by [BLP, Th. 1.10].

Let $D_0 \in S(H)$. For $n \in \mathbb{N}$ and $\varepsilon > 0$ we define
$$U(\varepsilon, n, D_0) := \{D \in S(H) \mid \|\phi_n(D) - \phi_n(D_0)\| < \varepsilon\} .$$
This is an open neighbourhood of D_0 in $\mathfrak{S}_n(H)$.

Let $(a, b) \subset \mathbb{R}$ be in the resolvent set of $\phi_n(D_0)$. Then there is $\varepsilon > 0$ such that (a, b) is in the resolvent set of $\phi_n(D)$ for all $D \in U(\varepsilon, n, D_0)$. Hence $\phi_n^{-1}((a, b))$ is in the resolvent set of D for all $D \in U(\varepsilon, n, D_0)$. Furthermore if $\mu \in \phi_n^{-1}((a, b))$, $\mu > 0$, then also $-\mu \in \phi_n^{-1}((a, b))$ and $U(\varepsilon, n, D_0) \to B(H)$, $D \mapsto 1_{[-\mu, \mu]}(D)$ is continuous.

This implies the following lemma, which will be used for the definition of the spectral flow:

Lemma 1.8. *If $D_0 \in SF_n(H)$ and $\mu \in (0, \frac{1}{n})$ is such that $\pm\mu$ are in the resolvent set of D_0, then there is $\varepsilon > 0$ such that $\pm\mu$ are in the resolvent set of D for all $D \in U(\varepsilon, n, D_0)$. Furthermore $1_{[-\mu, \mu]}(D)$ has finite-dimensional range for all $D \in U(\varepsilon, n, D_0)$ and the map*
$$\mathfrak{S}_n(H) \supset U(\varepsilon, n, D_0) \to K(H), \ D \mapsto 1_{[-\mu, \mu]}(D)$$
is continuous.

In particular all operators in $U(\varepsilon, n, D_0)$ are Fredholm.

Note that for a given $D_0 \in SF_n(H)$ a μ fulfilling the assumption of the lemma always exists since the spectrum of D_0 near zero is discrete.

Proposition 1.9.   1. *The identity induces a continuous map $S(H)_{\text{gap}} \to \mathfrak{S}(H)$.*
  2. *The space $SF(H)$ is open in $\mathfrak{S}(H)$.*
  3. *The identity induces a homeomorphism from $\mathfrak{S}(H) \cap SF(H)$ to $\mathfrak{S}\mathfrak{F}(H)$.*

Proof. The first assertion is a consequence of Lemma 1.3.

The second assertion follows from the previous lemma and the subsequent remark. Since the remark is in general wrong for a Hilbert C^*-module, we give another argument which also works for Hilbert C^*-modules: Let $D_0 \in SF_n(H)$ and let χ be a normalizing function for D_0 with $\text{supp}(\chi^2 - 1) \subset (-\frac{1}{n}, \frac{1}{n})$. Then $\chi(D_0)^2$ is invertible in $B(H)/K(H)$. Furthermore, since $\mathfrak{S}_n(H) \to B(H)$, $D \mapsto (\chi(D)^2 - 1)$

is continuous, also $\mathfrak{S}_n(H) \to B(H)/K(H)$, $D \mapsto \chi(D)^2$ is continuous. Hence there is an open neighbourhood U of D_0 in $\mathfrak{S}_n(H)$ such that $\chi(D)^2$ is invertible in $B(H)/K(H)$ for all $D \in U$. This implies that all $D \in U$ are Fredholm.

The proof of the third assertion is similar and left to the reader.   □

We denote the space of (not necessarily selfadjoint) Fredholm operators on H by $F(H)$. We identify $F(H)$ with a subspace of $SF(H \oplus H)$ via the injection

$$F(H) \to SF(H \oplus H), \; D \mapsto \begin{pmatrix} 0 & D^* \\ D & 0 \end{pmatrix} .$$

Note that if $D \in F(H)$ and $f : \mathbb{R} \to \mathbb{R}$ is an odd non-decreasing continuous function with $f^{-1}(0) = \{0\}$, then $f(D) \in F(H)$ is well defined.

The space $F(H)$ endowed with the subspace topology of $\mathfrak{S}\mathfrak{F}(H \oplus H)$ is denoted by $\mathfrak{F}(H)$.

For topological spaces X, Y we denote by $[X, Y]$ the set of homotopy classes of continuous maps from X to Y.

Theorem 1.10.

1. *The space $\mathfrak{S}\mathfrak{F}(H)$ represents the functor $B \mapsto K^1(B)$ from the category of compact spaces to the category of abelian groups.*
2. *The space $\mathfrak{F}(H)$ represents the functor $B \mapsto K^0(B)$ from the category of compact spaces to the category of abelian groups.*

Proof. We use the notation of [Jo]: Let $KC_{sa}(H)$ (where KC stands for "Kasparov cycles") be the space of selfadjoint bounded operators F on H with $\|F\| \le 1$ and $F^2 - 1 \in K(H)$ and endow it with the weakest topology such that the maps

$$KC_{sa}(H) \to H, \; F \mapsto Fx \;,$$

$$KC_{sa}(H) \to K(H), \; F \mapsto F^2 - 1$$

are continuous for all $x \in H$. The inclusion $KC_{sa}(H) \to \mathfrak{S}\mathfrak{F}(H)$ is continuous.

Let $KC(H)$ be the space of bounded operators F such that $\|F\| \le 1$ and $F^*F - 1$, $FF^* - 1 \in K(H)$. Consider $KC(H)$ as a subspace of $KC_{sa}(H \oplus H)$ as above.

By [Jo, Theorem 3.4], which is based on results of Bunke–Joachim–Stolz, the space $KC_{sa}(H)$ represents the functor K^1 and the space $KC(H)$ represents K^0.

(1) Let $D : B \to \mathfrak{S}\mathfrak{F}(H)$ be a continuous map. Let χ be a normalizing function for D and let $\chi_t(x) = (1 - t)x + t\chi(x)$. Then $B \to KC_{sa}(H)$, $b \mapsto \chi_1(D(b))$ and $[0, 1] \times B \to \mathfrak{S}\mathfrak{F}(H)$, $(t, b) \mapsto \chi_t(D(b))$ are continuous (here we use Prop. 1.1). It follows that the map $[B, KC_{sa}(H)] \to [B, \mathfrak{S}\mathfrak{F}(H)]$ induced by the inclusion $KC_{sa}(H) \to \mathfrak{S}\mathfrak{F}(H)$ is surjective.

For injectivity let $h : [0, 1] \times B \to \mathfrak{S}\mathfrak{F}(H)$ be a homotopy between continuous maps $B \to KC_{sa}(H)$, $b \mapsto h(i, b)$, $i = 0, 1$. Let χ be a normalizing function for h such that $\chi(1) = 1$ and let $\chi_t(x) = (1-t)x + t\chi(x)$. Since $\chi_t^2(1) - 1 = \chi_t^2(-1) - 1 = 0$, Lemma 1.5 implies that the map

$$B \to K(H), \; b \mapsto \chi_t(h(i, b))^2 - 1$$

is continuous for $i = 0, 1$. Furthermore $\chi_t(h(i, b))^2 - 1$ is continuous in t since $\chi_t^2 - 1$ depends continuously on t in $C([0, 1])$. Hence the map

$$([0, 1] \times \{0, 1\} \times B) \cup (\{1\} \times [0, 1] \times B) \to KC_{sa}(H), \ (t, x, b) \mapsto \chi_t(h(x, b))$$

is continuous and defines a homotopy in $KC_{sa}(H)$ between $\chi_0(h(0, \cdot)) = h(0, \cdot)$ and $\chi_0(h(1, \cdot)) = h(1, \cdot)$.

(2) The proof is analogous with the obvious modifications. $\square$

It follows that $\pi_0(\mathfrak{F}(H)) \cong \mathbb{Z}$. As usual an isomorphism is given by the index map. The results in the following section will imply that an isomorphism $[S^1, \mathfrak{S}\mathfrak{F}(H)] \to \mathbb{Z}$ is given by the spectral flow.

The proof of the previous proposition carries over to the case where H is the standard Hilbert $\mathcal{A}$-module $H_\mathcal{A}$ of a unital C^*-algebra $\mathcal{A}$ implying that $\mathfrak{S}\mathfrak{F}(H_\mathcal{A})$ is a representing space of the functor $B \mapsto K_1(C(B, \mathcal{A}))$ from the category of compact spaces to the category of abelian groups and $\mathfrak{F}(H_\mathcal{A})$ is a representing space for $B \mapsto K_0(C(B, \mathcal{A}))$. The corresponding statements for $SF(H_\mathcal{A})_{\mathrm{gap}}$ have been proven in [Jo].

In the following we give two examples of maps with values in $SF(H)$ that are continuous in $\mathfrak{S}\mathfrak{F}(H)$ but not gap continuous. Both arise from elliptic differential operators on a noncompact manifold.

Example. Let $H = L^2(\mathbb{R})$ and let $f \in C^\infty(\mathbb{R})$ be nonconstant, real-valued and bounded below by some $c > 0$. Set $f_t(x) = f(tx)$ for $t \in [0, 1]$. We define $D(t)$ on $L^2(\mathbb{R})$ to be the multiplication by f_t. The path $D : [0, 1] \to SF(H)$ is not gap continuous at $t = 0$, but it is continuous as a path in $\mathfrak{S}\mathfrak{F}(H)$.

Even if the resolvents are compact, they need not depend in a continuous way of t:

Example. Let $H = L^2(\mathbb{R}, \mathbb{C}^2)$. Let $f \in C_0^\infty(\mathbb{R})$ be a nonnegative function and let $g \in C^\infty(\mathbb{R})$ with $g \geq 0$, $g(0) = 1$, $g(1) = 0$ and $g(x) = 1$ for $|x| \geq 2$. Define $\psi_t(x) := \frac{g(tx)}{f(x)} + 1$ for $t \in [0, 1]$. Note that $\psi_t(x)$ is continuous in t and x. Define $D(t)$ to be the closure of

$$\begin{pmatrix} 0 & \psi_t(1 - \partial_x^2) \\ (1 - \partial_x^2)\psi_t & 0 \end{pmatrix} : C_c^\infty(\mathbb{R}, \mathbb{C}^2) \to L^2(\mathbb{R}, \mathbb{C}^2) .$$

Since $\frac{1}{\psi_t} \in C_0(\mathbb{R})$, the operator $\frac{1}{\psi_t}(1 - \partial_x^2)^{-1}$ is compact on $L^2(\mathbb{R})$ for any t, hence $D(t)^{-1}$ is compact for any t. Furthermore $D(t)^{-1}$ is uniformly bounded. Thus $[0, 1] \to \mathfrak{S}\mathfrak{F}(H)$, $t \mapsto D(t)$ is continuous. It is easy to check that $D(t)^{-1}$ is not continuous in t at $t = 0$. Hence D is not gap continuous.

Note that these examples have in common that the coefficients are continuous as maps from $[0, 1]$ to $C_{loc}(\mathbb{R})$ but not continuous (in the second example even not well defined) as maps from $[0, 1]$ to $C(\mathbb{R})$.

See [W, §6] for criteria for the continuity in $\mathfrak{S}\mathfrak{F}(H)$ of families of elliptic operators on noncompact Riemannian manifolds and families of well-posed boundary value problems.

2. Spectral flow

In the following we generalize the definition of the spectral flow in [BLP], which is based on the approach of [P], to continuous paths in $\mathfrak{SF}(H)$.

Definition 2.1. Let $(D_t)_{t\in[a,b]}$ be a continuous path in $\mathfrak{SF}(H)$ and assume that there is $\mu > 0$ such that $\pm\mu$ are in the resolvent set of D_t for all $t \in [a,b]$ and $1_{[-\mu,\mu]}(D_t)$ has finite-dimensional range. We define

$$\mathrm{sf}((D_t)_{t\in[a,b]}) = \dim \mathrm{Ran}(1_{[0,\mu]}(D_b)) - \dim \mathrm{Ran}(1_{[0,\mu]}(D_a)) \ .$$

If $(D_t)_{t\in[a,b]}$ is a general continuous path in $\mathfrak{SF}(H)$, then we define its spectral flow by cutting the path into smaller pieces to which the previous situation applies and adding up the contributions of the pieces. (This is always possible by Lemma 1.8 and the subsequent remark.)

Well-definedness can be proven as in [P].

The spectral flow has the following properties:

1. It is additive with respect to concatenation of paths.
2. For any non-decreasing continuous function $f : \mathbb{R} \to \mathbb{R}$ with $f^{-1}(0) = \{0\}$

$$\mathrm{sf}((D_t)_{t\in[a,b]}) = \mathrm{sf}((f(D_t))_{t\in[a,b]}) \ .$$

3. If $(U_t)_{t\in[a,b]}$ is a strongly continuous path of unitaries on H, then

$$\mathrm{sf}((U_t D_t U_t^*)_{t\in[a,b]}) = \mathrm{sf}((D_t)_{t\in[a,b]}) \ .$$

4. If D_t is invertible for all $t \in [a,b]$, then $\mathrm{sf}((D_t)_{t\in[a,b]}) = 0$.
5. If $(D_{(s,t)})_{(s,t)\in[0,1]\times[a,b]}$ is a continuous family in $\mathfrak{SF}(H)$ such that $D_{(s,a)}$ and $D_{(s,b)}$ are invertible for all $s \in [0,1]$, then

$$\mathrm{sf}((D_{(0,t)})_{t\in[a,b]}) = \mathrm{sf}((D_{(1,t)})_{t\in[a,b]}) \ .$$

6. If $(D_{(s,t)})_{(s,t)\in[0,1]\times[a,b]}$ is a continuous family in $\mathfrak{SF}(H)$ such that $D_{(s,a)} = D_{(s,b)}$ for all $s \in [0,1]$, then

$$\mathrm{sf}((D_{(0,t)})_{t\in[a,b]}) = \mathrm{sf}((D_{(1,t)})_{t\in[a,b]}) \ .$$

The proof of the first three properties is not difficult and is left to the reader. The fourth property follows from the fact that by Lemma 1.8 and by compactness of $[a,b]$ there is $\delta > 0$ such that $[-\delta, \delta]$ is a subset of the resolvent set of D_t for all $t \in [a,b]$ if D_t is invertible for all $t \in [a,b]$.

The following proposition implies the last two properties, namely homotopy invariance:

Proposition 2.2. *If* $(D_{(s,t)})_{(s,t)\in[0,1]\times[a,b]}$ *is a continuous family in* $\mathfrak{SF}(H)$, *then*

$$\mathrm{sf}((D_{(0,t)})_{t\in[a,b]}) + \mathrm{sf}((D_{(s,b)})_{s\in[0,1]}) - \mathrm{sf}((D_{(1,t)})_{t\in[a,b]}) - \mathrm{sf}((D_{(s,a)})_{s\in[0,1]}) = 0 \ .$$

Proof. Let $n \in \mathbb{N}$ be such that the family $(D_{(s,t)})_{(s,t)\in[0,1]\times[a,b]}$ is continuous in $\mathfrak{SF}_n(H)$.

If there is $\mu \in (0, \frac{1}{n})$ such that $\pm\mu$ are in the resolvent set of $D_{(s,t)}$ for all $(s,t) \in [0,1] \times [a,b]$, then $1_{[-\mu,\mu]}(D_{(s,t)})$ has finite-dimensional range for all (s,t) and the assertion follows from the definition of the spectral flow.

In general we find, by compactness of $[0,1] \times [a,b]$ and by Lemma 1.8, an $n \in \mathbb{N}$ such that each of the rectangles $[\frac{(m_1-1)}{n}, \frac{m_1}{n}] \times [a+(b-a)\frac{m_2-1}{n}, a+(b-a)\frac{m_2}{n}]$ with $m_1, m_2 = 1, 2 \ldots n$ has the following property: There is a $\mu \in (0, \frac{1}{n})$ such that $\pm\mu$ are in the resolvent set of $D_{(s,t)}$ for all points (s,t) of the rectangle. Hence for each of the rectangles an analogue of the formula holds by the previous argument. Since for fixed n these rectangles constitute a subdivision of $[0,1] \times [a,b]$, the formula follows from the additivity of the spectral flow with respect to concatenation. □

We draw some conclusions in the following two propositions. See [Le, §3] for similar results.

If P, Q are projections with $P - Q \in K(H)$, then $QP : P(H) \to Q(H)$ is Fredholm with parametrix PQ. Let

$$\text{ind}(P,Q) := \text{ind}(QP : P(H) \to Q(H)) .$$

It is well known that

$$\text{sf}((t(2P-1) + (1-t)(2Q-1))_{t\in[0,1]}) = \text{ind}(P,Q) .$$

Proposition 2.3. *Let* $(P_t)_{t\in[0,1]}$, $(Q_t)_{t\in[0,1]}$ *be strongly continuous paths of projections on* H *such that* $P_t - Q_t$ *is compact and continuous in* t. *Then*

$$\text{ind}(P_0, Q_0) = \text{ind}(P_1, Q_1) .$$

Proof. First we prove that the family $(F_{(s,t)})_{(s,t)\in[0,1]^2}$ defined by $F_{(s,t)} := t(2P_s - 1) + (2t-1)(2Q_s - 1)$ is continuous in $\mathfrak{S}\mathfrak{F}(H)$: Clearly $F_{(s,t)}$ depends in a strongly continuous way on (s,t). Hence, by Prop. 1.1, the operators $(F_{(s,t)} \pm i)^{-1}$ depend in a strongly continuous way on (s,t) as well. Furthermore $F_{(s,t)} - (2P_s - 1)$ is a compact operator depending continuously on (s,t). This and Lemma 1.2 imply that $\phi_n(F_{(s,t)}) - \phi_n((2P_s - 1))$ is a compact operator depending continuously on (s,t) for any $n \in \mathbb{N}$. From $\phi_n((2P_s - 1)) = 0$ it follows that $\phi_n(F_{(s,t)})$ is a compact operator depending continuously on (s,t). This shows the continuity.

Now by homotopy invariance

$$\text{sf}((t(2P_0-1) + (1-t)(2Q_0-1))_{t\in[0,1]}) = \text{sf}((t(2P_1-1) + (1-t)(2Q_1-1))_{t\in[0,1]}) .$$

□

The following technical lemma, which is an immediate consequence of [Le, Prop. 3.4] and Lemma 1.2, is needed for the proof of the subsequent proposition:

Lemma 2.4. *Let* $D \in S(H)$ *and let* K *be a symmetric operator with* $\text{dom}\, D \subset \text{dom}\, K$ *and such that* $K(D+i)^{-1}$ *is compact. Then* $f(D+K) - f(D) \in K(H)$ *for each function* $f \in C(\mathbb{R})$ *for which* $\lim_{x\to\infty} f(x)$ *and* $\lim_{x\to-\infty} f(x)$ *exist.*

Proposition 2.5. *Let $(D_t)_{t \in [a,b]}$ be a continuous path in $\mathfrak{SF}(H)$ with invertible endpoints and assume given a path of symmetric operators $(K_t)_{t \in [a,b]}$ with $\operatorname{dom} D_t \subset \operatorname{dom} K_t$ for all $t \in [a,b]$ such that $K_t(D_t + i)^{-1}$ is compact and continuous in t and such that $D_t + K_t$ is invertible for each $t \in [a,b]$.*
Then

$$\operatorname{sf}((D_t)_{t \in [a,b]}) = \operatorname{ind}(1_{\geq 0}(D_b), 1_{\geq 0}(D_b + K_b)) - \operatorname{ind}(1_{\geq 0}(D_a), 1_{\geq 0}(D_a + K_a)) \ .$$

Proof. Let n be such that $(D_t)_{t \in [a,b]}$ is a continuous path in $\mathfrak{SF}_n(H)$.

Lemma 1.2 implies that $\phi_n(D_t) - \phi_n(D_t + K_t)$ is compact and continuous in t. In particular $(D_t + K_t)_{t \in [a,b]}$ is a continuous path in $\mathfrak{SF}_n(H)$. Since each $D_t + K_t$ is invertible, by property (4) of the spectral flow

$$\operatorname{sf}((D_t + K_t)_{t \in [a,b]}) = 0 \ .$$

Let $\psi \in C(\mathbb{R})$ with $\psi|_{(-\infty, \frac{1}{3}]} = 0$ and $\psi|_{[\frac{2}{3}, \infty)} = 1$.

By homotopy invariance and additivity with respect to concatenation the spectral flow of the path $(D_t)_{t \in [a,b]}$ equals the spectral flow of the path $(\tilde{D}_t)_{t \in [a-1,b+1]}$ with $\tilde{D}_t = D_a + \psi(t - a + 1)K_a$ for $t \in [a-1, a]$, $\tilde{D}_t = D_b + (1 - \psi(t - b))K_b$ for $t \in [b, b+1]$ and $\tilde{D}_t = D_t + K_t$ for $t \in [a, b]$. Furthermore by additivity with respect to concatenation and since $\operatorname{sf}((D_t + K_t)_{t \in [a,b]}) = 0$,

$$\operatorname{sf}((\tilde{D}_t)_{t \in [a,b]}) = \operatorname{sf}((\tilde{D}_t)_{t \in [a-1,a]}) + \operatorname{sf}((\tilde{D}_t)_{t \in [b,b+1]}) \ .$$

We calculate $\operatorname{sf}((\tilde{D}_t)_{t \in [a-1,a]})$: Let $\chi \in C^\infty(\mathbb{R})$ be a normalizing function for $(\tilde{D}_t)_{t \in [a-1,a]}$ such that $\chi(D_a) = 2 \cdot 1_{\geq 0}(D_a) - 1$ and $\chi(D_a + K_a) = 2 \cdot 1_{\geq 0}(D_a + K_a) - 1$. By the previous lemma $\chi(D_a) - \chi(D_a + K_a) \in K(H)$. Then

$$
\begin{aligned}
\operatorname{sf}((\tilde{D}_t)_{t \in [a-1,a]}) &= \operatorname{sf}((\chi(\tilde{D}_t))_{t \in [a-1,a]}) \\
&= \operatorname{sf}(((1-t)\chi(D_a) + t\chi(D_a + K_a))_{t \in [0,1]}) \\
&= \operatorname{ind}(1_{\geq 0}(D_a + K_a), 1_{\geq 0}(D_a)) \ ,
\end{aligned}
$$

where the second equality follows from homotopy invariance and the third from the equation preceding Prop. 2.3. Analogously $\operatorname{sf}((\tilde{D}_t)_{t \in [b,b+1]}) = \operatorname{ind}(1_{\geq 0}(D_b), 1_{\geq 0}(D_b + K_b))$. □

Under slightly more restricted conditions (since the previous lemma has not been proven for Hilbert C^*-modules – the author did not check whether the rather complicated proof of [Le, Prop. 3.4] carries over) the statement of the proposition makes sense on a Hilbert C^*-module and was used as a definition of the noncommutative spectral flow in [W].

In the following we express the spectral flow in terms of a winding number.

Let $S^1 = [0,1]/_{0 \sim 1}$ with the standard smooth structure.

Let $\mathcal{U}(H) \subset B(H)$ be the group of unitaries and let

$$\mathcal{U}_K(H) = \{U \in \mathcal{U}(H) \mid U - 1 \in K(H)\} \ .$$

There is an isomorphism

$$w : \pi_1(\mathcal{U}_K(H)) \cong \mathbb{Z}$$

extending the classical winding number. In fact, if $s : S^1 \to \mathcal{U}_K(H)$ fulfills $(x \mapsto s(x) - 1) \in C^1(S^1, l^1(H))$, where $l^1(H) \subset B(H)$ is the ideal of trace class operators endowed with the trace class norm, then

$$w(s) = \frac{1}{2\pi i} \int_0^1 \mathrm{Tr}(s(x)^{-1} s'(x)) \, dx \ .$$

Proposition 2.6. *Let* $(D_t)_{t \in [0,1]}$ *be a continuous path in* $\mathfrak{SF}(H)$ *with invertible endpoints. Let* $\chi \in C(\mathbb{R})$ *be a normalizing function for the map* $t \mapsto D_t$ *such that* $\chi(D_0)$ *and* $\chi(D_1)$ *are involutions. Then*

$$\mathrm{sf}((D_t)_{t \in [0,1]}) = w([e^{\pi i (\chi(D_t)+1)}]) \ .$$

If $(D_t)_{t \in [0,1]}$ *is a continuous path in* $\mathfrak{SF}(H)$ *with* $D_0 = D_1$ *(not necessarily invertible), then the equation holds for any normalizing function of* $t \mapsto D_t$.

Proof. The term on the right-hand side is well defined by Lemma 1.5. We make use of the space $KC_{sa}(H)$, which was defined in the proof of Theorem 1.10.

Let $(D_t)_{t \in [0,1]}$ be a continuous path in $\mathfrak{SF}(H)$ with invertible endpoints. Since for any normalizing function χ of $t \mapsto D_t$

$$\mathrm{sf}((D_t)_{t \in [0,1]}) = \mathrm{sf}((\chi(D_t))_{t \in [0,1]}) \ ,$$

we only have to show that for any continuous path $(F_t)_{t \in [0,1]}$ in $KC_{sa}(H)$ such that F_0, F_1 are involutions

$$\mathrm{sf}((F_t)_{t \in [0,1]}) = w([e^{\pi i (F_t + 1)}]) \ . \qquad (*)$$

Both sides of this equation remain unchanged if we replace $(F_t)_{t \in [0,1]}$ by $(F_t \oplus I)_{t \in [0,1]}$, where I is an involution on H with infinite-dimensional eigenspaces. Hence we may assume that the eigenspaces of F_0, F_1 are infinite-dimensional. Then there is a unitary U with $F_0 = U F_1 U^*$. Furthermore by the contractibility of $\mathcal{U}(H)$ there is a continuous path $(U_t)_{t \in [1,2]}$ of unitaries, unique up to homotopy, with $U_1 = 1$ and $U_2 = U$. Define $G_t = F_{2t}$ for $t \in [0, \frac{1}{2}]$ and $G_t = U_{2t} F_1 U_{2t}^*$ for $t \in [\frac{1}{2}, 1]$. The path $(G_t)_{t \in [0,1]}$ is a loop in $KC_{sa}(H)$ with

$$\mathrm{sf}((F_t)_{t \in [0,1]}) = \mathrm{sf}((G_t)_{t \in [0,1]})$$

and

$$w([e^{\pi i (F_t + 1)}]) = w([e^{\pi i (G_t + 1)}]) \ .$$

Thus it is enough to prove equation $(*)$ for loops in $KC_{sa}(H)$. This will also show the second assertion of the proposition.

By homotopy invariance of the winding number and of the spectral flow for loops (see property (6)) and by $[S^1, KC_{sa}(H)] \cong K^1(S^1) \cong \mathbb{Z}$ it is sufficient to verify equation $(*)$ for some loop in $KC_{sa}(H)$ whose class generates $K^1(S^1)$. For example one can use the loop $(G_t)_{t \in [0,1]}$ arising as above from $F_t = -\cos(\pi t)P + (1 - P)$, where P is a projection whose range has dimension one. In this case equation $(*)$ is well known since $(G_t)_{t \in [0,1]}$ is a norm-continuous path. $\qquad \square$

References

[A]   M.F. Atiyah, *K-theory*, W.A. Benjamin, Inc., 1967.

[APS] M.F. Atiyah & V.K. Patodi & I.M. Singer, *Spectral asymmetry and Riemannian geometry III*, Math. Proc. Camb. Philos. Soc. **79** (1976), 71–99.

[AS]   M.F. Atiyah & I.M. Singer, *Index theory for skew-adjoint Fredholm operators*, Inst. Hautes Études Sci. Publ. Math. **37** (1969), 5–26.

[BLP] B. Booss-Bavnbek & M. Lesch & J. Phillips, *Unbounded Fredholm operators and spectral flow*, Canad. J. Math. **57** (2005), 225–250.

[HR]   N. Higson & J. Roe, *Analytic K-homology*, Oxford Mathematical Monographs. Oxford University Press, 2000.

[J]   K. Jänich, *Vektorraumbündel und der Raum der Fredholm-Operatoren*, Math. Ann. **161** (1965), 129–142.

[Jo]   M. Joachim, *Unbounded Fredholm operators and K-theory*, High-dimensional manifold topology, World Sci. Publishing, 2003, 177–199.

[Le]   M. Lesch, *The uniqueness of the spectral flow on spaces of unbounded self-adjoint Fredholm operators*, Spectral geometry of manifolds with boundary and decomposition of manifolds, Contemp. Math. 366, Amer. Math. Soc., 2005, 193–224.

[P]   J. Phillips, *Self-adjoint Fredholm operators and spectral flow*, Canad. Math. Bull. **39** (1996), 460–467.

[W]   C. Wahl, *On the noncommutative spectral flow*, J. Ramanujan Math. Soc. **22** (2007), 135–187.

Charlotte Wahl
Leibniz-Archiv
Waterloostr. 8
30169 Hannover, Germany
e-mail: ac.wahl@web.de